Topology of Gauge Fields and Condensed Matter

Topology of Gauge Fields and Condensed Matter

Michael Monastyrsky

Institute of Theoretical and Experimental Physics
Moscow, Russia

Translated from Russian by
Oleg Efimov

SPRINGER SCIENCE+BUSINESS MEDIA, LLC

Library of Congress Cataloging-in-Publication Data

Monastyrskii, Mikhail Il'ich.
Topology of gauge fields and condensed matter / Michael Monastyrsky ; translated from the Russian by Oleg Efimov.
p. cm.
Includes bibliographical references and index.

DOI 10.1007/978-1-4899-2403-2
1. Algebraic topology. 2. Gauge fields (Physics)--Mathematics. 3. Condensed matter--Mathematics. 4. Mathematical physics.
I. Title.
QC20.7.T65M65 1993
530.1'435--dc20 92-43571
CIP

Originally published by Plenum Press, New York in 1993
MyCopy version of the original edition 1993

Preface

For a long time, books on algebraic topology were of one of two kinds. Either they ended with the Klein bottle or else they were written in the style of a personal letter to Norman Steenrod.

Gian-Carlo Rota

The rapid penetration of new, or rather non-standard, mathematics into theoretical physics, such as the theories of field, condensed matter, and gravitation, has been a feature of recent years. This is not only because of the traditional interaction between mathematics and physics but also because of the advanced state of physical theory.

At present, there are common features in describing various physical phenomena. Quantum fluctuations in a vacuum, vortices in superfluid ^{3}He and neutron stars, singularities in gravitation are just a few examples.

This diversity of observations is describable by a relatively few number of fundamental principles. Gauge invariance and hidden symmetry are the basic unifying ideas. Another aspect of modern research is to go beyond perturbation theory to obtain numerical results that can contrast theory with experiment.

This approach requires the use of powerful modern mathematical methods. Physicists have in some instances mastered elements of the necessary mathematical theories and actually resorted to them implicitly. Only in the last decade has the necessity of applying the latest mathematical results to physics become obvious. Today, this realization has been felt by both mathematically inclined physicists and most theorists.

This situation requires a literature to help the working physicist orient himself in unusual branches of mathematics. Not only are surveys of existing results, so too are books discussing the apparatus needed, including various branches of algebraic topology, differential and algebraic geometry, and group theory which were treated at various levels of complexity, but almost always centered on mathematicians. The style of such a book differs from those on physics, which makes it difficult for the physicist to study modern mathematics. The treatment is often excessively general, and the motivations and examples follow proofs of general theorems, etc. The appearance of such brilliant books as *Ordinary Differential Equations* by Arnold [A2], *Modern Geometry* by Dubrovin, Fomenko, and Novikov [DFN], *Topology and Geometry for Physicists* by Nash and Sen [NS] shows that the way of presenting mathematics through physics may be particularly useful in introducing the subject.

This book is based on the same principles. It discusses several branches of algebraic topology, group theory, and differential geometry as they are applicable to modern research in field theory and the theory of condensed matter. The applications mostly touch upon the topological structure of monopole and instanton solutions to the Yang-Mills equations, the description of phases in superfluid ^{3}He, and the topology of singular solutions in ^{3}He and liquid crystals. With the exception of topological charge construction, the remarkable applications of topology to the theory of gravitation are omitted, as they are covered by Hawking and Ellis [HE], Penrose [P], etc. Even though they are the first steps in the field, these topological applications are impressive.

The discussion, background material included, presented the most difficulties. The rigorous treatment of any branch of mathematics must be based on set theory. A consistent treatment, starting from the basics, calls for definitions familiar to the mathematician, but not always to the physicist. Any attempt to include in a small monograph all the important applications of topology and group theory together with their complete analysis is infeasible. Nevertheless, this book is not merely a guide of the literature. A compromise is reached by the requirement that the background material with definitions and examples of manifolds, Lie groups, etc., contain arguments, few illustrations, precise definitions, and references where proofs are given. Results not easily found but important for latter work are discussed in detail. The approach remains the same in Chapter 3. Applications are treated with more attention, as they mostly come from original papers and surveys, and several were obtained quite recently.

It should be stressed that applications based on comparatively simple topological methods are only discussed in the monograph.

The modern theory of monopoles, instantons, and quantum strings rely heavily on the latest mathematical advances (The Atiyah-Singer index theory for a family of operators, Chern-Simons characteristic classes, twistors), and even induced new major results in topology (Donaldson's theorem for 4-manifolds, classification of fibre bundles over projective spaces CP^N, quantum groups, etc.). These branches of mathematics are characterized by extensive research, the importance of whose results will be fully realized soon.

From this point of view, this book can be regaded as a first step in studying more complicated branches of topology.

Besides the physicist, the monograph is also designed for the mathematician. To make the book useful for the latter category, I have also included a short survey of physical problems.

The reader is expected to have the equivalent of a course of theoretical physics given at a Faculty of Mathematics. More information can be acquired from the literature referred here.

The idea for this book occurred (1977) after giving a course of lectures on topology, as suggested by I. Shapiro, for theorists at the Institute of Theoretical and Experimental Physics in Moscow. An outline was discussed in detail with I. Lifshits and S. Novikov, whose support played a decisive role in the preparation of the monograph.

The advice of a number of other mathematicians and physicists was sought and I would particularly like to thank F. Bogomolov, K. Boreskov, V. Golo, E. Kats, G. Margulis, S. Natanson, V. Retakh, M. Solovyev, P. Polyakov, Yu. Rudyak. I'm thankful to V. Kirsanov, I. Solodukhina, G. Rybchinskaya, and L. Subbotina for their help in preparing the manuscript.

Contents

How to Use This Book

The numbering is by chapter, section, and subsection. Formulas are numbered by chapter. Theorems, lemmas, propositions, examples, and remarks are numbered similarly. The difficulties due to the use of the same letters to denote different concepts in mathematics and physics do not lead to confusion because of textual heterogeneity.

Since mathematics alternates with physics in the book, the bibliography is accordingly minimized. I have given preference to mathematical sources closest to the physicist in style or to my own textbooks when I was a student. The physical literature includes monographs and surveys with an accessible introduction to the subject. References to original work are given when directly used, or when the treatment is the best. I did not pursue the goal of strictly indicating priority, however I tried not to ascribe authorship unjustifiably. Basically, the references are for educative purposes.

Chapter 1
Preliminaries in Mathematical Setting. Basics

1.1 MANIFOLDS

1.1.1 Definition of a Manifold. Examples. Basic Properties

The simplest objects on which functions, vector and tensor fields can be defined are n-dimensional vector spaces R^n and their domains $\mathcal{U}$, open sets. Their main property is in the introduction of one system of coordinates.

This enables us, e.g., to represent functions $f(x)$ on $R^n (x \in R^n)$ or on $\mathcal{U}$ as those of n coordinates $x^1, \ldots, x^n$. We can do the same with vector fields. However, even very simple examples show that this is not always true, e.g., global coordinates cannot be introduced on the unit sphere $S^{n-1} \subset R^n$ given by the equation

$$\sum_{i=1}^{n} (x^i)^2 = 1.$$

Consider the sphere S^2 in more detail.

Example 1.1 $S^2 \subset R^3$ is determined by the equation

$$(x^1)^2 + (x^2)^2 + (x^3)^2 = 1$$

in spherical coordinates

$$x^1 = \cos\varphi \sin\theta, \quad x^2 = \sin\varphi \sin\theta, \quad x^3 = \cos\theta.$$

It is easy to see that the spherical coordinate system is single-valued and has no singularities for $0 < \theta < \pi$, $0 < \varphi < 2\pi$. However, since the points $\theta = 0, \pi$ are singular, which means that a single-valued function of coordinates θ and φ cannot be given on the sphere in the spherical coordinates. It can be shown that a system of global coordinates on the sphere is lacking. However, if at least one point is removed from it, then such a system can be constructed.

Consider the set $S^2 \setminus x_1 = \mathcal{U}_1$ (x_1 is the North pole). The domain $\mathcal{U}_1$ is homeomorphic to the Euclidean plane R^2, which means that $\mathcal{U}_1$ can be continuously mapped onto R^2 in a one-to-one fashion. Let $U_1 \to R^2$ be the stereographic projection from x_1 $(\theta = 0)$.

Under the projection, the point x on the sphere with coordinates θ and φ is mapped into the point $\tilde{x} \in R^2$ with polar coordinates $(r, \varphi) = (2\cot\theta/2, \varphi)$. A single-valued coordinate system for the domain $\mathcal{U}_2 = S^2 \setminus x_2$ can be obtained similarly (x_2 is the South pole).

Thus, to give a function on the sphere S^2, it suffices to define the function on its two domains $\mathscr{U}_1$ and $\mathscr{U}_2$. Meanwhile, with two systems of coordinates (u_1^1, u_1^2, u_1^3) and (u_2^1, u_2^2, u_2^3) acting, the transition from one system to the other on the intersection $\mathscr{U}_1 \cap \mathscr{U}_2$ and *vice versa* is specified by smooth (i.e., differentiable) functions, which means the existence of the nonvanishing Jacobian

$$\det(\partial u_1^i / \partial u_2^j) \neq 0.$$

The following definition of a manifold generalizes the above structure.

Definition 1.1 A set M^n is called an *n-dimensional manifold* if there exist (a) such a covering with domains $\mathscr{U}_i \subset M^n$, $\cup \mathscr{U}_i = M^n$ and (b) maps

$$\varphi_i : \mathscr{U}_i \to \mathscr{V}_i \subset R^n$$

such that if the intersection $\mathscr{U}_i \cap \mathscr{U}_j \neq \varnothing$, then the maps

$$\varphi_i \circ \varphi_j^{-1} : \varphi_j(\mathscr{U}_i \cap \mathscr{U}_j) \to \varphi_i(\mathscr{U}_i \cap \mathscr{U}_j) \tag{1.1}$$

are homeomorphisms between the domains in R^n.

A domain in R^n along with the map $\varphi_i : \mathscr{U}_i \to \mathscr{V}_i \subset R^n$ is called a chart, and a collection of the charts is an *atlas for the manifold.* Two charts $(\varphi_1, \mathscr{U}_1)$ and $(\varphi_2, \mathscr{U}_2)$, $\mathscr{U}_1 \cap \mathscr{U}_2 \neq \varnothing$, are said to be *compatible* if condition (1.1) is fulfilled.

In accordance with the restrictions imposed on the class of mappings and the topology of the space M^n, we obtain manifolds with different structures and different topological properties.

Definition 1.2 A manifold M^n is said to be *differentiable* if the maps $\varphi_i \circ \varphi_j^{-1}$ as functions of local coordinates $u^1, \ldots, u^n$ are differentiable functions.

We consider differentiable manifolds of class C^k, depending on the class of smoothness of the differentiable functions. The questions of smoothness will not be of importance for this book, and the existence on a manifold of an appropriate differentiability, e.g., of class C^∞, or analyticity, will always be implied. In the latter case, the manifold is said to be *analytic.*

Definition 1.3 If the mappings φ_i are complex-valued analytic functions and the transition maps $\varphi_i \circ \varphi_j^{-1}$ are holomorphic, then the manifold is said to be *complex.*

We now introduce some other definitions connected with the topology of manifolds.

To avoid various mathematical pathology not arising in most physical applications, we impose two natural restrictions on the topology of M^n.

(1) **Separability.** Any two points $x, y \in M^n$ possess disjoint neighborhoods (M^n is *Hausdorff*).

(2) There exists an atlas for M^n, consisting of at least countably many of the charts.

Definition 1.4 A manifold M^n is said to be *compact* if a finite covering can be selected from any open covering of M^n.

Example 1.2 The sphere S^2 is obviously compact. Prove the compactness of the sphere S^n.

An example of a noncompact manifold is the space R^n itself.

Consider another example of a compact manifold, i.e., the two-dimensional projective plane RP^2. The set RP^2 can be defined in several equivalent ways.

Example 1.3 Consider the set of points $x = (x^1, x^2, x^3) \in R^3$ with the following conditions for equivalence: Two points $x, x' \in R^3 \setminus \{0\}$ (where $\{0\} = (0, 0, 0)$ is the origin of coordinates) determine the same point $\bar{x} \in RP^2$ if there exists a nonzero real scalar λ such that $x' = \lambda x$, or $(x^{1\prime}, x^{2\prime}, x^{3\prime}) = (\lambda x^1, \lambda x^2, \lambda x^3)$. The coordinates x^1, x^2, x^3 of the point x are said to be *homogeneous*, the points $\bar{x} \in RP^2$. Let $\mathcal{U}_i$ be a neighborhood in RP^n, consisting of points whose ith homogeneous coordinate is not zero, i.e., $\mathcal{U}_i = \{\bar{x}, x^i \neq 0\}$, $i = 1, 2, 3$. This open set is connected, and a chart can be specified on it, namely,

$$\varphi_i : \mathcal{U}_i \to R^2,$$

by assigning the point $y = (x^1/x^i, x^j/x^i)$ to the point $\bar{x}(x^1, x^2, x^3)$. It is obvious that the whole projective plane can be covered with three charts. It remains to prove the compatibility, e.g., of the charts $(\varphi_1, \mathcal{U}_1)$ and $(\varphi_2, \mathcal{U}_2)$. Suppose that $\bar{x} \in \mathcal{U}_1 \cap \mathcal{U}_2$ and that the map $\varphi_{12}(y) = \varphi_2(\varphi_1^{-1}(y))$ of the domain $\{y^1, y^2 : y^1 \neq 0\}$ into the domain $\varphi_{21}(z) = \varphi_1(\varphi_2^{-1}(z)) = \{z^1, z^2, z^1 \neq 0\}$ is defined, where $y = \varphi_1(\bar{x})$, $z = \varphi_2(\bar{x})$. The maps φ_{12} and φ_{21} are diffeomorphisms on $\mathcal{U}_1 \cap \mathcal{U}_2$. Indeed,

$$z^1 = (y^1)^{-1}, \quad z^2 = y^2(y^1)^{-1}; \quad y^1 = (z^1)^{-1}, \quad y^2 = z^2(z^1)^{-1}. \tag{1.2}$$

It follows from formulas (1.2) that the transformations φ_i specify the structure of an analytic manifold. We have thereby proved that the real projective plane is a compact analytic manifold.

The representation of the projective plane in homogeneous coordinates may be given a simple geometric interpretation, namely, RP^2 is the set of nonorientable lines in the three-dimensional space R^3, passing through the origin.

There is an alternative topologically equivalent representation. Consider the unit sphere S^2 and identify the diametrically opposite points. We obtain one more representation of RP^2 if the diametrically opposite points of the boundary circle of the hemisphere S^2_+ are identified.

The results obtained can be easily extended to an n-dimensional space, namely, the projective space RP^n.

The definition of the projective plane as a hemisphere with the boundary circle points being identified involves a new object, namely, a *manifold with boundary.*

Manifolds with boundary arise naturally in boundary problems in the theory of differential equations, e.g., Dirichlet and Neumann problems.

The simplest example is the disk homeomorphic to the hemisphere.

The definition of a manifold with boundary leads to the corresponding modification of Definition 1.1.

Denote the half-space in R^n, i.e., the set of points $(x^1, \ldots, x^n)$, distinguished by the condition

$$x^1 \leqslant 0,$$

by R^n_-.

The hyperplane R^{n-1}

$$x^1 = 0$$

is called the *boundary* of R^n_-.

Definition 1.5 A manifold M^n is called a *manifold with boundary* if each point x admits a neighborhood $\mathcal{U}^n$ which is homeomorphic to the domain $\mathcal{V}^n$ in R^n_-. The points x mapped into the boundary are said to be *boundary points*. It is known that the concept of boundary is topologically invariant. The boundary of a manifold is denoted by ∂M^n. The dimension (dim) of ∂M^n is $n-1$.

A compact manifold M^n without boundary ($\partial M^n = 0$) is said to be *closed*.

To assign a meaning to the notion of the dimension, a condition that mappings into coordinate spaces are homeomorphisms is necessary.

It follows from the classical Brouwer theorem that two homeomorphic topological spaces are of the same dimension. It is easy to see that, without any of the properties involved in the definition of a homeomorphism, it is impossible to preserve a dimension under a mapping, e.g., a square can be mapped onto a cube in a one-to-one fashion or onto a line-segment continuously (by projecting the former).

We now introduce another important concept enabling us to separate manifolds into two classes, namely, manifold orientation.

Definition 1.6 Let M^n be an n-dimensional manifold without boundary, with atlas $\cup \mathcal{U}_i = M^n$. Orientation is defined on M^n if, for any two compatible charts $(\varphi_i, \mathcal{U}_i)$, $(\varphi_j, \mathcal{U}_j)$, $\mathcal{U}_i \cap \mathcal{U}_j \neq \varnothing$, the transformation Jacobian of one chart into the other is positive. If such an atlas can be given for M^n, then the manifold is said to be *orientable*. Otherwise, it is *non-orientable*.

An orientable manifold obviously has two orientations sign det $(\mathcal{U}_1^i / \mathcal{U}_2^j)$. It is easily shown that the property of orientability is a topological invariant, i.e., it is independent of the way in which the local covering is given or of properties of the ambient manifold.

The concept of orientation is preserved for manifolds with boundary. We show that orientation of the boundary ∂M^n is then compatible with that of the manifold M^n itself.

Let $\mathcal{U}^n$ be a neighborhood in M^n, whose intersection with ∂M^n is nonempty, i.e., $\mathcal{U}^{n-1} = \mathcal{U}^n \cap \partial M^n \neq \varnothing$.

With respect to local coordinates, the equation of the neighborhood $\mathcal{U}^{n-1}$ is of the form $x^1 = 0$. Therefore, the coordinates of $\mathcal{U}^{n-1}$ are $x^2, \ldots, x^n$. Similarly, we select another domain $\mathcal{V}^n \subset M^n$ with coordinates $(y^1, \ldots, y^n)$ so that $\mathcal{V}^{n-1} = \mathcal{V}^n \cap \partial M^n$ is nonempty. Consider the intersection $\mathcal{U}^n \cap \mathcal{V}^n \neq \varnothing$. In the common part $\mathcal{U}^n \cap \mathcal{V}^n$ of the neighborhoods, we have

$$y^j = y^j(x^1, \ldots, x^n), \quad j = 1, \ldots, n,$$

and, for $x^1 = 0$,

$$y^j = y^j(0, x^2, \ldots, x^n), \quad j = 2, \ldots, n.$$

Since $y^1(0, x^2, \ldots, x^n) = 0$, we obtain (on $\mathcal{U}^{n-1} \cap \mathcal{V}^{n-1}$)

$$\left| \frac{\partial y^i}{\partial x^j} \right| = \frac{\partial (y^1, \ldots, y^n)}{\partial (x^1, \ldots, x^n)} = \frac{\partial y^1}{\partial x^1} \cdot \frac{\partial (y^2, \ldots, y^n)}{\partial (x^2, \ldots, x^n)}. \tag{1.3}$$

Since the Jacobian on the left-hand side of (1.3) is nonzero and $\partial y^1/\partial x^1 > 0$, the sign of the Jacobian $|\partial y^i/\partial x^j|$ $(i, j = 2, \ldots, n)$ coincides with that of $|\partial y^i/\partial x^j|$ $(i, j = 1, \ldots, n)$, i.e., orientation of the boundary is consistent with that of the manifold itself.

We now consider some examples.

Example 1.4 (a) The space R^n is clearly orientable.

(b) The space S^n is orientable (for proof, resort to local coordinates specified by the stereographic projection).

(c) The complex n-dimensional space C^n is orientable, since each analytic map f: $C^n \to C^n$ determines a map f: $R^{2n} \to R^{2n}$ with a positive Jacobian.

(d) M_C^n is a complex orientable n-dimensional manifold. This follows from (c) and the definition of a complex manifold.

(e) An example of a non-orientable manifold is RP^2.

We now calculate the Jacobian of a map for the intersection of two charts. The formulas for transition are given in (1.2).

In the domain $\mathscr{U}_1$, $y^1 = x^2/x^1$, $y^2 = x^3/x^1$, in $\mathscr{U}_2$, $z^1 = x^1/x^2$, $z^2 = x^3/x^1$; in $\mathscr{U}_1 \cap \mathscr{U}_2$ $(x^1 \pm 0, x^2 \pm 0)$, $y^1 = (z^1)^{-1}$, $y^2 = z^2(z^1)^{-1}$. The Jacobian of the transition functions is

$$\det \begin{pmatrix} -1/(z^1)^2, 0 \\ -z^2/(z^1)^2, 1/(z^1) \end{pmatrix} = -1/(z^1)^3.$$

Thus, in transition from the domain $z^1 > 0$ to the domain $z^1 < 0$, the Jacobian changes its sign. The projective plane is therefore a nonorientable manifold.

The Jacobian $J = -1/(z^i)^{n+1}$ yields similar results for RP^n. Hence, for odd n, the Jacobian maintains its sign in transition from the domains $z^i > 0$ into those $z^i < 0$. Therefore, if $n = 2k + 1$, the manifold RP^n is orientable (see also Exercise 2.8); in particular, RP^1 is diffeomorphic to the circle S^1, whereas RP^3 to the group of orthogonal matrices of order 3 with the determinant equal to 1.

(f) The complex projective space CP^n.

We define the complex n-dimensional projective space CP^n as the set of all complex lines passing through the origin of C^{n+1}. We prove that CP^n is an n-dimensional complex manifold. Note that, in analogy to the real projective space RI^3, we can define CP^n by specifying an equivalence relation between pairs of points in C^{n+1}. Two points z and $z' \in C^{n+1}$ are said to be *equivalent* and denoted by $z' \sim z$ if there exists a complex scalar $\lambda \neq 0$ such that $z' = \lambda z$. Thus, points z can be regarded as equivalence classes of proportional tuples $(z^1, \ldots, z^{n+1})$ consisting of $n + 1$ complex numbers z. Similarly to the real case, we can introduce homogeneous coordinates and an atlas $(\varphi_i \mathscr{U}_i)$ in CP^n, e.g., in the neighborhood of $\mathscr{U}_1$, where $z_1 \neq 0$, the coordinates of a point $z = (z^1, \ldots, z^{n+1})$ are $w^i = z^i/z^1$, $i = 1, \ldots, n + 1$. It is easy to see that, when passing from coordinates in domain $\mathscr{U}_i$ to those in $\mathscr{U}_j$, the Jacobian in the domain $\mathscr{U}_i \cap \mathscr{U}_j$ is specified by complex analytic functions, i.e., CP^n is an n-dimensional complex manifold.

Since the system of neighborhoods $\mathscr{U}_i$ is a finite open cover of CP^n, we also obtain the compactness of the manifold CP^n when considering that $CP^n = \bigcup_{i=1}^{n} K_i$, where $K_i = \varphi_i^{-1}(K)$. Here K is a polydisk in

$$C^n : \{ w \in C^n, \mid w^i \mid \leqslant 1 \},$$

and φ_i are determined in analogy to the case of RP^n. This compactness is also due to the fact that complex lines intersect the unit sphere $S^{2n+1} : \mid z^1 \mid^2 + \ldots + \mid z^{n+1} \mid^2 = 1$, $S^{2n+1} \subset C^{n+1}$. It follows from the equivalence $z \sim z'$ that $\mid \lambda \mid = 1$, i.e., CP^n is obtained by factorizing S^{2n+1} with respect to the action of the group $U(1) = \{\exp(i\varphi)\}$.

We now define the concept of direct product which enables us to construct examples of new manifolds from those already available.

Definition 1.7 We call the manifold $M_1^k \times M_2^n$ of dimension $k + n$ consisting of pairs (x, y), where x, y are points in two manifolds M_1^k and M_2^n of dimensions k and n, respectively, the *direct product* of M_1^k and M_2^n, respectively.

The atlas consists of the neighborhoods

$$\mathscr{U}_i \times \mathscr{V}_j \xrightarrow{\mathscr{L}} R^k \times R^n \sim R^{k+n}.$$

The local coordinates of a point $z = (x, y)$ are the tuples $(x^1, \ldots, x^n, y^1, \ldots, y^n)$, i.e., the local coordinates of the points x and y.

If M_1^k and M_2^n are two oriented manifolds, then the direct product $M_1^k \times M_2^n$ is obviously an oriented manifold. If at least one of the factors is nonoriented, then the direct product is obviously nonoriented.

If M_1^k has a boundary ∂M_1^k and M_2^n is a close manifold, then the boundary of $M_1^k \times M_2^n$ is $\partial M_1^k \times M_2^n$.

That a manifold can be represented as the direct product of manifolds of fewer dimensions enables us to simplify studying its properties considerably.

Example 1.5 (1) An n-dimensional torus T^n is an orientable manifold. The proof is immediate from the representation $T^n = \underbrace{S^1 \times \ldots \times S^1}_{n}$ (where S^1 is a circle).

(2) The group $SO(4)$ is an orientable manifold. First, the reader should show that $SO(4)$ is the direct product $S^3 \times SO(3)$. The $SO(4)$ orientability now follows from that of the group $SO(3) \sim RP^3$ and S^3. We shall show below (see Subsec. 1.2.3) that all manifolds endowed with a Lie group structure are orientable.

The problem naturally arising in studying manifolds of given dimension is their classification. By the classification, we mean introduction of (finitely or countably many) invariants, enabling us to clear up the question whether or not two manifolds M_1 and M_2 are homeomorphic (or diffeomorphic). In fact, this is the main problem in topology, and all methods and efforts are directed to its solution. Unfortunately, there is no answer in this general formulation. Even for 4-dimensional manifolds it is impossible to say, whether two manifolds are homeomorphic if no additional restrictions are placed. However, confining ourselves to narrower classes, we see that the answer is "yes" (e.g., such a classification of all simply-connected manifolds M^n, $n \geqslant 5$, was obtained by Browder and Novikov [Bro]; for $n = 4$, by Freedman [Fr]).

The simplest manifolds are one-dimensional.

Exercise 1.1 Prove that all closed, connected and one-dimensional manifolds are diffeomorphic to a circle; nonclosed ones, to a straight line; those with boundary, to an interval or a half-interval.

The classification of all 2-manifolds is known. Of great interest are 2-manifolds arising in various physical problems.

Here is their classification.

(1) First, consider closed 2-manifolds, and distinguish the class of orientable ones. To obtain them, we proceed as follows. Take the sphere S^2. Cut out a hole, i.e., a domain homeomorphic to the circle. Denote the circle boundary by S_1^1, and attach to the domain a handle, which is an object with topological structure of a torus with a hole cut out. The hole boundary is S_2^1.

The procedure of attaching consists in identifying the boundary circle S_1^1 with S_2^1. We obtain a figure homeomorphic to the torus. Continuing, we see that the manifold's general form after p such steps is a sphere with p handles, homeomorphic to a p-pretzel (Fig. 1). The number of handles is a topological invariant of the surface and is called its *genus*.

The reader will easily prove that any two surfaces with a different number of handles are nonhomeomorphic, making use of the topological considerations in Subsec. 2.5.3.

Exercise 1.2 Prove that a sphere with p handles is orientable.

(2) Nonorientable surfaces. Closed nonorientable manifolds are constructed in a similar manner. The handle's part is played by another standard manifold, the favourite character of popular science literature, the *Möbius strip*.

Möbius strip. Let $ABCD$ be a rectangle with the sides AB and CD identified after rotating AB through 180°. The Möbius strip's boundary is homeomorphic to the circle S^1.

Nonorientable surfaces are obtained from the sphere S^2 by consecutively cutting holes, and attaching the Möbius strips to the latter. The number of the attached Möbius strips is an invariant of the surface.

Exercise 1.3 Show that the projective plane RP^2 is homeomorphic to the sphere with a Möbius strip attached.

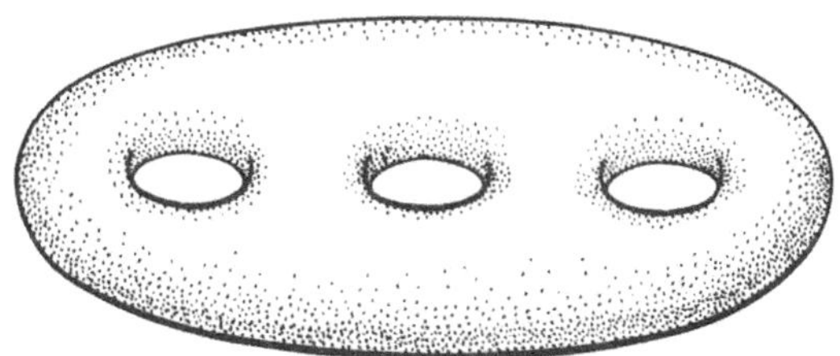

Fig. 1. Surface of genus p ($p = 3$)

Klein's bottle. The *Klein bottle* can be defined as follows. Take a cylinder and attach its boundaries by identifying boundary circles S_1^1 and S_2^1 and preliminarily reflecting one of them with respect to the diameter.

Exercise 1.4 Prove that the Klein bottle is a nonorientable surface. To which surface in the above construction of a nonorientable surface is it homeomorphic?

A manifold with boundary is obtained by the following construction.

Cut q holes out of S^2 and attach Möbius strips to p of them. We obtain a manifold M with boundary ∂M^2, consisting of q-p circles.

Exercise 1.5 Prove that the manifold obtained by attaching p handles and $q > 0$ Möbius strips to a sphere is homeomorphic to the sphere with $q + 2p$ Möbius strips attached.

The complete classification of two-dimensional surfaces includes the following two assertions.

(1) The proof that two-dimensional surfaces are pairwise nonhomeomorphic is quite simple. Nonorientable surfaces are not homeomorphic to orientable ones, since the orientability is a topological invariant. The nonhomeomorphism in each class follows from the existence of the topological invariant p, the genus of the surface.

(2) To prove that any closed 2-manifold is homeomorphic to one of the above M^2 is much more complicated, e.g., see the proof in [Mas 2].

An orientable 2-manifold M^2 can be realized as a surface in the space R^3 or as the Riemann surface of the algebraic function $w^2 = f(z) = P^{2n}(z)$, where $P^{2n}(z)$ is a polynomial of degree $2n$.

The relation between the degree n and genus of the surface is given by the formula $p = n - 1$.

Since non-orientable manifolds have no complex structure, they cannot be regarded as Riemann surfaces. The question naturally arises whether no-norientable manifolds can be regarded as surfaces in a Euclidean space.

The problem is closely related to another question of principle. We defined a manifold without restricting ourselves to positioning it in a fixed space R^n. On the other hand, we can always construct a manifold of particular form as the intersection of level lines of the functions $f_i(x^1, \ldots, x^n) = c_i$ (c_i = const), i.e., as a surface in R^n.

This brings up the question: is the notion of manifold more general, i.e., can we regard it as a submanifold embedded in R^N, where N is sufficiently large? The reply is given by Whitney in his fundamental theorem, to formulate which we need the precise definitions of a submanifold and of an embedding.

Definition 1.8 A subset $\mathscr{V} \subset M^n$ is called a *submanifold* if each point $x \in \mathscr{V}$ has such a neighborhood $\mathscr{W}$ in M^n and such a chart $\varphi: \mathscr{W} \to \mathscr{U} \subset R^n$ that $\varphi(\mathscr{W} \cap \mathscr{V})$ is a domain in a subspace $R^k \subset R^n$. $\mathscr{V}$ has the structure of the manifold M^n, and the charts $\mathscr{W}' = \mathscr{W} \cap \mathscr{V}$.

Definition 1.9 A manifold M^k is said to be *embedded* in a space M^l $(k < l)$ via a smooth map f if the rank of the Jacobian of the f is k for any $x \in M^k$, and f maps M^k onto $f(M^k)$ in a one-to-one fashion.

Theorem 1.1 (Whitney) *Each smooth manifold M^n can be embedded as a submanifold in a space R^N, where $N > 2n$.* In the general case, $N = 2n + 1$. However, for concrete spaces, more exact estimates can sometimes be obtained, e.g., two-dimensional nonorientable surfaces can be (without self-intersections) embedded in R^4. The proof can be found in [Rh] (the crux of the difficulty is to show that $N = 2n + 1$).

The result shows that the abstract concept of manifold as a matter of principle contains no objects different from surfaces in a Euclidean space of an appropriate number of dimensions. However, in many cases, especially in studying global properties, the abstract definition of a manifold, not involving the coordinates of the ambient space, is more convenient.

The concepts introduced enable us to extend the samples of manifolds strongly.

Example 1.6 Consider the set of all nonsingular real matrices $GL(n, R)$. It is easy to see that this space has the structure of N-manifold. The dimension of this manifold is $N = n^2$. Consider the following subsets in $GL(n, R)$:

(1) $O(n, R)$, the group of all orthogonal matrices.

(2) $SO(n, R)$, the group of all orthogonal matrices A with $\det A = 1$.

(3) The set of symmetric matrices with $\det A \neq 0$.

(4) The group of diagonal matrices with $\det A \neq 0$.

Is the set of all matrices of order n (singular ones included) a manifold?

Show that the sets in (1), (2), (3), and (4) are submanifolds of the manifold $GL(n, R)$.

It is wise to consider the embedding $GL(n, R) \subset GL(n, C)$, where $GL(n, C)$ is the group of all nonsingular complex matrices.

1.1.2 Functions and Vector Fields on Manifolds

We now study the behavior of functions and vectors under mappings of one manifold to another.

Let M_1^l and M_2^n be two differentiable manifolds and

$$\varphi : M_1^l \to M_2^n$$

a differentiable map. Then a function $f(x)$, $x \in M_2^n$, is transformed as a function of a function, i.e., for any f defined on a subset $\mathscr{V} \subset M_2^n$, a map

$$\varphi_* (f)(x) = f \circ \varphi(y)$$

is defined in M_2^n, where $y \in M_1^l$. The definition shows that φ_* acts reversely to φ and maps the space of functions $f(M_2^n)$ to the space $f(M_1^l)$.

Example 1.7 A differential mapping (or a function) of the interval $I = [0, 1]$ such that $f(t_0) = x_0$ is called a *(parametrized) curve* in a manifold M, emanating from the point x_0. The parametrized curve may not be a submanifold of M, since self-intersections of the curve, cusps, etc., are admissible.

1.1.2.1 Tangent space

We now define an object generalizing the notion of the vector tangent to a curve. The definition to be given of a tangent space does not depend on the choice of a local coordinate system.

Definition 1.10 Let M be a manifold. A *class of equivalent curves emanating from a point* x is called a *vector* v *tangent to* M at x. Meanwhile, two curves $\gamma_1 : I \to M$ and $\gamma_2 : I \to M$ passing through x are equivalent if they have the same derivatives at the point $\varphi(x)$ in any chart with x and, therefore, in any other chart with x, i.e., this definition does not depend on the system of coordinates. Now, consider the space of all curves passing through x and factorized with respect to the equivalence relation. Let the space be TM_x. An element of this space is just what is called the tangent vector at x.

Seemingly abstract and unusual, this definition completely coincides with the normal definition of the tangent vector to a curve. The equivalence condition means that, for all expansions into a Taylor series of functions f in the vicinity of a point x, the equivalent functions have the same linear parts, or the coefficients of $\partial f/\partial x^i/x$ coincide. Therefore, the tangent vector is determined by the collection of the first derivatives with respect to their local coordinates. The condition for equivalence shows that the tangent vector does not depend on the choice of a local coordinate system.

The set of tangent vectors has the structure of the vector space R^n and is denoted by M_x or TM_x. The dimension of the set coincides with $\dim M^n$.

We now give the formula for a tangent vector with respect to usual local coordinates.

Let $x \in W \subset M^n$. Any vector v tangent to M at the point x is specified by the set of components with respect to a local coordinate system. If $\gamma : I \to M^n$ is the curve emanating from x in the direction of v at a moment t_0, then

$$v^i = (d/dt)\, x^i(\gamma(t))\big|_{t = t_0}.$$

Consider the curve $\gamma(t)$ emanating from the point $x = (x^1, \ldots, x^n)$ along the coordinate line $x^i(\gamma_j(t)) = \delta^i_j(x^j + t)$. Then the tangent vector is $v^i = (\delta^1_i, \ldots, \delta^n_i)$. The tangent vectors v^i_x to the coordinate lines emanating from x are denoted by $\partial/\partial x^i/x$ and form the basis for the space TM_x.

Manifold of tangent vectors. Consider the space $TM^n = \bigcup_x TM^n_x$, the union of the tangent spaces at all points $x \in M^n$. Note that the tangent spaces determined at different points are disjoint. The set TM^n has the natural structure of a smooth manifold; charts of the manifold TM^n are the neighborhoods $\mathscr{W} \supset \mathscr{V} \times \mathscr{U}$ and the maps $\varphi : \mathscr{W} \to R^{n+n}$, where $\mathscr{V}$ is a neighborhood of x and $\mathscr{U}$ that of a vector $v \in TM_x$. We then have

$$\varphi : T\mathscr{V} \to R^{2n}, \qquad \varphi(v) = (x, v).$$

Different charts of TM, corresponding to different charts of the atlas for M, are compatible. Let $y^1, \ldots, y^n$ be another local coordinate system and $\eta^1, \ldots, \eta^n$ the components of the vector with respect to the system. Then

$$y^i = y^i(x^1, \ldots, x^n), \qquad \eta^i = \frac{\partial y^i}{\partial x^j} v^j.$$

Remark concerning terminology. The manifold TM is called a *tangent bundle* (more precise definition is a *total space of the tangent bundle*), since TM is one of the key examples of fiber bundles playing an important role in topology. Fiber bundles are defined in Sec. 1.4.

1.1.2.2 Vector field

We define a vector field on a manifold irrespective of the choice of local coordinates.

Definition 1.11 A *vector field* v *on* M is a smooth map $v: M \to TM$ such that the map $p \circ v: M \to M$ is the identity, i.e.,

$$p: TM \to M \Rightarrow p(v(x)) = x.$$

It is easy to see that the familiar definition of a vector field as the velocity field of phase flows, borrowed from the theory of ordinary differential equations, is quite equivalent for domains in a Euclidean space to the above abstract definition.

Let $x^1, \ldots, x^n$ be coordinates for R^n and g_t be the phase flow given by the $M^n \to \mathcal{U}^n \subset R^n$ 1-parameter group of diffeomorphisms of the space M^n. The velocity vector v of g_t is defined by the well-known formula

$$v(x) = \left.\frac{dg_t(x)}{dt}\right|_{t=0},$$

i.e., each point x is assigned the velocity vector given on the curve passing through the point. It is obvious that the correspondence satisfies Definition 1.11. Since a unique orbit for the flow can be (locally) drawn through each point, we obtain a vector field on the smooth manifold M^n. If we select an orthonormal basis $e_i = \partial/\partial x^i$ for the tangent space TM_x, then each vector field v_x can be uniquely represented as

$$v_x = J_i(x)\frac{\partial}{\partial x^i},$$

where $J_i(x)$ are functions of local coordinates.

The definition of a vector field in invariant terms (irrespective of the choice of the coordinate system) immediately enables us to apply to manifolds the whole theory of ordinary differential equations, constructed in domains in a Euclidean space. Meanwhile, the difference is revealed only in investigating global properties.

Let us see how vector fields are transformed under manifold mappings.

Let M and N be two smooth manifolds and f a smooth map. Since, under a smooth mapping, equivalence classes of smooth curves on M are carried into equivalence classes of those on N, the induced map

$$f^*: TM_x \to TN_y, \quad y = f(x),$$

arises. f^* is called the *differential* of f (also denoted by df).

Lemma 1.1 (a) *f^* is a linear map $TM_x \to TN_y$ of spaces.*

(b) *If we select a local coordinate basis $\partial/\partial x^i$ in TM_x and $\partial/\partial y^i$ in TN_y, then*

f^* *can be written as*

$$f^*(\partial/\partial x^i)_x = \left(\frac{\partial}{\partial x^i} y^j\right) \frac{\partial}{\partial y^j}\bigg|_y, \quad y = f(x).$$

Proof. Since each tangent vector in TM_x is of the form $X^i(x)\,\partial/\partial x^i$ and each tangent vector in TN_y is $y^j(y)\,\partial/\partial y^j$, it suffices to verify the linearity of f^*: $\partial/\partial x^i \to \dfrac{\partial y^j}{\partial x^i}\dfrac{\partial}{\partial y^j}$ for the basis vectors. It is obvious that the transformation is linear.

The condition (b) can be proved as simply as (a).

The Jacobian of the tangent bundle is of the form

$$\begin{pmatrix} A & 0 \\ H & A \end{pmatrix}, \quad A = \left(\frac{\partial x^i}{\partial y^j}\right), \quad H = \left(\frac{\partial^2 x^i}{\partial y^j \partial y^k}\right),$$

and a determinant of the Jacobian is $(\det A)^2 > 0$. Hence,

Corollary 1.1 *A tangent bundle to a manifold is an orientable manifold* (*even if the manifold itself is nonorientable*).

Exercise 1.6 Construct the tangent bundle to the projective plane and to the Klein bottle. To what manifolds are the tangent bundles homeomorphic?

Exercise 1.7 Consider the submanifold of the tangent bundle, formed by the pairs (x, e), where x is a point in the manifold M^n and e the unit tangent vector. The corresponding fiber bundle is an orientable $(2n - 1)$-dimensional manifold.

Such a fiber bundle is said to be *spherical.* Find the spherical bundle for S^2 and RP^2. Are the corresponding bundles homeomorphic?

Operations with vector fields. Transformation of vector fields. Let M_1^l, M_2^m, and M_3^n be three differentiable manifolds, φ_1 a differentiable map $M_1 \to M_2$, φ_2 a differentiable map $M_2 \to M_3$, and $TM_{1(p)}$, $TM_{2(q)}$, $TM_{3(r)}$ the tangent spaces at the points

$$p \in M_1, \quad q = \varphi_1(p) \in M_2, \quad r = \varphi_2(q), \quad r \in M_3.$$

Then the induced mapping of tangent spaces obeys the so-called *chain rule*

$$(\varphi_2 \circ \varphi_1)^* = \varphi_2^* \circ \varphi_1^*$$

(where $\circ$ means the composition of mappings).

This follows from the definition of φ^* and the familiar rule for differentiating a function of a function, i.e., $\dfrac{\partial z^k}{\partial x^i} = \dfrac{\partial z^k}{\partial y^j}\dfrac{\partial y^j}{\partial x^i}$ (with respect to the repeated summation indices).

On vector fields, besides the operations of addition and multiplication by a scalar usual for vector spaces, the *commutation* $X_3 = [X_1, X_2]$ of vector fields can be defined.

The *commutator* of two vector fields X_1 and X_2 is a vector field

$$X_3 = X_1 \circ X_2 - X_2 \circ X_1 = [X_1, X_2].$$

If the vector fields are given in the chart $(x) = (x^1, \ldots, x^n)$ as

$$X_1 = \xi^i(x)\,\partial/\partial x^i, \quad X_2 = \eta^j(x)\,\partial/\partial x^j,$$

then the field $X_3 = [X_1, X_2]$ is of the form $\xi^i(x)\dfrac{\partial \eta^j}{\partial x^i} - \eta^j(x)\dfrac{\partial \xi^i}{\partial x^j}$. In classical mechanics, X_3 is called Poisson's brackets of two vector fields.

The commutation has the following three properties: (1) antisymmetry, (2) bilinearity, and (3) fulfillment of the *Jacobi identity* for commutators of three vector fields:

$$[X_1 [X_2, X_3]] + [X_2 [X_3, X_1]] + [X_3 [X_1, X_2]] = 0.$$

One useful property of Poisson's bracket is formulated as follows.

Exercise 1.8 Two phase flows g_t, h_t generated by vector fields commute if and only if their Poisson's bracket is zero.

The commutation operation of two vector fields is a particular case of the Lie operator, or the directional differentiation of a tensor field with respect to a vector field.

We define the Lie derivative in two cases:

1. The Lie derivative of a function f with respect to a vector field v.

Definition 1.12 The *Lie derivative* of a function f: $\mathscr{U} \to R$ with respect to a field v at each point $x \in \mathscr{U} \subset M^n$ is the new function $L_v f$: $\mathscr{U} \to R$ whose value at x is equal to the derivative of f in the v direction:

$$(L_v f)(x) = L_{vx} f\big|_x.$$

In other words, if v is of the form $v = v^i \partial/\partial x^i$ in coordinates $x = (x^1, \ldots, x^n)$, then $(L_v f)(x) = v^i \partial f/\partial x^i$.

The operator L_v has the properties:

(1) $L_v(f + g) = L_v(f) + L_v(g)$,

(2) $L_v(f \circ g) = f(x) L_v g + L_v(f) g(x)$,

(3) $L_{v+w} = L_v + L_w$,

and

(4) $L_{fv} = f L_v$.

2. Let a vector field X generate a one-parameter group of transformations g_t. Locally, this is always so (corollary of the unique existence theorem for a system of ordinary differential equations).

We define the action of the one-parameter group $g_t(p)$ on a vector field Y as

$$g_t^*(p) : Y = g_t^*(Y)_p$$

(here g^* is the differential of the map g).

The infinitesimal transformation

$$L_X Y = \lim_{t \to 0} \frac{g_t^*(Y)_p - g_0^*(Y)_p}{t}$$

is called the *Lie derivative* of Y with respect to X.

We can show that the operator $L_X Y$ coincides with the commutator $[Y, X]$. The proof can be found in any textbook on differential geometry (e.g., [KN1, Ste]) and we omit it here.

1.1.2.3 Differential forms

Along with vector fields on a manifold, we can also define the dual object, *differential forms of degree one,* or simply *differential 1-forms.*

Differential 1-forms can be represented as $g_i df^i$ (where g_i and f^i are arbitrary functions). If we select local coordinates $x^1, \ldots, x^n$ in the chart of a manifold M^n, then the 1-form can be written as $a_i(x)\,dx^i$.

In transition from one local coordinate system to another, the form's coefficients change as components of the covector

$$a_i(x)\,dx^i \to a_i \frac{\partial x^i}{\partial y^j} dy^j,$$

i.e., the 1-form $a_i dx^i$ in coordinates $y^1, \ldots, y^n$ is $\tilde{a}_i(y)\,dy^i$, where

$$\tilde{a}_i(y) = a_j \frac{\partial x^j}{\partial y^i}.$$

If we introduce a basis $dx^1, \ldots, dx^n$ for the space R^{*n} of 1-forms, having defined the value of the basis vectors $(\partial/\partial x^1, \ldots, \partial/\partial x^n)$ on the vector space R^n by the formula

$$\left\langle \frac{\partial}{\partial x^i} \middle| dx^j \right\rangle = \delta_i^j, \tag{1.4}$$

then from the obvious properties of linearity it follows that R^{*n} and R^n are dual spaces relative to scalar product (1.4) and that the 1-forms make up the space of linear functionals on the tangent space.

By analogy to a tangent bundle, we can define the cotangent bundle of 1-forms as

$$\bigcup_x T_x^* M = T^* M.$$

A cotangent bundle, as well as a tangent bundle TM, has the structure of a differentiable bundle. In the Hamiltonian formalism of classical mechanics, a cotangent bundle plays the role of the phase space.

Proceeding from 1-forms, by means of exterior multiplication, we can construct forms of higher degree of p-forms.

Exterior (wedge) product of forms. We define exterior (wedge) product on basis forms, since the operation can be extended to arbitrary forms because of linearity. The exterior product is denoted by $\wedge$.

The operation $\wedge$ satisfies the associative, distributive, and anticommutative laws valid for Grassmann algebras, namely,

(1) $dx^i \wedge dx^j = -dx^j \wedge dx^i \Rightarrow dx^i \wedge dx^i = 0$

and

(2) $a \wedge dx^i = dx^i \wedge a = adx^i$,

where a is a function. Hence, any p-form can be represented as

$$\omega = a_{i_1 \ldots i_p} dx^{i_1} \wedge \ldots \wedge dx^{i_p} \tag{1.5}$$

in a given local coordinate system.

It is easy to show that this representation is unique. It also follows from (1) that forms of degree higher than n vanish on the manifold M^n.

We now formulate a number of properties of differential forms and additional definitions (see their proof, say, in [DFN1, Rh]).

(1) The form vanishes at x if all the form coefficients vanish at the point.

(2) Form's support. The support of a form is the smallest closed set, outside which $\omega = 0$.

(3) If α and β are forms of degree p and q, respectively, then

$$\alpha \wedge \beta = (-1)^{pq} \beta \wedge \alpha.$$

(4) The following representation of a form ω is convenient. We define coefficients for arbitrary systems of values of indices $i_1, \ldots, i_p$ and require that the antisymmetry condition be fulfilled, i.e., put $\alpha_{i_1 \ldots i_p} = 0$ if not all i_k are different and put $\alpha_{j_1}, \ldots, _{j_p} = \pm \alpha_{i_1 \ldots i_p}$ if $j_1 \ldots j_p$ is a permutation of indices $i_1 \ldots i_p$. The plus sign is selected in case of even permutation parity, the minus sign otherwise. Then the form ω is written as

$$\omega = (1/p1)\, \alpha_{i_1} \ldots {}_{i_p} dx^{i_1} \ldots dx^{i_p},$$

where all indices range from 1 through p independently.

(5) We define the *exterior differentiation* $\omega \rightarrow d\omega$ of a form ω. The form ω of degree p is sent into a form of degree $p + 1$. In the case of 0-forms ω^0 (i.e., functions), the operation coincides with taking the differential.

In the general case, d is specified by the formula

$$d\omega = d\alpha_{i_1} \ldots {}_{i_p} dx^{i_1} \wedge \ldots \wedge dx^{i_p}. \tag{1.6}$$

The operator d can be easily seen to satisfy the properties:

(a) $d(\omega_1 + \omega_2) = d\omega_1 + d\omega_2$ (linearity),

(b) $d(\omega_1 \wedge \omega_2) = d\omega_1 \wedge \omega_2 + (-1)^p \omega_1 \wedge d\omega_2$,

where p is the degree of the form ω_1, and, which is extremely important,

(c) $d \circ d\omega = 0$.

It follows that the exterior differentiation does not depend on the choice of local coordinates.

The property (c) generalizes the results, familiar from vector analysis, namely, rot grad $\varphi = 0$, div rot $A = 0$.

(6) The *Lie derivative* of a differential form is

$$L_X \omega = \lim_{t \to 0} \frac{\overset{*}{g}_t(\omega) - \overset{*}{g}_0(\omega)}{t},$$

where g_t is the local one-parameter group generated by the field X. $L_X\omega$ possesses the properties:

(a) $L_X(\omega_1 \wedge \omega_2) = L_X\omega_1 \wedge \omega_2 + \omega_1 \wedge L_X\omega_2$,

(b) $L_X d\omega = d(L_X\omega)$,

and

(c) $L_X\omega = dJ_X\omega + J_X(d\omega)$,

where J_X is the interior multiplication $\omega^p \to \omega^{p-1}$ which maps the form ω of degree p. J_X acts by the formula

$$(J_X\omega) = \sum_{i=1}^{p} (-1)^{i+1} a(x) L_X(x^i)\, dx^1 \wedge \ldots \wedge \widehat{dx^i} \wedge \ldots \wedge dx^p \tag{1.7}$$

on the basis form $\omega = a(x)\, dx^1 \wedge \ldots \wedge dx^p J_X$, where $\wedge$ means that the corresponding differential is omitted. Making use of differential forms, we show how the Hamiltonian equations can be written on a manifold M.

Now, we introduce an important definition.

Definition 1.13 A manifold M is said to be *symplectic* if there exists on it a nonsingular 2-form ω such that $d\omega = 0$ (we call this form *closed*).

An example of a symplectic manifold is given by the space of the cotangent bundle T^*M. A closed 2-form can be represented as

$$\omega = dp_i \wedge dq^i.$$

A vector field X on M is said to be *Hamiltonian* if $L_X\omega = 0$ (here ω is a closed 2-form).

We now turn to the covector field $W = J_X\omega$. It follows from property (6) that X is Hamiltonian if and only if the form W is closed. Since W is closed, the form can be locally represented as dF, where F is a smooth function. F is the Hamiltonian of the system and only locally one-valued. The condition for F to be a globally one-valued function on a manifold M is determined by the topology on the phase space. The corresponding results are given in Ch. 2.

The formalism of differential forms with values in vector spaces proves to be convenient for the theory of gauge fields in particular, in the Lie algebra of gauge groups.

1.2 LIE GROUPS

In this section, we shall set the foundations for studying the theory of Lie groups, to which a great variety of books are devoted. I would recommend the books [He, Sem, Ze], which can satisfy the requirements of any reader.

Definition 1.14 A manifold G endowed with a group structure is called a *Lie group* if the map $g_1 \circ g_2 \to g_1 g_2^{-1}$, $g_1, g_2 \in G$, is differentiable.

As it turns out, for G to be endowed with the structure of an infinitely differentiable, and even analytic, Lie group it suffices to require that the group operator be continuous. This very difficult theorem, the solution of the Hilbert fifth problem, was proved by Gleason, Montgomery, and Zippin in 1952.

The main examples of Lie groups are:

(1) The general linear group $GL(n, R)$ of all nonsingular transformations of the space R^n. The group operations are given by rational coordinate functions.

(2) Closed subgroups of a Lie group are also Lie groups. We thus obtain the examples of the group $O(n)$ of orthogonal matrices and of the group $SO(n)$ of orthogonal matrices with det = 1.

(3) A physical example: the isospin group $SU(2)$.

Definition 1.15 A *Lie subgroup* is a subgroup which is both a Lie group and a submanifold.

Example 1.8 Let $G = T^2 = S^1 \times S^1$. Consider an orbit in a torus, with an irrational rotation number. The corresponding trajectory is dense in the torus and therefore is not a Lie subgroup.

Proposition 1.1 *Let G be a Lie group and H a normal subgroup of G. The quotient group G/H is also a Lie group.*

If H is only a subgroup of G, then the quotient space G/H, the set of (e.g., left) cosets $\{gH\}$, $g \in G$, of G relative to the subgroup H can be formed. G/H has an interesting geometric interpretation.

1.2.1 Lie Groups as Groups of Transformations of a Manifold M

Let G be a Lie group, and M a C^∞-manifold. G acts on M on the left if there exists a C^∞-map $\varphi : G \times M \to M$, i.e., each pair of points (g, m), $g \in G$, $m \in M$, is assigned a motion $\varphi(g, m) = gm$ satisfying the conditions:

(a) For any $g \in G$, the map $g: M \to M : (g, m) \to gm$ is a diffeomorphism (it suffices to assume that g is a homeomorphism).

(b) $(gg_1)m = g(g_1 m)$ for all $g, g_1 \in G$, $m \in M$.

The group G acts on the right, $m \to mg$, on M if, instead of (b), (b′) holds: $m(gg_1) = (mg_1)g$. If a right action $g: m \to mg^{-1}$ is given, then the condition $(mg \cdot g_1) = (mg)g_1$ holds.

G acts *transitively* on M if for any two points m_1 and m_2 there exists a transformation from G, sending m_1 into m_2.

G acts *effectively* on M if a unique element from G, leaving all points m fixed, i.e., $\{g \in G/gm = m\}$, is the identity element $e \in G$.

G acts *freely* (without fixed points) on M if for each $m \in M$ the subgroup $\{g \in G, gm = m\}$ is trivial, i.e., $\{e\}$.

Example 1.9 Each Lie group is the group of transformations of itself (relative to a right or left translation).

Proposition 1.2 *Let G act transitively on M. Then $M \sim G/H$, where H is the isotropy group of a certain point m_0 (the set of all elements from G, which leave m_0 fixed).*

An isotropy group is also called a *stability group*, or the *point m_0 stabilizer.*

Proof. Let m_1 and m_2 be two points in M and $m_2 = gm_1$. Associate m_2 with the coset gH under a map $G/H \to M$. (Here H is the isotropy group of m_1.) It follows from the transitivity of the action of G on M that $G/H \to M$ is a mapping "onto". ∎

For a wide class of spaces (if, say, G/H is compact), the mapping is a homeomorphism.

Spaces on which Lie groups act transitively are said to be *homogeneous.*

Example 1.10 (1) $G = SO(3)$. Let m be the North pole. The stability group is the group $SO(2)$ of rotations about the vector $\mathbf{n} = (0, 0, 1)$. The space $M \sim SO(3)/SO(4)$.

(2) The projective plane $RP^2 = SO(3)/SO(2) \times Z_2$.

(3) The complex projective space $CP^n \sim S^{2n+1}/S^1$.

(4) Stiefel manifolds.

(a) We specify an orthonormal k-frame $(0 < k < n)$ in R^n, i.e., the set of k unit and pairwise orthogonal vectors. The set of all such k-frames forms a space called a *Stiefel manifold* $V_{n,k}$. We show that the space is homogeneous both for the group $O(n)$ and $SO(n)$. It is obvious that $O(n)$ acts transitively on the set of k-frames, since $O(n)$ preserves the length of vectors and their pairwise orthogonality. We select a fixed k-frame v_0^k. The group leaving v_0^k fixed is the $O(n-k)$ group of orthogonal rotations in the subspace orthogonal to all vectors in V_0^k. We can identify $V_{n,k}$ with $O(n)/O(n-k)$. If we now turn to the rotation subgroup $SO(n)$ isomorphic to $O(n)/Z_2$, then, as can be easily seen, we obtain the same Stiefel manifold $V_{n,k} = SO(n)/SO(n-k)$. The space $V_{n,k}$ can also be realized as that of $n \times k$ matrices A such that $AA^t = I$ (I being the unit matrix, and A^t the transpose of A). The condition distinguishes from the space of all $n \times k$ matrices a submanifold $V_{n,k}$ of dimension $nk - k(k+1)/2$. It immediately follows from the representation $V_{n,k} = SO(n)/SO(n-k)$ that $V_{n,k}$ is an orientable manifold. We investigate other properties of the Stiefel manifold below.

(b) In analogy to a real Stiefel manifold, we can consider a complex one $V_{n,k}^C$. $V_{n,k}^C$ is the collection of orthonormal k-frames in the complex space C^n. Similarly to the real case, we can show that $V_{n,k} = U(n)/U(n-k)$, where $U(n)$ is the group of unitary transformations of C^n. $U(n)$ is as a topological manifold diffeomorphic to $U(1) \times SU(n)$. $SU(n)$ is a unitary group with det $= 1$. The diffeomorphism

$\psi: SU(n) \times U(1) \to U(n)$ associates a pair of matrices $(u, \exp(i\varphi)) \in SU(n) \times U(1)$ with a matrix $\tilde{u} \in U(n)$; $\tilde{u}$ is obtained from u by multiplying the first row by $\exp(i\varphi)$.

Certainly, the diffeomorphism is not an isomorphism of the groups $SU(n) \times U(1)$ and $U(n)$, since $U(n)$ is not the direct and only semidirect* product of the groups $SU(n)$ and $U(1)$. As in the real case, we can show that $V^C_{n,k} = SU(n)/SU(n-k)$.

The manifold $V^C_{n,k}$ can be realized as the space of $n \times k$-complex matrices with the additional condition $AA^+ = I$ (A is a complex $n \times k$ matrix and A^+ the Hermitian adjoint), distinguishing a hypersurface in the space $C^{n^2} \sim R^{2n^2}$,

$$\dim_R V^C_{n,k} = 2nk - k^2.$$

An important point is that a complex Stiefel manifold is, generally speaking, not a complex manifold from the standpoint of existence of a complex structure on it. For example, it is obvious that, for odd k and even n, Stiefel manifolds are odd-dimensional and cannot have a complex structure. Nevertheless, the established terminology has a right to exist, since it reflects the fact that $V^C_{n,k}$ is defined over the field of complex numbers. Quaternion Stiefel manifolds can be considered similarly.

(5) Grassmann manifolds.

(a) Real Grassmann manifolds. Consider all k-dimensional linear subspaces (k-dimensional linear planes passing through the origin) of the space R^n. Denote the space of all k-planes in R^n by $G_{n,k}$.

Definition 1.16 The space $G_{n,k}$ is called a *real Grassmann manifold* or a *Grassman space* or simply *Grassmanian.*

To clarify, we show that $G_{n,k}$ is a homogeneous space and not only a manifold.

Note that any element of the group $O(n)$ transforms a k-plane into another k-plane and $O(n)$ operates on $G_{n,k}$ naturally. It is obvious that the action is transitive. To find the stability subgroup of a fixed point $x \in G_{n,k}$, let x be a fixed point in $G_{n,k}$, i.e., a fixed k-plane $R^k \subset R^n$. The stability subgroup H of x is the direct product $O(n-k) \times O(k)$. In fact, the group $O(n-k)$ leaves each vector from R^k fixed and the group $O(k)$ carries the plane R^k onto itself.

Hence, $G_{n,k} \simeq O(n)/O(n-k) \times O(k)$.

As an exercise, the reader can explicitly construct a system of neighborhoods on $G_{n,k}$ and prove that the latter is a differentiable manifold, $\dim G_{n,k} = nk - k^2$. For $k = 1$, we obtain a manifold $G_{n,1} = RP^{n-1}$. It can be seen from the example that $G_{n,k}$ is not, generally speaking, orientable.

It is natural to consider the space $\tilde{G}_{n,k}$ of orientable k-planes in an n-dimensional space $(0 < k < n)$.

The manifold $\tilde{G}_{n,k}$ is the homogeneous space $SO(n)/SO(k) \times SO(n-k)$, also called a Grassmannian. It is shown in Subsec. 1.4.1 that $\tilde{G}_{n,k}$ is a two-sheeted covering of $G_{n,k}$.

(b) Complex Grassmannian. Let $G^C_{n,k}$ be the space of complex k-planes in the n-dimensional complex space C^n. $G^C_{n,k}$ is called a *complex Grassmannian.* In analo-

* A semidirect product is denoted by $\dot{\times}$.

gy to the real case, it is easy to show that $G^C_{n,k} = U(n)/U(n-k) \times U(k)$. For $k = 1$, we obtain the complex projective space CP^{n-1}. The manifold $G^C_{n,k}$ is orientable even-dimensional (in the sense of real dimension); further, it admits a complex structure.

Stiefel and Grassmann manifolds play a key role in the theory of fibrations. In the subsequent chapters, we consider the manifolds from various points of view.

Many examples of homogeneous spaces arise throughout the book. In physics, homogeneous spaces mostly appear either as sets of extrema of potentials or as phase and configuration spaces of mechanical systems.

1.2.2 Lie Algebras

Consider a vector field X on a Lie group. Generally speaking, if we consider all vector fields on the group, then it is obvious that their space is infinite-dimensional and weakly related to the structure of a Lie group. We define the class of vector fields invariant under translations of a group G on itself. For definiteness, consider X on G, invariant under left translations. (Clearly, everything in the sequel is valid for right translations, since the transformation $g \to g^{-1}$ carries a left action into a right one, i.e., $gx \to xg^{-1}$.)

Definition 1.17 A vector field invariant under the differentials of left translations is said to be left-invariant on G, i.e., if $L_g : G \to G$ acts according to the formula

$$L_g(h) = gh,$$

then the field X is left-invariant, provided $dL_g X(h) = X(gh)$ for any $g, h \in G$.

It follows from the condition of left-invariance that X is defined on G globally.

It is easy to verify that the sum of two left-invariant fields is a left-invariant field. A left-invariant field also remains left-invariant when multiplied by a number. Poisson's bracket of two left-invariant fields is a left-invariant field and satisfies the Jacobi identity.

It follows from the invariance under translations that the field X is determined by its value at the point e. It is easy to see that a left-invariant vector field is determined by the tangent vector at $e \in G$ to the one-parameter group $g_t(e)$, $g_0(e) = e$. The number of linearly independent tangent vectors coincides with the dimension of the Lie group. Thus, the space of all left- (right) invariant vector fields on a Lie group forms a finite-dimensional vector field $\mathscr{G}$ with an additional operation, Poisson's bracket, or commutator, satisfying the properties

(a) $[x, y] = -[y, x]$ (skew-symmetry),

(b) $[x, [y, z]] + [y, [z, x]] + [z [x, y]] = 0$ (the Jacobi identity)

which is already known.

Definition 1.18 A vector space $\mathscr{G}$ with the structure of a commutator [,] is called a *Lie algebra*.

To determine a structure of the Lie algebra on the linear n-dimensional space R^n, it suffices to specify pairwise commutators for any two basis vectors e_i, e_j, i.e., the coefficients c_{ij}^k in the expansion

$$[e_i, e_j] = c_{ij}^k e_k. \tag{1.8}$$

c_{ij}^k are called the *structural constants* of the Lie algebra. It follows from (a) that $c_{ij}^k + c_{ji}^k = 0$. Therefore, it suffices to have $n^2(n-1)/2$ structural constants $c_{ij}^k (i < j)$.

Remark 1.1 The Jacobi identity in terms of structural constants is written as

$$c_{is}^p c_{jk}^s + c_{js}^p c_{ki}^s + c_{ks}^p c_{ij}^s = 0. \tag{1.9}$$

Thus, the variety of Lie algebras of dimension n can be regarded as the manifold given by $n^2(n-1)(n-2)/6$ quadratic equations (1.9) in $n^2(n-1)/2$ independent variables.

Lie algebras can obviously be considered irrespective of Lie groups. However, we are just interested in the relations between Lie algebras and Lie groups.

Consider some examples.

Example 1.11 (1) A full matrix group is the set $GL(n, R)$ of all nonsingular n-square matrices. The Lie algebra $\mathscr{G}L(n, R)$ is the set of all n-square matrices. The commutator of two matrices $A, B \in \mathscr{G}L(n, R)$ is defined as $[A, B] = AB - BA$.

(2) An orthogonal group $O(n, R)$ is the set of all matrices acting in R^n and preserving scalar product. The matrices satisfy the condition $AA^t = A^tA = I$. The Lie algebra $\mathscr{O}$ is the algebra of all skew-symmetric matrices $a : a^t = -a$. The commutator is determined as in (1). The group $O(n)$ has the subgroup $SO(n)$ of orthogonal matrices with det $= 1$. The Lie algebra $s\mathscr{O}(n)$ of the group $SO(n)$ coincides with $\mathscr{O}(n)$.

The result is a particular case of the general fact to be considered below (namely, that two groups isomorphic in a neighborhood of the identity, i.e., locally isomorphic groups, have the same Lie algebras).

(3) The unitary group $U(n)$ is the group of complex matrices with the condition $AA^+ = A^+A = I$, $A, A^+ \in U(n)$, $U(n) \subset GL(n, C)$ $(A^+ = \bar{A}^t$ is the Hermitian adjoint of A). The Lie algebra $u(n)$ is the algebra of all skew-Hermitian $\tilde{a}$, i.e., $\tilde{a} = -\tilde{a}^+$.

The $U(n)$ group contains a continuous subgroup $SU(n)$ of unitary matrices with det $= 1$. It is obvious that the latter condition leads to distinguishing in $u(n)$ the subalgebra $su(n)$ of skew-Hermitian matrices with tr $\tilde{a} = 0$.

(4) The commutative (or Abelian) group is $R^d \times T^q$ (T^q is a q-dimensional torus). The Lie algebra is also an Abelian algebra and isomorphic to the space $R^d \times R^q$ with the commutator $[A, B] = 0$.

This Lie algebra is obviously isomorphic to that of the group R^{d+q}.

(5) The group N of upper (lower) triangular n-square matrices with units on the principal diagonal. N is an example of a nilpotent group.

Let $A = \begin{pmatrix} 1 & & * \\ & 1 & \\ & & \ddots \\ 0 & & & 1 \end{pmatrix} \in N$. Then each element $\tilde{a}$ of the Lie algebra n is representable as $\tilde{a} = \begin{pmatrix} 0 & & * \\ & 0 & \\ & \ddots & \\ 0 & & 0 \end{pmatrix}$, i.e., n consists of upper (lower) triangular matrices with zeroes on the principal diagonal. This algebra (group) is said to be *Heisenberg-Weyl.* Its irreducible representations are called *commutation relations* and are of importance for the method of second quantization and the theory of coherent states (cf. Subsec. 1.3.3).

In conclusion, we formulate the concepts of Lie subalgebra, ideal and homomorphism of Lie algebras. The definitions are parallel to the corresponding ones for Lie groups.

Definition 1.19 A *Lie subalgebra* $\mathscr{G}_1$ *of a Lie algebra* $\mathscr{G}$ is the set of elements $g \in \mathscr{G}$ forming a subspace $\mathscr{G}_1$ of $\mathscr{G}$ such that, for any $g_1, g_2 \in \mathscr{G}_1$, $[g_1, g_2] \in \mathscr{G}_1$.

The *ideal* $\mathscr{I}$ is the subalgebra of $\mathscr{G}$ such that for any $g_1 \in \mathscr{G}$, $j \in \mathscr{I}$, $[g_1, j] \subset \mathscr{I}$.

Definition 1.20 A *homomorphism* Hom: $\mathscr{G} \to \mathscr{H}$ is the linear transformation $\varphi : \mathscr{G} \to \mathscr{H}$ preserving commutators.

1.2.3 Relation of a Lie Group to the Lie Algebra

Consider two Lie groups $SO(3)$ and $SU(2)$. Construct the homomorphism $SU(2) \to SO(3)$. We specify $SU(2)$ as the set of matrices of the form $u = \begin{pmatrix} \alpha, & \beta \\ -\bar{\beta}, & \bar{\alpha} \end{pmatrix}$ with $\det u = 1$. Hence, $SU(2) \sim S^3 \subset R^4 \sim C^2$. We consider $SO(3)$ as the group of rotations of the two-dimensional sphere S^2, not writing out its explicit parametrization (say, in terms of the Eulerian angles).

Consider S^2: $x^2 + y^2 + z^2 = 1/4$ of radius $r = 1/2$, mapping it via the stereographic projection from the North pole $(0, 0, 1/2)$ onto the plane W tangent to S^2 at the South pole $(0, 0, -1/2)$. We introduce complex coordinates $w = w_1 + iw_2$ on W. It is easy to see that each point (x, y, z) on S^2 is associated with the point $w = (w_1, w_2)$ with coordinates $\left(\frac{x}{-1/2 + z}, \frac{y}{-1/2 + z}\right)$. Each linear fractional transformation u on W completed with ∞, associated with the group $SU(2)$, is given as

$$w \to \frac{\alpha w + \beta}{-\bar{\beta} w + \bar{\alpha}}.$$

The transformation is associated with a rotation $g \in SO(3)$ of S^2. It is obvious that each g is associated with two matrices u and $-u$. Thus, $SO(3) \sim SU(2)/Z_2$, where $Z_2 = \left\{\begin{pmatrix} 1 & 0 \\ 0 & 1 \end{pmatrix}, \begin{pmatrix} -1 & 1 \\ 0 & -1 \end{pmatrix}\right\}$. $SU(2)$ is isomorphic to $SO(3)$ in a sufficiently small neighborhood of the identity. Therefore, these groups are said to be *locally isomorphic.* Globally, they are not. However, their Lie algebras are isomorphic, since Lie algebras are only defined in a neighborhood of the identity of a Lie group.

We now formulate the corresponding proposition showing the relationship between a Lie group and its Lie algebra.

Theorem 1.2 *Let G be a Lie group. Then there exists a one-to-one correspondence between connected subgroups of G and subalgebras of the Lie algebra.*

We now study in more detail the relation of a Lie algebra to its Lie group. Consider the space of matrices $\mathscr{G}L(n, R)$, already encountered, and define the exponentiation

$$\exp A = \sum_{0}^{\infty} (1/k!) A^k, \quad A^0 = I.$$

The series converges in norm in the space R^{n^2} with which the space of $n \times n$ matrices is identified.

Consider the one-parameter subgroup $\exp(tA)$. It is obvious that

$$\exp(t + s) A = \exp(tA) \exp(sA), \quad \exp(A) \exp(-A) = I.$$

Therefore, the curve $s \rightarrow \exp(sA)$ is a one-parameter subgroup of $GL(n, R)$. The left-invariant vector field is determined by the tangent vector at the identity element of the group G, i.e., at the point I: $\left.\dfrac{de^{sA}}{ds}\right|_{s=0} = A$. Hence, for a matrix group, the exponential map defines a map of a Lie algebra into its Lie group. The general definition models this situation.

Definition 1.21 Let G be a Lie group, $X \in \mathscr{G}$, and γ_X an integral curve of the field X, starting at the identity element of G. Then the *exponential map* $\mathscr{G} \rightarrow G$ is the map which sends X to the point $\gamma_X(1) = \exp X$ and coincides with $t \rightarrow \exp(tX)$.

It is easy to see that two groups G_1 and G_2 such that $G_1/C \sim G_2$ (C is a discrete normal subgroup of G_1) have the same Lie algebras.

Remark 1.2 Generally speaking, the exponential map does not map a Lie algebra onto the whole Lie group. This is characteristic only of special, so-called *exponential*, groups, in particular, containing nilpotent and compact Lie groups.

In the sequel, we mostly consider compact Lie groups.

We conclude with two simple but important characteristics of Lie groups.

1. A Lie group is an orientable manifold. The proof is obvious.

2. A Lie group is parallelizable, which means that n linearly independent vector fields can be selected on an n-dimensional Lie group.

The proof is extremely simple. Let $\mathscr{G}$ be the Lie algebra of a group G, i.e., the tangent space at $e \in G$, generating the frame of n vector fields at the point e.

Consider the transformation $G \rightarrow G$ via left translations $g : x \rightarrow gx$. The vector field is left-invariant at e, and, under the map $e \rightarrow x$, the tangent field at e is carried into the tangent field at x. Since the multiplication operation is continuous, we obtain a continuous field of frames.

The parallelizability property of manifolds is strong but does not hold in any case. For example, among the spheres S^n, only S^1, S^3, S^7, and S^{15} are parallelizable. We show below that parallelizability means that the tangent bundle is trivial.

Representation of a Lie group.

Definition 1.22 A homomorphism of G into the matrix group $GL(n, R)$ is called a *representation of the Lie group G*. A homomorphism of $\mathscr{G}$ into the Lie algebra of $GL(n, R)$ is called a *representation of the Lie algebra* $\mathscr{G}$.

Regular and adjoint representations are of particular interest.

(1) A regular representation is one obtained by left translations on G, i.e.,

$$T_g\colon f(h) \to f(g^{-1}h), \quad g,\ h \in G.$$

(2) Adjoint representations. For this we need the general concept of the automorphism of a Lie group and Lie algebra.

A continuous isomorphism of a Lie group G onto itself is called an *automorphism* of G. The set of all automorphisms of G forms a Lie group aut (G). Each $j \in$ aut (G) generates an automorphism dj of the Lie algebra G and the diagram

$$\begin{array}{ccc} \mathscr{G} & \xrightarrow{dj} & \mathscr{G} \\ \exp\downarrow & & \downarrow\exp \\ G & \xrightarrow{j} & G \end{array}$$

is commutative.

Since dj is a nonsingular linear transformation of $\mathscr{G}$, the map $j \to dj$ carries aut (G) into the group of linear transformations of $\mathscr{G}$ and is obviously a homomorphism since $d(j \circ k) = dj \circ dk$, i.e., a representation of aut (G) on the linear space $\mathscr{G}$ is obtained. When G is connected, the representation is faithful, i.e., its kernel is trivial.

Definition 1.23 The set of interior automorphisms of a group G is a subgroup of aut (G). Each element $x \in G$ sends y into xyx^{-1}, i.e., $j_x : y \to xyx^{-1}$. The mapping Ad $G \to Gl(\mathscr{G})$ defined as Ad $(x) = dj_x$ is called the *adjoint representation.*

Proposition 1.3 Ad *is the representation of G on the linear space* $\mathscr{G}$ (e.g., see the proof in [BC]).

The differential $d(\mathrm{Ad}) =$ ad of the adjoint representation is the adjoint representation of the algebra G and is given as

$$\mathrm{ad}\ (X) : Y = [X, Y], \quad X,\ Y \in \mathscr{G}.$$

The kernel of the adjoint representation of $\mathscr{G}$, i.e., the set of elements $X \in \mathscr{G}$ such that $[X, Y] = 0$ for all Y, is called the *center of the algebra* $\mathscr{G}$.

Consider the set C of elements from G commuting with all elements from G. It is obvious that C, called the *center of the group G*, is a normal subgroup of G. If G is connected, C is the adjoint representations's kernel. Thus, the adjoint representation induces a faithful representation of the group G/C (if G is connected), in particular, the groups $SU(2)$ and $SO(3)$ have isomorphic adjoint representations.

$SU(2)$ and $SO(3)$ belong to the class of semi-simple Lie groups. Semi-simple groups arise in the theory of gauge fields mostly as groups of internal symmetries.

Definition 1.24 A Lie algebra $\mathscr{G}$ is said to be *semi-simple* if it has no Abelian ideals different from $\{0\}$.

This follows from the natural relation of a Lie group to its Lie algebra.

Definition 1.25 A Lie group G is said to be *semi-simple* if it contains no nontrivial, connected, and commutative normal subgroups. A simple Lie group can be defined as one containing no nontrivial, connected normal subgroups as in the case of abstract groups. Accordingly, a simple Lie algebra $\mathscr{G}$ is the Lie algebra without ideals other than $\{0\}$ or $\mathscr{G}$.

We can prove that a semi-simple Lie algebra can be decomposed into the direct sum of simple Lie algebras. The corresponding proposition is also valid for Lie groups (if a finite center is neglected). Hence, the study of semi-simple Lie algebras (groups) can be reduced to that of the simple Lie algebras (groups). One important property follows from the semisimplicity of a Lie algebra.

Proposition 1.4 *A semi-simple Lie algebra has a trivial center* $\{0\}$.

Accordingly, a semi-simple Lie group has a discrete center.

We can isolate a subclass of compact semi-simple (simple) Lie groups K from semi-simple (simple) Lie groups G, assuming K to be a compact manifold.

A semi-simple compact group has a finite center.

The topological structure of a compact semi-simple group is especially important since, from the topological viewpoint, the nontrivial structure of an arbitrary Lie group is determined by its maximal compact semi-simple subgroup [Sem].

1.2.4 Differential Forms with Values in a Lie Algebra

Differential forms with values in a vector space V naturally generalize real- and complex-valued forms. V-valued forms turn out to be a convenient notation for investigating the global properties of gauge fields. For instance, it will be seen that the basic structural equations of the connection and curvature of manifolds can be written in terms of differential forms.

We are mostly interested in forms with values in a Lie algebra.

Definition 1.26 Let M^n be an n-dimensional manifold. A V-valued differential form Ω of degree k is a linear mapping of the space $\wedge^k T_xM \to V$ defined for each $x \in M^n$ (and depending differentiably on x). In other words, Ω is an element of the space $V \otimes \wedge^k T_x^*M$ for any $x \in M$. The set of all V-valued k-forms yields the modulus over the ring of differentiable functions. The space $\wedge_V(M)$ of all the forms is the direct sum of the spaces of V-valued forms of all degrees, namely,

$$\wedge_V(M) = \Sigma \wedge_V^k(M).$$

V-valued differential forms can be defined by selecting an appropriate basis in the space $V = (e^1, \ldots, e^n)$ such that the differential forms have coefficients which are vectors in V, i.e.,

$$\Omega = \omega_i \otimes e^i,$$

where ω_i are ordinary differential forms.

The corresponding operations on V-valued forms in the given basis are

(1) $d\Omega = d\omega_i \otimes e^i$.

(2) Let $l: V_1 \to V_2$ be a linear map. Then there exists a linear map of spaces, $l_\# : \wedge_{V_1}(M) \to \wedge_{V_2}(M)$. It is not difficult to verify that none of the operations depend on the choice of basis.

(3) $l_\#$ commutes with all operations on the forms, e.g.,

$$l_\# d\Omega = dl_\# \Omega, \quad l_\#(L_X\Omega) = L_X(l_\# \Omega),$$

etc.

Let V_1, V_2, and V_3 be three vector spaces and ϱ a bilinear map $V_1 \times V_2 \to V_3$. We define the bilinear map $\varrho_\#: \wedge_{V_1}(M) \times \wedge_{V_2}(M) \to \wedge_{V_3}(M)$ as follows. We select a basis $e^1, \ldots, e^k$ in V_1 and $f^1, \ldots, f^n$ in V_2 and write the forms $\omega = \omega_i e^i \in \wedge_{V_1}(M)$, $\Omega = \Omega_i f^i \in \wedge_{V_2}(M)$. Suppose that

$$\varrho_\#(\omega, \Omega) = \omega_i \wedge \Omega_j \otimes \varrho(e^i, f^j). \tag{1.10}$$

It is easy to see that the definition is independent of the choice of basis (i.e., coordinates).

It follows from formula (1.10) that

$$d\varrho_\#(\omega, \Omega) = \varrho_\#(d\omega, \Omega) + (-1)^q \varrho_\#(\omega, d\Omega) \tag{1.11}$$

if the form ω is to power q and

$$L_X \varrho_\#(\omega, \Omega) = \varrho_\#(L_X\omega, \Omega) + \varrho_\#(\omega, L_X\Omega)$$

for any vector field X on M.

If $\varphi: M \to N$ is a mapping of manifolds, then

$$\varphi^* \varrho_\#(\omega, \Omega) = \varrho_\#(\varphi^*\omega, \varphi^*\Omega).$$

In the theory of gauge fields, the role of the space V_i is played by the Lie algebras of gauge groups. Let $V = V_1 = V_2 = V_3$ be a Lie algebra G, in which case ϱ is the commutator (Lie bracket). We denote $\varrho_\#(\omega_1, \omega)$ by $[\omega_1 \wedge \omega_2]$. If ω_1 and ω_2 are two forms to powers p and q respectively, then $[\omega_1 \wedge \omega_2] = (-1)^{pq+1}[\omega_2 \wedge \omega_1]$ (the factor $(-1)^{pq}$ is due to the law of multiplication of forms in $\wedge_V(M)$ and (-1) to the anticommutativity of the Lie bracket).

The Jacobi identity is written as

$$(-1)^{pr}[\omega_1 \wedge [\omega_2 \wedge \omega_3]] + (-1)^{pq}[\omega_2 \wedge [\omega_3 \wedge \omega_1]] + (-1)^{qr}[\omega_3 \wedge [\omega_2 \wedge \omega_1]] = 0, \tag{1.12}$$

where ω_1 is a form to power p, $\omega_2 \to q$, and $\omega_3 \to r$.

In terms of vector-valued differential forms, the Maurer-Cartan equations for Lie groups can be written in a very convenient way.

We define a (left) invariant $\mathscr{G}$-valued fundamental form ω on a Lie group G which associates each vector $v \in T_g G$ with an element from $\mathscr{G}$, which is regarded as a (left) invariant field and takes the value v at the point g.

If $e_1, \ldots, e_n$ is a basis of the Lie algebra of a group G, then ω is represented as

$$\omega = \omega^i \otimes e_i, \quad \omega^i(e_j) = \delta^i_j,$$

where ω^i is a basis of left-invariant 1-forms. Then the relation

$$d\omega^i = \sum_{j<k} C^i_{jk} \, \omega^j \wedge \omega^k$$

holds, so that

$$d\omega = \sum_{j<k} C^i_{jk} \, \omega^j \wedge \omega^k \otimes e_i. \tag{1.13}$$

On the other hand, since $[e_k, e_j] = C^i_{jk} e_i$, taking into account formula (1.10), we obtain

$$[\omega \wedge \omega] = \omega^k \wedge \omega^j \otimes C^i_{kj} e_i. \tag{1.14}$$

Comparing (1.13) with (1.14), taking into consideration the summation over all i, j, k, and exchanging j and k, we obtain the *Maurer-Cartan equation*

$$d\omega = (-1/2)[\omega \wedge \omega]. \tag{1.15}$$

Exterior differentiation (1.15) leads to the Jacobi identity

$$[\omega \wedge [\omega \wedge \omega]] = 0.$$

Similar reasoning is also valid for right-invariant $\mathscr{G}$-valued forms α. Consider the operation of the group G on itself by right translations. Note that under the map $g \to g^{-1}$, $g \in G$, a left-invariant form ω is sent to a right-invariant form $-\alpha$. Therefore, relation (1.15) is transformed to

$$d\omega = (1/2)[\alpha \wedge \alpha]. \tag{1.16}$$

Later when describing connection forms and gauge fields, it will be convenient to use the following notation for G-invariant forms. We denote by dg the identity endomorphism of the tangent space T_g and consider it an element of the space $T_g \otimes T_g^*$. We have

$$\omega = (L_{g^{-1}})_* dg = g^{-1} dg, \tag{1.17}$$

where $(L_{g^{-1}})_*$ only acts on the first factor in $T_g \otimes T_g^*$. The notation of ω in (1.17) is nothing but the shortening of the left-invariant form. Similarly, a right-invariant form can be written as $\alpha = dgg^{-1}$. However, if the group G can be realized in a matrix form, then the expression $\omega = g^{-1}dg$ is meaningful.

Example 1.12 Let $G = GL(n, R)$ and $\mathscr{G}$ be the set of all $n \times n$ matrices, $\omega = A^{-1}dA$, where $A \in GL(n, R)$ and the matrix product is used.

1.3 ACTION OF GROUPS

1.3.1 Orbits of Groups

In Subsec. 1.2.1 a Lie group was defined as a group of transitive transformations of manifolds. In physics the study of the action of groups in a variety of spaces of states of physical systems seems to be important, e.g., the action of the Lorentz group in invariant relativistic equations or the action of crystallographic groups in a Euclidean space.

We give here general definitions of the action of groups that have applications in the context of the thermodynamic phases of ^{3}He, gauge fields, etc.

Definition 1.27 Let M be a space on which a Lie group G acts and $x \in M$. The subspace $G(x) = \{gx \in M$, where g ranges over all elements from $G\}$ is called the *orbit of the point x with respect to the action of G*. If $g \cdot x = h \cdot y$ for some $g, h \in G$, $x, y \in M$, then, for any g', $g'x = g'g^{-1}gx = g'g^{-1}hy$, i.e., $G(x) \subset G(y)$. The inverse inclusion is also obvious.

We have thereby proved that any two orbits are either disjoint or coincident.

We can define the *orbit space* $\Omega = M/G$ whose elements are orbits $x^* = G(x)$. Two points x^* and y^* from Ω coincide if and only if they are in one orbit.

We define the mapping

$$\pi: M \to M/G$$

associating each point x with its orbit.

Ω is assigned a natural topology (i.e., $\mathscr{U} \subset \Omega$ is open if and only if $\pi^{-1}(\mathscr{U})$ is open in M).

For noncompact groups G, the orbit space may have the trivial nondiscrete topology, an example being an irrational winding number of a torus.

However, for a compact group G, the orbit space is "better" arranged.

We now formulate the basic properties of the orbit space of a compact group G as follows.

Proposition 1.5 *If G is a compact group, then the map* $\varphi: G/G_x \to G(x)$ *is a homeomorphism for any x, where* G_x *is the stability subgroup at the point x.*

For the proof see [Br].

It is obvious that the definition of the orbit generalizes that of the transitive action of a Lie group G on M. In fact, G in this case acts on the space G/H by left translations.

By its definition, the transitive action is that with one orbit.

1.3.2 Crystallographic Groups

Remarkable examples of orbits generated by discrete groups are crystallographic lattices (domains). It is well known that *crystallographic domains* in R^n are "regular" sets in R^n, i.e., domains $\mathscr{U}$ of finite volume that can "tesselate" the whole of R^n. The exact meaning of the term consists in the following: Let $\mathscr{U}$ be a domain in R^n whose closure $\bar{\mathscr{U}}$ is compact in R^n. $\mathscr{U}$ is said to be a *fundamental domain* in R^n if the whole of R^n can be obtained by translations $\Gamma(\bar{\mathscr{U}})$, where Γ is a discrete subgroup of the group of motions $E(R^n)$, the sets $\gamma_i(\bar{\mathcal{U}}) \cap \gamma_j(\mathcal{U})$ being disjoint given $\gamma_i \neq \gamma_j$, γ_i, $\gamma_j \in \Gamma$. Recall that a subgroup Γ of a group G is said to be *discrete* if Γ is a discrete set in G. Under the action of G on a manifold M, Γ acts discretely, i.e., the set $\gamma_i x \in M$ forms a discrete set in M. We can now give a mathematically strict definition.

Definition 1.28 A *crystallographic group* Γ of R^n is a discrete subgroup of the group E of motions on R^n, with a finite volume of E/Γ.

To relate the definition to the geometric description of crystallographic lattices in the space R^n, we specify the action of Γ directly in R^n. Note that E is a semi-direct product $O(n) \times A$, where $O(n)$ is the group of orthogonal transformations, A the group of translations of R^n, and $A \sim R^n$. This representation of E follows from the well-known theorem that any Euclidean motion in R^n can be obtained as the composition of an orthogonal transformation and a translation. A is a normal subgroup of E, while $O(n)$ is not. R^n can be identified with the homogeneous space $E/O(n)$, where $O(n)$ is the stability subgroup of the point O. Γ acts on the space $E/O(n)$ according to the formula

$$\gamma : Hg \to \gamma Hg, \quad \gamma \in \Gamma, \tag{1.18}$$

where $H \sim O(n)$, $g \in E$. Such action can be given for arbitrary Lie groups G, H, and Γ, where H is a closed subgroup of G and Γ a discrete one. The coset space $\Gamma \setminus G/H$ obtained is called the *double coset space.* If the subgroup H is compact, then the space $\Gamma \setminus G$ is compact together with $\Gamma \setminus G/H$.

A subgroup $H \subset G$ is said to the *uniform* if the space G/H is compact.

We can now make a statement characterizing groups acting discretely in R^n.

Proposition 1.6 (1) *A group $\Gamma \subset E$ is discrete and uniform if and only if Γ acts on R^n discretely with the compact coset space.*

(2) Γ *acts freely on R^n if and only if Γ has no elements of finite order (no torsion).*

L. Bieberbach [Wf] is responsible for the basic theorems on the structure of crystallographic groups.

Theorem 1.3 *If $\Gamma \subset E$ is a crystallographic group, then $\Gamma \cap R^n = \Gamma'$ is a normal subgroup in R^n of a finite index in Γ. Any minimal set of generators of Γ' is a basis for R^n. With respect to this basis for $O(n)$, the components of elements of Γ have integral coordinates.*

Theorem 1.4 *For any integer n, there exist only finitely many classes of crystallographic groups isomorphic to each other and acting on R^n. Two crystallographic groups on R^n are isomorphic if and only if they are conjugate in the affine group A^n.*

Recall that an affine group A^n is the semi-direct product $GL(n, R) \dot{\times} A$, where A is the group of translations in the space R^n, A being naturally isomorphic to R^n.

An exact estimate of the number of non-isomorphic crystallographic groups cannot be obtained from the finiteness theorem. In the general case, to find all crystallographic groups which are not isomorphic to each other for $n > 3$ is a difficult unsolved problem.

However, in cases of dimensions 2 and 3, which are of primary interest to real physics, the answer is known. If $n = 3$ (real crystal), the number of different classes of crystallographic groups is 230. This classical result is due to Fyodorov and

Schoenflies. If $n = 2$, there are 17 different crystallographic groups [CM]. It is interesting that, in liquid crystals shaped like disks (see Subsec. 6.2.1), some two-dimensional crystal lattices are realized.

Another interesting class of orbits in infinite-dimensional spaces arises in quantum mechanics when constructing systems of coherent states.

1.3.3 Coherent States

I begin with the classical definition of coherent states for the simplest physical system, a harmonic oscillator with one degree of freedom.

Let Q be the Hermitian coordinate operator and P the momentum operator satisfying the commutation relation

$$[Q, P] = i\hbar,$$

where $\hbar = h/2\pi$ and h is Planck's constant.

We define the creation a^+ and annihilation a operators as follows:

$$a = \frac{1}{\sqrt{2\hbar\omega}}\left(\frac{P}{\sqrt{m}} - iQ\right), \quad a^+ = \frac{1}{\sqrt{2\hbar\omega}}\left(\frac{P}{\sqrt{m}} + iQ\right) \tag{1.19}$$

(ω is the angular frequency for the oscillator. We select the system of units such that $\omega = m = \hbar = 1$.)

It follows from (1.19) that

$$[a, a^+] = 1. \tag{1.20}$$

We can now give the first definition.

Definition 1.29 The eigenvectors* $|\alpha\rangle$ of the annihilation operation a are called *coherent states*

$$a|\alpha\rangle = \alpha|\alpha\rangle, \quad \alpha \in C.$$

The spectrum of a fills the complex plane C. We can show that the system of coherent states forms a complete (and even overcomplete) system of states, in terms of which an arbitrary state $|\psi\rangle$ can be decomposed.

Coherent states possess a number of useful properties which make them indispensible for investigating the statistical properties of physical systems (e.g., multiphoton processes in lasers) [KS]. However, we are now only interested in the mathematics.

We proceed with the harmonic oscillator. Let H be the Hamiltonian of an oscillator with zero vacuum expectation value $|\psi\rangle = |0\rangle$, i.e.,

$$\langle 0|Q|0\rangle = \langle 0|P|0\rangle = 0.$$

---

* Here, we use Dirac's notation for vectors (states) and operators conventional in quantum mechanics.

We define the unitary operator $U(p, q) = \exp[i(pQ - qP)]$ (where p and q are c-numbers). The action of U on the vacuum $|0\rangle$ is

$$|p, q\rangle = U(p, q)|0\rangle. \tag{1.21}$$

The vector $|p, q\rangle$ is the ground state of the oscillator with the coordinate shifted by q and the momentum by p. Thus, the expectation is

$$\langle p, q|Q|p, q\rangle = q, \ \langle p, q|P|p, q\rangle = p.$$

We express U in terms of the creation and annihilation operators as $U(\alpha) = \exp[(\alpha a^+ - \bar{\alpha}a)]$, the action on the vacuum being

$$U(\alpha)|0\rangle = |\alpha\rangle. \tag{1.22}$$

The law of multiplication for $U(\alpha)$ is

$$U(\alpha_2)U(\alpha_1) = \exp[i \ \mathrm{Im}(\alpha_2\bar{\alpha}_1)U(\alpha_2 + \alpha_1)] \tag{1.23}$$

($\mathrm{Im}\ \alpha_2\bar{\alpha}_1 = \mu$ is the area of the cell in C constructed on two vectors α_2 and α_1).

It follows from (1.23) that the operators $\exp(2\pi i t)U(\alpha)$ form a W_1 *group of commutation relations in the Weyl representation.*

The operators a, a^+, and I form a Lie algebra with commutation relation (1.20). If we write the general element of the Lie algebra as

$$tI + i(\alpha a^+ - \bar{\alpha}a), \tag{1.24}$$

then, applying the exponential mapping, we obtain W_1, where each element g is given by a pair (t, α) (t is real and α is complex). The law of composition is

$$g \cdot g_1 = (t, \alpha)(s, \beta) = (t + s + \mathrm{Im}\ \alpha\bar{\beta}, \alpha + \beta). \tag{1.25}$$

The operators $\exp(it)\ U(\alpha)$ define a representation of W_1 with matrix representation which is already known as the nilpotent group

$$W_1 = \begin{pmatrix} 1 & a & b \\ & 1 & c \\ 0 & & 1 \end{pmatrix}.$$

The same construction also holds for the algebra of commutation relations of an n-dimensional oscillator. The corresponding nilpotent group W_n has $2n + 1$ generators.

The classical coherent states are related to the simplest nilpotent Lie group W_n. Because of the great many Lie groups among dynamic symmetry groups, the construction of coherent states for arbitrary Lie groups is of interest.

The corresponding coherent states were introduced in [Per1]. This construction generalizes the result of [R] where the coherent states for rotation groups were built. The generalization due to Perelomov is related to the following observation: Retrace our steps to representation of the coherent states by shift operators $U(\alpha)$. In the Weyl representation the system of coherent states is obtained from the vacuum vector $|0\rangle$ under the action of $U(\alpha)$ defining a representation of the group W_n. From the classical Stone-von Neumann theorem [Per2] it follows that this representation is irreducible. Let G be an arbitrary Lie group, $T(g)$ an irreducible representation of G in a Hilbert space $\mathscr{H}$, and $|\psi_0\rangle$ a vector in $\mathscr{H}$. The orbit of $|\psi_0\rangle$ in $\mathscr{H}$, i.e., the set of vectors $\{|\psi\rangle\} = \{T(g)\psi_0\}$, $g \in G$, is called the *system of coherent states of type* $\{T, |\psi_0\rangle\}$ for G.

The system of generalized coherent states $\{T, |\psi_0\rangle\}$ has all the properties of conventional coherent states and is convenient in considering various physical problems.

Let us consider in more detail the class of orbits determining coherent states.

Let $\{|\psi_0\rangle\}$ be the set of vectors in $\mathscr{H}$, where $|\psi_g\rangle = T(g)|\psi_0\rangle$ and $T(g)$ is an irreducible unitary representation of G, $g \in G$. We assume that two vectors $|\psi_{g_1}\rangle$ and $|\psi_{g_2}\rangle$ define the same state if they differ by the phase factor exp (i, γ), i.e., $\exp(i\gamma)|\psi_{g_1}\rangle = |\psi_{g_2}\rangle$, which is equivalent to the condition $T(g_2^{-1}g_1)|\psi_0\rangle = |\psi_0\rangle$.

The set of elements $\{g\} \in G$ satisfying the condition forms a closed subgroups H. We call H the *stability subgroups* of the vector $|\psi_0\rangle$ (for $\gamma \equiv 0$, this coincides with the conventional definition). It follows from the definition that vectors $\{|\psi_g\rangle\}$ from the same coset differ by the phase factor and determine the same state, i.e., $\{|\psi_g\rangle\}$ only depends on a point $x \in M$ of the homogeneous space $M = G/H$, $|\psi_{g(x)}\rangle = |x\rangle$.

It follows from the irreducibility of the representation $T(g)$ that $\{|\psi_g\rangle\}$ forms a complete (or overcomplete) system of vectors in $\mathscr{H}$. By selecting complete subsystems related to subgroups of the group G we obtain a number of remarkable relations among coherent states. Of special interest is when G is semi-simple and the subgroups Γ are discrete, with compact or with finite volume coset, $\Gamma \setminus G$ [MP1, Mo].

The latter concluding remark closes this interesting subject. The reader may become acquainted with the contemporary state of this branch of mathematical physics in [Per2, Mo].

We now retrace our steps to describe orbit space.

1.3.4 Strata

All points along one orbit obviously have conjugate stability subgroups. However, the converse does not hold. The action of a group G on a manifold M can have several orbits of the same type, i.e., those with isomorphic stability subgroups. Such orbits are of the same type and form a stratum. A *stratum* is the union of all orbits of the same type. Partial ordering of all subgroups of a given group modulo conjugation occurs. In turn, it corresponds to (reverse) ordering in strata. The set of fixed points (with the maximal stability subgroup) forms the minimal stratum. If there are no fixed points under the action of G on M, then there may be several minimal strata. In physical applications, it often suffices to describe strata under the actions of groups or of exhibiting representatives of each stratum.

Example 1.13 (1) The group $SO(n)$ of rotations of the n-dimensional space R^n. The action is rotation.

There are two strata relative to the action: the first consists of spheres of nonzero radius and the second originates from zero.

(2) The Lorentz $SO(3, 1)$ group of motions in Minkowski space.

There are: (a) the spacelike region, (b) the timelike region, (c) the light cone, $t > 0$, (d) the light cone, $t < 0$, and (e) the point zero.

(3) $G = SO(3)$, $S^2(R^3)_0$, the space of real symmetric 3×3 matrices with trace 0, $\dim S^2(R^3)_0 = 5$. More complicated situations arise in studying thermodynamic phases in ^{3}He (see Subsec. 5.2.2).

The action of the group G on $S^2(R^3)_0$ is

$$g: m = g \cdot m g^{-1}, \quad g \in SO(3), \quad m \in M. \tag{1.26}$$

Since each symmetric matrix can be reduced by orthogonal transformation (1.26) to diagonal form, it is obvious that two matrices belong to the same orbit if and only if they have the same values of λ_i. Describe the corresponding orbits. Since $\lambda_1 + \lambda_2 + \lambda_3 = 0$, there are two independent eigenvalues. For definiteness, let them be λ_1 and λ_2. Ordering them, we have $\lambda_1 \geqslant \lambda_2 \geqslant \lambda_3$.

1. Let all λ_i be different, i.e.,

$$\lambda_1 > \lambda_2 > \lambda_3.$$

We now find the stability subgroup H of the point

$$m_0 = \begin{pmatrix} \lambda_1 & & 0 \\ & \lambda_2 & \\ 0 & & \lambda_3 \end{pmatrix} \in M.$$

H consists of the matrices

$$\begin{pmatrix} \pm 1 & & 0 \\ & \pm 1 & \\ 0 & & \pm 1 \end{pmatrix}$$

and is isomorphic to the group $Z_2 \oplus Z_2$, since $\det g = 1$.

2. Let $\lambda_1 = \lambda_2 > \lambda_3$. H of the point

$$m_1 = \begin{pmatrix} \lambda_1 & & 0 \\ & \lambda_2 & \\ 0 & & \lambda_3 \end{pmatrix}$$

consists of the matrices

$$G \ni g = \begin{pmatrix} * & & 0 \\ & & 0 \\ 0 & 0 & \pm 1 \end{pmatrix}$$

forming the normalizer of the subgroup $SO(2)$ of $SO(3)$. The orbit $G/H = SO(3)/SO(2) \times Z_2$ is isomorphic to RP^2.

The case $\lambda_1 > \lambda_2 = \lambda_3$ can be considered in a similar way. Besides, there is a minimal stratum $\lambda_1 = \lambda_2 = \lambda_3$. The stability subgroup of this point coincides with the whole of $SO(3)$.

We now present the elegant geometric interpretation of orbits given by Lawson [La] (see also [HL]). We need to introduce several general notions which are of independent interest.

Each orbit of a group G has quite definite volume, i.e., that of the corresponding submanifold in an induced metric. The volume V of a given orbit Ω is extremal among close orbits, i.e., those with conjugate stability subgroups if

$$\left. \frac{dV(\Omega_t)}{dt} \right|_{t=0} = 0$$

for all smooth families of the orbits Ω_t ($|t| < \varepsilon$, $\Omega_0 = \Omega$) on M. If Ω is isolated, i.e., if there is no other orbit with a conjugate stability subgroup in the vicinity of Ω, then the volume $V(\Omega)$ is obviously extremal. The following is due to Hsiang.

Theorem 1.5 *Let G be a compact group of motions in a Riemannian space M. Then any orbit Ω in M of extremal volume in the vicinity of close orbits of the same type is the minimal submanifold in M, i.e., of zero mean curvature.*

We now retrace our steps to a realization of orbits in the space $S^2(R^3)_0$. The scalar product on $S^2(R^3)_0 \sim R^5$ is defined by the formula $\langle A | B \rangle = \mathrm{Tr}\, AB$ (a particular case of the Killing metric; we took into account the fact that the matrices A and B are symmetric). The general definition of the Killing metric is given in Subsec. 1.5.8. Computation of the volume function $V(\Omega)$ for an orbit in general position yields $V(A) = c(\lambda_1 - \lambda_2)(\lambda_2 - \lambda_3)(\lambda_3 - \lambda_1)$, where λ_i are the eigenvalues of A. If we restrict the function V to the unit sphere S^4: $\lambda_1^2 + \lambda_2^2 + \lambda_3^2 = 1$ in $S^2(R^3)_0$, then $V(A)$ attains its maximum at the point $(\lambda_1, \lambda_2, \lambda_3) = (1/\sqrt{2}, 0, -1/\sqrt{2})$. The subset of matrices $\{A\} \subset S^2(R^3)_0$ with $|A| = 1$ and rank 2 forms a hypersurface in S^4. The orbit in the general position is $\Omega \sim SO(3)/Z_2 \oplus Z_2$ and has an eight-sheeted covering by the sphere S^3. The embedding of S^3 in S^4 is thereby given rise with the metric of nonconstant curvature induced on S^3.

The singular orbits ($\lambda_1 = \lambda_2$ or $\lambda_2 = \lambda_3$) are isolated from orbits of the same type; therefore, by the Hsiang theorem, they are minimal. Each singular orbit is the projective plane with constant Gaussian curvature minimally embedded in S^4.

The embedding can be given in the following explicit form. Consider the embedding of S^2 placed in R^3 and of radius $\sqrt{3}$, in the unit sphere S_1^4, namely,

φ: $S^2_{\sqrt{3}} \to S^4_1$, where

$$\varphi(x, y, z) = \left\{xy, xz, yz, \frac{1}{2}(x^2 - y^2), \quad \frac{1}{2\sqrt{3}}(x^2 + y^2 - 2z^2\right\} \tag{1.27}$$

and

$$S^2_{\sqrt{3}} = \{x, y, z \in R^3, x^2 + y^2 + z^2 = 3\}.$$

Since $\varphi(-x, -y, -z) = \varphi(x, y, z)$, the mapping φ specifies the minimal isometric embedding of RP^2 in S^4_1, the curvature of RP^2 being 1/3. Manifold (1.27) is called a *Veronese surface*.

Later we shall study the action of groups in a variety of situations: gauge groups of the Yang-Mills equations, orbits of invariance groups for potentials of free energy of liquid crystals, ^{3}He, etc.

The internal symmetry groups encountered in physics are compact. Noncompact ones basically arise as groups of space-time transformations or the second quantization method. The compactness property of internal symmetry groups is mostly related to the discreteness of quantum numbers (i.e., the spectrum) characterizing the system (the eigenvalues of Casimir operators generate a finite-dimensional space). In the next example, Cartan's theorem is proved thus providing a geometrical description of a compact groups's orbits.

In physical problems, group orbits often arise as the extremum points of G-invariant functions, say, minima of potentials in field-theory equations or as a Ginzburg-Landau potential in the theory of phase transitions.

We now formulate the corresponding results [MR].

Proposition 1.7 *Let G be a compact group acting differentiably on a real compact manifold M and f a real-valued smooth G-invariant function on M. Then f has at least one critical point for each connected component of any minimal stratum.*

That orbits arise in investigating extrema of G-invariant functions is quite natural.

Indeed, let $f(x)$ have an extremum at a point x.

It is obvious that $f(x)$ assumes the same value on the manifold $G(x)$, i.e., on the orbit of x.

1.3.5 Cartan's Theorem

Here, we discuss Cartan's theorem [C] on the orbits of compact Lie groups used in classification of thermodynamic phases. The proof involves a number of concepts of differential geometry and the theory of symmetric spaces [He], [EF] and can be ommitted in first reading.

As usual, we begin with examples.

Let G be the group $SO(3)$. Consider the action of G on $S^2 \subset R^3$. It is obvious that $S^2 = G/H$, where $H \sim SO(2)$ are rotations about the z-axis. The space S^2 can be embedded in $SO(3)$ as a submanifold. Cartan's theorem supplies the answer as to whether any orbit of the group can be embedded in G as a submanifold.

Definition 1.30 A manifold M such that a geodesic γ in N with ends in M wholly lies in M is called a *totally geodesic submanifold* of N.

Theorem 1.6 *Any orbit of a group G can be embedded in G as a totally geodesic submanifold.*

Proof. Consider the map Φ

$$\Phi: G/H' \to G,$$
$$\Phi: gH \to \sigma(g)g^{-1},$$

where σ is an involutive automorphism of the group G, $\sigma^2 = e$, $\sigma \neq e$, $(G_\sigma)_0 \subset H \subset G_\sigma$, G_σ is a set of fixed points of σ, and $(G_\sigma)_0$ is the component of the identity of G_σ.

We prove that Φ is the diffeomorphism G/H onto the closed totally geodesic submanifold

$$M_\sigma = \{g \in G \mid \sigma(g)g = e \tag{1.28}$$

(e is the identity element of G).

We recall the properties of σ necessary below.

Let d_{σ_e} be the differential of σ at the point e. d_{σ_e} is the involutive automorphism of the Lie algebra $\mathscr{G}$ of G. The operator d_{σ_e} separates the space G into two subspaces

$$\mathfrak{h} = \{x \in \mathscr{G} / d_{\sigma_e}(X) = X\}, \quad \varrho = \{X \in \mathscr{G} \mid d_{\sigma_e}(X) = -X\}$$

corresponding to the eigenvalues ± 1 of d_{σ_e}.

We can show that $\mathfrak{h}$ is the Lie algebra of H and ϱ the subspace orthogonal to $\mathfrak{h}$ in $\mathscr{G}$ (relative to the scalar product on $\mathscr{G}$).

ϱ is invariant to the adjoint representation Ad (H) [He].

The proof is reduced to verifying three statements:

1. Show that M_σ is a closed submanifold in G and dim M_σ = dim G/H.

Consider a map $\psi: G \to G$, i.e.,

$$g \to \psi(g) = \sigma(g)g. \tag{1.29}$$

The space M_σ is the kernel of ψ. To find dim M_σ we calculate the kernel $\operatorname{Ker} T_g = gk_g$ of the tangential map $T_g\psi: T_gG \to T_{\sigma(g)g}G$ at the point g, where

$$k_g = \{X \in \mathscr{G} \mid d_{\sigma_e}(X) + \operatorname{Ad}(g)X = 0\}, \tag{1.30}$$

i.e., Ker $T_g\psi = \varrho$, the rank of ψ is dim $\mathscr{G}$ − dim Ker ψ = dim $\mathfrak{h}$ and is independent of the choice of g. It follows from the theorem on a map with constant rank that M_σ is a closed submanifold in G and that dim M_σ = dim G/H.

2. We prove that M_σ is a totally geodesic submanifold in G. It is known that geodesics in the space G/H (relative to the G-invariant metric) are orbits of one-parameter subgroups of the group G [He]. Let γ be a geodesic in G tangent to

M_σ at the point g:

$$\gamma = g \exp(tX),\ X \in \mathscr{G}, \quad g \cdot X \in T_g M_\sigma = \operatorname{Ker} T_g \psi = g k_g.$$

We have

$$\sigma(\gamma_t)\gamma_t = \sigma[\ g \exp(tX)]g \exp(tX) = \sigma(g) \exp(td\sigma_e(X))\ g \exp(tX)$$

and, taking (1.30) into account,

$$\sigma(g) \exp(td\sigma_e(x))\ g \exp(tX) = \sigma(g) \exp(-\operatorname{Ad}(g)tX)\ g \exp(tX).$$

Since $\exp(\operatorname{Ad}(g)X) = g \exp(Xg^{-1})$, we have finally

$$\sigma(\gamma_t)\gamma_t = \sigma(g)\ g \exp(-tX)\ g^{-1}\ g \exp(tX) = \sigma(g)g = e.$$

3. It remains to prove that $M_\delta = \operatorname{Im}\Phi$. (Here $\operatorname{Im}\Phi$ means the image of Φ.) The inclusion $M_\sigma \supset \operatorname{Im}\Phi$ is obvious. We must prove that $M_\sigma \subset \operatorname{Im}\Phi$. Let g_0 be a point in M_σ. There exist finitely many points g_i in M_σ and geodesics γ_i in M_σ connecting g_{i-1} to g_i, $1 \leqslant i \leqslant N$, such that $g_N = \sigma(\tilde{g}_N)\tilde{g}_N^{-1} \in \operatorname{Im}\Phi$.

Let $g_{N-1} = g_N \exp(tX)$, where $X \in k_{g_N}$ for a value of the parameter t. Show that $g_{N-1} = \sigma(\tilde{g})\tilde{g}^{-1}$. We have

$$\begin{aligned} &\sigma\left[\exp\left(-\frac{1}{2}tX\right)g_N\right]\left[\exp\left(-\frac{1}{2}tX\right)g_N\right]^{-1} \\ &\quad= \exp\left[-\frac{1}{2}td_{\sigma_e}(X)\right]\sigma(g_N)g_N^{-1}\exp\left(\frac{1}{2}tX\right) \\ &= \exp\left[\frac{1}{2}t\operatorname{Ad}(g_N)X\right]g_N\exp\left(\frac{1}{2}tX\right) = g_N\exp(tX) = g_{N-1}. \end{aligned}$$

Repeating the procedure consecutively for g_i, $0 \leqslant i \leqslant N-1$, we obtain that $g_0 \in \operatorname{Im}\Phi$.

Remark 1.3 Since for each compact subgroup $H \subset G$ there exists an involutive automorphism σ leaving H fixed, all orbits of G can be realized as totally geodesic manifolds.

1.4 FIBER BUNDLES

Here, we provide some basic information about the theory of fibrations, i.e., the geometry of the theory of gauge fields. The theory of fibrations is compared with gauge-fields theory in the Chapter "Physical Structures".

1.4.1 Definition of a Fiber Bundle. Examples

To clarify the reason for introducing fiber bundles, we compare two examples.

Consider the Möbius strip and the cylinder. We know that they are not homeomorphic (the latter being orientable and the former not). However, if we select small neighborhoods of an arbitrary point on the cylinder and Möbius strip, then the spaces are easily seen to be arranged in a similar manner locally. Such a neighborhood admits a representation as the union of pairs of points x, y, where $x \in S^1$, $y \in I$ (I is an interval) (Fig. 2).

If we make x range over the whole circle S^1 and make I depend on x, then for the cylinder case I turns out to preserve orientation, returning to the initial point x; in the Möbius strip case, I reverses the orientation and turns through 180°. Thus, though the Möbius strip and cylinder locally are made similarly, they are different globally. The topologist will say that the cylinder is the *trivial* bundle and the Möbius strip *nontrivial.*

Another familiar example of a fiber bundle is the set of tangent vectors to an oriented two-dimensional surface.

We now give a formal definition of a *fiber space* (or *bundle*, or *fibration*; also called a *twisted product*, a *Steenrod bundle*, a *locally trivial fiber space*).

Definition 1.31 A fiber space $\xi = (E, p, F, B)$ is a composite object with the following elements:

1. The *total space E.*
2. The *base space B.*
3. A continuous map $p: E \to B$ (onto B) called the *projection.*
4. The fiber F. The full inverse image $F_x = p^{-1}(x) \subset E$ can be defined over each point $x \in B$. The set F_x is called the *fiber* over x. The requirement that fibers over different points should be homeomorphic is valid. Thus, the concept of the fiber F is meaningful, irrespective of the point of the base space.

Finally, another property showing that a fiber bundle must be locally equivalent to the direct product holds.

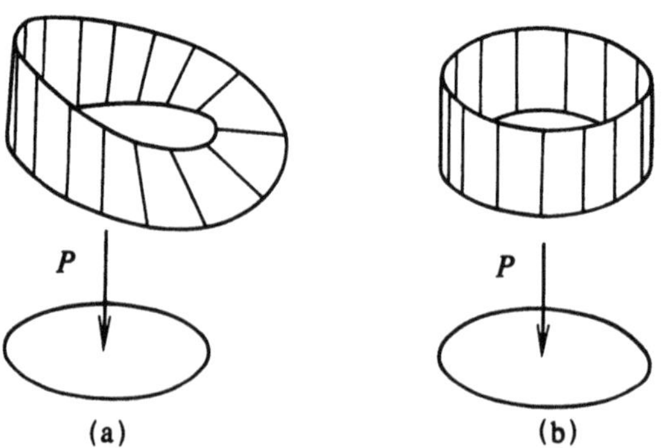

Fig. 2. The (a) Möbius strip and cylinder (b)

5. For any point $x \in B$ a neighborhood $\mathscr{V}$ and a homeomorphism $\varphi: \mathscr{V} \times F \to p^{-1}(\mathscr{V})$ exist so that

$$p\varphi(x', y) = x', \quad x' \in \mathscr{V}, \quad y \in F.$$

The most interesting examples are related to fiber bundles with the group G of homeomorphisms of the fiber acting in it in a fixed way.

Consider the Möbius strip, an example already familiar to us. The fiber bundle base space is the circle S^1, the fiber is the interval I. It is easy to see that the group $G = Z_2$ acts in the fiber

$$g: y \to \exp(i\pi)y, \quad g \neq e, \quad ey = e.$$

We now sharpen the definition of a fiber bundle, including the properties of G acting in the fiber. The action of the group is assumed to be free (i.e., $gy = y$, $y \in F$, entails $g = e$).

6. We specify the family of open subsets $\mathscr{V}_j$ covering all B and indexed by j. These *neighborhoods* are said to be *coordinate*. We define the homeomorphism

$$\varphi_j: \mathscr{V}_j \times F \to p^{-1}(\mathscr{V}_j)$$

(i.e., coordinate functions). The coordinate functions should satisfy the following conditions:

7. $p\varphi_j(x, y) = x$, where $x \in \mathscr{V}_j$, $y \in F$.

8. If the map $\varphi_{j,x}: F \to p^{-1}(x)$ is determined by the formula

$$\varphi_{j,x}(y) = \varphi_j(x, y),$$

then, for any two elements $i, j \in J$ and any $x \in \mathscr{V}_i \cap \mathscr{V}_j$, the homomorphism

$$g_{ji}(x) = \varphi_{j,x}^{-1}\varphi_{i,x}: F \to F$$

is generated by an action (i.e., an element) of the group G, which is unique due to the freeness of the action of G.

9. The map $g_{ji}: \mathscr{V}_i \cap \mathscr{V}_j \to G$, where $g_{ji}(x) = \varphi_{j,x}^{-1}\varphi_{i,x}$, is continuous.

The functions g_{ji} are called *coordinate transformations*. The properties below follow from their definition:

10. $g_{ij}g_{jk} = g_{ik}$ or $g_{ij}g_{jk}g_{ki} = e$ *for* $x \in \mathscr{V}_i \cap \mathscr{V}_j \cap \mathscr{V}_k$.

11. $g_{ij}g_{ji} = e$.

An analog of the concept of the graph of a function is a section of the fiber bundle.

Definition 1.32 Given a fiber bundle $E \xrightarrow{p} B$, a *cross-section* or *section* s of the fiber bundle is a continuous map $s: B \to E$ such that $ps(b) = b$ for any $b \in B$.

The existence of global sections imposes severe restrictions on the type of fiber bundles (see Subsec. 2.7.5).

An important example of fiber bundles are *principal bundles* (P, G, B).

Definition 1.33 A fiber bundle is said to be *principal* if the fiber is a group G. G acts on itself by translations. For definiteness, we make use of left translations.

Principal bundles (P, G, B) are related to free actions of groups on manifolds. The following precise statement is equivalent to the definition of a principal bundle.

Proposition 1.7′ *A group G acts freely (by right translations) on the space P. The orbits of action of G on P are in one-to-one correspondence with points of the base space B, i.e., if q_1 and q_2 are any two points from P, then $p(q_1) = p(q_2)$ if and only if $q_1 g = q_2$ for $g \in G$.*

The *proof* is not complicated and is left to the reader (see also [DFN, Steel]). Examples of principal bundles are repeatedly used below.

To make the set of examples more complete, consider some others; most were encountered above.

Example 1.14 (a) Coordinate functions of the Möbius strip. We cover the base space $B = S^1$ with two open sets $\mathscr{V}_1$ and $\mathscr{V}_2$. The intersection $\mathscr{V}_1 \cap \mathscr{V}_2$ consists of two components $\mathscr{V}$ and $\mathscr{W}$. We put $g_{12} = h$ (where h is a nontrivial element of the group Z_2), $g_{11} = e$, $g_{22} = e$, $g_{21} = [g_{12}]^{-1} = h$.

(b) Coverings. Fiber bundles with discrete fibers are called *coverings*, or *covering fiber bundles*. The simplest example is the covering of the circle S^1 with a real straight line R^1. The corresponding map $p: R^1 \to S^1$ is exponential. The fiber is the set of integers Z. We have $S^1 \sim R^1/Z$. We form the direct product of n circles, i.e., the n-dimensional torus T^n. The covering space for T^n is R^n.

The covering of the real projective plane RP^2 is S^2 (the fiber consisting of two points), which immediately follows from the construction of RP^2. A similar result also holds for RP^n. The covering space is S^n. The classical examples of covering spaces are the n-sheeted coverings of Riemann surfaces. However, coverings are not fiber bundles because of the existence of branch points. Transition to covering spaces in many cases simplifies the investigation of the topology of the original space.

Classification of coverings is given in Subsec. 2.1.4.

Example 1.15 Let G be a Lie group and H a closed subgroup. Consider the quotient space G/H. Under what conditions does the triple G, H, G/H form a fiber bundle? The answer is formulated as a theorem.

Theorem 1.7 *The triple G, H, G/H can always be turned into the fiber bundle*

$$G \xrightarrow{H} G/H$$

with the base space G/H and fiber H. The projection p associates any element $g \in G$ with its coset gH. The group of H acts by left translations.

This theorem admits the following useful generalization:

Let two closed subgroups H and K be embedded in a Lie group G, so that $G \supset H \supset K$. Consider quotient spaces G/K and G/H.

Theorem 1.8 *The space G/K is a fiber bundle over the base space G/H with fiber H/K and structure group H/K acting as the group of left translations, where K_0 is the largest normal subgroup of H belonging to K.*

These theorems are extremely important for applications, since most fiber bundles arising in physical problems are similar in type.

That homogeneous spaces can be fibered enables us to calculate their topological invariants. by using the algebraic technique of exact sequences (see Subsec. 2.1.1).

Consider several particular cases:

(a) $G = SU(n)$, $H = SU(n-1)$, $G/H = SU(n)/SU(n-1)$.

If each unitary matrix of order n with $\det A = 1$ is associated with its first row, then we obtain a fiber bundle $SU(n) \xrightarrow{SU(n-1)} S^{2n-1}$ whose fiber is the group $SU(n-1)$.

Similarly, we can consider the cases:

$$G = SO(n), \quad H = SO(n-1), \quad B = G/H \sim S^{n-1},$$
$$G = \mathrm{Sp}(n), \quad H = \mathrm{Sp}(n-1), \quad B = G/H \sim S^{4n-1},$$

where Sp is the standard notation of a real compact symplectic group [He].

(b) $G = SU(n)$, $H = SU(n-1)$, $K = SU(n-k)$, $V^C_{n,k} = G/K = SU(n)/SU(n-k)$ is the Stiefel manifold.

We have the fiber bundle

$$SU(n)/SU(n-k) \xrightarrow{SU(n-1)/SU(n-k)} SU(n)/SU(n-1) \tag{1.31}$$

and see that $V^C_{n,k}$ is that over S^{2n-1} with the Stiefel manifold of lesser dimension as a fiber. We can consider the sequence of fiber bundles

$$SU(n) \to V^C_{n,n-1} \to \dots \to V^C_{n,1} = S^{2n-1}, \tag{1.32}$$

the fiber of each type $V^C_{n,n-k} \to V^C_{n,n-k-1}$ being the sphere S^{2n-1} and the structure group being $SU(n)$.

The fiber bundle whose fiber is S^{2n-1} and the structure group the unitary group $SU(n)$ is called a $(2n-1)$-dimensional sphere bundle.

(c) $G = O(n)$, $H = O(n-1)$, $K = O(n-k)$.

In the real case, the situation is similar. We have the fiber bundle

$$V_{n,k} = O(n)/O(n-k) \xrightarrow{O(n-1)/O((n-k) \sim V_{n-1,k}} O(n)/O(n-1) = S^{n-1}, \tag{1.33}$$

where $V_{n,k}$ is a real Stiefel manifold.

Similarly to (1.3.2), we can consider the sequence of fiber bundles

$$V_{n,n} = O(n) \to V_{n,n-1} \to \dots \to V_{n,1} = S^{n-1}.$$

The corresponding fiber bundle $V_{n,n-k} \to V_{n-k-1}$ is called the $(n-1)$-dimensional sphere bundle.

In the real case, the fiber bundle

$$SO(n)/SO(n-k) \to SO(n)/SO(n-1)$$

can be naturally considered as well.

However, this coincides with (1.33). Indeed, the fiber bundle $V_{n,n} \to V_{n,n-1}$ has the zero-dimensional sphere S^0 as its fiber. The fiber bundle is not a two-sheeted covering, though $V_{n,n}$ does consist of two components (only one of them $SO(n)$ is a subgroup of $O(n)$, the other with det $= -1$ not forming a group). Since $g_1 \pm e \in O(i)$, det $g_1 = -1$, $V_{n,k} \to V_{n,n-1}$ topologically maps each component of $O(n)$ onto $V_{n,n-1}$. Therefore, we can identify $V_{n,k}$ with $\tilde{V}_{n,k} = SO(n)/SO(n-k)$, i.e., Stiefel manifolds are orientable.

If we turn to Grassmann manifolds, then the situation is different. Real Grassmannians were defined as spaces of k-dimensional planes in an n-dimensional space or as homogeneous spaces

$$G_{n,k} = O(n)/O(n-k) \times O(k).$$

Since one of the subgroups $O(n-k)$ and $O(k)$ contains a transformation with the negative determinant of the coefficients, the projection $O(n) \to G_{n,k}$ maps $SO(n)$ onto $G_{n,k}$. Consider the rotation subgroups $SO(n-k)$ and $SO(k)$.

The oriented Grassmannian is

$$\tilde{G}_{n,k} = SO(n)/SO(n-k) \times SO(k).$$

It is easy to see that $\tilde{G}_{n,k}$ is a two-sheeted covering of $G_{n,k}$ (since both spaces are connected, and the fiber is the zero-dimensional sphere).

1.4.2 Operations on Fiber Bundles

Here we introduce basic operations on fiber bundles, enabling us to construct new fiber bundles related to a given one. Suppose a principal bundle $P \xrightarrow{G} B$ is given. What fiber bundles are there with the same base space B and structure group G?

Such fiber bundles can obviously be determined by specifying the fiber F and action of G in it. The corresponding bundles are said to be *associated* with P.

All real Stiefel manifolds $V_{n,k}$ (see Example 1.10) are, for instance, associated with the principal bundle

$$O(n) \xrightarrow{O(n-1)} S^{n-1}.$$

Definition 1.34 Two bundles ξ and ξ' with the same base space, fiber and group are said to be *equivalent* if

(a) the corresponding map $h: E \to E'$ preserves the fibers and

(b) the coordinate transformations are *conjugate*.

The exact meaning of (a) is that if h maps the fiber F_x homeomorphically onto the fiber F'_x, then

(1) the diagram

$$\begin{array}{ccc} E & \xrightarrow{h} & E \\ p \downarrow & & \downarrow p' \\ B & \xrightarrow{\bar{h}} & B \end{array}$$

is commutative, where $\bar{h}$ is the map of the base space B onto itself induced by h, and

(2) the map $h_x: F'_x \to F_x$ generated by $\bar{h}: x' \to \bar{h}(x)$ induces the group transformation of the fiber on itself, $\bar{g}_{kj}(x)$, $x \in U_j \cap \bar{h}^{-1}(U'_k)$,

$$\bar{g}_{kj}(x) = \varphi'^{-1}_{k,x'} h_x \varphi_{j,x} = p'_k \bar{h}_x \varphi_{j,x}. \tag{1.34}$$

Transformation (1.34) is generated by an element g of G and is continuous.

Two coordinate transformations g and $\bar{g}$ are said to be *conjugate* if they satisfy the relationships

$$\bar{g}_{kj}(x) g_{ji}(x) = \bar{g}_{ki}(x), \quad x \in U_i \cap U_j \cap \bar{h}^{-1}(U'_k), \tag{1.35}$$

$$g_{ik}(\bar{h}(x)) \bar{g}_{kj}(x) = \bar{g}_{ij}(x), \quad x \in U_j \cap \bar{h}^{-1}(U'_k \cap U'_i). \tag{1.36}$$

Note that $\bar{h}$ can be regarded as the identity, and the equivalent fiber bundles only differ in coordinate transformations (1.35) and (1.36) (see the proof in [Stee]).

Making use of the concept of equivalence, we can formulate the notion of an associated bundle.

We first need the definition of an adjoint bundle. A principal bundle P with the base space B and fiber G is said to be *adjoint* to $\xi = (E, p, B, F)$ (with the group G). The group G acts by translations, $G \to G$, $g(g_0) = gg_0$, where g_0 is fixed.

Definition 1.35 Two bundles ξ and ξ' with the same base space B are said to be *associated* if the principal bundles adjoint to them are equivalent.

Sections of fiber spaces. Existence of nontrivial sections in a fiber bundle is an important global characteristic of the fiber space, e.g., among closed orientable two-dimensional manifolds, only the tangent bundle over a torus has a nontrivial section. To calculate the corresponding topological invariants, the topological technique is developed in Ch. 2.

We only consider here the following. Let a fiber bundle $p: E \to B$ have a nontrivial section. Can the coordinate functions be transformed so that the fiber bundle is equivalent to the direct product? We formulate the answer as the theorem below.

Theorem 1.9 *A principal bundle with a group G is equivalent to the direct product if and only if the bundle has a section.*

Proof. Given a section $s: B \to E$. For $x \in \mathscr{U}_j$, put $\lambda_i(x) = p_i s(x)$. If follows from the relationship $g_{ji}(p(x) p_i(x) = p_j(x)$ that

$$g_{ji}(x) = \lambda_j(x) \lambda_i^{-1}(x), \quad x \in \mathscr{U}_i \cap \mathscr{U}_j, \tag{1.37}$$

which means that the fiber bundle is equivalent to the direct product.

Conversely, if E is equivalent to the direct product, then the coordinate transformations are of form (1.37). Put $s_i(x) = \varphi_i(x, \lambda_i(x))$. Then $s_i(x) = s_j(x)$ if $x \in \mathscr{U}_i \cap \mathscr{U}_j$, therefore, s defined on $\mathscr{U}_i$ as s_i is a section over B. ■

The associated bundle is equivalent to the direct product if the former's adjoint principal bundle is equivalent to it. However, it does not follow, generally speaking, from the existence of a section of a principal bundle that a section of the associated bundle exists. There only exists a family of sections of the associated bundle, with pairwise disjoint ranges of values filling the total space.

Induced bundles. We define an *induced bundle* as the one determined by a mapping $f: N \to M$ of two topological spaces.

Proposition 1.8 *Given a principal bundle $P(M, G)$ and a mapping f. There exists a unique (up to an isomorphism) principal bundle $Q(N, G)$ with the mapping $\bar{f}: Q \to P$ induced by f and with the automorphism of G corresponding to the identity mapping.*

The *proof* follows from the following construction. We consider the direct product $N \times P$ and distinguish the subset Q of points $(y, u) \in N \times P$ such that $f(y) = p(u)$. The projection $p_1: Q \to N$ is determined by the equality $p_1(y, u) = y$. Thus, the diagram

$$\begin{array}{ccc} Q & \xrightarrow{\bar{f}} & P \\ {\scriptstyle p_1}\downarrow & & \downarrow{\scriptstyle p} \\ M & \xrightarrow{f} & M \end{array}$$

is commutative. G acts on Q as

$$g(y, u) \to (y, gu), \quad g \in G, \quad (y, u) \in Q.$$

We specify a local covering $(\mathscr{U}_i, \varphi_i)$ on (M, G). We put $\bar{\mathscr{U}}_i = f^{-1}(\mathscr{U}_i)$ and define the coordinate functions $\varphi_i' : \bar{\mathscr{U}}_i \times G \to p^{-1}(\mathscr{U}_i)$ on $Q(N, G)$ by the formula $\varphi'(y, u) = (y, \varphi(f(y), u))$. Then $(\bar{\mathscr{U}}_i, \varphi_i')$ form a local covering of the fiber bundle $Q = \bar{f}P$. Hence, the induced bundle is also a locally trivial fiber space; in particular, if a bundle is trivial, then the induced bundle is also trivial.

Note that the same construction also holds for vector bundles and arbitrary associated ones.

The construction of an induced bundle is extremely important, since, by induction, the main properties of the original bundle are preserved. We now list some of them. The proofs are simple and left to the reader.

1. The bundles induced by equivalent fiber bundles are equivalent.
2. The bundles induced by associated bundles are associated.
3. If a space B can be continuously contracted to a point x_0, then the fiber bundle over B induced by the map $f: B \to x_0$ is equivalent to the direct product. As a corollary, we immediately obtain that the bundle over R^n or disk D^n: $\{\vec{x} \in R^n, \ |\vec{x}| \leqslant a\}$ is trivial. We make use of this result in Subsec. 2.6.4 (Lemma 2.10).

Induced bundles are used in constructing the characteristic classes of fiber bundles (see Sec. 2.7). It turns out that for each class of fiber bundles with a compact simple Lie group there exist universal bundles $EG \to BG$. Other bundles are obtained by the induction operation.

Besides principal bundles, a special class of associated bundles of vector bundles $\xi = (E, F, B)$ with the fiber R^n is essential. Including the tangent bundle TM already encountered above, they admit a number of algebraic operations enabling us to construct fiber bundles.

Direct product. Let ξ_1 and ξ_2 be two vector fiber bundles with projections $p_i: E_i \to B_i$, $i = 1, 2$. The fiber bundle with the total space $E_1 \times E_2$, base space

$B_1 \times B_2$, and the projection $p = p_1 \times p_2 : E_1 \times E_2 \to B_1 \times B_2$, where each fiber $(p_1 \times p_2)^{-1}(b_1, b_2) = F_{b_1} \times F_{b_2}$, $b_1 \in B_1$, $b_2 \in B_2$, is also endowed with the structure of a vector space, is called the *direct product* $\xi_1 \times \xi_2$.

Example 1.16 The tangent bundle $T(T^2)$ to the torus T^2 is the direct product $T(S^1) \times T(S^1)$. This can be obviously carried over to T^n by induction.

The Whitney sum. We now define an analog of the direct sum of two vector bundles ξ and η over the same base space B. Consider the diagonal embedding $d: B \to B \times B$, i.e., $b \to (b, b)$. The fiber bundle $\tilde{d}(\xi \times \eta)$ over B induced by d is called the *Whitney sum* of ξ and η and denoted by $\xi \oplus_W \eta$. Each fiber $F_b(\xi \oplus \eta)$ of $\xi \oplus_W \eta$ is canonically isomorphic to the direct sum $F_b(\xi) \oplus F_b(\eta)$ of the spaces.

The Whitney sum enables us to carry over to the total space the concepts of the subspace and its orthogonal complement. We now formulate the corresponding definitions.

Definition 1.36 Let ξ and η be two vector fiber bundles over B and such that $E(\xi) \subset E(\eta)$. Then ξ is called a *sub-bundle* of η if each fiber of $F_b(\xi)$ is a vector subspace of $F_b(\eta)$. The following is easy to prove.

Let ξ_1 and ξ_2 be two sub-bundles of a fiber bundle η and such that each fiber $F_b(\eta)$ equals the direct sum $F_b(\xi_1) \oplus F_b(\xi_2)$. Then η is isomorphic to the Whitney sum $\xi_1 \oplus_W \xi_2$.

If the base space of a fiber bundle is a (not necessarily compact) manifold, each vector bundle can be equipped with the Euclidean metric. The concept of the *orthogonal complement to the sub-bundle* ξ in η can then be introduced. It suffices to take the subspace $F_b(\xi^\perp)$ orthogonal to the subspace $F_b(\xi)$ in each fiber $F_b(\eta)$. We construct the fiber bundle $\xi^\perp$ by glueing all fibers of $F_b(\xi^\perp)$ when the point b ranges over the base space B. We have the fiber bundle $\xi^\perp \subset \eta$.

The following is valid.

Proposition 1.9 *A fiber bundle η is isomorphic to the Whitney sum $\xi \oplus_W \xi^\perp$.*

The *proof* is not complicated and given in [MSt].

The importance of the Whitney sum becomes evident when studying characteristic classes.

Tensor product of fiber bundles. Since the tensor product is defined for two vector spaces R^n and R^m, we can attempt to construct the tensor product of two fiber bundles ξ and η on the same base space B.

The simplest way is to define the tensor product $F_b^n \oplus F_b^m$ of two vector spaces, the fibers F_b^n and F_b^m at a point b of the base space B and then construct the total space by glueing (uniting) all spaces $F_b^n \oplus F_b^m$, where b runs over all B. We obtain the fiber bundle $\xi \oplus \eta$. The proof of its local triviality is obtained in a standard way [MSt]. Note that the group $GL(n, R) \oplus GL(m, R)$, i.e., the tensor product of the groups acting in R^n and R^m, acts in the fiber of the tensor product. The group can be extended to the full linear group $GL(mn, R)$ of the space $F^n \oplus F^m \sim R^{mn}$.

If additional structures exist on vector spaces R^n, then they can be carried over to the corresponding fiber bundles.

1. The structure of a complex space C^n can be introduced on R^{2n} and a complex n-dimensional vector bundle ξ_C with the fiber C^n can be considered.

2. The automorphism $z \to \bar{z}$, $z \in C^n$, acts on the complex space C^n. The corresponding fiber bundle $\bar{\xi}_C$ is said to be *conjugate* to ξ_C.

3. The Hermitian metric can be introduced on C^n and carried over to complex fiber bundles over manifolds (it suffices for us to require that the manifolds should be paracompact for the metric to exist).

4. Similar definitions can also be given both for quaternionic fiber bundles with fiber $H^n \sim R^{4n}$ and by introducing on R^4 the structure of a quaternion algebra.

The topology of fiber spaces is studied in Ch. 2. Now we turn to the differential geometric characteristics of manifolds and to their fiber bundles.

1.5 CONNECTION IN A FIBER BUNDLE

A connection in a fiber bundle is introduced to define such important geometric concepts as a parallel translation of a vector along a curve, the curvature of a manifold, etc. On the other hand, as shown in Ch. 3, Sec. 3.1, the introduction of the connection is equivalent to specifying a gauge field and the field strength tensor coincides with the curvature tensor of the corresponding connection.

The classical definition of the connection in the two-dimensional case is recalled in Subsec. 1.5.1.

A brief summary of the main definitions in the theory of connections in fiber bundles is given in invariant form below. Besides the usual advantages of notational invariance (the theory of characteristic classes, see Sec. 2.7), the treatment based on differential forms is necessary to study the relations between differential geometric and topological properties of manifolds.

1.5.1 Connection on Two-Dimensional Manifolds

Let x be a point in $M^2 \subset R^3$ and $v(x)$ the vector tangent to M^2 at x. Consider a parallel translation of $v(x)$ along a curve $u(t)$ with coordinates $u^1(t)$, $u^2(t)$, so that $v(x(t_0)) = v(u^1(t_0), u^2(t_0)) = v(x)$. In transition of v from $x(t_0)$ into the infinitely close point $x(t_0 + dt)$, the tangent vector v is sent into the vector $v + dv$, generally speaking, now no longer tangent at $x(t_0 + dt)$. We translate the vector $v + dv$ to x and split it into two components, one in the tangent space and the other normal to the surface,

$$\langle n | v + dv \rangle | n \rangle, \tag{1.38}$$

where n is the unit vector of the normal. Since v is tangent to M^2 at x, we have $\langle n | v \rangle = 0$, therefore (1.38) equals

$$\langle n | dv \rangle | n \rangle. \tag{1.39}$$

The tangential component is

$$| v + dv \rangle_{\tan} = | v + dv \rangle - \langle n | dv \rangle | n \rangle. \tag{1.40}$$

This vector is already tangent at x. The difference between the vectors $|v + dv\rangle_{\tan}$ and v is called the *absolute differential* of v in the transition from x to x' and denoted by

$$Dv = |dv\rangle - \langle n|dv\rangle |n\rangle. \tag{1.41}$$

Associating each vector $v(x)$ with the vector $Dv(x)$ specifies a *connection* on the manifold. Since the vector field determines the tangent bundle on M^2, the specifying of a connection on M^2 determines actually the connection in TM^2, or if we confine ourselves to unit vectors, determines the tangent bundle, i.e., that with the fiber S^1.

Let us see what the conidition $Dv = 0$ means.

In projecting the vector $v + dv$ onto the tangent plane, the value of the vector at the infinitely close point x' equals its value at x, i.e., the vector $v + dv$ is obtained from v by a parallel translation into the infinitely close point x'. The equations for covariant differentiation can be represented as follows:

Let $v(t) = \{v^1(t), v^2(t)\}$. At any point t, vector $v(t)$ can be decomposed as

$$v = v^1 r_1 + v^2 r_2$$

in terms of the coordinate vectors r_1 and r_2 of the surface at the same point t.

Omitting intermediate calculations familiar from the course in classical differential geometry, we give the final answer

$$\begin{aligned} Dv^1 &= dv^1 + \Gamma^1_{\alpha\beta} v^\alpha du^\beta, \\ Dv^2 &= dv^2 + \Gamma^2_{\alpha\beta} v^\alpha du^\beta, \end{aligned} \tag{1.42}$$

where $\Gamma^\gamma_{\alpha\beta}$ are Christoffel symbols.

We have already considered the concept of parallel translation for the infinitesimal displacement of a vector. A similar operation can be carried out for a parallel translation of a vector along a path $r(t)$. The result depends on the path and not only on $\Gamma^\gamma_{\alpha\beta}$.

The possibility of a parallel translation along $r(t)$ emanating from a point x easily follows from solvability of the system of differential equations

$$\frac{dv^1}{dt} = \Gamma^1_{\alpha\beta} v^\alpha \frac{du^\beta}{dt},$$
$$\frac{dv^2}{dt} = \Gamma^2_{\alpha\beta} v^\alpha \frac{du^\beta}{dt}.$$

However, it is clear that the value of v at each point x' can be totally different, depending on the path r joining x' and x. There are manifolds for which the parallel translation does not depend on r. Such are developable surfaces in the two-dimensional case.

In what follows, we give an invariant definition of a connection in a principal bundle and in the associated ones.

1.5.2 Connection in a Principal Bundle

Let P be a principal bundle with base space M, structure group G, and projection $p: P \to M$.

Let P_x be the tangent space to P at a point $x \in P$ and V_x the subspace of P_x, tangent to the fiber passing through x. The subspace can be identified with the Lie algebra $\mathscr{G}$ of G.

Definition 1.37 A *connection* Γ *on* (P, G, M) is the correspondence associating each point $x \in P$ with a tangent subspace H_x if

(1) P_x is the direct sum of V_x and H_x;

(2) for any two elements $g \in G$, $x \in P$, H_x is invariant under right translations, namely,

$$dR_g H_x = H_{xg};$$

(3) the field H_x smoothly (C^∞) depends on x.

The latter condition means that if a field X given on P admits decomposition (1) at each point $x \in P$ and is smooth (differentiable), then the corresponding component is also smooth, so is V_x.

Given a connection Γ, V_x is called the *vertical subspace* (which explains the notation) at a point x and H_x the *horizontal subspace*. The vertical component of X is denoted by vX and the horizontal one by hX.

The projection p of the total space P generates a linear map $dp: P_x \to M_{p(x)}$ (onto the tangent space of the base space).

It follows from the existence of Γ that the space $M_{p(x)}$ is isomorphic to H_x under dp; therefore, dim H_x = dim $M_{p(x)}$.

For a given vector field X on M we can define the lift of X into a fiber bundle P, called the *horizontal lift* of X and denoted by $\tilde{X}$.

Definition 1.38 A *lift* $\tilde{X}$ of X is the unique horizontal field $\tilde{X}$ on P covering X, i.e., $p\tilde{X} = X$.

The unique existence of $\tilde{X}$ follows from what is written two lines above. To prove the differentiability of $\tilde{X}$, it suffices to notice that the proof can be obtained in a neighborhood of a point $x \in M$, where the fiber bundle can be regarded as the direct product.

For a lift $\tilde{X}$, the following is valid:

(a) $dR_g\tilde{X} = \tilde{X}$, $g \in G$,

(b) $\tilde{X} + \tilde{Y} = \widetilde{X + Y}$,

(c) $h[\tilde{X}, \tilde{Y}] = [\tilde{X}, \tilde{Y}]$.

The concept of the lift of a vector field is necessary in Subsec. 1.5.6 to define a parallel translation of a vector along fibers. We now give the dual formulation of a connection in terms of differential forms.

1.5.3 Connection Form ω

Given a connection Γ on P, we identify the Lie algebra $\mathscr{G}$ of a group G with the Lie algebra $\mathscr{G}_V$ of (left) invariant vector fields on P as follows.

At a point x, consider the decomposition $P_x = H_x + V_x$ and put $\mathscr{G}_V$ equal to the tangent space to G at x. Since G acts freely in a principal bundle, $\mathscr{G} \sim \mathscr{G}_V$. We define the connection 1-form ω with the value in $\mathscr{G}$ as a linear map under which the image of an element $a^* \in \mathscr{G}_V = V_x$ is the corresponding element $a \in \mathscr{G}$, while the image of $h^* \in H_x$ is the zero element in $\mathscr{G}$. The form ω vanishes on horizontal vectors. Such a form is said to be *vertical*. If a form ω_1 vanishes on vertical vectors, then it is said to be *horizontal*.

We now make a statement showing equivalence of the definitions of the connection both in terms of the distribution of horizontal spaces and in terms of connection forms.

Proposition 1.10 (1) *The form ω is smooth*; (2) *ω is equivariant, i.e., $(R_g^*\omega)X = \mathrm{Ad}\ (g^{-1})\omega(X)$, for any vector field X on P.*

Property (1) follows from the similar one for a horizontal distributions. It suffices to prove (2) for a vertical field, since for a horizontal one, hX, $R_g hX$ is also a horizontal field, on which ω vanishes. If X is vertical, then we use the relation $dgX = \mathrm{Ad}\ (g^{-1})X$. We have

$$(R_g^*\omega)X = \omega(R_g \cdot X) = \mathrm{Ad}\ (g^{-1})X = \mathrm{Ad}\ (g^{-1})\omega(X). \ \blacksquare$$

1.5.4 Parallel Translation

We show that the concept of connection enables us to introduce in fiber bundles the notion of parallel translation of a vector along certain curves. The concepts are basically equivalent.

Let γ be a broken curve of class C^∞ in M. Then its (horizontal) lifting is a curve $\tilde{\gamma}$ in P such that

1. $\tilde{\gamma}$ is horizontal, i.e., its tangent vector is horizontal, and
2. $p\tilde{\gamma} = \gamma$.

The fundamental unique existence theorem for the lifting of a curve holds.

Theorem 1.10 *Let γ be a curve in M, $\gamma: I \to M$, $I = [0, 1]$, and $x \in p^{-1}(\gamma(0))$. Then γ can be lifted in a unique way so that γ (0) $= x$* (for a proof see [BC]).

Here is an important corollary.

Corollary 1.1 *Let Γ be a connection on (P, M, G). Then we can define a parallel translation on $p^{-1}(\gamma(0)) \to p^{-1}(\gamma(1))$ of the fiber F_γ over the point $p^{-1}(\gamma(0))$ along a curve γ. It is obvious that the transformation of F_γ commutes with right translations R_g: $F_\gamma R_g = R_g F_\gamma$ and are homomorphisms of the group of paths, i.e., $F_{\gamma\sigma} = F_\gamma \circ F_\sigma$, if the curves γ and σ are such that $\sigma(0) = \gamma(1)$.*

We leave the proof to the reader (see also [BC]).

Existence of a connection can be easily derived from the possibility of parallel translation of the fiber along a curve γ. Indeed, suppose that a given parallel translation commutes with a right one of the fiber $p^{-1}(\gamma(0))$ along γ. Then we can define the horizontal subspace H_x in P_x (where $x = p^{-1}(0)$) as the tangent space at the point 0 of the curve $\gamma \to F_{\gamma_u} x$, the parameter u on γ varying from $\gamma(0)$ to $\gamma(t)$.

The connection generated by an infinitesimal parallel displacement and expressed with respect to local coordinates coincides with the classical definition of the connection in terms of the Christoffel coefficients. We shall perform the corresponding calculations in Subsec. 1.5.6, turning now to the invariant treatment of curvature forms.

1.5.5 Curvature Form

For each connection 1-form ω we can define a $\mathscr{G}$-valued 2-form Φ by the following rule. Put $\Phi = D\omega$, where

$$D\omega = d\omega \circ H, \tag{1.43}$$

and H is a horizontal field. The operator D is called the *covariant differential* of ω.

Formula (1.43) means that

$$D\omega(t_1, t_2) = d\omega(ht_1, ht_2), \tag{1.44}$$

where t_1 and t_2 are two vectors from P_x and ht is the horizontal component of the vector t.

It is obvious that Φ is horizontal. It is easy to verify that Φ is also equivariant and therefore defined on the whole fiber bundle.

The main result relating connections to a curvature form is the following structural equation due to Cartan.

Proposition 1.11 *Let ω be a connection form and Φ a curvature. Then*

$$d\omega = -(1/2)[\omega, \omega] + \Phi. \tag{1.45}$$

(The proof can be found in numerous books on differential geometry [BC], [DFN], [Ste], [KN1].)

The classical *Bianchi identity* $D\Phi = 0$ follows from (1.45).

Proof. Apply the operator D to both sides of relation (1.45), namely,

$$Dd\omega = -(1/2)D[\omega, \omega] + D\Phi.$$

We have

$$dd\omega(ht_1, ht_2, ht_3) = 0,$$

since $d^2 = 0$; in turn $D[\omega, \omega] = 0$, because $[\omega, \omega]$ is a vertical form. We obtain $d\Phi = 0$. ■

1.5.6 Connection and Curvature in a Tangent Bundle. Local Coordinates

Introducing the covariant derivative of a vector field along a curve enables us to establish equivalence of the definitions of the connection given in terms of the

Christoffel symbols Γ^k_{ij} and by the invariant approach in the preceding subsection.

First, we notice that the classical concept of the connection (or of the linear connection) is related to introduction of the tangent bundle TM associated with the principal bundle (P, G, M), where G is the group of all nonsingular linear transformations of the space R^n, $G = GL(nR)$.

The principal bundle $(P, GL(n, R), M^n)$ is called the *bundle of frames* and is of the following structure. Let u be the frame at a point $x \in M^n$. The set of frames $\{u\}$ at all $x \in M^n$ forms a bundle P. We specify P with respect to local coordinates. The frame is $u = (Y_1, \ldots, Y_n)$, where $\{Y_i\}$ is the basis of tangent vectors at x. Each element $g \in GL(n, R)$ is represented by a nonsingular matrix g_i^j $(i, j = 1, \ldots, n)$ and acts on P (on the right), namely,

$$u \cdot g \Rightarrow g_j^i y_i, \qquad u \in P,$$

i.e., $u \cdot g$ is a new frame at the same point x.

The projection $p: P \to M^n$ is defined as $p(u) = x$. If $x^1, \ldots, x^n$ are the local coordinates of a point $x \in \mathscr{V} \subset M^n$, then each frame $u \in p^{-1}(\mathscr{V})$ is written as

$$u = (Y_1, \ldots, Y_n), \; Y_i = Y_i^k \frac{\partial}{\partial x^k}, \tag{1.46}$$

where Y_i^k is a nonsingular matrix. Each nonsingular matrix Y_i^k defines u by (1.46). Thus, x^i and Y_i^k define local coordinates in the fiber bundle P_{GL}. It is easy to verify that $(P, GL(n, R), M)$ is a smooth manifold.

Let F be an n-dimensional vector space with a fixed basis $\xi = (\xi^1, \ldots, \xi^n)$, $F \sim R^n$. The group G acts in the fiber F as

$$g \cdot \xi = g_j^i \xi^j.$$

Consider the tangent bundle TM with fiber $T_xM \sim F$, associated with the principal bundle P_L. Each frame $u = (Y_1, \ldots, Y_n)$ can be regarded as a linear map F onto T_x, $x = p(u)$ such that $u \cdot \xi^i = Y_i$, $i = 1, \ldots, n$.

A connection Γ in the principal bundle P determines a connection in the associated bundle TM according to the following rule. If $\tau = x_t$ $(a \leqslant t \leqslant b)$ is a curve in M and $\tilde{\tau} = u_t$ a lift in P, then, for each fixed $\xi \in R^n$, the curve $\tau' = u_t\xi$ is, by definition, a lift τ in TM.

Let φ be a section in TM defined over the curve τ so that $p \circ \varphi\,(x_t) = x_t$ for all t (where p is the projection $TM \to M$). We denote by $\dot{x}_t$ the vector tangent to $\tau = x_t$ at a point t.

Definition 1.39 The limit

$$\nabla_{\dot{x}_t}\varphi = \lim_{\varepsilon \to 0} [\tau^{t+\varepsilon}(\varphi(x_{t+\varepsilon}) - \varphi(x_t))], \tag{1.47}$$

where $\tau^{t+\varepsilon}: p^{-1}(x_{t+\varepsilon}) \to p^{-1}(x_t)$ is a parallel translation of the fibre $p^{-1}(x_{t+\varepsilon})$ along τ from $x_{t+\varepsilon}$ to x_t, is called the *covariant derivative* of the vector field (section) φ in the direction of $\dot{x}_t$ (for each fixed t).

The field $\nabla_{\dot{x}_t}\varphi \in p^{-1}(x_t)$ defines for each t a section in TM along τ.

The section φ is parallel, i.e., the curve $\varphi(x_t)$ is horizontal in TM, if and only if $\nabla_{\dot{x}_t}\varphi = 0$ for all t. The covariant derivative $\nabla_{\dot{x}_t}\varphi$ satisfies the equalities

$$\begin{aligned} \nabla_{\dot{x}_t}(\varphi + \psi) &= \nabla_{\dot{x}_t}\varphi + \nabla_{\dot{x}_t}\psi, \\ \nabla_{\dot{x}_t}(\lambda \cdot \varphi) &= \lambda \cdot \nabla_{\dot{x}_t}\varphi + (\dot{x}_t\lambda)\varphi(x_t), \end{aligned} \tag{1.48}$$

where λ is a real function defined on τ.

We define the covariant derivative of the field φ for any vector $X \subset T_xM$. Let $\tau = x_t$ be a curve in M, $-\varepsilon \leqslant t \leqslant \varepsilon$ such that $X = \dot{x}_o$. Put $\nabla_X\varphi = \nabla_{\dot{x}_t}\varphi$. It is clear that $\nabla_X\varphi$ does not depend on the choice of τ.

Let X, $Y \in T_xM$ and φ, ψ be two sections in TM, defined in a neighborhood of $x \in M$.

We have

$$\begin{aligned} \nabla_{X+Y}\varphi &= \nabla_X\varphi + \nabla_Y\varphi, && \text{(a)} \\ \nabla_X(\varphi + \psi) &= \nabla_X\varphi + \nabla_X\psi, && \text{(b)} \\ \nabla_X(\lambda\varphi) &= \lambda \cdot \nabla_X\varphi + (X\lambda) \cdot \varphi, && \text{(c)} \\ \nabla_{\lambda X}(\varphi) &= \lambda \cdot \nabla_X\varphi, && \text{(d)} \end{aligned} \tag{1.49}$$

where λ is a real function defined in a neighborhood of x.

The proofs (only (a) being rather difficult) can be obtained by the reader or found in any differential geometry reference (e.g., see [KN1]).

The properties (a)-(d) characterize a connection uniquely and can be taken as its definition.

We now can easily obtain the classical definition by writing (a)-(d) with respect to local coordinates.

Let $x^1, \ldots, x^n$ be local coordinates defined in a neighborhood of a point $x \in \mathcal{U} \subset M^n$. We denote by ∂_k coordinate vector fields $\partial_k = \partial/\partial x^k$. Then each vector field X on $\mathcal{U}$ is uniquely written as

$$X = v^k\partial_k,$$

where v^k are real functions on $\mathcal{U}$, in particular,

$$\nabla_{\partial_i}\partial_j = \Gamma_{ij}^k\partial_k.$$

The functions Γ_{ij}^k determine the connection completely. Indeed, for any two fields $X = v^i\partial_i$ and $Y = w^j\partial_j$, the field

$$\nabla_X Y = (v^i w^k{}_{,\,i})\partial_k \tag{1.50}$$

is given according to (a)-(d), where the symbol $w^k{}_{,\,i}$ denotes the function

$$w^k{}_{,\,i} = \partial_i w^k + \Gamma_{ij}^k w^j$$

(recalling that tensor notation, i.e., summation over repeated indices, is used). Conversely, if for any smooth functions Γ_{ij}^{k}, $\nabla_X Y$ is determined by formula (1.50), then properties (1.49) are obvious. The equivalence of the connection's definition by the covariant derivative ∇ to that by the Christoffel symbols Γ_{ij}^{k} has thereby been proved.

Remark 1.4 In the literature, the term "affine connection" instead of "linear connection" is often used. As a matter of fact, a larger group $\tilde{G} = GL(n, R) \times L(n)$, where $L(n)$ is the group of translations of R^n, isomorphic to R^n, can act on a tangent bundle and not only $GL(n, R)$. A connection in a tangent bundle where each fiber is isomorphic to R^n and endowed with the structure of an affine space may be said to be affine. It is easy to see that there exists a natural correspondence between the linear and affine connections, which can be established by the group homomorphism $\tilde{G} \to G$ (e.g., see the details in [KN1]). In the sequel, we restrict ourselves to the linear connection.

We now turn to connection forms ω in a principal fiber space $P = (P, G, M)$. Since P is locally trivial, for each neighborhood $\mathscr{U}$ of the covering M, there exists a map $\psi_{\mathscr{U}}: p^{-1}(\mathscr{U}) \to \mathscr{U} \times G$. Let z be the coordinate in $p^{-1}(\mathscr{U})$, namely,

$$\psi_{\mathscr{U}}(z) = \{p(z) = x,\ s_{\mathscr{U}}(z)\}.$$

The group acts freely in P (left action being selected for definiteness). Therefore,

$$s_{\mathscr{U}}(g \cdot z) = g \cdot s_{\mathscr{U}}(z), \quad z \in p^{-1}(\mathscr{U}), \quad g \in G. \tag{1.51}$$

If $z \in p^{-1}(\mathscr{U} \cap \mathscr{V})$, then it follows from (1.51) that $s_{\mathscr{U}}(gz)^{-1} s_{\mathscr{V}}(gz) = s_{\mathscr{U}}(z)^{-1} s_{\mathscr{V}}(z) = g_{\mathscr{U}\mathscr{V}}$ and $g_{\mathscr{U}\mathscr{V}}$ only depends on $p(z)$ (and not on $g \in G$).

The coordinate functions $g_{\mathscr{U}\mathscr{V}}: \mathscr{U} \cap \mathscr{V} \to G$ satisfy conditions (10) and (11) in the definition of a fiber space (Subsec. 1.4.1). Thus, in the fiber bundle over the neighborhoods $\mathscr{U}_i$, we can introduce coordinates $z = (x, g)$. We specify connection forms ω_i on P at first in each local neighborhood $\mathscr{U}_i$ and then, after defining the identification rule on the intersection of $\mathscr{U}_i$ and $\mathscr{U}_j$, we obtain the connection form defined globally on P.

Let $H_z \subset T_z$. By definition, ω belongs to the vertical subspace V_z^* of the cotangent space T_z^*, which is equivalent to defining a $\mathscr{G}$-valued form ω whose restriction to the fiber is $ds_{\mathscr{U}} s_{\mathscr{U}}^{-1}$. Since the action of G is left, ω is right-invariant. Locally, ω is given as

$$\omega(z) = ds_{\mathscr{U}} s_{\mathscr{U}}^{-1} + \theta_{\mathscr{U}}(x, s_{\mathscr{U}}, dx). \tag{1.52}$$

The condition for equivalence (see Proposition 1.10) for the above is

$$\omega(gz) = \mathrm{Ad}(g)\omega(z). \tag{1.53}$$

It follows from (1.53) that $\omega(z)$ is representable locally as

$$\omega(z) = ds_{\mathscr{U}} s_{\mathscr{U}}^{-1} + \mathrm{Ad}(s_{\mathscr{U}})\theta_{\mathscr{U}}(x, dx), \tag{1.54}$$

where $\theta_{\mathscr{U}}(x, dx)$ is a $\mathscr{G}$-valued 1-form on $\mathscr{U}$.

The rule for glueing $\theta_{\mathscr{U}}$ together, which can serve as another definition of a connection, follows from (1.53).

Let $\mathscr{U}$ and $\mathscr{V}$ be two neighborhoods in the base space M, $\mathscr{U} \cap \mathscr{V} \neq \varnothing$. The forms $\omega_{\mathscr{U}}$ and $\omega_{\mathscr{V}}$ defined in $p^{-1}(\mathscr{U})$ and $p^{-1}(\mathscr{V})$, respectively, must coincide on $p^{-1}(\mathscr{U} \cap \mathscr{V})$. Making use of the relation $s_{\mathscr{U}}(z)g_{\mathscr{U}\mathscr{V}}(z) = s_{\mathscr{V}}(z)$, we obtain from (1.54) by a simple calculation the transformation rule for forms θ, namely,

$$\theta_{\mathscr{U}} = dg_{\mathscr{U}\mathscr{V}}g_{\mathscr{U}\mathscr{V}}^{-1} + \mathrm{Ad}(g_{\mathscr{U}\mathscr{V}})\theta_{\mathscr{V}}(x, dx). \tag{1.55}$$

Transformation $\theta_{\mathscr{U}} \to \theta_{\mathscr{V}}$ (1.55) is called a *local transformation of the connection* in differential geometry. In the field theory, it corresponds to a gauge transformation (see Subsec. 3.1.1).

We now turn to the definition of a connection form ω on vector bundles associated with the principal one. Let $\xi = (E, p, F, G, M)$ be the bundle associated with the principal bundle (P, G, M). The associated connection form ω_E is uniquely constructed by using ω. The construction is as follows. Since G acts in the fiber F, each element $g \in G$ determines the vector field on F, generated by the Lie algebra $\mathscr{G}$. Suppose a point $y \in F$ and the tangent vector τ at y are given. We now define an analog of (1.52). If suffices to obtain the formula for ω_E in a neighborhood of the point $(p_E(y), y)$. We define ω_F^0 on F by putting $\omega_F^0(y, \tau) = \tau$. The restriction of ω_E on each fiber must coincide with the value ω_F^0. ω_E should be of the form

$$\omega_E = \omega_F^0 + \theta(x, y, dx), \quad x = p_E(y), \tag{1.56}$$

on the direct product $p_E^{-1}(\mathscr{U}_i) = \mathscr{U}_j \times F$ with coordinates x, y. The form $\theta(x, y, dx)$ in a coordinate neighborhood of a point (x, y) can be represented as $\theta = \theta_\mu(x, y)\, dx^\mu$, $\theta_\mu(x, y)$ being the tangent vector to the fiber at the point $y \in E$. The element of $\mathscr{G}$ coinciding with $\theta_\mu(x, y)$ in θ determines a vector field η on F. We select the value of the field $\theta_\mu(x, y) = \eta(y)$, $x = p(y)$, at y, which is just the tangent vector to the fiber at y. By definition, the forms dx^μ vanish on any vector τ tangent to F. Therefore, the restriction of ω_E to the fiber is ω_F^0.

The equation $\omega_E = 0$ specifies the horizontal direction R_x^n for points $x \in E$ in the associated bundle. Formula (1.56) invariantly defines ω_E.

Formulas (1.56) are simple and explicit for the bundle associated with the principal one P with $G = GL\,(n, R)$. Let F be a vector space R^n and the group $GL(n, R)$ operate in F linearly. Then elements η of the Lie algebra $\mathscr{G}L(n, R)$ can be regarded as matrices $A_\eta\colon R^n \to R^n$. We introduce coordinates ζ^i $(i = 1, \ldots, n)$ in R^n. If $A_\eta = (a_j^i)$, then $\eta(\zeta) = \eta^i = a_j^i\zeta^j$.

The field $\theta_\mu(x) = (\theta_\mu(x)_j^i) = \theta_{\mu j}^i$ is of matrix form, the matrix acting in R^n. In other words, a connection is locally given by the matrix $\theta_{\mu j}^i$ $(i, j = 1, \ldots, n,$ $\mu = 1, \ldots, m, m = \dim M)$ in the vector bundle with the fiber R^n or by the matrix-valued form $\theta = \theta_{j\mu}^i dx^\mu$.

For the tangent bundle TM, the coefficients $\theta_{\mu j}^i$ coincide with the Christoffel symbols $\Gamma_{j\mu}^i$.

We have thereby established the equivalence of the definition of a connection by form, parallel translation, and covariant differentiation along a path.

Example 1.17 Compute a connection in another important case of a Stiefel bundle, a principal bundle with the group $O(q)$ [Ch]. This bundle is universal for those with the structure group $O(q)$ (see Subsec. 2.6.4).

Let R^{q+N} be an Euclidean space. We select a basis $e_a^b = (e_a^1, \ldots, e_a^{q+n})$, $1 \leqslant a \leqslant q + N$ such that the matrix $X = e_a^b$ is orthogonal. $O(q + N)$ can be identified with the space of all orthogonal frames e_a (or all orthogonal matrices).

Put

$$de_a = \alpha_{ab} e^b. \tag{1.57}$$

Then, for $\alpha = (\alpha_{ab})$, we have

$$\alpha = dXX^{-1} = -\alpha^t,$$

where α^t is the transpose of α.

As noted in Subsec. 1.2.2, Stiefel manifolds $V_{N+q,q} = O(N + q)/O(q)$ can be identified with that of all orthonormal q-frames and the Grassmann manifold

$$G_{N+q,q} = O(q + N)/O(N) \times O(q)$$

with q-planes generated by $e_1, \ldots, e_q$. The matrices $\alpha_N = (\alpha_{ab})$, $1 \leqslant a, b \leqslant N$, define a connection in the fiber bundle

$$V_{N+q,q} \xrightarrow{O(N)} G_{N+q,q}.$$

In fact, these α_N determine a form with values in the Lie algebra $\mathscr{O}(N)$, themselves defined on $O(q + N)$, but obtained from forms on $O(q + N)/O(q)$, since they are invariant under Ad g, $g \in O(q)$. After multiplying on the left, the α_N are transformed with respect to the adjoint representation of the group $O(N)$; therefore, the forms determine a connection in $V_{N+q,q}$. A connection for the complex bundle is defined by a similar formula:

$$V_{N+q,q} \xrightarrow{U(N)} G_{N+q,q}.$$

1.5.7 Curvature of a Linear Connection

We now obtain explicit formulas for the curvature of a linear connection. We make use of two parallel and equivalent approaches. One resorts to the concept of covariant derivative, the other and dual, to the curvature 2-form Φ already defined.

We start with the latter approach and show that the curvature form is locally expressed as

$$\Phi = \mathrm{Ad}\,(s_{\mathscr{U}}^{-1})\Omega_{\mathscr{U}}, \tag{1.58}$$

where $\Omega_{\mathscr{U}} = d\theta_{\mathscr{U}} + (1/2)[\theta_{\mathscr{U}}, \theta_{\mathscr{U}}]$ and $\theta_{\mathscr{U}}$ is the form on the base space determined by (1.54).

The proof of (1.58) is based on the following.

Lemma 1.2 *Let θ be a $\mathscr{G}$-valued 1-form on $\mathscr{U} \subset M^n$, $s \in G$, and $\alpha = s^{-1}dS$ a left-invariant $\mathscr{G}$-valued 1-form on G. Then*

$$d(\mathrm{Ad}\,(s^{-1})\theta) = \mathrm{Ad}\,(s^{-1})d\theta + [\mathrm{Ad}\,(s^{-1})\theta, \alpha]$$

on $\mathscr{U} \times G$.

The proof for a matrix-valued form α is not complicated, namely,

$$\begin{aligned} d(s^{-1}\theta s) &= ds^{-1}\theta s + s^{-1}d\theta s + s^{-1}\theta ds \\ &= \mathrm{Ad}\,(s^{-1})d\theta + s^{-1}\theta s \cdot s^{-1}ds - s^{-1}ds \cdot s^{-1}\theta s \\ &= \mathrm{Ad}\,(s^{-1})d\theta + [\mathrm{Ad}\,(s^{-1})\theta, \alpha]. \end{aligned}$$

The proof in the general case is left to the reader (or see [Ch]).

We apply the operator D to ω and use the lemma. We obtain

$$\Phi = D\omega = d\omega + (1/2)[\omega, \omega] = \mathrm{Ad}\,(s^{-1})\Omega_{\mu}. \tag{1.59}$$

Due to (1.51 and 1.59), two forms $\Omega_{\mathscr{U}}$ and $\Omega_{\mathscr{V}}$ are related so that

$$\Omega_{\mathscr{V}} = \mathrm{Ad}\,(g_{\mathscr{U}\mathscr{V}}^{-1})\Omega_{\mathscr{U}} \tag{1.60}$$

on the intersection of domains $\mathscr{U}$ and $\mathscr{V}$ in the base space.

Since forms θ are defined on the base space, the curvature forms can be locally written as

$$\Omega = \Omega_{\mu\nu}dx^{\mu} \wedge dx^{\nu},$$

by taking (1.58) into account, where

$$\Omega_{\mu\nu} = \partial_{\mu}\theta_{\nu} - \partial_{\nu}\theta_{\mu} + [\theta_{\mu}, \theta_{\nu}],\ \theta = \theta_{\mu}dx^{\mu}.$$

The curvature form Φ is immediately defined on the principal bundle P. We fix the principal bundle $P_{GL} = (P, GL(n, R), M)$. One would like the construction of the curvature form $\tilde{\Phi}$ of a linear connection to be as follows. We consider the form $\tilde{\omega}$ or a linear connection in the fiber bundle T^*M associated with P_{GL}. By applying D to $\tilde{\omega}$ we obtain the curvature form $\tilde{\Phi}$ of the linear connection; however, this procedure cannot be carried out directly. The forms ω are defined in $\mathscr{G}L(n, R)$ of dimension n^2, while T^*M has the dimension n of the fiber F. Nevertheless, we obtain structural equations for the forms generated by the connection on introducing an additional 1-form θ, which takes its values in $F \sim R^n$. The equations are similar to (1.45).

We specify forms ω for the fiber bundle P_{GL} as a matrix-valued form ω_j^i with values in the space of $n \times n$ matrices (i.e., in the Lie algebra $\mathscr{G}L(n, R)$). We define the 1-form $\tilde{\theta} = (\theta^1, \ldots, \theta^n)^t$ taking values in the set of column vectors of $n \times 1$ matrices.

$\tilde{\theta}$ is determined by the equality

$$p^*(u_x) = \theta^i(u_x)u_i, \tag{1.61}$$

where $u_x \in P_x$ and u_i is the frame $\{u_1, \ldots, u_n\}$ at the point $p(u) = x$. In local coordinates (1.46), $\theta^i = Z^i_j dx^j$, where the matrix $Z^i_j = (Y^i_j)^{-1}$.

Note that $\tilde{\theta}$ is horizontal, and, therefore, independent of the choice of a connection. Sometimes, $\tilde{\theta}$ is called a *solder form* [BC].

$\tilde{\theta}$ specifies the dual structure on P, called a *horizontal* (or *ground*) *vector field* A^* on P.

A^* is defined as a (unique) horizontal field on P such that $\tilde{\theta}(A^*(X)) = X$ at each point of the space P. The basis for the space of horizontal fields generates fields A_i^* assuming on the frame $\{X_1, \ldots, X_n\}$ values coinciding with the components of vector X_i under "projecting" p_x onto the base space M^n. It is obvious that the fields A^* depend on the connection in P.

The fundamental vector fields A^i_j and horizontal ones A_i^* form a basis in the tangent space TP. Similarly, the forms ω^i_j and θ^i make up a basis for T^*P. In this case, the pairs ω^i_j, A_i^* and θ^i, A^i_j form mutually complementary pairs. This means that $\omega^i_j(A^l_k) = 0$ if $(i, j) = (l, k)$ and

$$\omega^i_j(A^i_j) = 1, \quad \omega^i_j(A_k^*) = 0, \quad \theta^k(A^i_j) = 0, \quad \theta^i(A_j^*) = \delta^i_j,$$

A^i_j acting on the vector $\xi_k \in F$ as $A^i_j\xi_k = \delta^i_j\xi_k$.

The torsion form Θ is an important characteristic of the bundle of frames and the tangent bundle, which is related to the form $\tilde{\theta}$.

Definition 1.40 The 2-form $D\tilde{\theta} = d\tilde{\theta}\,(hX, hY)$ is called the *torsion form* Θ of a given linear connection.

Θ is specified on P and takes the value in the fiber F. It possesses the following property:

$$R_g\Theta = g^{-1}\Theta,$$

which is easily verifiable and is a consequence of a similar fact for $\tilde{\theta}$. Θ satisfies the structural equation

$$d\tilde{\theta}(x, y) = (1/2)\{\omega(Y)\cdot\tilde{\theta}(X) - \omega(X)\cdot\tilde{\theta}(Y)\} + \Theta(X, Y), \tag{1.62}$$

where $\omega\cdot\tilde{\theta}$ is the result of the action $\omega \in \mathscr{G}$ on $\tilde{\theta} \in F$.

In terms of components, (1.62) is written as

$$d\theta^i = -\omega^i_j \wedge \theta^j + \Theta^i,$$

where

$$\tilde{\theta} = (\theta^1, \ldots, \theta^n)^t, \ \Theta = (\Theta^1, \ldots, \Theta^n)^t.$$

Along with (1.62), we consider the structural equation

$$d\omega(X, Y) = (1/2)\{\omega(X)\cdot\omega(Y) - \omega(Y)\cdot\omega(X)\} + \Omega \tag{1.63}$$

for the curvature form, where $\cdot$ is a matrix multiplication. In componentwise notation, the equation takes the form

$$d\omega_k^i = -\omega_j^i \wedge {}_k^j + \Omega_k^i. \tag{1.64}$$

Both relations are proved by checking the validity of formulas (1.62) and (1.63) for the particular cases of horizontal fields X, Y, vertical fields X, Y, horizontal field X and vertical field Y as given in [KN1].

We show that the vertical component of a field $[Y, Z]$, where Y, Z are two horizontal fields, satisfies the relation

$$\omega(v[Y, Z]) = -2\Omega(Y, Z). \tag{1.65}$$

If $\tilde{\theta}$ is constant on Y and Z, i.e., Y and Z are linear combinations of the ground fields A_i^*, then

$$\tilde{\theta}([Y, Z]) = -2\Theta(Y, Z). \tag{1.66}$$

Proof. (a) Formula (1.65) immediately follows from the Maurer-Cartan equation and structural equation (1.63), which in this case is equivalent to the definition of the form Ω, namely, $2d\omega(Y, Z) = Y\cdot\omega(Z) - Z\cdot\omega(Y) - \omega([Y, Z])$. Since $\omega(Y) = \omega(Z) = 0$, then $2d\omega(Y, Z) = 2\Omega(Y, Z) = -\omega([Y, Z])$.

(b) Making use of structural equation (1.62) and the Maurer-Cartan equation, we obtain

$$2d\tilde{\theta}(Y, Z) = -\tilde{\theta}([Y, Z]) = 2\Theta(Y, Z). \tag{1.67}$$

Compare the approach by means of forms with the classical tensor treatment. Recall the definition of a tensor field.

Let F be the vector space over a field R, $\dim F = n$, and F^* the conjugate space. The multilinear mapping

$$\underbrace{F \times F \times \ldots \times F}_{r} \times \underbrace{F^* \times F^* \times \ldots \times F^*}_{s} \to R$$

is called an s-times contravariant and r-times covariant *tensor of type* (s, r) *on* F.

The set of all tensors of type (s, r) forms a vector space on R denoted by T_r^s. If each point x in the manifold M^n is associated with a tensor of a type (s, r) defined on T_xM, then we obtain a tensor field on M^n. We denote a tensor field of type (s, r) by $t_r^s \subset T_r^s$. To define a smooth structure on the space of all tensor fields is not difficult. The set of tensor fields is then turned into a differentiable fiber bundle over M^n. Important particular cases of tensors are vectors, or elements of the space T_0^1, differential forms, or elements of T_1^0, and metrics, or tensors of type (0, 2).

We show that there exists a one-to-one correspondence between 2-forms Θ and Ω on P and tensor torsion fields T and curvature fields R on the base space M^n.

We start with the following lemma.

Lemma 1.3 *There exists a one-to-one correspondence between the set of vector fields X on M^n and that of differentiable functions f on P assuming values in a vector space F and such that*

$$f(u \cdot g) = g^{-1}f(u)$$

for any $u \in P$, $g \in G$.

The correspondence is established as follows:

$$f(u) = \tilde{\theta}_u(\tilde{X}),$$

where *$\tilde{X}$ is the lift of X relative to an arbitrary linear connection.*

Proof. If $\tilde{X}$ is a lift of a field X, then, by the definition of the solder form $\tilde{\theta}$,

$$\tilde{\theta}_u(\tilde{X}) = u^{-1}p(\tilde{X}) = u^{-1} \cdot X_{p(u)},$$

i.e., $f(u) = u^{-1} \cdot X_{p(u)}$ is an arbitrary differentiable function. The equivariance property obviously holds, namely,

$$\tilde{\theta}_{ug}(\tilde{X}) = (ug)^{-1}p(\tilde{X}) = g^{-1}u^{-1}X_{p(ug)} = g^{-1}u^{-1}X_{p(u)} = g^{-1}f(u),$$

since $R_g^*(\tilde{\theta}) = g^{-1} \cdot \tilde{\theta}$, $R_g\tilde{X} = \tilde{X}$.

Conversely, any function $f: P \to F$ generates a vector field X on M if $f(ug) = g^{-1}f(u)$. X is determined by the equality $X_x = u \cdot f(u)$ and irrespective of the choice of u if $p(u) = x$. The correspondence is one-to-one. ∎

Similarly to Lemma 1.3, we state Lemma 1.4 by establishing a correspondence between horizontal 1-forms α on P and tensor fields of type (1.1) on the base space M.

Lemma 1.4 *There exists a one-to-one correspondence between tensor fields t_1^1 on M and the set of equivariant 1-forms α on P, namely,*

$$R_g^*\alpha = g^{-1} \cdot \alpha$$

for any $g \in G$.

The proof is similar to that of Lemma 1.3. We only note that α is defined as

$$\alpha_u(\tilde{X}) = u^{-1} \cdot t_1^1(p\tilde{X}).$$

The form $\tilde{\theta}$ satisfies the conditions of Lemma 1.3. The corresponding tensor field

$$t_1^1(X) = u\tilde{\theta}_u(\tilde{X}) = u \cdot (u^{-1}p(\tilde{X})) = X$$

consists of the identity transformations of the tangent space at each point in the manifold M.

By analogy to Lemmas 1.3 and 1.4, we make the corresponding statement for horizontal equivariant 2-forms.

Lemma 1.5 *There exists a one-to-one correspondence between forms β on P and tensor fields t_2^1 on M, namely,*

$$t_2^1(X, Y) = -t_2^1(Y, X), \tag{1.68}$$

the former being defined as $t_2^1(X, Y) = u\beta_u(\tilde{X}, \tilde{Y})$.

We take the form $2\tilde{\theta}$ as β. The corresponding field $t_2^1(X, Y) = T(X, Y)$ is called the *field of the torsion tensor.*

Lemma 1.6 *There exists a one-to-one correspondence between the set of tensor fields t_3^1 such that*

$$t_3^1(X, Y) = -t_3^1(Y, X) \tag{1.69}$$

and the set of $\mathscr{G}$-valued horizontal 2-forms γ on P so that

$$R_g^*\gamma = \mathrm{Ad}\,(g^{-1})\gamma;$$

t_3^1 defined for any two vector fields (X, Y) is of type (1.1).

t_3^1 corresponding to the curvature 2-form 2Ω is called the *tensor curvature field with $R(X, Y) = -R(Y, X)$.*

Show that tensor torsion fields T and tensor curvature fields R are defined by covariant differentiation ∇ as

$$\text{(a) } T(X, Y) = \nabla_X Y - \nabla_Y X - [X, Y], \tag{1.70}$$

$$\text{(b) } R(X, Y) = \nabla_X \nabla_Y - \nabla_Y \nabla_X - \nabla_{[X, Y]}\,. \tag{1.71}$$

Proof. (a) Let $\tilde{X}$, $\tilde{Y}$ be two lifts of vector fields X and Y, respectively, and

$$(\nabla_X Y)_x = u(\tilde{X}_u\tilde{\theta}(\tilde{Y})), \quad p(u) = x.$$

Note that $h[\tilde{X}, \tilde{Y}] = [\widetilde{X, Y}]$. We make use of structural equation (1.62) and the definition of $T(X, Y)$. We have

$$T_x(X, Y) = u \cdot 2\Theta_u(\tilde{X}, \tilde{Y}) = u \cdot (2d\tilde{\theta}(\tilde{X}, \tilde{Y})) = u \cdot (\tilde{X}_u \cdot \tilde{\theta}(\tilde{Y}) - \tilde{Y}_u \cdot \tilde{\theta}(\tilde{X}) \\ - \tilde{\theta}_u[\tilde{X}, \tilde{Y}]) = \nabla_X Y_x - \nabla_Y X_x - [X, Y]_x. \ \blacksquare$$

(b) We select an arbitrary vector field Z on M and consider the function $f = \tilde{\theta}(\tilde{Z})$. Since $[\tilde{X}, \tilde{Y}] = h[\tilde{X}, \tilde{Y}] + v[\tilde{X}, \tilde{Y}]$, applying (1.71) to Z, we obtain

$$([\nabla_X, \nabla_Y]Z - \nabla_{[X, Y]}Z)_x = u \cdot (\tilde{X}_u\tilde{Y}f - \tilde{Y}_u\tilde{X}f - (h[\tilde{X}, \tilde{Y}]_u f) \\ = u \cdot (v[\tilde{X}, \tilde{Y}])_u \cdot f.$$

We select an element $A \in \mathcal{G}$ such that the fundamental field is $A_u = (v[\tilde{X}, \tilde{Y}])_u$. According to (1.65),

$$2\Omega_u(X, Y) = -\omega_u(v[\tilde{X}, \tilde{Y}]) = -A.$$

To complete the proof of formula (1.71), it remains to recall that $f = \tilde{\theta}(Z)$ satisfies the conditions of Lemma 1.3 and the property

$$A_u f = -A \cdot f(u), \tag{1.72}$$

where the left-hand side is the result of applying A_u to f and the right-hand side of applying an element $A \in \mathcal{G}$ to $f(u) \in F$. In fact, let g_t be the one-parameter subgroup in G generated by A. Then

$$A_u f = \lim_{t \to 0} \frac{f(ug_t) - f(u)}{t} = \frac{g_t^{-1} f(u) - f(u)}{t} = -Af. \blacksquare \tag{1.73}$$

With respect to local coordinates, the points $x = (x^1, \ldots, x^n)$ in the manifold M, the components of torsion tensors T and curvature tensors R are

$$T(\partial_i, \partial_j) = T^k_{ij}\partial_k, \ T^k_{ij} = \Gamma^k_{ij} - \Gamma^k_{ji}, \tag{1.74}$$
$$R(\partial_i, \partial_j)\partial_k = R^l_{ijk}\partial_l, \ R^l_{ijk} = \partial_i\Gamma^l_{ki} - \partial_k\Gamma^l_{ji} + \Gamma^m_{ik}\Gamma^l_{jm} - \Gamma^m_{ji}\Gamma^l_{km}.$$

The formulas easily follow from (1.70) and (1.71) when written out in the basis (∂_i).

We now write structural equations (1.62) and (1.63) in terms of the components of T and R.

We specify linear forms $\bar{\omega}^i_j \bar{\theta}^k$ and $\mathcal{V} \subset M^n$ so that

$$\bar{\omega}^i_j = \sigma_* \omega^i_j, \ \theta^k = \sigma_* \bar{\theta}^k, \tag{1.75}$$

where ω^i_j and θ^k are defined on p. 67 and σ is a section of the bundle of frames $\sigma: \mathcal{V} \to P$, $x \in \mathcal{V} \to (X_1, \ldots, X_n)$.

Since Ω and Θ are tensor forms, they can be expressed in terms of 1-forms θ^k and functions T^i_{jk}, R^i_{ikl}.

Put

$$\Theta^i = (1/2) T^i_{jk} \theta^j \wedge \theta^k, \quad T^i_{jk} = -T^i_{kj}, \tag{1.76}$$

and

$$\Omega^i_j = (1/2) R^i_{jkl} \theta^k \wedge \theta^l, \quad R^i_{jkl} = -R^i_{jlk}. \tag{1.77}$$

The forms Ω^i_j and Θ^i are carried over to the base space M^n by means of the section σ, namely,

$$\bar{\Omega}^i_j = \sigma_* \Omega^i_j, \quad \bar{\Theta}^i = \sigma_* \Theta^i.$$

Cartan's structural equations are written in terms of coordinates $\bar{\omega}_j^i$ and $\bar{\theta}^k$ on the base space M^n as

$$d\bar{\theta}^i = -\bar{\omega}_j^i \wedge \bar{\theta}^j + \bar{\Theta}^i, \tag{1.78}$$
$$d\bar{\omega}_j^i = -\bar{\omega}_l^i \wedge \bar{\omega}_j^k + \bar{\Omega}_k^i. \tag{1.79}$$

If we choose local coordinates in a neighborhood $\mathscr{V}$ of a point $x = (x^1, \ldots, x^n)$, $x \in \mathscr{V} \subset M$ and the vector fields are $X_i = \partial_i$, then the basis is $\bar{\theta}^i = dx^i$, $\bar{\omega}_j^i = \Gamma_{jk}^i dx^k$.
Equation (1.78) is written as

$$\bar{\Theta}^i = \bar{\omega}_j^i \wedge \bar{\theta}^j = \Gamma_{jk}^i dx^k \wedge dx^j = (1/2) T_{jk}^i dx^j \wedge dx^k,$$

i.e., taking into account (1.76), $T_{jk}^i = \Gamma_{jk}^i - \Gamma_{kj}^i$. We similarly derive relation (1.74) for R_{ijk}^l from (1.79).

In the general case, additional terms arise in (1.74) for the fields X_i and X_j with the relation $[X_i, X_j] = c_{ij}^k X_k$, e.g., $T_{jk}^i = \Gamma_{jk}^i - \Gamma_{kj}^i - c_{jk}^i$.

It is easy to see that the pair involving curvature form Ω and torsion tensor R as well as the pair consisting of curvature form Θ and torsion tensor T completely determine each other. To prove that curvature and torsion determine a connection is somewhat more complicated (e.g., see [KN1]).

Remark 1.5 It is interesting to see how arbitrary the choice of a connection is in a principal bundle P if the curvature form Φ is fixed. The answer has been obtained quite recently. We turn to the problem in Ch. 6 (Subsec. 6.1.1).

1.5.8 Connection and a Metric

Consider a manifold M on which a metric g_{ik} and connection Γ_{ij}^k are given. Generally speaking, a connection and metric are different structures not related to each other. However, there exists an important class of connections corresponding to the metric, i.e., Riemannian connections.

Here, we restrict ourselves to the study of manifolds with a Riemannian metric g_{ik}. Recall its definition.

Definition 1.41 A *metric* g_{ik} is said to be *Riemannian* if it is given by a positive definite symmetric quadratic form $g_{ik}dx^i dx^k$, where g_{ik} is a tensor of rank (0, 2), $\det \|g\| > 0$.

The Riemannian metric generates a scalar product $\langle X, Y \rangle$ in each tangent space T_xM, $x \in M$, X, $Y \subset T_xM$. The Riemannian metric exists on any differentiable manifold with a countable basis [Stee].

Along with a positive definite Riemannian metric, it would be interesting to investigate spaces with a pseudo-Riemannian metric in whose definition the condition for positive definiteness is omitted, e.g., the Minkowski metric

$$ds^2 = dt^2 - [(dsx^1)^2 + (dx^2)^2 + (dx^3)^2]$$

on R^4 or its generalization

$$ds^2 = \sum_{i=1}^{k} (dx^i)^2 - \sum_{k+1}^{n} (dx^i)^2$$

to space R^n.

Another example is the canonical Killing metric on a noncompact semi-simple Lie group.

Example 1.18 The Killing metric on a group G.

Let $\mathscr{G}$ be the Lie algebra of G. For any two elements $X, Y \in \mathscr{G}$, we define $\langle X, Y\rangle = \mathrm{Tr}\,(\mathrm{ad}\, X \circ \mathrm{ad}\, Y)$, the *Killing form on* $\mathscr{G}$. $\langle X, Y\rangle$ is obviously bilinear and symmetric on $\mathscr{G}$. It is not hard to check that $\langle X, Y\rangle$ is invariant under Ad G. Cartan's theorem [Sem] supplies a criterion for the nonsingularity of a Killing form.

Theorem 1.11 *A form $\langle X, Y\rangle$ is nonsingular if and only if the algebra $\mathscr{G}$ is semi-simple.*

Let $\mathscr{G}$ be a semi-simple algebra. We realize it as the tangent space T_eG to the identity element of the group G. We define the scalar product

$$g_e\langle X, Y\rangle = \langle X, Y\rangle = -\mathrm{Tr}\,(\mathrm{ad}\, X \circ \mathrm{ad}\, Y), \quad X, Y \in \mathscr{G} \tag{1.80}$$

in T_eG, defined at an arbitrary point $x \in G$ by left translations of G as

$$g_x\langle X, Y\rangle = \langle L_{x^{-1}}X, L_{x^{-1}}Y\rangle. \tag{1.81}$$

The metric $\langle\ ,\ \rangle$ is bi-invariant on G. That the metric is left-invariant follows from the definition, whereas the invariance under right translations follows from that of the Killing form under adjoint representation Ad G. Indeed, we must show that $g_x\langle R_xX, R_xY\rangle = g_e\langle X, Y\rangle = \langle X, Y\rangle$ for all $X, Y \in \mathscr{G}$ and $x \in G$. But $\langle R_{x^{-1}}X, R_{x^{-1}}Y\rangle = \langle L_xR_{x^{-1}}X, L_xR_{x^{-1}}Y\rangle$ and $L_xR_{x^{-1}} = \mathrm{Ad}\, x$. Our statement just follows from the invariance of $\langle X, Y\rangle$ under Ad G.

Differentiation of

$$\langle \mathrm{Ad}\,(g)\, X, \mathrm{Ad}\,(g)\, Y\rangle = \langle X, Y\rangle,$$

namely,

$$\frac{d}{dt}\langle \mathrm{Ad}\, g(t)\, X, \mathrm{Ad}\, g(t)\, Y\rangle_{t=0} = \frac{d}{dt}\langle X, Y\rangle,$$

yields

$$\langle(\mathrm{ad}\, Z)X, Y\rangle + \langle X, (\mathrm{ad}\, Z)Y\rangle = 0, \tag{1.82}$$

where $Z = \dot{g}(t)|_{t=0}$.

If we confine ourselves to the class of compact semi-simple Lie groups, then according to Weyl (see [Sem]), the Killing metric is positive definite. For the group $O(n)$, this follows from explicitly calculating $\mathrm{Tr}\,(\mathrm{ad}\, X \circ \mathrm{ad}\, Y)$. In the general case, it is proved that any compact Lie group is contained in $O(n)$ for an appropriate choice of a positive definite invariant form on the space V with a given representation of the group G. In the sequel, we mostly consider only Riemannian metrics, though most results are also valid in the pseudo-Riemannian case.

Definition 1.42 A linear connection ∇_X is said to *correspond to a metric g* if a parallel translation of tangent vectors along any curve in M preserves their scalar product.

Meanwhile, if the torsion tensor vanishes (i.e., $\Gamma_{ij}^k = \Gamma_{ji}^k$), then such a *connection* is said to be *symmetric*, or *Levi-Civita.*

Theorem 1.12 (Fundamental theorem of Riemannian geometry) *A symmetric linear connection corresponding to the Riemannian metric is unique.*

The proof can be found in any differential geometry text-book; we especially recommend the simple treatment in [Mil4].

We now give the final expression for the Christoffel coefficients

$$\Gamma_{ij}^l = (1/2)(\partial_i g_{jk} + \partial_j g_{ik} - \partial_k g_{ij})g^{kl}$$

in terms of the metric g_{ik}, where g^{kl} is the matrix inverse of g_{kl}.

1.5.9 Geodesic Lines

The existence of a linear connection enables us to give a definition of a geodesic (line) on a manifold.

Definition 1.43 Let x_t, $t \in (a, b)$, $-\infty \leqslant a, b \leqslant \infty$, be a curve of class C^∞ in M^n. x_t is called a *geodesic* if the velocity field $X = \dot{x}_t$ is translated along the curve, in other words, x_t is a geodesic if $\nabla_X X = 0$ for all $t \in (a, b)$.

If we write the equation $\nabla_X X = 0$ with respect to local coordinates, then we obtain the classical equation for geodesics

$$\frac{d^2x^k}{dt^2} = \Gamma_{ij}^k \frac{dx^i}{dt} \frac{dx^j}{dt} = 0.$$

Note that the definition of a geodesic essentially depends on parametrization. It is easy to see that only affine transformations of the parameter $t \to \alpha t + \beta$ are admissible, leaving the geodesic invariant.

We establish the relation between geodesics on a manifold M and curves in the bundle of frames P over M.

Theorem 1.13 *A projection onto M of any integral curve in the standard base field given on P is geodesic in M. Any geodesic in M can be obtained in this way.*

In particular, the theorem shows that finding a geodesic (or an equation of the second degree) in M can be reduced to the definition of a vector field on P, i.e., the solution of simultaneous equations of the first order.

Geodesics are defined if affine transformations of the natural parameter t, or "time", are neglected and said to be *canonical.*

Definition 1.44 A linear connection on M is said to be *complete* if any geodesic x_t can be extended to the whole range of t.

Definition 1.45 A geodesic γ: $[a, b] \to M$ is said to be *minimal* if it is no longer than any piecewise differentiable path joining the endpoints. It is not difficult to show that each sufficiently small geodesic segment is minimal. However, a geodesic may be not minimal as a whole, e.g., all arcs of a great circle on a sphere are geodesics, but an arc length greater than πR is obviously not minimal. The example shows that there can be infinitely many minimal geodesics. However, we can prove that two points in one small neighborhood are joined by a unique minimal geodesic.

If we define the distance between two points ϱ (p, q), $p, q \in M$, as the infimum of the lengths of all piecewise differentiable curves joining p and q together, the space $\check{M}$ becomes metric. The important theorem due to Hopf and Rinow [BC] states that the geodesic completeness of a manifold M is equivalent to its metric completeness (i.e., each fundamental sequence of points in M converges).

The trivial examples of complete spaces are R^n, Minkowski's pseudo-Riemannian space $\check{M}^4$ and spheres S^n.

We select only two results from the elegant theory of geodesics on manifolds which shall be needed below. One is related to the well-known property of a geodesic as the shortest line between two points, the other, to the definition of a geodesic on a Lie group.

Assume that a manifold M is Riemannian. We denote the length of a vector $v \in TM_x$ by $|v| = \langle v, v\rangle^{1/2}$. The action for a path γ from a point a to a point b, $a, b \in M$, is defined as

$$S_a^b(\gamma) = \int_a^b \left|\frac{d\gamma}{dt}\right|^2 dt.$$

The action S is a function on the space Ω (p, q, M) of all piecewise differentiable paths γ joining a to b.

Proposition 1.12 (Least action principle) *Let M be a complete space and the distance $\varrho(p, q)$ be d. Then the action function S: $\Omega(p, q, M) \to R$ attains its minimum d^2 exactly on the set of minimal geodesics joining p and q together.*

Geodesics on a Lie group can be defined as trajectories on G. This and Definition 1.43 coincide for Lie groups. We now prove a more general statement.

Theorem 1.14 *One-parameter subgroups $g(t)$, and only they, of a Lie group G are geodesics on G. There exists a unique connection corresponding to the Riemannian metric on G, or the Killing metric $\langle\ ,\ \rangle$; $g(t)$ are geodesics in $\langle\ ,\ \rangle$.*

Proof. We will consider G with a bi-invariant metric. Let X, Y be two left invariant vector fields on G. Construct the connection on G corresponding to the metric. We notice that $\nabla_X X = 0$ for any left invariant field X on G since integral curves in the field are left translations of one-parameter groups and, therefore, geodesics. Hence,

$$\nabla_{X+Y}(X + Y) = \nabla_{X+Y}X + \nabla_{X+Y}Y = \nabla_Y X + \nabla_X Y = 0, \quad \nabla_X Y = -\nabla_Y X.$$

On the other hand, $T(X, Y) = \nabla_X Y - \nabla_Y X - [X, Y] = 0$. Therefore, $\nabla_X Y = \nabla_Y X + [X, Y]$, i.e., the connection $\nabla_X Y$ on G, corresponding to the metric, is

$$\nabla_X Y = (1/2)[X, Y] = (1/2)L_Y X,$$

where L_Y is the Lie derivative.

It immediately follows that one-parameter groups $g(t) = \exp(tX)$ on G generated by X are geodesics, namely,

$$\nabla_{\dot{g}(t)}X = (1/2)[X, X] = 0.$$

Since a parametrized curve with any tangent vector can be issued from the identity element of the group G, it follows by the uniqueness theorem from the local coincidence of the geodesic and the one-parameter group that the former is a one-parameter subgroup of G. ■

Remark 1.6 In proving the theorem, we have made use of the Killing metric without explicitly mentioning it. We show that the connection $\nabla_X Y = (1/2)[X, Y]$ corresponds to the Killing metric $\langle X, Y\rangle = -\mathrm{Tr}\,(\mathrm{ad}\,X \circ \mathrm{ad}\,Y)$, i.e.,

$$\partial_X\langle Y, Z\rangle = \langle \nabla_X Y, Z\rangle + \langle Y, \nabla_X Z\rangle,$$

or

$$\partial_X\langle Y, Z\rangle = \langle (1/2)[X, Y], Z\rangle + \langle Y, (1/2)[X, Z]\rangle. \tag{1.83}$$

Since $[Y, Z]$ is invariant under translations along the curve, $\partial_X[Y, Z] = 0$. We prove that the right-hand side of (1.83) is zero too. This follows from the relation $\langle (\mathrm{ad}\,X)Y, Z\rangle = -\langle Y, (\mathrm{ad}\,X)Z\rangle$, because ad X is a linear operator skew-symmetric relative to the Killing metric.

We now turn to the last item of the short digest on differential geometry, namely, the holonomy group.

1.5.10 Holonomy Group

As we know, the existence of a connection in a principal bundle P enables us to introduce the concept of parallel translation of the fiber along a curve $x_t \in M$. We select a fixed point x_0 in the base space M and consider all closed paths $x_i(t)$ starting and ending at x_0. We call them *loops with the base point* x_0. Consider the set $C(x)$ of all such curves. The operation of multiplication (i.e., composition) of two curves is naturally defined. The loop x_3 obtained by consecutively travelling over loops x_1 and x_2 is their product. The inverse element is defined in the obvious fashion by moving along the loop in the reverse direction. A parallel translation of the fiber $p^{-1}(x)$ along x_1 induces an isomorphism of the fiber with itself, i.e., there exists an element $g \in G$ such that $\tilde{x}_0^2 = g\tilde{x}_0^1$, where $\tilde{x}_0^1$ and $\tilde{x}_0^2$ are two elements of $p^{-1}(x_0)$.

Definition 1.46 The set of all isomorphisms H of the fiber $p^{-1}(x_0)$ with itself forms the holonomy group of a connection ω on P. As can easily be seen, a holonomy group is the natural generalization of the monodromy group of a vector field (i.e., of a differential equation).

Definition 1.47 A subgroup H^0 of a holonomy group H generated by null-homotopic* loops is called the *restricted holonomy group*.

---

* The definition of a homotopy is given in Sec. 2.1.

A holonomy group H can be naturally embedded as a subgroup in the structure group G of a fiber bundle. In fact, we regard two points u and u_1 as equivalent ($u \sim u_1$) if they can be joined together by the horizontal curve. The group $H(u)$ (with fixed *base point* u) then coincides with the set of elements $g \in G$ such that $u \sim gu$. Since $gu \sim gu_1$ if $u \sim u_1$ for any $u, u_1 \in P$, $g \in G$, $H(u)$ forms a subgroup of G.

We can show that for a wide class of manifolds M, the groups H and H^0 are Lie (e.g., see the proof in [KN1]).

To study the holonomy group for fiber bundles with a connection is important due to several circumstances.

Firstly, the existence of a certain holonomy group enables us to reduce connections from a larger structure group to a smaller one and thereby obtain information about the existence (or nonexistence) of additional structures in the fiber bundle or base space such as metric, spin, complex, and other structures.

Secondly, the knowledge of a holonomy group enables us (by analogy with a monodromy group) to globalize certain local properties of manifolds. In this way Cartan first obtained the classification of globally symmetric spaces [He].

One of the main theorems related to the holonomy concept is due to Ambrose and Singer [KN1].

Theorem 1.15 *Let (P, G, M) be a principal bundle, ω a connection on P, and Ω the curvature form for ω. The Lie algebra $\mathscr{H}$ of the holonomy group H coincides with the subspace $V \subset \mathscr{G}$ generated by all elements $\Omega_v(X, Y)$, where $v \in P$, X and Y are arbitrary horizontal vectors at the point v.*

We define the linear holonomy group $HL(M)$ of a manifold M as the holonomy group of the tangent bundle TM.

As a simple corollary, we obtain from Theorem 1.15 a characteristic of flat manifolds. A *manifold M* is *flat* if its linear holonomy group is discrete. For compact manifolds, we can prove a more precise statement.

Proposition 1.13 *A compact and connected Riemannian manifold M is flat if and only if the group $HL(M)$ is finite.*

It is interesting to note that by the Auslander-Kuranishi theorem [Wf] each finite group can be the holonomy group of a compact and flat manifold.

It is with this result that we end this brief outline of holonomy group properties. How holonomy groups can be applied to field theory (i.e., the problem of global gauge) is discussed in Subsec. 6.1.2.

Chapter 2
Elements of topology. How two given manifolds can be differentiated

In this chapter, we briefly account two topological theories enabling us to solve this fundamental problem in principle, viz., homotopy and homology. .

The trick consists in constructing a system of invariants which are algebraic and analytic in nature and can be calculated according to some recipe. Two manifolds M_1 and M_2 are *non-homeomorphic* (*non-diffeomorphic*) if at least one of the invariants assumes different values on M_1 and M_2. The invariants can be either discrete or continuous; we have already encountered such simple cases as the dimension or orientation of a manifold.

Homotopy and homology methods can be compared if we resort to a physical analogy roughly as are the study of electric fields from inside and from outside, e.g., the distribution of the field due to charge over the surface of a sphere can be determined either by interaction with some known standard or independently, viz., by studying the structure (field) of the conductor itself.

An analog of the former method is a homotopy theory while that of the latter is a homology theory. Different non-trivial relations exist between the two.

2.1 HOMOTOPY THEORY

Homotopy theory models the first method, studying a manifold by introducing (mapping) a probe into it. The probe is an n-dimensional sphere. We now turn to exact formulations.

We take the unit n-dimensional cube I^n with the boundary $\partial I^n = \cup I^{n-1}$ (where I^{n-1} are $(n-1)$-faces) contracted to a point. We obtain the unit n-dimensional sphere S^n. Consider all maps of S^n to a manifold M^k.

The set of all the maps forms an infinite-dimensional group slightly related to the structure of M^k.

The set of classes of homotopy equivalent mappings becomes important for other reasons.

A homotopy can be intuitively imagined in the one-dimensional case as follows. Take two elastic threads fixed at two points and then place them in a domain, say, in a square. We will say that two curves, or "threads", are homotopic if one can be made coincident with the other.

It is easy to see that an equivalence relation can be introduced on the set of homotopic curves. If we replace the "elastic threads" with "films", i.e., the images of an n-cube, and speak of coincidence of the films instead, then we obtain a homotopy of n-dimensional objects.

2.1.1 Homotopy Groups

Definition 2.1 Let M^n be a connected manifold. Two maps $f_1, f_2: I^n \to M^n$ are said to be *homotopic* ($f_1 \sim f_2$) if there exists a continuous map $\tau: I^n \times I$ such that $\tau(I^n, 0) = f_1(x)$, $\tau(I^n, 1) = f_2(x)$, $x \in I^n$. (Actually, we can replace the cube I^n in the homotopy definition by any topological space N^k and speak of a homotopy of two maps of the space N^k to M^n.) It is obvious that a homotopy is an equivalence relation.

We now define the important concept of a homotopy group. We begin with the definition of the group operation of addition (or multiplication) of any two elements on the set of homotopy equivalent maps $S^n \to M^k$ and the identity and inverse elements.

Definition 2.2 The set of homotopy equivalent maps $f: S^n \to M^k$ together with the group operation is called an *n-dimensional homotopy group* and denoted by $\pi_n(M^k)$, whereas the class containing f by $\{f\}$.

(1) *Identity element* e of $\pi_n(M)$ is the class of mappings into a fixed point $x \in M$. We denote the element by zero for $n \geqslant 2$ if addition is regarded as the group operation.

(2) *Addition operation.* We define it for two representatives of classes $\{f_1\}$ and $\{f_2\}$, namely,

$$(f_1 + f_2)(t) = f_1(2t_1, t_2, ..., t_n) \qquad \text{if } t_1 \in [0, 1/2],$$
$$(f_1 + f_2)(t) = f_2(2t_1 - 1, t_2, ..., t_n) \qquad \text{if } t_1 \in [1/2, 1],$$
$$f_1(1, t_2, ..., t_n) = f_2(0, t_2, ..., t_n),$$

where $t = (t_1, ..., t_n)$ is an arbitrary point in the cube.

Geometrically, this means that we separate the cube into two sets, $t_1 \in [0, 1/2]$ and $t_2 \in [1/2, 1]$, first traversing the image of the map f by extending the interval $[0, 1/2]$ into $[0, 1]$ and then the image of f_2. It is assumed that maps with fixed points are considered, i.e., $f_i(0) = x$, $x \in M$ for all f_i.

It is easy to see that the homotopy class $\{f_1 + f_2\}$ only depends on $\{f_1\}$ and $\{f_2\}$; therefore, the formula

$$\{f_1\} + \{f_2\} = \{f_1 + f_2\}$$

determines addition in the set $\pi_n(M)$.

(3) *Inverse element* $\{f\}^{-1}$. The map

$$f^{-1} = f(1 - t_1, t_2, ..., t_n)$$

is the inverse of f. All group properties can be verified quite simply. That $\pi_n(M)$ for $n \geqslant 2$ is commutative (see below) is more complicated to prove, e.g., by showing that the zero element of π_n has the defining property

$$\{f\} + \{0\} = \{f\}.$$

It suffices to construct the homotopy sending the map $f + 0$ into f. Map the interval [0, 1/2] into [0, 1], while the interval [1/2, 1] into a point and define $f + 0$ by the formulas

$$f + 0 = \begin{cases} f(2t_1, \ldots, t_n), & 0 \leqslant t_1 \leqslant 1/2, \\ 0(2t_1 - 1, \ldots, t_n), & 1/2 \leqslant t_1 \leqslant 1. \end{cases}$$

Since the map 0 is contracted to a point, $f + 0 \sim f$.

(4) *Associativity* $\{F_1\{F_2, F_3\}\} = \{\{F_1, F_2\}F_3\}$.

Before turning to the investigation of a homotopy group's properties, we make two remarks.

Remark 2.1 When defining homotopy groups by a mapping of an n-dimensional cube with boundary contracted to a point, a fixed point x_0 arises in the image. Therefore, the homotopy group is often denoted by $\pi_n(M, x_0)$. It is easy to show that the homotopy groups of connected spaces with different base points are isomorphic.

Proof. Let $\pi_n(M, x_0)$ and $\pi_n(M, x_1)$ be two n-dimensional homotopy groups with different base points x_0 and x_1. Since the space M is connected, x_0 and x_1 can be joined together with a path γ.

We associate each element f of $\pi_n(M, x_0)$ with the element $\gamma \circ f \in \pi_n(M, x_1)$, the required isomorphism being thereby established. The map $\gamma f \gamma^{-1}$ is an automorphism of $\pi_n(M, x_0)$ with itself.

Below we omit the base point in the notation of $\pi_n(M, x_0)$ except for the cases where this is given explicitly.

Remark 2.2 The first homotopy group $\pi_1(M)$ is that of classes of homotopy equivalent paths or loops. It is also said to be the *fundamental* or *Poincaré* group. We shall use the first name. The second is also found in the mathematical literature; however, it can be a cause of confusion in a book for physicists as a Poincaré group in physics is the full group of motions of four-dimensional Minkowski space-time.

The main properties of homotopy groups:

1. $\pi_n(M)$ is commutative for $n \geqslant 2$.

Proof. Let f_1 and f_2 be two maps $S^n \to M$ and the point $s_0 \to x_0$ belongs to the equator $S^{n-1} \subset S^n = \left(\sum\limits_0^n (t^i)^2 = 1\right)$. We prove that the product $f_1 \cdot f_2$ is homotopic to $f_2 \cdot f_1$.

We restrict f_1 to the upper hemisphere $D^+ : t^0 \geqslant 0$ and f_2 to the lower $D^- : t^0 \leqslant 0$.

We select a point s_0 with coordinates $t^0 = 0$, $t^1 = 1$, $t^2 = t^n = 0$ and consider a map φ sending the upper hemisphere into the lower. φ is obtained by rotating S^n through π about the orthogonal complement to the plane $(t^0 = 0)$.

Consider the homotopy $f_1 \cdot f_2\varphi_t$, where φ_t are rotations through $0 \leqslant \varphi \leqslant \pi$. For $t = 0$, we obtain the map $f_1 \cdot f_2$, while for $t = 1$, $f_2 \cdot f_1$. Since the point s_0 remains fixed under the homotopy, the class of $\{f_1 \cdot f_2\}$ is homotopic to $\{f_2 \cdot f_1\}$.

The fundamental group $\pi_1(M)$ need not necessarily be commutative, e.g., $\pi_1(M^2)$ (where M^2 is a two-dimensional surface of genus $g > 1$) is a non-commutative group and is of the following structure.

Let $a_1, ..., a_n, b_1, ..., b_n$, $n = 2g$, be the generators of a group Γ with the defining relation

$$a_1 b_1 \dots a_g b_g a_{g+1}^{-1} b_{g+1}^{-1} \dots a_{2g}^{-1} b_{2g}^{-1} = 1. \tag{2.1}$$

We can show that Γ with $4g$ elements and one relation (2.1) is the fundamental group of a two-dimensional surface and non-commutative for $g > 1$ [Mas2].

2. $\pi_1(M, x_0)$ is commutative if M is a Lie group G. It is obvious that it suffices to consider the group $\pi_1(G, e)$, where e is the identity element.

We define multiplication in $\pi_1(G, e)$ as

$$(f_1 * f_2)(t) = f_1(t) * f_2(t)$$

for a topological group, where $f_1 * f_2$ is a group multiplication in G.

Let Hom (M, N) be the set of all continuous mappings of a topological space M to a topological space N. The following holds.

Proposition 2.1 *For any two maps f_1 and f_2 from* Hom (M, N), *the map $f_1 * f_2$ is homotopic in* Hom (M, N) *to $f_1 \times f_2$. In particular, this is valid for maps* $S^1 \to M$ and $S^n \to M$.

Let c be a loop starting at the point $e \in G$. Consider the induced map

$$c^{\#}: \pi_1(G, e) \to \pi_1(G, e),$$

where

$$c^{\#}\gamma = c\gamma c^{-1}, \quad \gamma \in \pi_1(G, e).$$

We prove that the automorphism of the group $\pi_1(G, e)$, induced by c, is the identity. It suffices to show that the *path* $c\gamma c^{-1}$ is homotopic to γ. This automatically follows from the lemma below.

Lemma 2.1 *Let c be the path in G joining e to g, $g \in G$. Then the isomorphism*

$$c^{\#}: \pi_n(G, e) = \pi_n(G, g),$$

where the action of π_1 on π_n is $c^{\#}\gamma = c\gamma c^{-1}$, $\gamma \in \pi_n(G, e)$, *coincides with the one induced by a left or right translation of the group G* by g.

It follows that the operation of π_1 with itself, given by inner automorphisms, is trivial, i.e., $c\gamma c^{-1} \sim \gamma$; therefore, $\tilde{c}\tilde{\gamma} = \tilde{\gamma}\tilde{c}$. It also follows that, for topological groups, the action of the fundamental group π_1 on π_n is trivial.

This is not the case for arbitrary manifolds. We consider the action of fundamental groups on higher homotopy ones in Subsec. 2.1.5.

3. The homotopy groups of the direct product are

$$\pi_i(M_1 \times M_2) = \pi_i(M_1) + \pi_i(M_2).$$

The proof is simple and left to the reader.

The main trick in calculating homotopy groups is the following. We represent the space under consideration as the product — naturally, not necessarily direct — of two other spaces whose homotopy groups are known and find the homotopy group of a more complicated space by algebraic relations between these groups. The appropriate algebraic tool is the method of exact sequences.

Algebraic excursion. Exact sequences of groups. Let A_1 and A_2 be two abstract groups and φ_1 a homomorphism of A_1 into A_2.

The *kernel of* $A_1 \xrightarrow{\varphi_1} A_2$ is the set of elements $\{x\} \in A_1$ such that, under φ_1, they are mapped into the zero element of A_2 (making use of additive notation). The kernel of φ is a group and denoted by Ker φ. The *image of* $\varphi: A_1 \to A_2$ is the set of elements $\{y\} \in A_2$ such that $\{y\} = \{\varphi(A_1)\}$ and denoted by Im φ. The concepts of image and kernel can be formulated for any groups.

In a manner similar to vector spaces, we now define the concept of an exact sequence of groups.

Definition 2.3 Given a set of groups A_i and maps $\varphi_i: A_i \to A_{i+1}$, the sequence of the groups $A_i \xrightarrow{\varphi_i} A_{i+1} \xrightarrow{\varphi_{i+1}} A_{i+2}$ is said to be *exact* in the terms A_i, A_{i+1}, A_{i+2} if Ker $\varphi_{i+1} = \text{Im}\,\varphi_i$.

In terms of the exact sequences, a number of properties of groups and maps can be formulated:

(a) $0 \to A \to 0$.

It is obvious that the exactness of the sequence means that the group A vanishes.

(b) $0 \to A \to B \to 0$.

Exercise 2.1 Show that the exactness is equivalent here to the condition that groups A and B should be isomorphic,

$$\text{(c) } 0 \to A \to B \to C \to 0. \tag{2.2}$$

Prove that the exactness of (2.2) entails the relation $C \simeq B/\text{Im}\,(A)$.

Calculations with exact sequences prove useful, since, knowing some terms, we can often draw a conclusion about others.

Remark 2.3 Recall that the definition of an exact sequence involves map, and not just group, properties.

Given an exact sequence of groups

$$0 \to Z_2 \to B \to Z_2 \to 0.$$

What follows from its exactness?

It is easy to see that there exist at least two exact sequences

$$0 \to Z_2 \to Z_4 \to Z_2 \to 0,$$

$$0 \to Z_2 \to Z_2 + Z_2 \to Z_2 \to 0, \tag{2.3}$$

if we confine ourselves to commutative groups. Therefore, to obtain the right answer, we have to know the structure of the corresponding maps and not only that of the nearby groups.

Exercise 2.2 Consider the groups Z of all integers and $2Z$ of all even integers. It is obvious that there exists the homomorphism

$$\varphi: Z \to 2Z$$

whose kernel is Ker $\varphi \simeq Z_2$.

Thus, Z and $2Z$ can be the terms of the exact sequence

$$0 \to Z_2 \xrightarrow{i} Z \xrightarrow{\varphi} 2Z \to 0$$

($2Z$ also being isomorphic to Z).

On the other hand, we can construct the exact sequence

$$0 \to Z_2 \xrightarrow{i} Z + Z_2 \xrightarrow{\varphi} Z \to 0.$$

φ is a projection map onto the first factor of the group $Z + Z_2$ and i the embedding of Z_2 in $Z + Z_2$.

2.1.2 Homotopy Groups of Fiber Bundles

We need one additional property of fiber bundles.

Proposition 2.2 *Let $p: E \to B$ be a fiber bundle with fiber $F = p^{-1}(b)$, X a topological space, $f_0: X \to E$ a map, $g_0 = pf_0: X \to B$ its projection, and $g_t: X \to B$, $t \in [0, 1]$, a homotopy starting at g_0. Then there exists a homotopy $f_t: X \to E$ starting at f_0 and covering g_0, i.e., $pf_t = g_t$* (see the proof in [Stee]).

This important property of fiber bundles defined in Subsec 1.4.1 is called the *covering homotopy theorem.* If we replace the local triviality condition of fiber bundles by this theorem, then the corresponding fiber bundles are called *Serre fiber spaces.* They are, generally speaking, a wider class, as, for instance, they can have non-homeomorphic fibers. However, for "good" spaces X (say, polyhedra), the concepts coincide (see [Stee] for the exact conditions on the topology of X).

The main result of the subsection is the proof of the following theorem.

Theorem 2.1 *The sequence* of *homotopy groups*

$$\to \pi_k(F) \xrightarrow{i_*} \pi_k(E) \xrightarrow{p_*} \pi_k(B) \xrightarrow{\partial_*} \pi_{k-1}(F) \to \dots \to \pi_0(B) \qquad (2.4)$$

is exact.

We start with definitions. The map i_* is defined as the induced map of embedding i of the fiber F into the fiber space E.* The map p_* is induced by the

*The map $f: A \to B$ induces $f_*: \pi_k(A) \to \pi_k(B)$, i.e., $g: S^k \to A$ goes to $g_1: S^k \to B$ where $g_1 = f \cdot g$.

projection map $p: E \to B$ of the total space. It is less easy to define the operator

$$\partial_*: \pi_k(B) \to \pi_{k-1}(F),$$

which is called a *boundary.*

Boundary operator ∂. Let I^k be the unit k-cube and $f: I^k \to B$. Consider a fixed face I^{k-1}. The cube's boundary is $\partial I^k = I^{k-1} \cup J^{k-1}$ (where J^{k-1} is the union of all the other faces). Taking the boundary $\partial I^{k-1} = I^{k-1} \cap J^{k-1}$, we show that each map $f: I^k \to B$ such that $f: \partial I^k \to b_0 \in B$ can be assigned a map $g: I^{k-1} \to F = p^{-1}(b_0)$ such that $g: \partial I^{k-1} \to e_0$ (e is the base point in F, $b_0 = p(e_0)$). If we turn to the corresponding homotopy classes, then the required map is obtained.

For the proof we need the following lemma.

Lemma 2.2 *Let $p: E \to B$ be a fiber bundle, let $e_0 \in E$ and $b_0 \in B$ be two base points, let $p(e_0) = b_0$, let a map $f: I^k \to B$ carry ∂I^k into b_0, and let I^{k-1} be a fixed k-face.*

There exists a map $\tilde{f}: I^k \to E$ covering f, i.e., there exists $p\tilde{f} = f$ such that $\tilde{f}/\partial I^k \setminus I^{k-1} = e_0$.

Note that it immediately follows from the definition of a (Serre) fiber space that $\tilde{f}/\partial I^k \subset p^{-1}(b_0) = F$.

Proof. The map $f: I^k \to B$ is defined as a homotopy f_t defined on I^{k-1} filling up I^k (Fig. 3) so that $f_0: \partial I^k \setminus I^{k-1} \to b_0$ and $f_1: I^{k-1} \to b_0$. We cover this homotopy by $\tilde{f}_t$, $\tilde{f}_0: \partial I^k \setminus I^{k-1} \to e_0$.

We identify $\tilde{f}_t$ with the required map f. Since $\partial I^{k-1} = I^{k-1} \cap J^{k-1}$, $\partial I^{k-1} \to e_0$ under the map $\tilde{f}$. ■

Thus, each map $g: I^k \to B$ sending the boundary ∂I^k into a point* can be associated with a homotopy $\tilde{f}: I^{k-1} \to F$ which sends the boundary into a point.

Now, turning to the equivalence classes of homotopic maps, we obtain the homomorphism

$$\partial_*: \pi_k(B) \to \pi_{k-1}(F).$$

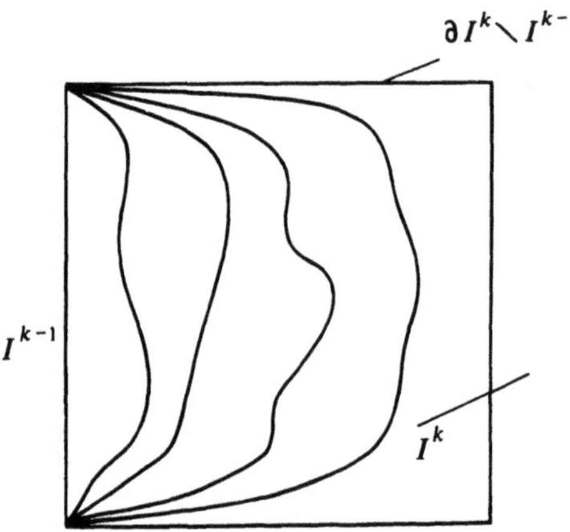

Fig. 3. Definition of homotopy f_t. Part of boundary $I^k \setminus I^{k-1}$ of cube I^k ($k = 2$) is in solid line

*Img is called a *k-dimensional spheroid in B,* since g maps $S^k \to B$.

Turning our attention to the exact sequence in (2.4), we see that the proof can be separated into several steps.

1. $\operatorname{Im} i_* = \operatorname{Ker} p_*$, which is trivial.

2. $\operatorname{Ker} \partial_* = \operatorname{Im} p_*$.

(a) $\operatorname{Ker} \partial_* \supset \operatorname{Im} p_*$. The set $\operatorname{Im} p_*$ consists of k-dimensional "spheroids" in B obtained from those in E by the projection (i.e., the fiber F is mapped into a point $b_0 \in B$), whereas the set $\operatorname{Ker} \partial_*$ consists of $(k - 1)$-dimensional spheroids contractible in F to a point. It follows from the structure of the operator ∂_* that $\operatorname{Ker} \partial_* \supset \operatorname{Im} p_*$. (b) To prove the inverse inclusion, we have to show that the spheroids in B, which are contractible in F to a point, are in the set obtained from spheroids in E under the projection $p: E \to B$, which is obvious.

3. $\operatorname{Ker} i_* = \operatorname{Im} \partial_*$.

Proof. 1. We show that $\operatorname{Ker} i_* \supset \operatorname{Im} \partial_*$, i.e., if $\partial\alpha \in \pi_{k-1}(F)$, then $i_* (\partial\alpha) \sim 0$. Indeed, it follows from the definition of ∂ that the map $f: I^k \to B$ from $\pi_k (B, b_0)$ can be associated with $g: I^{k-1} \to F = p^{-1}(b_0)$ such that I^{k-1} is sent into the base point e_0 $(p(e_0) = b_0)$. It is clear that, under the map i, the element g is contracted in E to the point e_0 by the corresponding homotopy.

2. Conversely, let α be in $\operatorname{Ker} i_*$, i.e., an element in $\pi_{k-1}(F)$ be contractible to a point in the fiber bundle E. Fixing e_0 and reversing the argument of the preceding item, we see that the element $\alpha \sim \partial\beta$, where $\beta \in \pi_k(B)$. ∎

Remark 2.4 If the space B is connected, then $\pi_0(B)$ vanishes (i.e., there is one-connected component) and the exact sequence in (2.4) ends in zero (recalling that we used additive notation).

Resorting to the properties of exact sequences and with a knowledge of the homotopy groups of the simplest spaces, we can calculate the homotopy groups of more complicated and interesting objects.

Example 2.1 1. The homotopy groups π_i $(i > 1)$ of the covering space $\tilde{M}$ coincide with those of M itself.

By definition of the covering space, $M = \tilde{M}/\Gamma$, where Γ is a discrete group acting on $\tilde{M}$. Since the triple $\tilde{M} \xrightarrow{\Gamma} M$ is a fiber bundle, we derive from the exact sequence

$$\pi_i(\Gamma) \to \pi_i(\tilde{M}) \to \pi_i(M) \to \pi_{i-1}(\Gamma)$$

that $\pi_i(\tilde{M}) = \pi_i(M)$ for $i > 1$, since $\pi_i(\Gamma) = 0$ for $i > 0$.

2. $\pi_i(S^1)$. The universal covering space of S^1 is R^1, therefore, $\pi_i(S^1) = 0$ for $i > 1$.

3. $\pi_i(M^2)$, where M^2 is a two-dimensional orientable surface of genus $g \geqslant 1$. Since the covering space of an orientable two-dimensional manifold (or the universal covering space) is a plane, $\pi_i(M^2) = 0$ for $i > 1$.

4. A non-orientable surface of genus $g > 1$, N_g^2. Non-orientable surfaces can be two-sheeted covered by corresponding orientable surfaces, doubles of non-orientable ones (see the corresponding constructions in Example 2.16 and [Mil7]). Thus, the surface $N_g^2 \sim M_{g+1}^2/Z_2$. It follows from this representation that $\pi_i(N_g^2) = 0$ for $i > 1$ and $\pi_i(N_g^2) = \pi_1(M_{g+1}) + Z_2$.

5. $\pi_1(S^1)$. Calculation of the circle's fundamental group requires direct reasoning.

We show that $\pi_1(S^1) = Z$. Indeed, we consider maps $f: S^1 \to S^1$ of the form z^n. We realize both circles as the sets of complex numbers $\{z, |z| = 1\}$. It is obvious that the result does not depend on the representation of a circle. We prove that two maps f_1 and f_2 with different degrees n_1 and n_2 are not homotopic, for which it suffices to show that the degree of a map f, $\deg f = n$, is a topological invariant. Postponing until Sec. 2.5 a discussion in detail of the concept of the degree of a map, we use the well-known definition of the degree of a map of a circle as the index of an analytic function

$$n = \frac{1}{2\pi i} \int_{S^1} \frac{df}{f}.$$

Two maps are homotopic if and only if they have the same indices. Thus, $\pi_1(S^1) = Z$.

6. $\pi_1(T^n)$. The fundamental group of the torus is

$$\pi_1(S^1 \times \dots \times S^1) = \underbrace{Z + \dots + Z}_{n}.$$

7. $\pi_i(S^n) = 0$ for $i < n$.

8. $\pi_1(T_1M^2)$ and $\pi_1(T_1N^2) = ?$, where T_1M^2 (T_1N^2) is the manifold of unit tangent vectors to the orientable surface M^2 and non-orientable one N^2.

9. Wedge of spheres. Let S_1^p and S_2^q be two spheres of dimensions p and q, respectively. We fix a point s_0 in S_1^p and a point s_1 in S_2^q. We consider the formal union of S_1^p and S_2^q identifying (i.e., gluing) s_0 and s_1. We obtain a new topological object called the *wedge* of two spheres S_1^p and S_2^q. The customary notation is $S_1^p \vee S_2^q$. It is obvious that the wedge of n spheres $S_1^{p_1} \vee S_2^{p_2} \vee \dots \vee S_n^{p_n}$ and also that of n topological spaces $X_1 \vee X_2 \vee \dots \vee X_n$ can be defined similarly.

The wedge of spaces can be identified with a subspace of a direct product as follows. For simplicity, take two spaces X and Y. Then the wedge $X \vee Y$ is identified with the subspace $(X \times y_0) \cup (Y \times x_0)$ of the direct product $X \times Y$ with the help of the map $k: (X \vee Y, u_0) \to (X \times Y, w_0)$, where u_0 is a point from $X \vee Y$ obtained by identifying x_0 with y_0, $w_0 = (x_0, y_0)$ and k is defined by the formula

$$k(u) = \begin{cases} (u, y_0) \text{ if } u \in X, \\ (x_0, u) \text{ if } u \in Y. \end{cases}$$

It is easy to see that $\pi_1(S_1^p \vee S_2^q) = Z * Z$ for $p = q = 1$ and $\pi_1(S_1^p \vee S_2^q) = 0$ for $p, q > 1$. For $p = 1$, $q = 0$, $\pi_1(S_1^1 \vee S_2^0) = Z$, where $*$ is the free product of the groups.

Exercise 2.3 1. Prove that $\pi_n(S_1^p \vee S_2^q) = \pi_n(S_1^p) + \pi_n(S_2^q)$ for any $n < p + q - 1$.

2.* For $n > 1$, $\pi_n(X \vee Y, u_0) = \pi_n(X, x_0) + \pi_n(Y, y_0) + \pi_{n+1}(X \times Y, X \vee Y, w_0)$.

* The definition of the relative homotopy group π_{n+1} is given in Subsec. 2.1.3.

The structure of the wedge of spaces has important applications in topology. We shall soon need it in defining a Whitehead product.

2.1.3 Relative Homotopy Groups

Homotopic invariants of manifolds can also be defined for manifolds with a boundary. They are relative homotopy groups $\pi_i(M, \partial M, x_0)$. The idea of their definition is simple. We will study mappings which are homotopic if their boundary elements are not taken into account.

Consider a more general definition that is convenient for more than manifolds with a boundary.

Definition 2.4 Let M be a topological space, A a subset in M, and x_0 a base point. Elements $\alpha \in \pi_i(M, A, x_0)$ are the equivalence classes of homotopic maps $D^i \to M$ under which $\partial D^i = S^{i-1} \to A$ and the marked point s_0 is sent into x_0, $s_0 \in S^{i-1}$. The definition is equivalent to the following. We consider a map $f\colon I^i \to M$ with the base point x_0 sending $\partial I^i \to x_0$. Let $\bar{f}$ be the restriction of f to $\underline{I}^{i-1}$. Since $\partial I^{i-1} = I^{i-1} \cap J^{i-1}$ and goes into the same point x_0 under the map $\bar{f}$, we have $f\colon I^{i-1} \to A$. Thus we have the set of maps $\mathscr{F}\colon (I^i, I^{i-1}, s_0) \to (M, A, x_0)$. The set of homotopy classes $\pi_i(M, A, x_0)$ of such maps is a group when $i \geqslant 2$.

We now list the main properties of relative homotopy groups.

(1) $\pi_i(M, A, x_0)$ for $i > 2$ is a commutative group. $\pi_2(M, A, x_0)$ may be non-commutative (e.g., it can be isomorphic to the absolute fundamental group $\pi_1(M, x_0)$).

The *proof* is similar to the case of absolute groups and is left to the reader.

(2) $\pi_i(M, A, x_0)$ for $A \to x_0$ coincides with an absolute homotopy group. The proof is obvious.

(3) Under continuous mappings of manifolds

$$\begin{array}{ccc} f\colon M & \to & N \\ \cap & & \cap \\ A & \to & B \\ \cap & & \cap \\ x_0 & \to & y_0 \end{array}$$

we obtain natural mappings of homotopy groups

$$f_*\colon \pi_i(M, A, x_0) \to \pi_i(N, B, y_0),$$

where $D^i \to M \xrightarrow{f} N$.

The homomorphisms remain unaltered under a homotopy of the map f such that $A \to B$, $x_0 \to y_0$.

For relative homotopy groups, we can define the boundary operator

$$\partial_*\colon \pi_i(M, A, x_0) \to \pi_{i-1}(A, x_0).$$

Since each map $f: D^i \to M$ defines $f': S^{i-1} = \partial D^i \to A$, we associate f with $f'|_{\partial D^i}$.

Under homotopies of f in the class $\alpha \in \pi_i(M, A, x_0)$ the map of the boundary varies in the class in $\pi_{i-1}(A, x_0)$. It is easy to verify that the map ∂_* is a homomorphism (for $i > 1$), i.e., carries a product into a product.

The main information about relative homotopy groups is contained in the following theorem.

Theorem 2.2 *The sequence of homotopy groups*

$$\pi_i(A, x_0) \xrightarrow{i_*} \pi_i(M, x_0) \xrightarrow{j_*} \pi_i(M, A, x_0) \xrightarrow{\partial_*} \pi_{i-1}(A, x_0) \longrightarrow \ldots$$

is exact, where i_* is induced by the embedding

$$A \subset X,\ j_*: (M, x_0) \subset (M, A, x_0).$$

The proof is similar to that of Theorem 2.1.

The following corollaries can easily be obtained.

(1) *If M is contractible, i.e., $\pi_i(M) = 0$, $i \geqslant 0$, then*

$$\pi_i(M, A) = \pi_{i-1}(A).$$

(2) *Let $M = D^n$, $A = S^{n-1}$. Then $\pi_n(D^n, S^{n-1}, x_0) = Z$, $\pi_i(D^n, S^{n-1}, x_0) = 0$ for $i < n$.*

Exercise 2.4 Calculate $\pi_i(M^2, A)$, where M^2 is a two-dimensional orientable surface and A the boundary of M, $A \sim \partial M$.

Physical examples of the application of homotopy groups are given in the subsection on the classification of surface defects in liquid crystals and superfluid phases in ^{3}He.

2.1.4 Homotopy Groups of Covering Spaces

Recall that the covering space $\tilde{M}$ of a space M is a fiber bundle over M with a discrete fiber.

Definition 2.5 A covering space with $\pi_1(\tilde{M}) = 0$ is said to be *universal.*

We classify coverings by means of the group $\pi_1(\tilde{M}, \tilde{x}_0)$.

Theorem 2.3 *Each connected covering $\tilde{M}$ determines the class of pairwise conjugate groups $\pi_1(M, x_0)$ which are the images of $\pi_1(\tilde{M}, \tilde{x}_0)$, $p(\tilde{x}_0) = x_0$.*

If a discrete group Γ is identified with the fiber F, then these coverings are said to be *regular.* Regular coverings $\tilde{M}$ of M correspond to normal subgroups of $\pi_1(M)$.

Proof. Let $\tilde{x}_1 \in p^{-1}(x_0)$ be a point in the fiber, generally speaking, other than $\tilde{x}_0$ and $\gamma: I \to \tilde{M}$ a path joining $\tilde{x}_0$ to $\tilde{x}_1$. Each such path induces the isomorphism

$$\gamma_*: \pi_1(\tilde{M}, \tilde{x}_1) \to \pi_1(\tilde{M}, \tilde{x}_0).$$

On the other hand, the covering map $p\gamma$ is a loop at x_0 and therefore determines an element $\sigma \in \pi_1(M, x_0)$. It follows from the definition of the homomorphisms

p_* and γ_* that for any element $\alpha \in \pi_1(\tilde{M}, \tilde{x}_1)$

$$p_*\gamma_*\alpha = \sigma p_*\alpha\sigma^{-1},$$

i.e., the group $p_*\pi_1(\tilde{M}, \tilde{x}_1)$ coincides with $\sigma p_*\pi_1(\tilde{M}, \tilde{x}_0)\,\sigma^{-1}$ which is conjugate to the group $p_*[\pi_1(\tilde{M}, \tilde{x}_0)] \in \pi_1(M, x_0)$. On the other hand, let σ be an arbitrary element of $\pi_1(M, x_0)$ and $\tau: I \to M$ an arbitrary loop from the class σ. By the covering homotopy property, there exists a path $\gamma: I \to \tilde{M}$ such that $\gamma(0) = \tilde{x}_0$ and $p\gamma = \tau$. In this case, the point $\tilde{x}_1 = \gamma(1)$ of the fiber $p^{-1}(x_0)$ only depends on σ. It can easily be checked that the equality $x_1 = x_0$ holds if and only if $\sigma \in p_*[\pi_1(\tilde{M}, \tilde{x}_0)]$.

We have thereby shown that each right coset of $\pi_1(M, x_0)$ relative to the subgroup $p_*[\pi_1(\tilde{M}, \tilde{x}_0)]$ is associated with the point $\tilde{x}_1 = p^{-1}(x_0)$ such that each path $\gamma: I \to \tilde{M}$ joining it to $\tilde{x}_0$ covers the loop $\sigma \in \pi_1(M, x_0)$ contained in the coset.

The class $\chi(\tilde{M}, \tilde{x}_0) = \{p_*[\pi_1(\tilde{M}, \tilde{x}_0)]\}$, $\tilde{x}_0 \in p^{-1}(x_0)$, of conjugate subgroups of $\pi_1(M, x_0)$ is said to be *characteristic of the covering of M at x_0*.

Each subgroup of $\chi(\tilde{M}, \tilde{x}_0)$ is isomorphic to $\pi_1(\tilde{M})$.

The class $\chi(\tilde{M})$ consists of only one group if and only if $p_*\pi_1(M, \tilde{x}_0)$ is a normal subgroup of $\pi_1(M)$. The result obviously does not depend on the point in the base space. The covering is then regular. Thus property of a regular covering can also be taken as its definiton. We can summarize the results as follows.

Proposition 2.3 *Simply connected coverings are a special case of fiber spaces, where the covering space $\tilde{M}$ is a fiber bundle with the structure group $\pi_1(M)$ and the discrete homogeneous space $\pi_1(M)/\varrho_*\pi_1(\tilde{M})$ is the fiber. Regular covering spaces correspond to principal bundles.*

Example 2.2 1. R^1 is a universal covering of S^1. The covering map is constructed by the exponential map $\exp: x \to \exp(2\pi i x)$, $x \in R^1$.

2. The covering maps are $S^1 \to S^1: z \to z^n$.

3. The sphere S^n can cover projective space RP^n as a two-sheeted covering. It is easy to calculate the homotopy groups of RP^n:

$$\pi_1(RP^n) = Z_2, \quad \pi_i(RP^n) = 0, \quad i < n, \quad \pi_n(RP^n) = Z.$$

4. The universal covering space $\tilde{M}^2$ of a closed non-orientable surface coincides with the covering R^2 of the orientable surface.

5. Lens spaces L^n.

The covering space is S^{2n-1}. We consider the action of the group Z_n on $S^{2n-1} \subset C^n$. We realize the sphere S by the equation

$$|z_1|^2 + \ldots + |z_n|^2 = 1,$$

and specify the action of Z_n on S^{2n-1} as that of the group of the nth roots of unity, namely,

$$(z_1, \ldots, z_n) \to \left(\exp\left(\frac{2\pi_i\alpha_1}{n}\right)z_1, \ldots, \exp\left(\frac{2\pi_i\alpha_n}{n}\right)z_n\right).$$

We immediately derive from $L^n \sim S^{2n-1}/Z_n$ that $\pi_1(L^n) = Z_n$, $\pi_i(L^n) = 0$, for $i < 2n - 1$.

6. Wedges of spheres.

(a) The covering space of the wedge $\tilde{M} = S_1^1 \vee S_2^1$. The wedge $S_1^1 \vee S_2^1$ is homotopy equivalent to the figure-of-eight in the plane; the covering $\tilde{M}$ is constructed as follows. We consider a universal covering R^1 over S_1^1, representing one of the circles as a helical line and gluing the circles $S_{x_1}^1, \ldots, S_{x_n}^1$ at points $x_1, \ldots, x_n$ (where x_i are integers) which lie over the point $s = (s_0 \sim s_1)$, $s_0 \in S_1^1$, $s_1 \in S_2^1$. We obtain a covering $\tilde{M}$ over $S_1^1 \vee S_2^1$. It is obvious that the universal covering $\tilde{M} \sim R^1 * R^1$. Therefore, $\pi_1(\tilde{M}) = 0$. Geometrically, $\tilde{M}$ can be regarded as an acyclic graph in the plane R^2. In fact, let a point o be the inverse image of the point s. Four edges (forming a cross) emanate from o as from a vertex so that two of them are carried into a cycle a generating S_1^1, while the other two into a cycle b generating S_2^1. Each subsequent vertex of any edge generates a new cross, etc. The corresponding infinite graph Γ is a universal covering space of the figure-of-eight. Γ has no cycles and is therefore contractible.

(b) The covering space for the wedge $M = S_2^1 \vee S_2^2$. It is easy to see that the universal covering space of M is $\tilde{M}$ obtained by attaching spheres S^2 at integral points $x_1, \ldots, x_n$ lying over s in the helical line $l \sim R^1$. Hence, $\pi_i(S_1^1 \vee S_2^2) = \pi_i(\ldots S^2 \vee S^2 \ldots \vee S^2 \ldots)$ for $i > 1$ (cf. Exs. 2.3).

We have considered coverings, assuming fiber isomorphism. However, there may be more complicated situations, e.g., Riemannian surfaces determined by algebraic functions with ramification points. Such coverings are said to be *ramified* and not considered here.

2.1.5 Action of the Fundamental Group π_1 on Higher Homotopy Groups

We have already seen that, for linearly connected spaces, the homotopy groups $\pi_k(M, x_0)$ are isomorphic for different x_0. However, as is shown by the example of the group $\pi_1(M, x_0)$, the isomorphism is established via the transformation

$$\pi_1(M, x_0) \ni \gamma \to \alpha_* \gamma \in \pi_1(M, x_1),$$

where α_* is induced by a path α joining the point x_0 to x_1. In particular, for one-connected manifolds, the isomorphism does not depend on a path. For a closed path $\gamma \in \pi_1(M, x_0)$, the correspondence $\beta \to \gamma_*(\beta)$ determines the action of $\pi_1(M, x_0)$ on $\pi_k(M, x_0)$ via group isomorphisms.

If $f: \tilde{M} \to M$ is the universal covering determined by the free action of a discrete group Γ on $\tilde{M}$, then (1) $\Gamma \sim \pi_1(M, x_0)$ and Γ acts on $\tilde{M}$ by motions $\tilde{M} \to \tilde{M}$. The action coincides with that of $\pi_1(M, x_0)$ on $\pi_k(M, x_0)$ as a group of operators without $\pi_1(M)$ depending on the point x_0 (since $\tilde{M}$ is one-connected). (2) If we do not fix the correspondence of points $s_0 \to x_0 \in M$, the classes of the free homotopy of maps $S^i \to M$ are in a natural one-to-one correspondence with the orbits of operators from $\pi_1(M, x_0)$ acting on $\pi_k(M, x_0)$.

This result is used when studying defects in liquid crystals (see Subsec. 5.1.2).

Example 2.3 The action $\pi_1(RP^2)$ on $\pi_i(RP^2)$: $\pi_1(RP^2) = Z_2$, $\pi_i(RP^2) = \pi_i(S^2)$ for $i > 1$.

Let g be an element of $\pi_1(RP^2)$, $g \neq$ of maps $S^i \to M$. The action $g: S^2 \to S^2$ reverses the orientation, namely, $g(x) = -x$. Therefore, the action of g on an element $a \in \pi_2(RP^2)$ is $g(a) = -a$. We define the action $\pi_1(RP^2)$ on $\pi_i(RP^2)$ for $i > 2$. The homotopy groups $\pi_i(RP^2) = \pi_i(S^2)$ are known. We now formulate one result valid for the homotopy groups of even-dimensional spheres.

Proposition 2.4 *For each even-dimensional sphere S^n, the group $\pi_m(S^n)$ is finite for any $m > n$ and different from $m = 2n - 1$. The group $\pi_{2n-1}(S^n)$ is the direct sum of the group Z and a certain finite group.*

For $n = 2$, $\pi_3(S^2) = Z$. All the other groups are finite. Since π_i are Abelian, π_i can be decomposed into the direct sum of cyclic groups. It is easy to see that the action of π_1 on π_i (for $i > 3$) is reduced to that on the generator $a \in Z_k$. If $ga = -a$, then $(-a)^k = e$. Hence, for groups Z_k with odd k, there is a non-trivial action of π_1 on π_i and there is not for even k. It follows from Table 3 that the first non-trivial homotopy group of odd order is $\pi_9(S^2) = Z_3$.

Since space $RP^3 \sim SO(3)$ is a Lie group, the action $\pi_1(RP^3)$ on $\pi_i(RP^3)$ is trivial. An example of the action of $\pi_1(M)$ on $\pi_i(M)$, when π_1 is not commutative, is considered in Subsec. 5.1.2.

2.1.6 Whitehead Product

Another operation introducing the structure of a graded algebra, or a Lie superalgebra, in modern physical terminology, can be defined on the set of homotopy groups. Called a Whitehead operation, it has interesting applications in the theory of defects in liquid crystals (see Subsec. 5.1.6.6).

Let M be the direct product of spheres $S^i \times S^j$ and A the wedge $S^i \vee S^j$ with identified base points s_0 and s_1, $s_0 \in S^i$, $s_1 \in S^j$. The point $s = (s_0, s_1)$ is the base point in $S^i \times S^j$. The map f

$$D^{i+j} = D^i \times D^j \xrightarrow{f} S^i \times S^j$$

is defined (contracting the disk boundaries ∂D^i and ∂D^j to the points s_0 and s_1, respectively), which sends the boundary $\partial D^{i+j} = (\partial D^i \cup D^j) \cup (\partial D^j \cup D^i)$ into the wedge of spheres, since the generator $\alpha \in \pi_i(S^i, s_0)$ carries ∂D^i into s_0 and $\beta \in \pi_j(S^j, s_1)$ carries ∂D^j into s_1. Thus, f represents an element of the group $\pi_{i+j}(S^i \times S^j, S^i \vee S^j, s)$ for which the boundary homomorphism

$$\partial: \pi_{i+j}(S^i \times S^j, S^i \vee S^j, s) \to \pi_{i+j-1}(S^i \vee S^j, s)$$

is defined.

By $\partial(f)$ we define the product of elements $a \in \pi_i(X, x_0)$ and $b \in \pi_j(X, x_0)$ in the group $\pi_{i+j-1}(S^i \vee S^j, s)$.

Construction. Let a and b be two elements representing classes of homotopy groups.

Definition 2.6 The *Whitehead product,* or *superproduct of two elements a and b,* is the element $[a, b] \in \pi_{i+j-1}(X, x_0)$ obtained as follows. Let

$$a: S^i \to X, \ b: S^j \to X$$

and the points $s = s_0 = s_1$ be identified in the wedge $S^i \vee S^j$.

Consider the sequence of maps $S^{i+j-1} \xrightarrow{\partial f} S^i \vee S^j \xrightarrow{a \circ b} X$.

Their composition $\partial f \circ (a \circ b)$ defines

$$[a, b] \in \pi_{i+j-1}(X, x_0).$$

The main properties of the Whitehead product are:

1. $[a, b] = (-1)^{ij}[b, a]$ (where i and j are the dimensions of the spheres S^i and S^j, respectively).

Proof. The orientation of the disk $D^i \times D^j$ with the frame (e^i, e^j) differs from that of $D^j \times D^i$ with (e^j, e^i) in the sign $(-1)^{ij}$. ∎

2. The product $[a, b]$ in the group $\pi_1(X, x_0)$ $(i, j = 1)$ coincides with the commutator $aba^{-1}b^{-1}$ in the group.

Proof. Let $a, \ b \in \pi_1(X, \ x_0)$. Consider the map of the disk $D^2 = D^1 \times D^1 \to X$ realized as a square in the plane R^2. The boundary ∂D^2 coincides with that of the square oriented clockwise. The base point s coincides with the origin of coordinates. A map f sends ∂D^2 into the wedge of two circles (one-dimensional spheres). Meanwhile, the boundary operator ∂f carries ∂D^2 into the element $aba^{-1}b^{-1}$ (see Fig. 4).

3. If $a \in \pi_1(X, \ x_0)$, $b \in \pi_q(X, \ x_0)$, $q > 1$, then $[b, \ a] = b - a(b)$ and $[a, b] = (-1)^q(b - a(b))$, $a(b)$ denoting the action of the element a on b.

4. Jacobi identity. Let $p, q, r > 1$ and

$$a \in \pi_p(X, x_0), \ b \in \pi_q(X, x_0), \ c \in \pi_r(X, x_0).$$

The relation

$$(-1)^{qr}[a[b, c]] + (-1)^{rp}[b[c, a]] + (-1)^{pq}[c[b, a]] = 0$$

holds. Along with the obvious bilinearity of the bracket $[a, b]$ the "graded form"

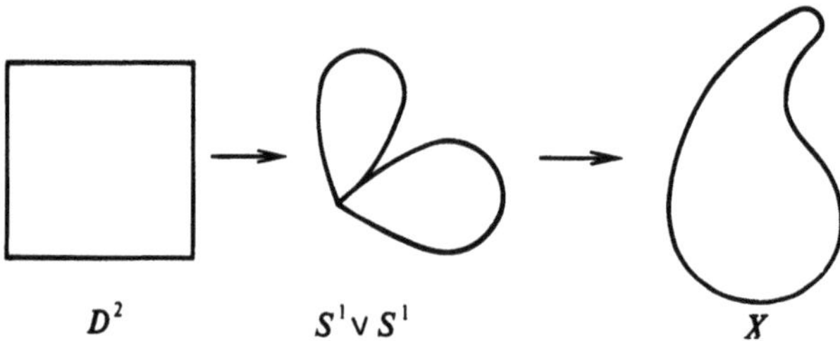

Fig. 4. Whitehead product $(i, \ j = 1)$

of the Jacobi identity enables us to define the graded Abelian group

$$\Pi(X, x_0) = \sum_{p=1}^{\infty} M_p,$$

where $M_p = \pi_{p+1}(X, x_0)$. The graduation is specified by the Whitehead product as $M_p \otimes M_q \xrightarrow{[\,]} M_{p+q}$.

5. For any path $\gamma: I \to X$ joining two points x_0 and x_1 together and for any elements $a \in \pi_p(X, x_0)$ and $b \in \pi_q(X, x_0)$

$$\gamma_{p+q-1}[a, b] = [\gamma_p(a), \gamma_q(b)],$$

where γ_{p+q-1} is a homomorphism of an element $a \in \pi_p(X, x_0)$ in motion along γ.

6. If φ is a continuous map $X \to Y$, then

$$\varphi_*[a, b] = [\varphi_*(a), \varphi_*(b)] \in \pi_{p+q-1}(Y),$$

where $a \in \pi_p(X)$ and $b \in \pi_q(X)$.

7. The Whitehead product for Lie groups is trivial.

Proof. Let G be a Lie group and the base point e the identity element. If $a, b \in \pi_1(G)$, then the proof is a consequence of the commutativity of $\pi_1(G)$, i.e., $[a, b] = aba^{-1}b^{-1} = e$. For $a \in \pi_1(G)$ and $b \in \pi_i(G)$, $i > 1$, the triviality of $[a, b]$ follows from that of the action of the fundamental group on higher homotopy groups of a Lie group. ∎

Let $i, j \geqslant 2$. The elements a and b are represented by the maps

$$\hat{a}: \quad D^i \to X, \ \partial D^i \to e, \ \hat{b}: D^j \to X, \ \partial D^j \to e.$$

Consider the map

$$\hat{a}\hat{b}: D^i \times D^j \to X,$$

putting $\hat{a}\hat{b} = \hat{a}(x)\hat{b}(y)$. It induces the element $[a, b]$ on the boundary $\partial(D^i \times D^j) = S^{i+j-1}$. Since $[a, b]$ is homotopic to an element $\hat{c} \in S^i \vee S^j$, $[a, b] = 0$.

2.1.7 Survey of Calculations of Homotopy Groups in Physical Applications

Here, we discuss calculations of the homotopy groups of several classes of manifolds encountered in physical applications.

The calculations are partly given in the book where appropriate, others can be obtained by the reader as an exercise. However, most results require a considerably more complicated technique than that in the monograph. I indicate in brackets physical problems and some original works where the corresponding groups are used.

2.1.7.1 Homotopy groups of Lie groups

(1) $\pi_2(G) = 0$ (see the proof in Sec. 2.4 and applications in the theory of monopoles in Subsec. 4.1.3).

(2) $\pi_3(G) \neq 0$ (see the proof in Sec. 2.4 for semisimple Lie groups and applications in the theory of instantons in Subsec. 4.2.5).

(3) $\pi_i(G), i > 3$. The theory of homotopy groups of classical Lie groups contains the remarkable Bott periodicity theorem (see [Mil4]) enabling us to calculate all "stable" homotopy groups. This result calls for some preliminary clarification.

Classical Lie groups fall into three categories: orthogonal $O(n)$, unitary $U(n)$, and symplectic groups $Sp(n)$. We consider each.

(a) $O(n)$. It suffices to restrict ourselves to the group $SO(n)$.

Proposition 2.5 *The inclusion homomorphism*

$$\pi_i(SO(n)) \subset \pi_i(SO(n+1))$$

is an isomorphism for $i \leqslant n-2$ *and an epimorphism for* $i = n-1$.

The group $\pi_1(SO(n))$ is Z for $n = 2$ and Z_2 for $n \geqslant 3$, $\pi_2(SO(n)) = 0$. $\pi_3(SO(2)) = 0$, $\pi_3(SO(3)) = Z$, $\pi_3(SO(4)) = Z + Z$, $\pi_3(SO(n)) = Z$ for $n \geqslant 5$.

(b) $U(n)$. The inclusion homomorphism $\pi_i(U(n)) \subset \pi_i(U(n+1))$ is an isomorphism for $i \leqslant 2n-1$ and an epimorphism for $i = 2n$. This follows from the exact homotopy sequence

$$\ldots \to \pi_{i+1}(S^{2n+1}) \to \pi_i(U(n)) \to \pi_i(U(n+1)) \to \pi_i(S^{2n+1}) \to \ldots$$

of the fiber bundle $U(n+1) \xrightarrow{U(n)} S^{2n+1}$. The groups $\pi_i(U(n)) \sim \pi_i(U(n+1)) \sim \pi_i(U(n+2))$ are isomorphic for $i \leqslant 2n$. The isomorphic groups $\pi_i(U(\infty))$ ($U(\infty) = \lim U(1) \subset U(2) \subset \ldots \subset U(n)$) are called *stable homotopy groups* and denoted by $\pi_i(U)$.

We can directly consider the homotopy groups $\pi_i(U)$, where $U = \lim U(n)$ ($n \to \infty$) is the direct limit of inclusions $U(1) \subset \ldots \subset U(n)$ [ES].

Theorem 2.4 *The stable homotopy groups* $\pi_i(U)$ *are periodic with period* 2, $\pi_0(U) = \pi_2(U) = \pi_4(U) = 0$ *and* $\pi_1(U) = \pi_3(U) = \pi_5(U) = \ldots = Z$.

$\pi_i(U)$ arises in the models of $1/N$ ($N \to \infty$) expansion in field theory (see [Ho2]).

We consider homotopy groups of stable orthogonal groups $\pi_i(O)$.

Like unitary groups, the homotopy groups of stable orthogonal groups also possess Bott periodicity: $\pi_i(O(n))$, $n \to \infty$, $\pi_i(O) = \pi_{i+8}(O)$.

Calculations of $\pi_i(O)$ for $i \leqslant 7$ and of symplectic groups are tabulated in Table 1.

TABLE 1 ($\pi_i(O)$, $\pi_i(S\pi)$, $1 \leqslant i \leqslant 7$)

i	1	2	3	4	5	6	7
$\pi_i(O)$	Z_2	0	Z	0	0	0	Z
$\pi_i(Sp)$	0	0	Z	Z_2	Z_2	0	Z

Applications. Stable homotopy groups $\pi_i(O)$ like unitary groups arise in the models of $1/N$ expansion. Depending on the dimension of a homotopy group (actually, of physical space), instantons either exist or do not exist in the models.

(c) Homotopy groups of symplectic groups Sp(n). The inclusion homomorphism $\pi_i(\mathrm{Sp}(n)) \to \pi_i(\mathrm{Sp}(n+1))$ is an isomorphism for $i \leqslant 4n+1$ and an epimorphism for $i = 4n+2$. The result follows from the exactness of the homotopy sequence of the fiber bundle $\mathrm{Sp}(n+1) \xrightarrow{\mathrm{Sp}(n)} S^{4n+3}$:

$$\ldots \to \pi_{i+1}(S^{4n+3}) \to \pi_i(\mathrm{Sp}(n)) + \pi_i(\mathrm{Sp}(n+1)) \to \pi_i(S^{4n+3}) \to \ldots$$

The homotopy groups $\pi_i(\mathrm{Sp}(n))$ are stabilized for $n \geqslant (i+2)/4$. We define a stable symplectic group in a similar manner as a unitary or orthogonal one, viz., as the direct limit of the inclusions $\mathrm{Sp}(1) \subset \mathrm{Sp}(2) \subset \ldots \subset \mathrm{Sp}(n) \subset \ldots$. We denote the ith homotopy group by $\pi_i(\mathrm{Sp})$. The corresponding Bott periodicity theorem is as follows.

Theorem 2.5

$$\pi_i(\mathrm{Sp}) = \pi_{i+q}(\mathrm{Sp}).$$

2.1.7.2 Homotopy groups of homogeneous spaces

1. Stiefel manifolds.

(a) Real Stiefel manifolds $V_{n,k}$.

Proposition 2.6 *A manifold $V_{n,k}$ is connected and $\pi_i(V_{n,k}) = 0$ for $i < n - k$:*

$$\pi_i(V_{n,k}) = \begin{cases} Z \text{ if } n-k \text{ is even or } k = 1, \\ Z_2 \text{ if } n-k \text{ is odd and } k > 1. \end{cases}$$

The generator of the group $\pi_{n-k}(V_{n,k})$ is determined by the mapping $f: S \xrightarrow{n-k} V_{n,k}$ given as follows. Let v_0 be a fixed orthogonal $(k-1)$-frame of an n-dimensional space and S^{n-k} be the unit sphere in the R^{n-k+1}-dimensional space orthogonal to v_0. The map f assigns each point $x \in S^{n-k}$ the orthogonal k-frame obtained from v_0 by adding the radius-vector of x as the first vector.

The proof is in [Stee]. The assertions partly follow from the exact sequence of the fiber bundle

$$SO(n) \xrightarrow{SO(n-k)} V_{n,k}.$$

Definition 2.7 A manifold M is said to be *k-connected* if $\pi_k(M^n) = 0$ for all $i \leqslant k$.

It follows from Proposition 2.6 that the manifold $V_{n,k}$ is $(n-k-1)$-connected.

Similarly to classical stable groups, we can define stable Stiefel manifolds $V_{\infty,k}$. The corresponding homotopy groups $\pi_i(V_{\infty,k})$ vanish for all i. Therefore, the spaces $V_{\infty,k}$ are infinitely connected.

(b) Complex Stiefel spaces $V^C_{n,k}$.

In a similar manner to Proposition 2.6, the following holds.

Proposition 2.7

$$\pi_i(V^C_{n,k}) = \begin{cases} 0 & \text{for } i < 2n - 2k + 1, \\ Z & \text{for } i = 2n - 2k + 1. \end{cases}$$

In the latter case, any fiber of the projection $V^C_{n,k} \to V^C_{n,k-1}$ is a representative of the generator of the group Z.

We could construct quaternionic Stiefel spaces similarly, replacing the field of complex numbers by a quaternion field. (The reader can carry out the corresponding constructions if necessary; see also [Stee].)

Stable complex Stiefel spaces are determined as $V^C_{\infty,k} = U(\infty)/U(\infty - k)$.

2. Grassmannians.

(a) Real oriented Grassmannians $G_{n,k}$.

Homotopy groups $\pi_i(G_{n,k})$ can easily be found if we notice that $G_{n,k}$ can be regarded as the base space for the Stiefel bundle $V_{n,k}$ with fiber $SO(k)$. We then obtain from the exactness of the homotopy sequence of the fiber bundle

$$V_{n,k} \xrightarrow{SO(k)} G_{n,k}$$

that the following holds.

Proposition 2.8

$$\pi_i(G_{n,k}) = \pi_{i-1}(SO(k)) \text{ for } i \leqslant n - k - 1.$$

It follows from the duality isomorphism $G_{n,k} \sim G_{n,n-k}$ that $\pi_i(G_{n,k}) = \pi_{i-1}(SO(n - k))$.

The universal Grassmannians $G_{\infty,k} = SO(\infty)/SO(\infty - k) \times SO(k)$ are constructed in a similar manner as universal Stiefel spaces.

Exercise 2.5 Calculate $\pi_i(G_{\infty,k})$.

(b) Complex Grassmannians $G^C_{n,k}$.

Exercise 2.6 Calculate $\pi_i(G^C_{n,k})$.

3. Homotopy groups of symmetric spaces.

The symmetric spaces $M_1 = SU(n)/SO(n)$ and $M_2 = SU(2n)/Sp(n)$ arise when constructing baryons from solitons (see [Wit2, 3]). Using the homotopy groups of M_1 and M_2, we can classify classical solutions assured by the corresponding Lagrangian.

We now list the results of calculating $\pi_i(M_1)$ and $\pi_i(M_2)$ for $i \leqslant 5$ obtained from Table 2.

4. We consider homotopy groups of spheres.

Calculating homotopy groups of spheres is the most complicated problem in topology and is still fully unsolved. However, for physical applications, a knowledge of the homotopy groups of relatively small dimensions is sufficient.

We start with general constructions necessary for calculating π_i.

We are coming now to the Freudenthal suspension.

Let f be a map $S^i \to S^n$. Consider the spheres S^i and S^n as the equators of the spheres S^{i+1} and S^{n+1}, respectively.

TABLE 2 $\pi_i(M_j)$, $j = 1, 2$, $i \leqslant 5$

i	1	2	3	4	5
$\pi_i(M_1)$	0	Z_2, $n \geqslant 3$	Z_2, $n \geqslant 4$	0, $n \geqslant 4$	Z, $n \geqslant 3$
$\pi_i(M_2)$	0	0	0	0	Z, $n \geqslant 3$

Note. The spaces M_1 and M_2 are examples of irreducible globally symmetric Riemannian spaces classified by Cartan; see [He] *for a detailed treatment of Cartan's theory. The results in the table can be obtained by analyzing the exact sequence of the homotopy groups of the fiber bundles $SU(n) \xrightarrow{SO(n)} M_1$ and $SU(2n) \xrightarrow{Sp(n)} M_2$. To calculate the group $\pi_4(M)$, we use Hurewicz's theorem (Theorem 2.15) and Borel's results concerning the cohomology of M_1 and M_2* [Bor]. *$\pi_{4n+r}(M_1)$, $r \leqslant 5$, were calculated by Harris* [Ha] *and $\pi_{4n+r}(M_2)$, $r \leqslant 5$, by Mimura* [Mim]. *The problem of calculating all $\pi_i(M_j)$ is unsolved since it is equivalent to finding the homotopy groups of spheres.*

The map

$$Ef\colon S^{i+1} \to S^{n+1}$$

coinciding with f on the equator and sending the upper (lower) hemisphere D_+^{i+1} (D_-^{i+1}) onto upper (lower) hemisphere D_+^{n+1} (D_-^{n+1}) of the sphere S^{n+1} is called the *suspension* of f and denoted by Ef (*Einhängung* according to Freudenthal). Ef maps the center of each hemisphere into that of the corresponding hemisphere and is linear on the radii. A homotopy of f into f' can be completed to a homotopy of Ef into Ef'.

Each homotopy class of a map $f\colon S^i \to S^n$ is thereby associated with that of $Ef\colon S^{i+1} \to S^{n+1}$, in other words, the map

$$E\colon \pi_i(S^n) \to \pi_{i+1}(S^{n+1})$$

called the *suspension mapping* or *Freudenthal suspension,* is defined.

Freudenthal's suspension theorem is now stated as follows.

Proposition 2.9 *Let $E\colon \pi_i(S^n) \to \pi_{i+1}(S^{n+1})$.*

(1) *If $i < 2n - 1$, then E is an isomorphism.*

(2) *If $i = 2n - 1$, then E is an epimorphism.*

The proof of (1) is the "easy" part of Freudenthal's theorem and can be found in [Stee] and that of (2) is the "difficult" part and is in [Hu].

Remark 2.5 Homotopy groups of low dimensions (see Table 3), in particular, $\pi_3(S^2)$, now have very interesting applications in physics. The Hopf invariant is related to $\pi_3(S^2)$ and repeatedly appears below.

The group $\pi_4(S^3) = Z_2$ is responsible for the interaction of solitons in the Witten model (see [Wit2, 3]).

TABLE 3 The homotopy groups of spheres $\pi_{n+k}(S^n)$, $2 \leqslant n \leqslant 7$, $1 \leqslant k \leqslant 7$

k \ n	2	3	4	5	6	7
1	Z	Z_2	Z_2	Z_2	Z_2	Z_2
2	Z_2	Z_2	Z_2	Z_2	Z_2	Z_2
3	Z_2	Z_{12}	$Z + Z_{12}$	Z_{24}	Z_{24}	Z_{24}
4	Z_{12}	Z_2	$Z_2 + Z_2$	Z_2	0	0
5	Z_2	Z_2	$Z_2 + Z_2$	Z_2	Z	0
6	Z_2	Z_3	$Z_2 + Z_{24}$	Z_2	Z_2	Z_2
7	Z_3	Z_{15}	Z_{15}	Z_{30}	Z_{60}	Z_{120}

Note. The homotopy groups $\pi_{n+k}(S^n)$, $k \leqslant 30$, and $\pi_{\infty+k}(S^\infty)$ are known at present [To], *where $\pi_{\infty+k}(S^\infty)$ is the stable homotopy group defined below. It follows from Freudenthal's suspension theorem that the groups $\pi_i(S^n) \sim \pi_{i+1}(S^{n+1})$ for $i < 2n - 1$ are isomorphic. Therefore, the suspension isomorphism in the series of $\pi_{n+k}(S^n)$ enables us to identify the groups for $n > k + 2$ with the only group, just called a stable homotopy group of the form $\pi_{\infty+k}(S^\infty)$.*

2.2 HOMOLOGY THEORY

Homology theory introduces a new connection between invariants of manifolds. Continuing the "physical" analogy, we say that a homology theory studies the intrinsic structure of a manifold by breaking it into a system of portions arranged simply, or, more precisely, in a standard way. Then, given certain rules for glueing the portions together, the theory obtains the whole manifold. The main problem consists in proving the resultant geometric quantities that are independent of the decomposition and glueing (i.e., proving the topological invariance of the characteristics).

We will use various approaches. Some are more convenient in stating and proving theorems, others, in practical calculations. We shall mostly confine ourselves to stating and illustrating the principal concepts by way of example. For a rigorous proof we refer the reader to the well-known references [ESt, DFN].

2.2.1 Cellular Decomposition

Cellular decomposition is a decomposition into cells, each of which is homeomorphic to a ball of a certain dimension.

Before turning to precise definitions, we consider three examples.

Example 2.4

(1) Sphere S^n. We select a point τ^0. The set $\tau^n = S^n \setminus \tau^0$ is then homeomorphic to the n-dimensional ball.

(2) Torus T^2. Let τ^0 be a point, l_1 a meridian, and l_2 a parallel so that

$$\tau_1^1 = l_1 \setminus \tau^0, \; \tau_2^1 = l_2 \setminus \tau^0, \; \tau^2 = T^2 \setminus (l_1 \cup l_2).$$

The cellular decomposition of the torus contains one zero-dimensional cell, two one-dimensional, and one two-dimensional cells.

(3) Projective plane RP^2. We select the projective line RP^1 in RP^2. We have $RP^2 \supset RP^1 \supset p^0$ (p^0 is a point). We put $\tau^0 = p^0$, $\tau^1 = RP^1 \setminus p^0$, $\tau^2 = RP^2 \setminus RP^1$. The cellular decomposition of the projective plane consists of the three cells τ^0, τ^1, τ^2.

We now give the precise definition of a cellular decomposition of a topological space K.

We regard K as Hausdorff, i.e., impose the following separability condition on it. For any two distinct points $x_1, x_2 \in K$, there should exist disjoint neighborhoods $\mathscr{U}_1 \ni x_1$, $\mathscr{U}_2 \ni x_2$, $\mathscr{U}_1 \cap \mathscr{U}_2 = \varnothing$. The condition is quite a weak constraint on the topology of the space, but it enables us to avoid various pathologies.

Definition 2.8 The decomposition of a space K into mutually disjoint cells $\{\tau^a\}$ is said to be *cellular* (K a *CW-complex* space) if the following three conditions are fulfilled.

A. Each cell τ^a is homeomorphic to a ball of some dimension, called the *dimension of the cell.*

For each τ^a there exists a continuous map $\varphi: D^a \to K$ of a closed ball D^a ($a = \dim \tau^a$) into the space K such that φ homeomorphically maps the interior of the ball onto τ^a. The map is said to be *characteristic* and corresponds to τ^a.

B. For any τ^a with $\dim \tau^a > 0$, its boundary $\partial\tau^a = \bar{\tau}^a \setminus \tau^a$ is contained in the union of finitely many cells of smaller dimension. If the manifold is non-compact, then another property should hold.

C. Any set $N \subset K$ for which the intersection $N \cap \tau^a$ is closed (for any cell τ^a) is closed. Spaces admitting cellular decompositions are called *cell complexes.* Manifolds of compact and non-compact type (with a countable base) are cell complexes. A particular case of cellular decompositions is a simplicial decomposition (or decomposition into simplexes).

2.2.2 Operations on Cell Complexes

A number of operations enabling us to pass to subsets and quotient sets of cell complexes can be introduced.

A. Subdivision of a complex.

Definition 2.9 Let $\{\tau^a\}$ be a cellular decomposition of a space K. A closed subset $L \subset K$ consisting of integral cells is called a *subcomplex* and the decomposition corresponding to L *a subdivision of K.*

Definition 2.10 The union of all cells of dimensions less than n or equal to n of a cell complex K is called its *n-dimensional skeleton* and is denoted by K^n.

It is obvious that each K^n of a cellular decomposition of K is a subdivision. The zero-dimensional skeleton K^0 is non-void and discrete.

B. Factorization of a cell complex.

Let K be a cellular decomposition and L its subdivision. We denote by K/L the subspace obtained from K by identifying L with one point σ and we denote by $\varphi: K \to L$ the identification map. Then the sets $\varphi(\tau)$ and σ, when τ runs over the cells not involved in L, generate, as can easily be seen, the cellular decomposition of the space K/L. The transition from K to the cellular decomposition of K/L is called the *factorization of K with respect to L.*

Example 2.5 Let K be a cell complex and n an integer. The cellular decomposition K^n/K^{n-1} consists of one zero-dimensional cell σ^0 and several n-dimensional cells (which are in one-to-one correspondence with the n-dimensional cells of K). It follows from the definition of K that the space K^n/K^{n-1} is the familiar wedge of n-dimensional spheres.

C. Attaching an n-dimensional cell.

This operation enables us to construct inductively the cellular decomposition of a complex. Let K be a cell complex, D^n an n-dimensional cell, and f a map $\partial D^n = S^{n-1} \to K$.

The space $K \cup_f D^n$ is said to *have an attached cell* (via f) if the space is obtained from the disjoint union of K and D^n by identifying each point $x \in S^{n-1}$ with the point $f(x) \in K$. For the zero-dimensional case, $n = 0$, D^0 is a point, S^{-1} is regarded as the empty set, and K with the attached point is the union of K with a distant point.

We can introduce a topology defining the cellular decomposition on $\check{K} = K \cup_f D^n$.

Here, certain details are omitted. An arbitrary map $f: S^{n-1} \to K$ may not be in K^{n-1}. However, we can prove that there is a homotopic map already possessing the property, the homotopy type of $K \cup_f D^n$ remaining unaltered. We can construct a cellular decomposition of any space by attaching cells since the operation applied to K^{n-1} consecutively yields K^n and the union of all skeletons is K.

D. Cellular maps.

A necessary tool in the study of cellular decompositions of arbitrary manifolds is the cellular map.

Definition 2.11 Let K and L be two cellular decompositions. A continuous map $f: K \to L$ is said to be *cellular* if, for any n, $f(K^n) \subset L^n$.

Theorem 2.6 *Any continuous map $f: K \to L$ is homotopic to a cellular one; if f has already been cellular on the subdivision $K^* \subset K$, then the homotopy can be regarded as constant on K^*.*

The proof is obtained in several steps. We only outline it (see the details in [Hu]).

We first consider the case of attaching one additional cell. Let K be obtained from a subdivision K' by attaching one n-dimensional cell τ^n and L from a subdivision L' by attaching one m-dimensional cell σ^m $(m > n)$.

Let a map $f_0: K \to L$ satisfy the condition $f_0(K') \subset L'$. Then there exists a homotopy $f_t: K \to L$ $(0 \leqslant t \leqslant 1)$ of f_0, constant on K' and such that $f_1(K') \subset L'$.

To prove this particular case, we construct a homotopy (constant on K') sending f_0 into a map f_* such that there are points in σ^m not belonging to the image $f_*(K)$.

This can be done by introducing coordinates into the cells τ and σ (via characteristic maps) and by approximating f_* by differentiable functions (Weierstrass' theorem).

Let the map f_* be constructed. Then there exists a point $y_0 \in \sigma^m$, $y_0 \notin f_*(K)$. We denote by $h: D^m \to L$ the characteristic map corresponding to σ^m and put $x_0 = h^{-1}(y_0)$. Furthermore, let ω_t be a deformation of the set $D^m \setminus x_0$ on itself so that it is constant on the boundary ∂D^m and satisfies the condition $\omega_1(D^m \setminus x_0) \subset \partial D^m$.

We put

$$f_t'(x) = \begin{cases} h\omega_t h^{-1} f_*(x) & \text{for } f_*(x) \in \sigma^m, \\ f_*(x) & \text{for } f_*(x) \notin \sigma^m. \end{cases}$$

We obtain a continuous deformation f_t' of the map $f_*: K \to L$ into $f_1': K \to L$ such that $f_1'(K) \subset L'$ and constant on K'. ■

The general case is a consequence if the following simple argument is employed. We attach zero-, one-, etc., dimensional cells not in K^* to K^* one by one and, by taking into account that $f(K) \subset L'$, we reduce the dimension of the image in L using the particular case already proved.

The procedure is correct, since under a continuous map $f: K \to L$ the image of any cell only intersects finitely many cells. ■

This theorem will enable us below to blur the distinction between continuous and cellular maps.

To study cellular maps and their images, it is convenient to introduce coordinates in cells and to consider maps of the cells in coordinate functions.

Let D_0^n be the unit ball in R^n $(\xi^1, ..., \xi^n)$, i.e., the set of points satisfying the inequality

$$(\xi^1)^2 + (\xi^2)^2 + ... + (\xi^n)^2 \leqslant 1.$$

The unit sphere S_0^{n-1} is the ball's boundary. An n-dimensional cell τ^n of a cellular decomposition K is said to be *coordinate* if the corresponding characteristic map $\varphi: D_0^n \to K$ is fixed. φ is called a *coordinate map.* We introduce coordinates ξ^1, ..., ξ^n in a coordinate cell τ^n, assuming that a point $x \in \tau^n$ has the same coordinates as its inverse image $\varphi^{-1}(x) \in D_0^n \setminus S_0^{n-1}$.

Assume that all cellular decompositions are given via a fixed coordinate mapping (i.e., coordinates are introduced in all the cells).

Let $f: K \to L$ be a cellular map. We select a point $x \in K^n \setminus K^{n-1}$ $(n \geqslant 1)$ whose image $y = f(x)$ is in the set $L^n \setminus L^{n-1}$. Let τ^n be an n-dimensional cell containing x and σ^n be an n-dimensional cell containing y. We introduce coordinates ξ^1, ..., ξ^n in τ^n and η^1, ..., η^n in σ^n. In the vicinity of x, the map f is given with respect to the coordinates as

$$\eta^i = f_i(\xi^1, ..., \xi^n).$$

Definition 2.12 A map f is said to be *regular* if the functions f_i have continuous partial derivatives of order one and the determinant of the Jacobian $\det I = \left| \frac{\partial \eta^i}{\partial \xi^j} \right|_x$ is nonzero.

If $\det I$ is positive at a point x, then we regard x as *positive*; if $\det I < 0$, then x is regarded as *negative*. If $n = 0$ and $x \in K^0$, then any cellular map $f: K \to L$ is regular at x. By definition, any point $x \in K^0$ is regarded as positive.

A cellular map $f: K \to L$ is said to be regular at a point $y \in L^n \setminus L^{n-1}$ $(n \geqslant 1)$ if the mapping is regular at each point of the inverse image $x \in f^{-1}(y) \cap K^n$.

The inverse image of y only has finitely many points in each n-dimensional cell of the decomposition K. If $y \in L^0$, then any cellular map $f: K \to L$ is regular at y. A *cellular map* $f: K \to L$ is said to be *regular* if there is a point y in any cell $\sigma^n \subset L$ at which the map is regular. Any cellular map turns out to be homotopic to a regular one. A more exact result is formulated as follows.

Proposition 2.10 *Let $f_0: K \to L$ be a cellular map. There exists a deformation f_t of f_0 so that each f_t $(0 \leqslant t \leqslant 1)$ is cellular and the map f regular.* (The proof involves approximation of continuous functions by differentiable ones, e.g., see [GG].)

E. Degree of a map.

We now introduce the concept of the degree of a map, which enables us to make the concept of adhesion of two cells meaningful. We formulate the necessary results.

Theorem 2.7 *Assume that for each cell of any cellular decomposition a coordinate map is fixed. Then each cellular map $f: K \to L$ and each pair of n-dimensional cells $\tau^n \subset K$ and $\sigma^n \subset L$ can be associated with a certain integer $[f\tau: \sigma]$ called the degree of map* f of τ^n onto σ^n with the following properties:*

(1) If $f_t: K \to L$ is a deformation such that for any t, $0 \leqslant t \leqslant 1$, the map f_t is cellular, then the degrees of the maps f_0 and f_1 are the same, namely, $[f_0\tau: \sigma] = [f_1\tau: \sigma]$ for any n-dimensional cells $\tau^n \subset K$ and $\sigma^n \subset L$ (property of *constancy of the degree under homotopy*).

(2) Let $f: K \to L$ be regular at a point $y \in L^n \setminus L^{n-1}$, σ^n a cell of the decomposition of L with y, τ^n an arbitrary n-dimensional cell in K, p the number of positive points of the inverse images $f^{-1}(y)$ in τ^n, and q that of negative ones. Then $[f\tau: \sigma] = p - q$. The degree of a map is determined uniquely. To construct it, we use Theorem 2.7.

Let f be a cellulary map. Deform it into a regular map f_1. We select a point y_0 in a cell $\sigma \subset L$ at which f_1 is regular. On the basis of property (2) in Theorem 2.7, the degree of the map $[f_1\tau: \sigma]$ can be uniquely determined as the difference of the corresponding values of p and q. Then, on the basis of property (1), $\deg f_1$ coincides with $\deg f$. The proof that the definition is correct, i.e., of the independence of $\deg f$ of the choice of regular values y_0, can be obtained by the reader (see also [Mil7]).

The degree of a map is one of the most important topological concepts. We shall consider different applications of this characteristic in more detail later.

We now formulate the property of the degree for a sequence of maps.

Lemma 2.3 *Let $f: k \to L$ and $g: L \to M$ be two cellular maps. Then the relation*

$$[(g \circ f)\tau: \varrho] = \sum_{\sigma} [f\tau: \sigma] \cdot [g\sigma : \varrho]$$

* We will also use the usual notation $\deg f$ if it is clear from the context which sets are mapped.

holds for any n-dimensional cells $\tau^n \subset K$ *and* $\varrho^n \subset M$, *where the summation is over all n-dimensional cells* σ^n *from* L.

The proof easily follows from the definition of the degree of a map and from the properties of the Jacobians of compositions of maps.

2.2.3 Cell Complex Homology

Here, we construct the homology groups of manifolds admitting a cellular decomposition, for which it is necessary to define several related algebraic objects.

We also assume that a cellular decomposition is given and that coordinates are fixed in each cell.

2.2.3.1 Chains. Chain groups

Definition 2.13 Formal (finite) sums of the form

$$k_1\tau_1^n + \ldots + k_n\tau_n^n,$$

where $\tau_1^n, \ldots, \tau_n^n$ are n-dimensional cells and k_i integers, are called *n-dimensional chains* of a cellular decomposition of K, i.e., chains are elements of an n-dimensional vector space. The addition operation can be given on the chain space, reducible to collecting similar terms, namely,

$$(k_1 + k_1')\tau_1^n + \ldots(k_n + k_n')\tau_n^n,$$

turning the space into an Abelian group denoted by $C_n(K)$.

Let $f\colon K \to L$ be a cellular map. If the cell $\tau^n \subset K$ is mapped onto the cells $\sigma_1^n, \ldots, \sigma_s^n \subset L$ with degrees $l_1, \ldots, l_s$ (i.e., $[f\tau^n\colon \sigma_1] = l_1$, etc.) and the other n-dimensional cells of L contain no points from $f(\tau^n)$, then τ^n is assumed to be sent into the chain $l_1\sigma_1^n + \ldots + l_s\sigma_s^n$. Generally speaking, if there is an n-dimensional chain $x_n = k_1\tau_1^n + \ldots + k_n\tau_n^n$ of K, then the chain is carried into

$$f_{\#}(x_n) = \sum k_i [f\tau_i\colon \sigma^n]\sigma^n$$

under f, where summation is over the subscripts i and all n-dimensional cells of L.

It is obvious that $f_{\#}$ is a homomorphism $C_n(K) \to C_n(L)$.

2.2.3.2 Chain's boundary

First, we define the boundary of a cell. The unit ball D_0^n admits a cellular decomposition $\sigma_0 \in S^{n-1}$, $\sigma^{n-1} = S^{n-1} \setminus \sigma_0$, $\sigma^n = D_0^n \setminus S^{n-1}$. For $n \geqslant 2$, we fix coordinates in the cells σ^n and σ^{n-1}, regard σ^{n-1} as the boundary of σ^n (this obviously coincides with the definition of the boundary of D_0^n as the set S^{n-1}), and write $\sigma^{n-1} = \partial\sigma^n$. For $n = 1$, the unit ball D_0^1 consists of the interval $[-1, 1]$ whose boundary S^0 consists of two points, 1 and -1. We denote the first one by σ_+^0 and the second by σ_-^0 and put $\partial\sigma^1 = \sigma_+^0 - \sigma_-^0$ (an oriented boundary). Thus, the boundary of the cell $\sigma^n = D_0^n \setminus S^{n-1}$ is defined for all $n \geqslant 1$. If τ^n is an arbitrary n-dimensional cell of the decomposition $K (n \geqslant 1)$ and $\chi: D_0^n \to K$ the corresponding coordinate map, then it is natural to assume that σ^n is carried into τ^n under χ and $\partial\sigma^n \to \partial\tau^n$, i.e.,

$$\partial\tau^n = \chi_{\#}(\partial\sigma^n) = \chi_{\#}(\sigma^{n-1}),$$

by definition.

The coefficient with which a given cell τ_i^{n-1} is involved in a chain $\partial\tau^n$ is called the *incidence number* for τ^n and τ_i^{n-1} and denoted by $[\tau^n : \tau_i^{n-1}]$. Thus, $\partial\tau^n = \sum_i [\tau^n : \tau_i^{n-1}]\tau_i^{n-1}$.

We now define the boundary of a chain by making use of additive properties and taking into account that

$$\partial : (k_1\tau_1^n + \dots + k_n\tau_n^n) = k_1\partial\tau_1^n + k_2\partial\tau_2^n + \dots + k_n\partial\tau_n^n$$

is a homomorphism.

We have thus introduced the boundary operator $\partial_n : C_n(K) \to C_{n-1}(K)$. Consider the map's kernel, i.e., the set of chains $\{C_n\}$ such that $\partial_n C_n = 0$. Elements of the group are called *cycles.* It is obvious that the set of cycles forms a group denoted by $Z_n(K)$.

The operator ∂ possesses the following fundamental property.

Lemma 2.4

$$\partial_{n-1} \circ \partial_n = 0. \tag{2.5}$$

This enables us to define a new homology group $H_n(K)$ of K.

Indeed, it follows from (2.5) that $\operatorname{Im} \partial_n \subset \operatorname{Ker} \partial_{n-1}$. Then

$$H_n(K) = Z_n(K)/B_n(K),$$

where $B_n(K) \subset Z_n(K)$: $B_n(K) = \{b_n(K) \in C_{n-1}(K)\}$, and $b_n(K) = \operatorname{Im} \partial_n(C_n(K))$.

We now prove Lemma 2.4. From the definition of a boundary, we have

$$\begin{aligned}\partial \circ \partial\tau^n &= \partial\left(\sum_i [\tau^n : \tau_i^{n-1}]\tau_i^{n-1}\right)\\ &= \sum_i [\tau^n : \tau^{n-1}]\tau_i^{n-1}\\ &= \sum \sum [\tau^n : \tau_i^{n-1}][\tau_i^{n-1} : \tau_j^{n-2}]\tau_j^{n-2} = 0,\end{aligned} \tag{2.6}$$

since the boundary τ_i^{n-1} of $(n-1)$-dimensional cells is involved twice with opposite signs. The properties hold for an arbitrary chain because the operation ∂ is a homomorphism.

2.2.3.3 Properties of homology groups

Exactness of a sequence of homology groups. For homology groups of a cellular complex, we have

Theorem 2.8 *Let K be a cell complex and L a subcomplex. Then the sequence*

$$H_n(L) \xrightarrow{i_*} H_n(K) \xrightarrow{j_*} H_n(K/L) \xrightarrow{\partial_*} H_{n-1}(L) \to \ldots \qquad (2.7)$$

is exact.

The proof is similar to that for the theorem on the exactness of a sequence of homotopy groups. Some of the changes are indicated below.

We define the corresponding homomorphisms.

1. Let $f: K \to L$ be a cellular map. It is easy to see that the homomorphisms f_* and ∂ commute, or that the diargam

$$\begin{array}{ccc} C_n(K) & \xrightarrow{\partial} & C_{n-1}(K) \\ f_* \downarrow & & \downarrow f_* \\ C_n(L) & \xrightarrow{\partial} & C_{n-1}(L) \end{array}$$

is commutative. Hence,

$$f_*(Z_n(K)) \subset Z_n(L), \ f_*(B_n(K)) \subset B_n(L)$$

and the homomorphism of the quotient groups $f_* H_n(k) \to H_n(L)$ is thus defined.

2. We give a definition of the maps i_* and j_*. Consider spaces K, L, K/L $(L \subset K)$ and maps

$$L \xrightarrow{i} K \xrightarrow{j} K/L,$$

where i is an embedding map and j the identification (or factorization).

They are associated with group homomorphisms

$$C_n(L) \xrightarrow{i_*} C_n(K) \xrightarrow{j_*} C_n(K/L).$$

The sequence can be made a part of an exact one

$$0 \to C_n(L) \xrightarrow{i_*} C_n(K) \xrightarrow{j_*} C_n(K/L) \to 0 \qquad (2.8)$$

by the definition of the corresponding chains. However, in the transition to homology groups, the exactness of (2.8) is violated. A non-trivial homomorphism $H_n(K/L) \to H_{n-1}(L)$ arises, which we denote by ∂_*.

Construction of ∂_*. We consider the commutative diagram with exact rows

$$\begin{array}{ccccccccc}
 & & & & C_{n+1}(K) & \xrightarrow{j_*} & C_{n+1}(K/L) & \to & 0 \\
 & & & & \partial\downarrow & & \downarrow\partial & & \\
0 & \to & C_n(L) & \xrightarrow{i_*} & C_n(K) & \xrightarrow{j_*} & C_n(K/L) & \to & 0 \\
 & & \downarrow\partial & & \partial\downarrow & & \downarrow\partial & & \\
0 & \to & C_{n-1}(L) & \xrightarrow{i_*} & C_{n-1}(K) & \xrightarrow{j_*} & C_{n-1}(K/L) & \to & 0 \\
 & & \downarrow\partial & & \partial\downarrow & & & & \\
0 & \to & C_{n-2}(L) & \xrightarrow{i_*} & C_{n-2}(K) & & & &
\end{array}$$

We select an element $\zeta \in H_n(K/L)$ associated with a cycle $z \in Z_n(K/L)$, a representative of ζ. Since j_* is an epimorphism, there exists an element $x \in C_n(K)$ satisfying the condition $j_*(x) = z$, $\partial(j_*(x)) = 0$, because z is a cycle. Since $(j_* x)$ is in the kernel of ∂, $j_*(\partial x) = 0$, due to the commutativity of the diagram, i.e., $\partial x \in \operatorname{Ker} j_* = \operatorname{Im} i_*$; therefore, there exists an element $y \in C_{n-1}(L)$ such that $i_*(y) = \partial x$.

Due to the diagram's commutativity, we derive from the condition $\partial(i_* y) = \partial(\partial x) = 0$ that $i_*(\partial y) = 0$. However, i_* is a monomorphism; therefore, $\partial y = 0$, i.e., $y \in Z_{n-1}(L)$. Let $\eta \in H_{n-1}(L)$ be the class defined by the cycle y. Put $\eta = \partial_* \zeta$. Travelling across the diagram, we can show by a similar argument that η does not depend on the representative of the homology class. Correctness in the construction of the operator ∂_* has thereby proved. The same constructions show that the homology sequence is exact, since in constructing ∂_* the kernel and image of the map were calculated. ∎

Note that in contrast with the homotopy group sequence, the left-hand end of homology sequence (2.7) must vanish, beginning with certain n ($n = \dim K + 1$), since there are no cells of dimensions greater than that of the manifold K in its cellular decomposition.

We now formulate without proof some properties of the homology groups of manifolds.

1. Group $H_0(S^0) = Z$. $H_{-k}(M) = 0$ (by definition).
2. Group $H_n(S^0) = 0$ for $n \neq 0$ (dimension property).
3. Given two maps $f: M_1 \to M_2$ and $g: M_2 \to M_3$ of spaces, $(g \circ f)_* = g_* \circ f_*$.
4. If f, $g: M_1 \to M_2$ are homotopic, then $f_* = g_*$.
5. Manifolds of the same homotopy type have isomorphic homology groups; in particular, if $f: M_1 \to M_2$ is a homomorphism, then f_* is an isomorphism, i.e., the homology groups coincide.

Remark 2.6 We used cellular decompositions and maps in considering homology groups. By the lemma on approximating continuous maps by cellular ones, a cellular map can be replaced by a continuous one in all statements on maps.

Remark 2.7 In the definition of a homology group, integral coefficients were used. We can consider any coefficient group G. Then the set of chains $k_i \tau_i$ with coefficients in G can be regarded as a chain group of the form $C(K) \otimes G$. The corresponding homology groups are denoted by $H_i(K, G)$. We will also encounter $G = Z_2, R, C$ besides group Z. All properties of the homology groups are preserved if $H_0(K, Z) = Z$ is replaced by $H_0(K, G) = G$.

Definition 2.14 Homology groups are finite-dimensional vector spaces $H_i(M, G)$. The ranks (or dimensions) of the corresponding groups of spaces are called the *Betti numbers*

$$b_i = \dim H_i(M, G).$$

2.2.3.4 Calculation of the homology groups of the simplest spaces

(1) Sphere S^n. Groups $H_i(S^n, Z)$.

Since a cellular decomposition of the sphere consists of two cells x_0 and $S^n \setminus x_0$, it is obvious that $H_i(S^n) = 0$ for $0 < i \leqslant n - 1$ and $H_0(S^0) = Z$, $H_n(S^n) = Z$ because an n-dimensional cell generates a cycle.

(2) Wedge of spheres $M = \vee S_j^n$. The groups $H_i(M, Z)$, $j = 1, ..., q$.

The wedge of n-dimensional spheres, joined at a point x_0, has the cellular decomposition consisting of one zero-dimensional cell x_0 and of q n-dimensional cells. Accordingly, the homology groups are

$$H_0(M, Z) = Z,$$
$$H_i(M, Z) = 0, \quad i \neq 0, n,$$
$$H_n(M, Z) = \underbrace{Z + ... + Z}_{q}.$$

(3) Torus T^2. The groups $H_i(T^2, Z)$.

The decomposition of a torus into cells is familiar (see Sec. 2.2.1, Fig. 5). It remains for us to find the boundaries of the cells σ_j^i $(i = 0, 1, 2)$. We have $\partial\sigma^2 = \sigma_1^1\sigma_2^1(\sigma_1^1)^{-1}(\sigma_2^1)^{-1} = 0$, $\partial\sigma_1^1 = \partial\sigma_2^1 = 0$, $\partial\sigma^0 = 0$.

Hence, $H_i(T^2)$ are of the form

$$H_0(T^2) = Z,$$
$$H_1(T^2) = Z + Z,$$
$$H_2(T^2) = Z.$$

(4) Klein bottle $H_i(K^2)$.

The cellular decomposition of the Klein bottle like that of the torus contains one zero-dimensional, two one-dimensional, and one two-dimensional cells.

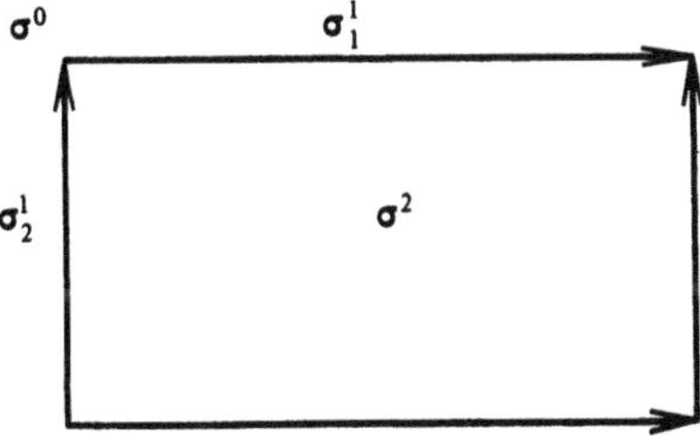

Fig. 5. Cellular decomposition of torus

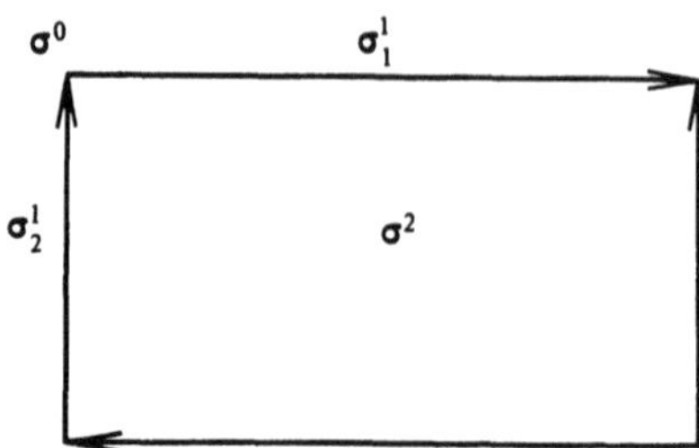

Fig. 6. Celluar decomposition of the Klein bottle

However, the boundaries of the chains have different incidence numbers (Fig. 6), namely, $\partial\sigma^0 = \partial\sigma_1^1 = \partial\sigma_2^1 = 0$, $\partial\sigma^2 = 2\sigma_1^1$. Hence, $H_2(K^2) = 0$, $H_1(K^2) = Z + Z_2$, $H_0(K^2) = Z$.

(5) Real projective space RP^n.

The cellular decomposition of RP^n is constructed in Sec. 2.1. To find the homology groups, it is necessary to determine the cell boundaries. Let the decomposition consist of cells $\sigma^0, \ldots, \sigma^n$. It is not hard to see that $\partial\sigma^0 = 0$, $\partial\sigma^1 = 0$, $\partial\sigma^2 = 2\sigma^1 \cdots$ $\cdot\, \partial\sigma^{2k+1} = 0$, $\partial\sigma^{2k+2} = 2\sigma^{2k+1}$. Hence,

$$\begin{aligned}
&H_0(RP^n, Z) = Z, && \\
&H_k(RP^n, Z) = 0 && \text{for } k = 2r, \\
&H_k(RP^n, Z) = Z_2 && \text{for } k = 2r + 1, \\
&H_n(RP^n, Z) = 0 && \text{for } n = 2r, \\
&H_n(RP^n, Z) = Z && \text{for odd } n.
\end{aligned}$$

The latter result yields another definition of the orientability of a manifold (as shown in Sec. 1.1, RP^n is orientable for odd n and nonorientable for even n).

(6) Complex projective space CP^n.

The canonical cellular decomposition of CP^n consists of the cells $\sigma^r = CP^r \setminus CP^{r-1}$, $\dim_R \sigma^r = 2r$, $0 \leqslant r \leqslant n$.

The characteristic map is determined for a cell by the composition of maps $p \circ i$:

$$R^{2r} \sim C^r \xrightarrow{p} CP^r \xrightarrow{i} CP^n,$$

where i is the embedding map $CP^r \to CP^n$ and p the quotient map $C^r \to CP^r$. It is easy to see that CP^n can be represented as the quotient space of the unit ball $D_0^{2n} \subset C^n$ relative to the decomposition whose elements are interior points of D_0^{2n} and circles cut on the boundary $\partial D^{2n} = S^{2n-1}$ by straight lines in the space C^n passing through the origin.

It is now easy to calculate the homologies of CP^n. The cellular space's decomposition consists of even-dimensional cells such that $\partial\sigma^{2r} = 0$; therefore, $H_i(CP^n, Z) = Z$ for $i = 2r$, $H_i(CP^n, Z) = 0$ for $i = 2r + 1$.

2.2.3.5 Relative homology groups

Like the construction of relative homotopy groups, we can also construct relative homology groups. The homology groups, or pairs K, L, $K \supset L$, have actually been already defined as homologies of the space K/L.

The construction of the homology groups of the three spaces $X \supset A \supset B$ are of no lesser interest. Calculation of the homology groups of spaces X, where $X = X_1 \cup X_2$, such that $X_1 \cap X_2 \neq \varnothing$, leads to this type of problems.

Homology sequence of a triple. Let $X \supset A \supset B$ and the embedding maps

$$i: (A, B) \subset (X, B), \quad j: (X, B) \subset (X, A)$$

be given. The maps are assumed to be admissible (i.e., with the product operation defined relative to which they form a group). The embedding maps and boundary operators are defined as follows:

$$\begin{array}{lll} i: A \to X, & j: X \to (X, A), & \partial: H_q(X, A) \to H_{q-1}(A), \\ i': B \to X, & j': X \to (X, B), & \partial': H_q(X, B) \to H_{q-1}(B), \\ i'': B \to A, & j'': A \to (A, B), & \partial'': H_q(A, B) \to H_{q-1}(B) \end{array}$$

(see [ESt]).

Another boundary operator $\bar{\partial}$ is defined as the composition of maps $j''_* \partial$, where j_* is the homology group mapping induced by the embedding j. The map $\bar{\partial}$: $H_q(X, A) \to H_{q-1}(A, B)$ is called the *boundary operator of the triple.*

In a similar manner to sequence (2.7), we can construct

$$H_q(A, B) \to H_q(X, B) \to H_q(X, A) \to H_{q-1}(A) \to \dots . \tag{2.9}$$

Theorem 2.9 *Sequence* (2.9) *is exact.*

The proof is a tedious repetition of Theorem 2.8 with necessary modifications (see the proof in [ESt]).

Another triple of spaces, useful in calculating the homology groups of complicated manifolds, is called a *triad* (X, X_1, X_2) and consists, by definition, of a space X and its two subspaces X_1 and X_2 such that, for X, X_1, X_2, $X_1 \cup X_2$, $X_1 \cap X_2$, and all their pairs, embedding maps are defined.

Certain technical restrictions are placed on a triad, which always hold, for example, for manifolds with a boundary and the corresponding submanifolds.

We now state the theorem on the exactness of the homology sequence of a triad (X, X_1, X_2) for $X = X_1 \cup X_2$, $A = X_1 \cap X_2$.

Theorem 2.10 *The sequence*

$$H_q(A) \xrightarrow{\psi} H_q(X_1) + H_q(X_2) \xrightarrow{\varphi} H_q(X) \xrightarrow{\Delta} H_{q-1}(A) \to \dots$$

is exact, where ψ, φ, Δ *are defined by the formulas* $\psi u = (h_{1*} u, -h_{2*} u)$, $u \in H_q(A)$, $\varphi(v_1, v_2) = m_{1*} v_1 + m_{2*} v_2$, $v_1 \in H_q(X_1)$, $v_2 \in H_q(X_2)$, $\Delta w = -\partial_1 k_{1*}^{-1} l_{1*} w = \partial_2 k_{2*}^{-1} l_{2*} w$, $w \in H_q(X)$.

Here all the homomorphisms h_{i*}, k_{i*}, l_{i*}, m_{i*} $(i = 1, 2)$ are induced by the embedding maps as in the diagram

$$
\begin{array}{ccccc}
 & & H_q(A) & & \\
 & h_{2*}\swarrow & & \searrow h_{1*} & \\
H_q(X_2) & & \downarrow & & H_q(X_1) \\
 & m_{2*}\searrow & & \swarrow m_{1*} & \\
 & & H_q(X) & & \\
 & l_{1*}\swarrow & & \searrow l_{2*} & \\
H_q(X, X_1) & & \downarrow j_* & & H_q(X, X_2) \\
 & j_{1*}\nwarrow & & \nearrow j_{2*} & \\
k_{1*}\uparrow & & H_q(X, A) & & \uparrow k_{2*} \\
 & i_{2*}\nearrow & & \nwarrow i_{1*} & \\
H_q(X_2, A) & & \downarrow \partial & & H_q(X_1, A) \\
 & \partial_1\searrow & & \swarrow \partial_2 & \\
 & & H_{q-1}(A) & &
\end{array}
$$

Calculation of homology groups. Using the exact sequence in (2.9), called the *Mayer-Vietoris sequence*, the homology groups of various spaces can be calculated by representing the spaces by glueing together those already known.

Exercise 2.7 Calculate

(a) the homology of a surface of genus g,

(b) the homology of a handle or a torus with a hole cut in it,

(c) the homology of a two-dimensional non-orientable surface,

(d) the homology of a sphere with a hole cut in it and with a Möbius strip attached, and

(e) the homology of fiber spaces.

To calculate the homology (and cohomology) groups of fiber spaces, Leray developed the powerful method of spectral sequences. The treatment is outside the scope of the book (see the accessible discussion in [Hu]). By means of spectral sequences, the homology of a fiber space can be calculated if those of the base space and fiber are known, in particular, the method of spectral sequences is convenient for finding the homology of classical Lie groups, Stiefel and Grassmann manifolds, etc.

For reference, we give the exact sequences of homology groups in two particular cases (see the proof in [Hs]).

Proposition 2.11 *Given a fibration $p: E \to B$ whose fiber is an n-dimensional sphere and that the base space B is one-connected. Then the sequence of homology groups*

$$\to H_{m+1}(B) \to H_{m-n}(B) \to H_m(E) \to H_m(B) \to H_{m-n+1}(B),$$

induced by the fibration is exact.

The sequence is said to be *Gysin*.

Given a fiber $p: E \to S^n$ of the n-dimensional sphere with fiber F, the following is valid.

Proposition 2.12 *The sequence of homology groups*

$$H_m(E) \to H_{m-n}(F) \to H_{m-1}(F) \xrightarrow{i_*} H_{m-1}(E) \to H_{m-n-1}$$

is exact.

Here i_* is induced by the fiber embedding $F \subset E$. The sequence is said to be *Wang*.

Example 2.6 Find $H_i(SU(3))$ with the help of the Wang sequence.

We have the fiber bundle $SU(3) \xrightarrow{SU(2)} S^5$ and the corresponding exact sequence of homology groups is

$$H_m(SU(2)) \to H_m(SU(3)) \to H_{m-5}(SU(2)) \to H_{m-1}(SU(2) \to H_{m-1}(SU(3)).$$

Since $SU(2) \sim S^3$, $H_0(S^3) = H_3(S^3) = Z$, $H_i(S^3) = 0$, $i = 1, 2$.

It immediately follows from the exactness of the above sequence that

$$\begin{aligned} H_m(SU(3)) &= Z \quad \text{for } m = 0, 3, 5, 8, \\ H_m(SU(3)) &= 0 \quad \text{for } m = 1, 2, 4, 6, 7. \end{aligned}$$

Remark 2.8 Originally, the homology was constructed for simplicial decompositions or decompositions into simplexes (such as a one-dimensional segment, two-dimensional triangle, etc.). Accordingly, homology theories (with mapping simplexes into manifolds) are said to be *simplicial*. The fundamental theorem of the homology theory concerning independence of homology groups from the kind of decomposition shows that manifold's decomposition selected is immaterial. However, for practical calculation, the cellular decompositions are more convenient, e.g., the cellular decomposition of T^2 contains one zero-dimensional, two one-dimensional, and one two-dimensional cells. Accordingly, a simplicial decomposition contains the minimal number 42 of simplexes (see [Mas2]).

The simplicial homology theory is left for the reader to construct.

We consider the properties of homological invariants and the relation with differential geometric property of manifolds later. For the present, we define another important invariant dual of a homology theory, namely, a cohomology theory. For manifolds, a cohomology group admits a remarkable formulation in terms of differential forms.

2.3 COHOMOLOGY THEORY

2.3.1 Definition and Basic Properties

Here, we outline the dual object to homology groups $H_i(K)$, namely, cohomology groups $H^i(K)$. Taking into account a certain parallelism with a homology theory, we briefly formulate the main properties of the cohomology groups.

2.3.1.1 Cochain group

Let K be a cell complex. The group of n-cochains $C^n(K, G)$ is the set Hom $(C_n(K), G)$, where $C_n(K)$ is the group of n-chains with coefficients in the group G. The set of cochains is that of linear functionals on the chain space. If the value $\langle c^1 | c_1 \rangle$ of the cochain c^1 is defined on the chain c_1, then we can define the operator δ adjoint of the operator ∂ and, accordingly, cocycle, coboundary, and cohomology groups.

First, the following general notation. Let φ be a group homomorphism $A \to B$. We can consider the dual one $\varphi^{\#}$: Hom $(B, G) \to$ Hom (A, G) defined by the equality $\varphi^{\#}(b) = b \circ \varphi$ for any $b \in$ Hom (B, G). $\varphi^{\#}$ possesses the following obvious property. Given homomorphisms $A \overset{\varphi_1}{\to} B \overset{\varphi_2}{\to} C$, for the adjoint homomorphisms

$$\varphi_1^{\#}: \text{Hom}(B, G) \to \text{Hom}(A, G), \quad \varphi_2^{\#}: \text{Hom}(C, G) \to \text{Hom}(B, G),$$
$$(\varphi_2 \circ \varphi_1)^{\#}: \text{Hom}(C, G) \to \text{Hom}(A, G),$$

the relation

$$(\varphi_2 \circ \varphi_1)^{\#} = \varphi_1^{\#} \circ \varphi_2^{\#}$$

holds.

If the homomorphism $\varphi: A \to B$ is zero, then $\varphi^{\#}$: Hom $(B, G) \to$ Hom (A, G) is also zero.

We now retrace our steps to the definition of a cohomology group.

2.3.1.2 Cohomology groups

We define the coboundary operator δ_n as $\partial_n^{\#}$ adjoint of ∂_n, where

$$\partial_n: c_n(K, G) \to \dot{c}_{n-1}(K, G).$$

Consider the sequence of homomorphisms

$$c^{n-1}(K, G) \overset{\delta_n}{\to} c^n(K, G) \xrightarrow{\delta_{n+1}} c^{n+1}(K, G).$$

The composition of maps $\delta_{n+1} \circ \delta_n$ is trivial since

$$\delta \circ \delta = \partial^{\#} \circ \partial^{\#} = 0. \tag{2.10}$$

We can now define cocycle and coboundary groups as the dual objects of cycle and boundary groups.

Definition 2.15 A cochain $c^n(K, G)$ is called a *cocycle* if $\delta c^n(K, G) = 0$ and is denoted by $z^n(K, G)$. In other words, a cocycle is the kernel of the map $f: c^{n-1}(K, G) \to c^n(K, G)$.

Definition 2.16 A *coboundary* $b^n(K, G)$, $n \geqslant 1$, is the image of a cochain $c^{n-1}(K, G)$ under a homomorphisms δ_n: $c^{n-1}(K, G) \to \dot{c}^n(K, G)$. For $n = 0$, the set $B^0(K, G) = \{b^0\}$ is by definition that of homomorphisms $c_0(K) \to G$ which take the same value at all vertices from K. It immediately follows from property (2.10) that $Z^n(K, G) \supset B^n(K, G)$. Therefore, the quotient group $Z^n(K, G)/B^n(K, G)$ can be formed.

Definition 2.17 The groups $H^n(K, G) = Z^n(K, G)/B^n(K, G)$ are called the nth *cohomology groups* of the space K with coefficients in G.

The same axioms as for homology groups with obvious modifications hold for cohomology ones. We prove in the next subsection that they also hold if the cohomology groups are realized as spaces of differential forms.

2.3.2 Cohomology Groups of Differentiable Manifolds

The cohomology groups of differentiable manifolds with coefficients in the fields of real or complex numbers can be introduced with the help of exterior differential forms. This deep result is due to de Rham [Rh].

Consider two closed differential forms ω^k of degree k on a manifold M^n. Let the space of closed forms be Z^k. The subspace of forms $\{\omega^k\} \subset B^k$ such that $\omega = d\tau$, where the operator d is exterior differentiation and τ is the space of all forms of degree $k-1$, is called the *space of exact forms.* Since $d \circ d = 0$, we can form the quotient group $H^k = Z^k/B^k$ called the *de Rham cohomology group* of M^n. De Rham's main theorem states that the groups H^k are isomorphic to the cohomology groups of the cellular decomposition of M^n. De Rham's theorem enables us to simplify strongly the calculation of the cohomology groups of manifolds, formulations of the cohomology group properties, and proofs of a number of classical theorems.

We have to underline the essential difference between the cohomology of compact and non-compact manifolds M. Although the main concepts are formulated independently of the compactness or noncompactness of M, certain results, e.g., the finite dimensionality of the space $H^i(M, R)$, are only valid for compact manifolds. The difference becomes clear if the properties of harmonic functions on the plane and sphere are compared. The concept of the cohomology group for non-compact manifolds should be sharpened. We can consider both cohomology $H^i(M)$ and cohomology with compact support $H^i_k(M)$. If in the definition of a cohomology group we consistently consider forms with compact support, then we obtain the definition of $H^i_k(M)$. We need $H^i_k(M)$ in stating Alexander's duality theorem (Theorem 2.14) as applied to the theory of links of defects (Subsec. 5.1.6). The usual cohomology is implied in the remaining text. In fact, the arising manifolds M are mostly compact. The reader may find the de Rham theory for non-compact manifolds in [Rh].

2.3.2.1 Properties of cohomology groups

Cohomology groups can be turned into a ring if the multiplication operation

$$\omega_1^p \wedge \omega_2^q = (-1)^{pq} \omega_2^q \wedge \omega_1^p$$

is introduced on cocycles via the exterior multiplication of forms.

Definition with the help of forms enables us to verify easily the main cohomology ring's properties.

1. For any manifold M^n, the group $H^0(M^n, R)$ is a linear space of dimension q, where q is a number of connected components of the space M^n.

Proof. By definition, $H^0(M^n, R) = \{f \mid df\} = 0$, i.e., f is a locally constant function on each connected piece. There are no exact forms. ∎

2. Using the familiar properties of forms, it is easy to see that, given a map $f: M_1 \to M_2$, a map of forms $\omega \to f_*(\omega)$ such that the diagram

$$\begin{array}{ccc} \omega & \xrightarrow{f_*} & f_*(\omega) \\ d\downarrow & & \downarrow d \\ \omega_1 & \xrightarrow{f_*} & f_*(\omega_1) \end{array}$$

is commutative is defined. The map $f_*: H^k(M_2, R) \to H^k(M_1, R)$ are thus defined since the equivalence classes pass into each other (closed forms into closed and exact ones into exact).

3. Let $f_1: M_1 \to M_2$, $f_2: M_1 \to M_2$.

If two maps f_1 and f_2 are homotopic, then the cohomology group's maps coincide (the proof can be found in [DFN]).

2.3.3 Integration of Differential Forms

The connection between cohomology and homology groups of differentiable manifolds can be established by the integration of differential forms in the fastest and most natural way and a global topological characterization of manifolds (such as characteristic classes) can be introduced, etc.

Integration of differential forms with detailed proofs is treated in a number of books. An especially refined treatment is given in [Whi] and [Rh]. We only formulate the main concepts and results to be used in subsequent chapters.

Let q^p be a cube in a Euclidean space R^p with edge length ε, i.e., $q^p = (t^1, \ldots, t^p)$, $0 \leqslant t^i \leqslant \varepsilon$, $i = 1, \ldots, p$. We construct a cube Q^p in a manifold M. Let f be a smooth mapping $f: q^p \to \mathcal{U} \subset M$ and $\mathcal{U}$ a coordinate neighborhood on M. By definition, Q^p is then equal to the image $f(q^p)$. The local coordinates of points from Q^p are functions of $t = (t^1, \ldots, t^p)$.

Consider a chain $c^p = \Sigma a_i Q_i^p$ (a_i are real numbers* and Q_i^p different cubes in M).

To define a chain's boundary ∂c^p, it suffices to define the boundary ∂Q^p of an individual cube Q^p as the image $f(\partial q^p)$. The boundary of the standard cube, $q^p \in R^p$, is naturally defined as

$$\partial q^p = \sum_{i=1}^{p} (-1)^i (q_{i_0}^p - q_{i\varepsilon}^p).$$

Summation is over all faces (here $q_{i_0}^p$ is the face $t_i = 0$ and $q_{i\varepsilon}^p$ the face $t_i = \varepsilon$). It is obvious that $\partial \circ \partial(q^p) = 0$. Note that the definition of Q-chains by means of cubes $f(q)$ is a version of the construction of the cohomology theory.

* We thereby consider a cohomology with real coefficients (see [ESt] for the relation between cohomology groups over different domains of coefficients).

Since it is proved that all (CW-complex, simplicial, and cubical) theories are equivalent for manifolds, it is clear that all cohomology operations are also valid for cubical decompositions. For instance, the boundary operator can be carried over to chains.

Consider compact and orientable manifolds. The method is unsuitable for non-orientable ones; for non-compact ones, it requires additional discussion.

A manifold M^n can be decomposed into cubes and thus turned into an n-cycle.

Let $\omega(x) = q_{i_1 \dots i_p} dx^{i_1} \wedge \dots \wedge dx^{i_p}$ be an exterior form of degree p, defined in a coordinate neighborhood $\mathscr{U} \subset M^n$ containing Q^p. A map $f: q^p \to Q^p$ enables us to associate $\omega(x)$ with the induced form $(f_* \omega)(t)$ obtained by replacing x with $x(t)$ given on $q^p \subset R^p$. We put

$$\int_{Q^p} \omega = \int_{q^p} f_*^* \omega. \qquad (2.11)$$

The integral does not depend on the choice of parameter t and, therefore, is defined correctly.

Thus, each ω in terms of (2.11) gives a linear functional, or the value of ω on the p-dimensional cube, i.e., each determines a cochain. Integral (2.11) on an arbitrary chain $c^p = \Sigma\, a_i Q_i^p$ is defined by linearity, namely,

$$\int_{c^p} \omega = \sum a_i \int_{Q_i^p} \omega.$$

The basic formula in the theory of integration of differential forms is the Stokes generalized theorem, including all classical versions such as Green's formula, Green's theorem, Stokes' formula, etc.

Proposition 2.13 (Stokes' formula) Let ω be a form of degree $p - 1$ and c^p a p-dimensional chain. Then

$$\int_{c^p} d\omega = \int_{\partial c^p} \omega$$

(see the proof in [Whi]).

Example 2.7 (Green's formula)

Let $\mathscr{U}$ be a domain in R^2, $\partial\mathscr{U}$ its boundary, and $\omega = a(x, y)dx + b(x, y)dy$ a 1-form.

We have

$$\int_{\partial\mathscr{U}} a\,dx + b\,dy = \int_{\mathscr{U}} \left(\frac{\partial a}{\partial y} - \frac{\partial b}{\partial x} \right) dx \wedge dy.$$

It immediately follows from Stokes' formula that $H^n(M^n, R) = R$.

Indeed, let ω^n be a closed n-form on M^n. Such a form $\hat{\omega}$ always exists, being an element of volume M^n. Since M^n is a cycle, the cohomology class is generated by the value of $\hat{\omega}$ on M^n. On the other hand, no n-form ω^n is cohomologous to zero. Assume the contrary, namely, that $\omega^n = d\tau^{n-1}$, where τ^{n-1} is a form of degree $n-1$. We have $\int_{M^n} d\tau^{n-1} = \int_{\partial M^n} \tau^{n-1} = 0$, therefore, $d\tau^{n-1} = 0$. ∎

2.3.4 Structure of the Algebra of Differential Forms

We can introduce various operations on the space of differential forms enabling us to carry the whole theory of differential operators given on R^n over arbitrary Riemannian manifolds. On the other hand, using differential forms we can give an analytic treatment of characteristic classes (see Sec. 2.7).

2.3.4.1 Scalar product in the space of forms

Let M^n be an n-dimensional real orientable compact n-dimensional manifold with the Riemannian metric $ds^2 = g_{ik}dx^i dx^k$.

We can introduce a scalar product, denoted g_{ik}, for two vectors ξ and η in the tangent space TM_x at a point x. The product is defined as

$$\langle \xi \mid \eta \rangle = g_{ij}\xi^i \eta^j. \tag{2.12}$$

Since tangent vectors vary as contravariant tensors and g_{ij} is a covariant tensor, expression (2.12) does not depend on the choice of a coordinate system on M^n.

We now define the scalar product of differential forms, starting with the forms of degree one. Let $\omega = a_k dx^k$ and $\varphi = b_j dx^j$. The scalar product of two forms is defined as

$$\langle \omega \mid \varphi \rangle = g^{kj} a_k b_j, \tag{2.13}$$

where g^{kj} is the matrix inverse to g_{jk}.

The coefficients of ω and φ are covariant tensors and (2.13), like (2.12), is independent of the coordinates in M^n. Since g^{ij} is positive definite and symmetric, $\langle \omega \mid \omega \rangle > 0$.

If the metric of M^n is Euclidean (e.g., if M^n is the torus T^n), then

$$\langle \omega \mid \varphi \rangle = \delta^{ij} a_i b_j.$$

The scalar product of two homogeneous forms ω^p and φ^p of degree p is defined as follows. Let

$$\begin{aligned} \omega^p &= \frac{1}{p!} a_{i_1 \ldots i_p} dx^{i_1} \wedge \ldots \wedge dx^{i_p}, \\ \varphi^p &= \frac{1}{p!} b_{j_1 \ldots j_p} dx^{j_1} \wedge \ldots \wedge dx^{j_p}. \end{aligned} \tag{2.14}$$

Then

$$\langle \omega^p \mid \varphi^p \rangle = \frac{1}{p!} g^{i_1 j_1} g^{i_2 j_2} \ldots g^{i_p j_p} a_{i_1 \ldots i_p} b_{j_1 \ldots j_p}. \tag{2.15}$$

It is easy to check that the form $\langle \omega^p \mid \varphi^p \rangle$ is positive definite (we transform g^{ij} to principal axes).

If the degrees of ω and φ are not equal, then we put $\langle \omega \mid \varphi \rangle_x = 0$. Now, the scalar product of two inhomogeneous forms is defined by linearity.

The scalar products of two forms given at a point $x \in M^n$ can be integrated across the whole of M^n and thereby introduced into the space $\wedge M$ of all differential forms.

The product is defined as

$$\int_M \langle \omega_x \mid \varphi_x \rangle dV, \tag{2.16}$$

where dV is the element of volume and $dV = G dx^1 \wedge \dots \wedge dx^n$ defined by the Riemannian metric. The function G is expressed in terms of the determinant of the metric tensor by the familiar equality

$$G = \sqrt{\det \| g_{ij} \|}.$$

Formula (2.16) possesses all the properties of a scalar product. Completing $\wedge M$ with respect to the norm turns it into a complete Hilbert space.

2.3.4.2 Operators in $\wedge M$

Three pairs of adjoint operators enabling us to write invariantly operators acting on forms can be introduced in the Hilbert space $\wedge M$.

1. Operator $*$. We first define its action on a so-called *basis forms* representable as

$$\omega = dx^1 \wedge \dots \wedge dx^p,$$

where $dx^1 \dots dx^p$ is a basis for p-dimensional subspace $T_y^{p*}(M^n)$ of the cotangent space $T_y^*(M^n)$.

The operator $*$ is determined by the following properties specifying it uniquely. Let ω be a basis form of degree p and $*\omega$ that of degree $n - p$ (n is the dimension of the manifold M^n).

(a) The space of $*\omega$ is orthogonal to that of ω.

(b) $(n - p)$-dimensional volume of $*\omega$ equals the p-dimensional volume of ω.

(c) $\omega \wedge *\omega > 0$ for the selected orientation of M^n.

Example 2.8 The operator $*$ generalizes the property of vector product. If a bilinear form $\omega = dx^1 \wedge dx^2$ is given in the three-dimensional Euclidean space R^3, then $*\omega$ is the vector product of vectors of ω.

For arbitrary ω, the form $*\omega$ is defined at each point in M^n as $(*\omega_x) = *(\omega_x)$.

We now list the main properties of $*$:

(1) $$**\omega = (-1)^{p(n-p)}\omega, \tag{2.17}$$

where ω is a homogeneous form of degree p.

Proof. By the linearity of $*$, it suffices to prove formula (2.17) for a basis form ω representable as

$$\omega = dx^1 \wedge \ldots \wedge dx^p.$$

The form $*\omega$ is written as

$$dx^{p+1} \wedge \ldots \wedge dx^n, \quad **\omega = (-1)^\gamma dx^1 \wedge \ldots \wedge dx^p.$$

The coefficient γ is determined by the relation $dV = dx^1 \wedge \ldots \wedge dx^n = (*\omega) \wedge$ $\wedge *(*\omega) = (-1)^\gamma dx^{p+1} \ldots dx^n \wedge dx^1 \ldots dx^p = (-1)^{\gamma - p(n-p)} dx^1 \wedge \ldots \wedge dx^n$, i.e., $\gamma =$ $= p(n-p)$. The result is naturally valid for any form given on M^n. ■

(2) The scalar product of two forms ω and φ can be written as

$$\langle \omega \mid \varphi \rangle = \int \omega \wedge * \varphi. \tag{2.18}$$

Proof. It suffices to prove the equality $\langle \omega_x \mid \varphi_x \rangle \, dV = \omega_x \wedge * \varphi_x$ at any point $x \in M^n$. Both sides are linear with respect to ω_x and φ_x, therefore, it suffices to verify the equation for basis forms $\omega = dx^{i_1} \wedge \ldots \wedge dx^{i_p}$ and $\varphi = dx^{j_1} \wedge \ldots \wedge dx^{j_p}$ selected at x.

We have $\langle \omega \mid \varphi \rangle = 1$ if $(i_1 \ldots i_p) = (j_1 \ldots j_p)$ and $\langle \omega \mid \varphi \rangle = 0$ otherwise.

If the form φ contains at least one differential in ω, then $\langle \omega \mid \varphi \rangle = 0$. However, if $\varphi = \omega$, then $\langle \omega \mid \omega \rangle = \omega \wedge * \omega = dV$. ■

The general case can easily be reduced to the above if we notice that, at an arbitrary point in a manifold, the Riemannian metric g_{ik} can be transformed linearly to principal axes. The reader can reconstruct the details by himself or resort to the extensive bibliography (cf. [Ch]).

(3) The operator $*$ is unitary.

Proof.

$$\langle *\omega \mid *\varphi \rangle = \int *\omega \wedge **\varphi = (-1)^{p(n-p)} \int *\omega \wedge \varphi = \int \varphi \wedge *\omega = \langle \varphi \mid \omega \rangle = \langle \omega \mid \varphi \rangle.$$

Since $*$ is unitary, the operator $*^{-1}$ is the adjoint of $*$. ■

2. The operator δ.

By means of $*$, we can construct the operator δ adjoint to the exterior differentiation d.

Let ω^{p-1} be a form of degree $p-1$ and φ^p that of degree p. Then

$$\langle d\omega \mid \varphi \rangle = \int d\omega \wedge *\varphi. \tag{2.19}$$

We transform the integrand by using the formula for differentiating the exterior product

$$d\omega \wedge *\varphi = d(\omega \wedge *\varphi) + (-1)^p \omega \wedge d*\varphi \tag{2.20}$$

(see Subsec. 1.2.3).

The form $d(*\varphi)$ is of degree $n - p + 1$. Using relation (2.17), we obtain

$$d(*\varphi) = (-1)^{(n-p+1)(p-1)} **d*\varphi. \tag{2.21}$$

We put

$$\delta = (-1)^{np+n+1} *d*\varphi. \tag{2.22}$$

Then expression (2.19) can be rewritten as

$$\langle d\omega \mid \varphi \rangle = \int_M d(\omega \wedge *\varphi) + \int_M \omega \wedge *\delta\varphi = \langle \omega \mid \delta\varphi \rangle,$$

since the integral with respect to $d(\omega \wedge *\varphi)$ vanishes by Stokes' formula.

We have thereby proved that $\delta: \wedge^k M \to \wedge^{k-1} M$ is the adjoint of the operator d and decreases the degree of a form by unity.

The main property of δ is that $\delta \circ \delta = 0$. Indeed,

$$\delta \circ \delta\varphi = \pm *d(*d*)\varphi = *(d \circ d)*\varphi = 0,$$

since $d^2 = 0$. ■

Remark 2.9 It follows from (2.22) that if n is even, then $\delta = -*d*$ and $d = *\delta*$. If n is odd, then $\delta = (-1)^p *d*$.

Example 2.9 We now calculate the operator δ for the simplest case of the Euclidean metric. The corresponding compact manifold is the torus T^n.

(a) Let ω be the 1-form

$$\omega = a_i dx^i.$$

Then

$$\delta\omega = -*d*\omega = -*d(a_i * dx^i) = -*\frac{\partial a_i}{\partial x^j} dx^j \wedge *dx^i$$

$$= -*\left(\frac{\partial a_i}{\partial x^i}\right) dx^1 \wedge \ldots \wedge dx^n = -\left(\frac{\partial a_i}{\partial x^i}\right) = \operatorname{div} \omega.$$

(b) Let ω be a p-form. We have

$$\omega = a(x^1, \ldots, x^n) dx^1 \wedge \ldots \wedge dx^p,$$

$$\delta\omega = (-1)^{np+n+1} * da(x) dx^{p+1} \wedge \ldots \wedge dx^n = (-1)^{np+n+1} * \frac{\partial a}{\partial x^j} dx^j \wedge dx^{p+1} \wedge \ldots \wedge dx^n$$

$$= (-1)^j \frac{\partial a}{\partial x^j} dx^1 \wedge \ldots \wedge \widehat{dx^j} \wedge \ldots \wedge dx^p.$$

3. Interior multiplication operator. A third pair of adjoint operators of the scalar product (2.16) and (2.18) are the exterior $\wedge$ and interior I multiplication of forms. $\wedge$ is the familiar operator of action of a form ω on a form φ as $\omega \wedge \varphi$. We denote the action by $e(\omega)\varphi$ and find the operator adjoint of $e(\omega)$. Let ω be a form of degree s and φ that of degree p. We have

$$e(\omega)\varphi \wedge *\psi = (-1)^{ps}\varphi \wedge e(\omega) * \psi = (-1)^{ps}\varphi \wedge **^{-1}e(\omega)*\psi,$$

i.e., $(-1)^{ps}*^{-1}e(\omega)*$ on forms of degree $p + s$ is the operator adjoint of $e(\omega)$ defined on forms of the same degree. We put $q = p + s$ and define interior multiplication $I(\omega) = (-1)^{s(q+s)}*^{-1}e(\omega)*$ on forms of degree q. $I(\omega)$ is the operator adjoint of $e(\omega)$. The exterior multiplication operator of degree s increases the form's degree by s, whereas the interior multiplication operator decreases by s. In the special case of $s = 1$, the operator I has been defined in Subsec. 1.1.2.3.

The operator $I(\omega)$ can be rewritten more conveniently as $I(\omega) = (-1)^{(q+s)s}*^{-1}e(\omega)*$ on forms of degree q. We replace the operator $*^{-1}$ by $(-1)^{(q+1)(n-q-1)}*$ and omit even addends in the exponent.

We have introduced interior multiplication operator for the sake of completeness, which is important in considering differential operators on manifolds. In particular, the relation of Lie operators to exterior and interior multiplication was discussed in Subsec. 1.1.2.3.

4. Laplacian. By means of the operators d and δ, we can construct the second-order self-adjoint operator

$$\Delta = (d + \delta)^2 = d \circ \delta + \delta \circ d \tag{2.23}$$

acting on forms ω and called a *Laplacian*, since, for manifolds with the Euclidean metric, it acts on functions in the same way as the usual Laplacian

$$\Delta = \frac{\partial^2}{\partial(x^1)^2} + \dots + \frac{\partial^2}{\partial(x^n)^2}.$$

The result is immediately apparent from the simple calculation:

$$\omega = f(x^1, \dots, x^n), \quad d\delta + \delta d = \delta d.$$

Let

$$\delta\left(\frac{\partial f}{\partial x^i}dx^i\right) = -\sum \frac{\partial^2 f}{\partial(x^i)^2} = -\Delta f.$$

The operator Δ is commutative by definition, namely,

$$\Delta\delta = \delta\Delta, \quad *\Delta = \Delta*, \quad \Delta d = d\Delta.$$

Definition 2.18 A form ω is said to be *harmonic* if $\Delta\omega = 0$.

Proposition 2.14 *A form ω is harmonic if and only if $d\omega = \delta\omega = 0$. If $d\omega = \delta\omega = 0$, then obviously $\Delta\omega = 0$. The converse follows from the relation*

$$0 = \langle \Delta\omega \mid \omega\rangle = \langle d\delta\omega + \delta d\omega \mid \omega\rangle = \langle \delta\omega \mid \delta\omega\rangle + \langle d\omega \mid d\omega\rangle.$$

The scalar product is positive definite, therefore, $\delta\omega = d\omega = 0$. The following can also be proved in a simple way.

Proposition 2.15 *Let $H_1 \subset \wedge M$ be the subspace of harmonic forms, H_2 the subspace of forms of the $d\alpha$ type, and H_3 that of the $\delta\beta$ type. Then*

(1) $H_i \perp H_j$ for $i \neq j$.

(2) If $\omega \perp H_i$ for all $i = 1, 2, 3$, $\omega = 0$.

The main result about the space of forms on a compact manifold consists in the following.

Proposition 2.16 *The space $\wedge M$ can be uniquely decomposed into the direct sum of spaces H_i as*

$$\wedge M = H_1 + H_2 + H_3, \tag{2.24}$$

$i = 1, 2, 3$ (see the proof in [Rh]).

The existence of orthogonal decomposition of forms enables us to derive an important statement on the finite dimensionality of the cohomology groups of compact manifolds. This is the fundamental theorem due to Hodge.

Theorem 2.11 *Any closed form ω is cohomologous to a unique harmonic form.*

Contrasting this with de Rham's theorem on the cohomology of differential manifolds, we obtain

$$H^p(M^n, R) \sim H_1^p,$$

where H_1^p is the space of harmonic forms of degree p.

As a corollary, we state *Poincaré's duality theorem.*

Theorem 2.12 *The space of cohomology $H^p(M^n, R)$ of a compact manifold M^n is isomorphic to $H^{n-p}(M^n, R)$.*

Proof. The space of pth cohomology of a manifold M^n is generated by harmonic forms of degree p. The operator $*$ acts on the forms as $\omega^p \to \omega^{n-p}$. Since Δ and $*$ commute, the harmonic forms ω^q are sent under the map $*$ into harmonic forms $\tilde{\omega}^{n-q}$. ∎

We put dim $H^p(M^n, R) = b^p$. b^p are called the *Betti numbers*. It follows from Theorem 2.12 that $b^p = b^{n-p}$, which coincides with the definition of the Betti numbers of homology groups b_p introduced above. This follows from another statement of Poincaré's duality theorem.

Theorem 2.13 *There exists the isomorphism*

$$D: H^{n-i}(M^n, R) \to H_i(M^n, R).$$

For a proof see [Ch]. We only outline its construction. It is convenient to define the action of the operator D directly on the spaces $H_* = \sum H_i(M^n, R)$ and $H^* = \sum_i H^i(M^n, R)$ adjoint to the natural scalar product $\langle x \mid \omega\rangle = \int_x \omega$ for $\omega \in H^i(M^n, R)$, $x \in H_i(M^n, R)$. Let $\omega_1 \in H^{n-i}(M^n, R)$. We define the linear operator

$D: \omega_1 \to y \in H_i(M^n, R)$ by the formula

$$\int_{D\omega_1} \omega = \int_{M^n} \omega_1 \wedge \omega = \langle \omega_1 \mid \omega \rangle,$$

where $\langle \omega_1 \mid \omega \rangle$ is the scalar product in the space of forms.

We have thereby constructed the map $D: H^{n-i}(M^n, R) \to H_i(M^n, R)$. Poincaré's duality theorem can be stated as follows.

Proposition 2.17 *The diagram*

$$\begin{array}{ccc} H^i(M^n, R) & \to & H^{n-i}(M^n, R) \\ \downarrow & & \downarrow \\ H_i(M^n, R) & \to & H_{n-i}(M^n, R) \end{array}$$

is commutative.

Remark 2.10 (a) The theorem remains valid if the coefficient field R is replaced by any coefficient domain Z, C.

(b) It has been assumed in Theorems 2.12 and 2.13 that M is a compact orientable manifold. However, if it is non-orientable, then the duality theorem can be stated as

$$H^i(M, Z_2) \simeq H_{n-i}(M, Z_2).$$

In conclusion, we formulate another duality relation due to Alexander [MS], which we need for the study of defect links (see Subsec. 5.1.6.4). Here it is in simplified form. Let K be a compact subset in the sphere S^n.

Theorem 2.14 (Alexander)

$$H^{i-1}(K, Z) \simeq H_{n-i}(S^n \setminus K, Z). \tag{2.25}$$

The restrictions on the set K are rather weak, e.g., K can be a finite cell complex, $\dim K < n$. The reader may find the exact formulations in [MS].

2.4 RELATION BETWEEN HOMOTOPY AND HOMOLOGY GROUPS

Homotopy and (co-)homology groups possess both similarities and radical natural differences (e.g., homotopy groups depend, generally speaking, on a point).

We start with the difference.

1. First and foremost, as the dimension of a homotopy group increases compared with that of the manifold, the group $\pi_i(M^n)$ is not trivialized for $i > n$. Their calculation becomes more and more complicated. On the contrary, the groups $H_i(M^n)$ and $H^i(M^n)$ vanish if $i > n$.

2. The excision axiom $H_q(X, A) \sim H_q(X \setminus \mathcal{U}, A \setminus \mathcal{U})$ is violated for homotopy groups.

We illustrate this effect by simple example.

Example 2.10 As the space X, take the sphere S^2 and break it by the equator S^1 into two hemispheres S^2_+ and S^2_-. We fix a point x and consider the mapping of homotopy groups

$$\pi_q(S^2_+, S^1) \to \pi_q(S^2, S^2_-)$$

relative to the point. Induced by the corresponding embeddings, the above is an excision map. However, as can easily be seen, it is isomorphic to Freudenthal's suspension map (see Subsec. 2.1.7) inducing the homomorphism

$$E: \pi_{q-1}(S^1) \to \pi_q(S^2),$$

which is obviously not an isomorphism here.

We now turn to the relation between homotopy and homology groups.

(1) Let $\pi_1(M)$ be the fundamental group and $H_1(M)$ the homology group of the manifold M. Then $H_1(M) = \pi_1(M)/[\pi_1, \pi_1]$, where $[\pi_1, \pi_1]$ is the commutator in π_1.

We have seen that $\pi_1(M)$ is non-Abelian in the general case, while H_1, as well as all H_i, is Abelian.

If π_1 is Abelian, then $\pi_1 = H_1$.

(2) *Hurewicz's theorem* [Hu]. This famous theorem is

Theorem 2.15 *Let $\pi_i(X) = 0$ for all $i < n$. Then*

$$\pi_n(X) = H_n(X).$$

Relations between homotopy and homology groups sometimes enable us to reduce the calculation of the homotopy to homology groups, which are easier to compute.

We illustrate by a number of theorems.

Theorem 2.16 *The fundamental group of a semi-simple and compact Lie group G is finite. Since $\pi_1(G)$ is Abelian, $\pi_1(G) = H_1(G)$.*

It suffices to show that $H^1(G) = 0$.

We prove that there exists no non-trivial harmonic 1-form ω^1 on G by resorting to the following auxiliary result of independent interest.

Lemma 2.5 *Let G be a compact Lie group and H its closed subgroup such that the space G/H is symmetric. Then any differential 1-form ω^1 invariant under the action of G is harmonic (relative to the invariant metric on G/H)* [Ch].

G is naturally a symmetric space. It suffices to show by Lemma 2.5 that there exists no bi-invariant nonzero 1-form ω^1 on G.

Assume that the invariant 1-form ω^1 exists. It does not vanish on a vector $v \in TG_e$ (e is the identity element in G), $\omega^1 = 0$ on an $(n-1)$-dimensional subspace $V \subset R^n \simeq T_eG$, dim $G = n$, invariant under the adjoint representation Ad G. The orthogonal complement $V^\perp$ is one-dimensional and therefore fixed relative to Ad G. Consider the mapping exp $V^\perp \to G$ specifying the one-parameter group in the center, which is contrary to the semisimplicity of G. ■

Remark 2.11 That G is semi-simple is essential since for compact non-semi-simple groups (e.g., tori, nilpotent groups, etc.) $\pi_1(G)$ can be the direct sum of cyclic groups.

Theorem 2.17 *For an arbitrary Lie group*

$$\pi_2(G) = 0.$$

The requirement for semisimplicity is now not necessary. However, the proof can be easily reduced to the case of a compact and semi-simple Lie group.

Indeed, the following statement due to Iwasawa [Sem] holds. Any simply connected Lie group G can be decomposed into the product KAN, where K is a compact subgroup, A a simply connected commutative subgroup, and N the solvable subgroup homeomorphic to the Euclidean space. Hence, $\pi_2(G) = \pi_2(K)$.

The universal covering space of a compact and solvable group K_2 is homeomorphic to the Euclidean space. Hence, $\pi_2(K) = \pi_2(K_1 \times K_2)$, where K is a compact semi-simple group.

Note that the homotopy groups of non-semi-simple and compact groups (e.g., tori) do not coincide with the homology ones, e.g., $\pi_i(T^n) = 0$ for $i > 1$, while $H_i(T^n) \neq 0$ for all $i \leqslant n$, which once again underlines the difference between them.

We now turn to the proof of Theorem 2.17, assuming that G is a compact semi-simple group.

Since $\pi_2(G) = \pi_2(\tilde{G})$, where $\tilde{G}$ is the universal covering group of G, by Hurewicz's theorem, it suffices to prove that $H_2(\tilde{G}) = 0$. It follows from Theorem 2.16 that $\tilde{G}$ is semi-simple and compact. Since $H_2(\tilde{G}) = H^2(\tilde{G})$, we show that there are no closed invariant 2-forms ω^2 on $\tilde{G}$, i.e., $H^2(\tilde{G}) = 0$.

Proof. Define the form $\Omega^2 = \lambda_1\omega^2 + \lambda_2 g$ on $\tilde{G}$, where g is a symmetric positive definite form generated by the Killing form on $\tilde{G}$ while ω^2 an arbitrary skew-symmetric 2-form. g is the Riemannian metric on $\tilde{G}$. Let V be the tangent space of $\tilde{G}$ at the point e.

The 2-form Ω^2 determines the map $V \to V^*$ (where V^* is the space adjoint of V), i.e., a linear functional $l(v)$, where $v \in V$. Consider the subspace V_1 on which the functional vanishes. Since Ω^2 is invariant under the adjoint representation $\mathrm{Ad}\,G$, V_1 is also invariant under $\mathrm{Ad}\,G$. Since G is semi-simple, its representation is irreducible. Therefore, V_1 is either $\{0\}$ or the whole space. The latter is impossible, since the form ω^2 is skew-symmetric and g symmetric. If $\Omega^2 = 0$, then $\omega^2 = \gamma g$, which is also impossible, a contradiction, and the theorem is thus proved.

Since $\pi_1(G) = \pi_2(G) = 0$ for semi-simple simply connected compact Lie groups, $\pi_3(G) = H_3(G)$.

Theorem 2.18

$$\pi_3(G) \neq 0.$$

Proof. Since $H_3(G, R) \simeq H^3(G, R)$, it suffices to construct a non-vanishing closed invariant 3-form. Put $\omega^3(X, Y, Z) = \langle [X, Y] \mid Z \rangle$, where X and Y are two left-invariant vector fields on G, $\langle X \mid Y \rangle$ a bi-invariant Riemannian metric on G, and $[X, Y]$ the commutator for the fields.

It is easy to see that the form ω^3 is bi-invariant and, therefore, closed and skew-symmetric in all the three variables. It thereby determines a nonzero element of the group $H^3(G, R)$. ∎

We can show that $\pi_3(G) = Z$ for simple groups G.

2.5 DEGREE OF A MAP AND INDICES OF VECTOR FIELDS

We have already encountered the concept of the degree of a map (in Example 2.1) when calculating the fundamental group of the circle. The definition was given in Subsec. 2.2.2. Here, we study the concept in more detail, in particular, establishing its relation to the principal homology and homotopy characterization of a manifold and giving different methods for finding the degree.

2.5.1 Definitions of the Degree of a Map. Examples

Consider the example of a continuous map $f: S^1 \to S^1$, $S^1 \subset C$, whose main properties can be summarized as follows:

(1) The degree of f (deg f) is the number of revolutions of the curve turning around the origin.

(2) Deg f is an invariant of continuous deformations.

(3) Deg f is the unique invariant of deformations: If two maps f_1 and f_2 have the same degree, then they are homotopic.

(4) There exist maps with a given degree.

We can indicate several methods for calculating the degree of a map. One, called *differential*, has already been mentioned. This is the representation

$$\deg f = 1/(2\pi i) \int \frac{df}{f}$$

familiar from complex analysis.

The definition is valid for differentiable functions; however, since each continuous function can be approximated by differentiable ones (*Weierstrass' theorem*), this also holds in the continuous case.

The second definition is as follows: consider a map $g = f/|f|: S^1 \to S^1$ and approximate it by a differentiable one, finding the number of points in the preimage of the general point, taking into account the signs of the points.

Both definitions admit a generalization to maps of n-dimensional manifolds.

We start with the second definition used in Sec. 2.2 when mappings of a cell σ into a cell complex were defined.

Definition 2.19 The *degree of a smooth map of connected oriented closed manifolds* is

$$\deg f = \sum \operatorname{sgn} \det \left(\frac{\partial y_0^\alpha}{\partial x_j^\beta} \right), \qquad y_0 \in N,\ x_j \in M,\ f(x_j) = y_0. \tag{2.26}$$

The correctness of the definition follows from the theorem below.

Theorem 2.19 *The degree of a map does not depend on the regular value y_0 and remains unaltered under homotopy.*

Proof. 1. First, we show that if f: $M \to N$ is homotopic to g: $M \to N$, then $\deg f = \deg g$ for any common regular value $y \in N$.

Consider a homotopy $F_t = M \times [0, 1] \to N$ such that $F_0(x) = f(x)$, $F_1(x) = g(x)$. Let $y \in N$ be a general regular point of the map F_t (always existing by the Sard theorem [GG]). Its preimage $F^{-1}(y)$ consists of finitely many smooth curves ending on the "sides" $M \times 0$ and $M \times 1$; the others are closed. The inverse image of the neighborhood of y are tubular neighborhoods of the inverse image of y. Since the contribution to the degree of a map is only given by the curves ending on $M \times 0$ and $M \times 1$, we consider only the tubes embracing the sides. The tubes can be broken into disks normal to the axial curve. Take one with base on $M \times 0$ and "move" along it. The base disk D is mapped into a neighborhood $\mathscr{U} \subset M$, say, with positive orientation. Since the orientation is discrete, the map is continuous, the curve γ consists of regular points, and the tubular neighborhood is small, the orientation of the disks cannot be reversed when moving along γ. If the curve goes back to $M \times 0$ when the parameter t equals unity, the corresponding disk is attached to the base with opposite orientation; if γ ends on $M \times 1$, then with the same orientation.

A similar picture is obtained for curves starting on $M \times 1$. Thus, the number of disks with positive orientation equals that with negative orientation on $M \times 0$ and $M \times 1$, coinciding with the number of tubes going from the "sides" (taking the signs of orientation into account), which means that $\deg f = \deg g$.

2. If y and z are two different regular points of the map f: $M \to N$, then we construct the diffeomorphism h: $N \to N$ isotopic to the identity and sending y into z. Then h preserves its orientation and $\deg(f, y) = \deg(h \circ f, z)$. f and $h \circ f$ are homotopic, therefore, $\deg(f, y) = \deg(f, z)$.

Remark 2.12 For non-orientable manifolds, the concept of a Jacobian's sign cannot be defined invariantly in a neighborhood of a regular point. The degree of a map is only defined as the residue mod 2.

We now define the degree of a map for manifolds with a boundary. Given maps $(M, \partial M) \to (N, \partial N)$ such that the boundary ∂M is sent into the boundary ∂N, the following theorem is true.

Theorem 2.20 *The degree of a map of the boundary coincides with that of the manifold itself, namely,*

$$\deg f \mid \partial M = \deg f.$$

The proof is based on the following observation. We can construct a smooth homotopy of a map $f: M \to N$ so that none of the interior points of M is mapped into ∂N. Now, we take a regular point y on ∂N and consider its full preimage $f^{-1}(y)$ in ∂M. If we shift y into a small tubular neighborhood of ∂N, then the corresponding points of the preimage are in the tubular neighborhood of ∂M. Since the number of preimages and their signs are unaltered under this deformation, $\deg f$ equals $\deg f \mid \partial M$ (for details see [DFN]).

We now turn to a generalization of the second definition of the degree of a map to the n-dimensional case.

Let M^n and N^n be two closed orientable manifolds of dimension n and f: $M^n \to N^n$ a smooth map of degree q: $\deg f = q$. Take a differential form Ω of rank

n on N^n, representable as $\Omega = \varphi_\alpha(y)dy^1_\alpha \wedge \ldots \wedge dy^n_\alpha$ in terms of local coordinates $y_\alpha \in N^n$, $y_\alpha = (y^i_\alpha)$. The integrals $\int_N \Omega$, $\int_M f_*(\Omega)$ are defined (locally) of the form $f_*(\Omega) = \varphi_\alpha(f(x))dx^1_\beta \wedge \ldots \wedge dx^n_\beta \cdot \det(\partial y^i_\alpha/\partial x^j_\beta)$ in the coordinate patch $x_\beta(x^j_\beta)$ such that $f(x^j_\beta) \subset (y^i_\alpha)$.

Proposition 2.18

$$\int_M f_*(\Omega) = \deg f \int_N \Omega. \tag{2.27}$$

The proof is left to the reader (see also [DFN]).

Consider the concrete computation of the degree of a map in certain particular cases of interest.

Degree of a vector field on a surface. Let M^n be a compact n-dimensional manifold with a boundary ∂M.

Definition 2.20 The map

$$f\colon \partial M \to S^{n-1}$$

is said to be *Gauss, spherical,* or *normal* if each point $x \in \partial M$ is associated with the unit vector normal to ∂M at x and pointing outwards.

We calculate the Gauss map's degree for a closed hypersurface Q in R^3. We specify Q in parametric form

$$x^\alpha = x^\alpha(u^1, u^2)$$

locally. A closed 2-form, the element of area, is defined on the sphere S^2 (being the volume form for S^n in the n-dimensional case)

$$\Omega = \frac{1}{4\pi}\,\frac{x^1dx^2 \wedge dx^3 - x^2dx^1 \wedge dx^3 + x^3dx^2 \wedge dx^1}{((x^1)^2 + (x^2)^2 + (x^3)^2)^{3/2}}. \tag{2.28}$$

The factor $1/(4\pi)$ is due to the normalization condition

$$\int_{S^2} \Omega = 1 \tag{2.29}$$

for the area of the sphere.

It follows from Proposition 2.18 and (2.28), (2.29) that for any vector field ξ on Q the degree of the Gauss map f is $\int f_*(\Omega)$.

Since the degree of a Gauss map does not depend on the form of the field (the nonsingularity assumption only being essential), it suffices to calculate $\int f_*(\Omega)$ for the normal unit field $n(x)$ pointing outwards of Q.

The reader can perform the calculation (with respect to local coordinates) independently or can turn to [DFN]. Note that the appropriate result can be obtained by using the general properties of a Riemannian manifold specialized for a two-dimensional surface. Since the Riemannian metric always exists on the surface and there is only one symmetric connection corresponding to the metric, the curvature form Ω_{12} is $d\omega_{12}$, where ω_{12} is the connection form. Under f, Ω is carried into Ω_{12}. The latter form is obviously proportional to the element of area of Q, namely, $\Omega_{12} = K\,dS$, where K is a scalar function. The coefficient K is the Gaussian curvature of Q.

Geodesic coordinates $ds^2 = dx^2 + g(x, y)\,dy^2$ can be introduced on the two-dimensional surface Q. Assuming sections of the cotangent bundle to be $\theta_1 = dx$ and $\theta_2 = g\,dy$, we obtain the orthonormal basis in a neighborhood of any point $x \in Q$.

The Maurer-Cartan equations for the connection forms

$$\begin{aligned} d\theta_1 &= \omega_{12} \wedge \theta_2, \\ d\theta_2 &= -\,\omega_{12} \wedge \theta_1 \end{aligned} \tag{2.30}$$

have a unique solution $\omega_{12} = (\partial g/\partial x)\,dy = g_x\,dy$ $(\partial g/\partial x = g_x)$. Hence, $\Omega_{12} = \partial^2 g/\partial x^2\,dx \wedge dy = -(g_{xx}/g)\,dS$.

The Gaussian curvature is expressed by the formula

$$K = -\,g_{xx}/g. \tag{2.31}$$

The final result is summarized as follows.

Proposition 2.19 *Neglecting the normalizing factor, the integral of the Gaussian curvature on a closed surface is equal to the degree of the Gauss map on the surface.*

The result is generalized to the n-dimensional case and only the scalar curvature of the manifold is calculated in place of the Gaussian curvature.

2.5.2. Index of a Vector Field

The concept of the degree of a map has remarkable applications in the study of singular points of vector fields on a manifold.

Let $\xi(x)$ be a vector field on a manifold M in the neighborhood of a point x_0. The point x_0 is said to be singular if $\xi(x_0) = 0$. x_0 is called an *isolated singular point* if in a small neighborhood of it $\xi(x) \neq 0$ for $x \neq x_0$ and *non-degenerate* if $J = \det\left(\frac{\partial \xi^i}{\partial x^j}\right)\Big|_{x_0}$ does not vanish in the neighborhood.

The following statement is a consequence of the implicit-function theorem: *if x_0 is non-degenerate, then it is isolated.*

Definition 2.21 The *index of a non-degenerate singular point* x_0 is

$$\operatorname{Ind} x_0 = \operatorname{sgn}\det\left(\frac{\partial \xi^i}{\partial x^j}\right)_{x = x_0} = \operatorname{sgn}\det\begin{pmatrix} \lambda_1 & & \\ & \ddots & \\ & & \lambda_n \end{pmatrix},$$

where $\lambda_1, \ldots, \lambda_n$ are the eigenvalues of the Jacobian matrix.

For a gradient vector field $v = \operatorname{grad} f$, Ind $v = (-1)^{l(x_0)}$, where $l(x_0)$ equals the number of negative squares in the canonical representation of the quadratic form d^2f/x_0. The number is called the *Morse index* of the function f at a non-degenerate critical point x_0. We discuss the Morse index in Subsec. 2.5.3.

We now define the index in general and as applied to non-degenerate singular points.

Let $Q_\varepsilon = S_\varepsilon^{n-1}$ be a sphere of small radius $\varepsilon > 0$ surrounding a singular point x_0 so that the vector field does not vanish on the sphere. The (spherical) Gauss map

$$f_\varepsilon: Q_\varepsilon \to S^{n-1}$$

is defined for Q_ε.

Definition 2.22 The *index of an isolated singular point* x_0 of a vector field ξ is the degree of the Gauss map f, namely,

$$\operatorname{Ind}_{x_0}(\xi) = \deg f. \tag{2.32}$$

For non-degenerate points, this definition coincides with the above.

Outline of the proof. It suffices to note that, in a small neighborhood of x_0, ξ is homotopic to the linear vector field $\dot{x} = Ax$, in homotopy, all fields $\xi(x, t)$ vanishing only at x_0. Assume that such a homotopy is constructed. It then suffices to calculate the degree of the map A. On the unit sphere, deg A is ± 1 and is determined by the sign of det A, namely, if det $A > 0$ (the orientation is preserved), then deg $A = 1$; if det $A < 0$, then deg $A = -1$ (see [Ar] for more detail).

The classification of singular points in terms of the eigenvalues of the matrix $A = \left.\dfrac{\partial \xi^i}{\partial x^i}\right|_{x_0}$ leads to the classification of singular points of vector fields familiar from a course of differential equations (e.g., [Ar]).

The basic theorem characterizing the deep relation between the indices of singular points of vector fields and the topology of manifolds was discovered by Poincaré for 2-manifolds and by Hopf in the n-dimensional case.

We formulate several versions of the corresponding theorems.

Theorem 2.21 *Let ξ be a vector field on an n-dimensional closed hypersurface Q pointing outwards and $x_1, \ldots, x_k$ be isolated singularities of ξ in Q. Then the index sum $\Sigma \operatorname{Ind} x_i$ equals the degree of the Gauss map $f_\varepsilon: Q \to S^{n-1}$.*

Proof. We cut out a ball of radius ε. We obtain a new manifold with a boundary around each zero of ξ. The map $\bar{\xi} = \xi(x)/|\xi(x)|$ sends the manifold into S^{n-1}. The degrees of the restrictions of $\bar{\xi}$ to different boundary components sum to zero. But $\bar{\xi}|Q$ is homotopic to f_ε.

In fact, consider the field $\dfrac{(1-t)\bar{\xi} + tf_\varepsilon}{|(1-t)\bar{\xi} + tf_\varepsilon|}$. The denominator vanishes nowhere since the direction of one of the fields is not reverse to that of the other at any point $x \in Q$. On the other hand, the sum of the degrees of the restrictions to the other boundary components is $-\Sigma \operatorname{Ind} x_i$. The minus sign is due to the orientation of the small sphere S_ε^{n-1} as of the boundary of the manifold $D_\varepsilon^n(x)$ in $X \setminus D_\varepsilon^n(x)$

being opposite to the orientation of $S_\varepsilon^{n-1}(x)$ as of the boundary of $D_\varepsilon^n(x)$. Thus, we have

$$\deg f_\varepsilon(x) - \Sigma \operatorname{Ind} x_i = 0.$$

The topological nature of the index theorem for vector fields becomes clear after defining a new topological invariant, i. e., the Euler characteristic. In the next subsection, we establish its relation to the index sum of vector fields on a manifold.

We introduce the characteristic in several ways.

2.5.3 Euler Characteristic

The quantity

$$\chi(M^n) = \Sigma(-1)^i \operatorname{rank} H_i(M^n),\ i = 0, \ldots, n,\ n = \dim M,$$

is called the *Euler characteristic* of the manifold M^n.

The geometric definition requires several facts from the intersection theory.

Let P^r and Q^s be two closed submanifolds of M^n of dimensions r and s, respectively. By Theorem 1.1 (Whitney), we can always regard M as a Euclidean space of a sufficiently large dimension.

P^r is said to *intersect transversally to* Q^s (or to be in general position) if, at any point $x \in P^r \cap Q^s$, the tangent spaces T_xP^r and T_xQ^s generate the space tangent to M; in particular, it follows that, in general position, the intersection $P^r \cap Q^s$ is a smooth $(r + s - n)$-dimensional submanifold.

Example 2.11 A straight line P^1 and the plane Q^2 in three-dimensional space intersect transversally if P^1 does not meet Q^2 at a zero angle, i. e., is not in Q^2.

If the sum of dimensions r and s is n, then, in general position, they intersect at one or more points. If M, P, and Q are oriented, then each intersection point x_i is given a sign by the following rule: Let τ_j^r be the orienting tangent frame to P^r at a point x_j and τ_j^s the orienting frame to Q^s at x_j. The point x_j is assigned a plus sign if the union frame $\tau = (\tau_j^r, \tau_j^s)$ is orienting for M^n at x_j. Otherwise, a minus sign is given. The sign is denoted by sgn $x_j(P \circ Q)$.

Definition 2.23 The *intersection number* of two manifolds, P^r and Q^s, is the integer

$$\operatorname{Ind}(P \circ Q) = \sum_{j=1}^{m} \operatorname{sgn}_{x_j}(P \circ Q),$$

where m is the number of intersection points.

In the non-orientable case, $\operatorname{Ind}(P \circ Q)$ is defined as the residue modulo 2 of the number of intersection points.

Properties of the intersection number

1. $\operatorname{Ind}(Q \circ P) = (-1)^{rs} \operatorname{Ind}(P \circ Q)$.

The *proof* is obvious.

2. *If two submanifolds Q_1 and Q_2 are homotopic, then their intersection numbers*

coincide for any P:

$$\text{Ind}(Q \circ P) = \text{Ind}(Q_2 \circ P).$$

The *proof* is very similar to that of the invariance theorem for the degree of a map under homotopic transformations (see [DFN]). We obtain the result as a consequence of a more general theorem, whose simple corollary is as follows.

Proposition 2.20 *The intersection number for two closed submanifolds P and Q in a Euclidean space is zero.*

Proof. Translate Q by a vector a so that $Q + a$ does not intersect P, which is possible since Q is compact, $Q + a$ homotopic to Q, and $\text{Ind}(Q + a) \circ P = 0$.

The bulk singularity of a vector field. Given a tangent vector field ξ on a manifold P^r, consider the space of linear elements $N = (x, \xi)$, where x is a point in P^r and ξ a vector at x. The vector field determines an embedding $f: P^r \to N$ by the rule $F_\xi(x) = (x, \xi(x))$. We denote the embedding image by $P(\xi)$. The manifold $P(0)$ associated with the null vector field is identified with P.

A vector field ξ is said to be in *general position* if the manifolds P and $P(\xi)$ are. Then ξ only has isolated singular points (t-regularity). If P is oriented by τ^r at a point x, then so is N by the frame (τ^r, τ^r) at the points (x, ξ).

Lemma 2.6 *All singular points x_j of a field ξ in general position are non-degenerate. The sign of x_j as of a point in the intersection $P(0) \cap P(\xi)$, involved in the definition of an intersection number, coincides with* $\text{Ind}\,\xi = \text{sgn det}\left(\frac{\partial \xi^\alpha}{\partial x^\beta}\right)_{x_j}$ *of a singular point of ξ.* (See the proof in [DFN].)

It follows that the sum of the indices of a tangent vector field is equal to the intersection number for manifolds P and $P(\xi)$. To prove a similar assertion for an arbitrary vector field η in general position, it suffices to show that η can be homotopized into ξ, which is easy to implement since any field can be homotopized, viz., into the null field

$$\xi(x, t) \to (1 - t)\xi(x).$$

Hence, the embeddings $P \subsetneq P(\xi)$ and $P \subsetneq P(\eta)$ are homotopic and, by the (intersection number) theorem, $\text{Ind}\,(P \circ P(\xi)) = \text{Ind}(P \circ P(\eta))$.

We can derive a number of interesting corollaries.

Proposition 2.21 *If* $\dim P^r$ *is odd, then the sum of the indices of a vector field on the closed orientable manifold P^r is zero.*

Proof. It follows from property (1) that $\text{Ind}\,(P(0) \circ P(\xi)) = (-1)^{r^2}\text{Ind}\,(P(\xi) \circ P(0)) = -\text{Ind}\,(P(\xi) \circ P(0))$. On the other hand, since the vector fields 0 and ξ are homotopic, $\text{Ind}\,(P(0) \circ P(\xi)) = \text{Ind}\,(P(\xi) \circ P(0))$. ∎

Corollary 2.1 *For any smooth function f with non-degenerate singular points x_j on a closed orientable manifold M^n, the expression $\sum_{x_j}(-1)^{i(x_j)}$ does not depend on the choice of f and is zero if* $\dim m$ *is odd, where $i(x_j)$ is the index of a singular point x_j, or the number of squares with a minus sign in the form $d^2f|_{x_j}$.*

This follows from the definition of the index of a singular point of the field grad f.

We now investigate the properties of functions said to be *Morse* with non-degenerate critical points.

An important property pointing to the relation between indices of vector fields and the topology of a manifold is the equality

$$\sum_{x_m}(-1)^{i(x_m)}c_{i(x_m)} = \chi(M), \tag{2.33}$$

where $c_{i(x_m)}$ is the number of critical points of index $i(x_m)$.

The proof reduces to establishing the equivalence of the definition of the Euler characteristic in terms of homology groups and the following definition.

Definition 2.24 Let $P^r \subset N$ be a submanifold embedded in the space of linear elements as the null field and P_ε a translated submanifold so that P and P_ε intersect transversally.

The *number* $\chi(P) = \operatorname{Ind}(\Delta \circ \Delta)$ *of self-intersection* of the diagonal $\Delta(x, x)$ in the direct product $P \times P$ is called the *Euler characteristic* of the manifold P.

The proof of (2.33) separates into two steps.

Step 1. It is required to prove that the definition of the Euler characteristic $\chi(M)$ as of the intersection number of the diagonal is equivalent to the definition of the index of a tangent field.

Step 2. $\chi(M)$ defined as the alternating sum of the ranks of homology groups coincides with $\chi(M)$ found in terms of Morse indices.

Proof of Step 1. Let ξ be the tangent field given at a point $x \in M$. We define a small translation of the diagonal Δ in $M \times M$: $(x, x) \to (x, x')$, where $x' = x + \xi\, dt$, with the help of ξ.

The self-intersection points of Δ are the singular points of the vector field ξ. It is now easy to find the index of Δ as it is equal to the intersection number of the manifolds $y = x$ and $y = x + Ax$. On the other hand, $\operatorname{Ind}\Delta = \operatorname{Ind}\xi = \operatorname{Ind}\tilde{\xi}$, the latter for the field $\tilde{\xi} = x' = Ax$, which is 1 if $\det A > 0$ and -1 if $\det A < 0$, coinciding with the index of ξ.

Proof of Step 2.

Theorem 2.22 (Morse) *Let $f(x)$ be a function of M with non-degenerate critical points. The number of critical points of $f(x)$ of index i is c_i. Then*

$$\sum_i (-1)^i c_i = \chi(M).$$

Proof. Consider the gradient field grad f on M in a neighborhood of a point $x = x_0$, namely, $f(x) = f(x_0) - (x^1)^2 - \dots - (x^i)^2 + (x^{i+1})^2 + \dots + (x^n)^2$, $\operatorname{Ind} \operatorname{grad} f = (-1)^i$, where x_0 is a critical point of index i. But, the sum of indices of critical points of $v = \operatorname{grad} f$ is $\chi(M)$. ∎

In proving the theorem, the existence of f with non-degenerate critical points has been assumed. The proof of existence on each compact manifold of such a function is also due to Morse [Mil4].

Usually, the Euler characteristic $\chi(M)$ is defined as

$$\Sigma(-1)^i \dim H_i(M, R). \tag{2.34}$$

The proof of the equality

$$\Sigma(-1)^i \dim H_i = \Sigma(-1)^i c_i$$

follows from the remarkable surgery theorem of Morse.

We only state it (see the proof and a number of topological consequences in [Mil4]).

We need another concept.

Let f be a real function on M. We denote the set $f^{-1}\ [-\infty, a] = \{x \in M: f(x) \leqslant a\}$ be M^a.

Theorem 2.23 (Morse's surgery theorem) *Let f be a Morse function on M. If each M^a is compact, then M has the homotopy type of a cell complex, in which each critical point of index i is associated with one cell of dimension i.*

The results of Subsection 2.5.2 admit various applications and generalizations. We illustrate this by a number of examples, some we shall use in subsequent chapters and consider in greater detail.

Example 2.12 We begin with a classical result. The *fundamental theorem of algebra*: (a) Any algebraic equation $P(z) = 0$, $z \in C$, has at least one root, (b) $P(z) = a_n z^n + \ldots + a_1 z + a_0 = 0$ has n roots. (Show that the map $S^2 \to S^2$ given by the polynomial $P(z)$ is homotopic to $P_1(z) = a_n z^n$ and find its degree.)

Example 2.13 The Euler characteristic of a two-dimensional closed orientable surface with genus g is $\chi(M^2) = 2 - 2g$. Hence, on any surface other than T^2, a tangent vector field has at least one singular point.

Similar argument enables us to prove the following fact.

Example 2.14 Inside a limit cycle on a plane, there is always an equilibrium position.

Example 2.15 On a closed odd-dimensional manifold M^{2k+1}, there always exists a field without singularities.

We first prove that $\chi(M^{2k+1}) = 0$.

Let f be a Morse function on M^{2k+1}. Then $\chi(M^{2k+1}) = c_0 - c_1 + \ldots + c_{2k+1}$. However, for the function $-f$, we have $c_{2k+1} - \ldots - c_0 = -\chi(M^{2k+1})$, i. e., $\chi(M^{2k+1}) = 0$. Therefore, there exists at least one non-singular field grad f on M^{2k+1}.

Example 2.16 Vector field on a manifold with a boundary.

Let $M^m \subset R^n$ be a smooth manifold with boundary ∂M. Consider tangent vector fields on M with the following properties $(*)$. The field has finitely many zeros and is on the boundary pointing outwards. We would like to find an analog of the Poincaré-Hopf theorem for manifolds M such that $\partial M \neq \varnothing$, for which it is useful to introduce the double $\tilde{M}$ of M. $\tilde{M}$ is obtained from M by attaching a second copy of M along the boundary ∂M. In doing so each point in ∂M is identified with its copy in the boundary of the second copy and we obtain an m-dimensional manifold $\tilde{M}$ without a boundary.

We construct a field ν on $\tilde{M}$ with the above properties (*). It is then clear that

$$\chi(\tilde{M}) = 2\,\Sigma\,\mathrm{Ind}\,\nu \qquad \text{if } \dim M \text{ is even,}$$
$$x(\tilde{M}) = 2\,\Sigma\,\mathrm{Ind}\,\nu - \chi(\partial M) = 0 \qquad \text{if } \dim M \text{ is odd.}$$

Before turning to the next example, we should introduce two (quite familiar [H]) definitions.

Let γ be a closed curve bounding a domain $\mathscr{U} \subset R^2$ and ξ a vector field in $\bar{\mathscr{U}}$, $\mathscr{U}$ having no singular points of ξ and ξ not vanishing on γ.

Definition 2.25 The index of a field ξ on γ (Ind ξ) is the degree $\deg n(x)$ of the Gauss map $n(x) = \xi(x)/|\xi(x)|$.

Ind ξ equals the degree of the mapping $S^1 \to S^1$.

If there are singular points in $\mathscr{U}$, then the definition should be modified naturally. The singular points of ξ are to be removed and the domain $\mathscr{U} \setminus \varepsilon(x_i)$ considered, where $\varepsilon(x_i)$ are disks of radius ε surrounding the points.

Definition 2.26 Let γ be a closed smooth curve surrounding a simply connected domain $\mathscr{U}$ and let a field ξ touch the boundary γ only at finitely many points $y_1, \ldots, y_n$. A point y_i is said to be of *internal* (*external*) *tangency of the orbit* $y(t)$ if the solution of the equation

$$\dot{y} = \xi, \quad y(0) = y_i$$

is in $\mathscr{U}$ (in $R^2 \setminus \bar{\mathscr{U}}$) for $\varepsilon > |t| > 0$.

Theorem 2.24 *Let ξ be a smooth vector field on the plane R^2, L be a smooth contour in R^2, I be the number of internal tangency points for ξ and L, and E be the number of external tangency points. Then*

$$\mathrm{Ind}\,\xi = 1 + \frac{I - E}{2}.$$

This entails an interesting corollary.

Corollary 2.2 *Let x_0 be a unique critical point of the gradient field grad f (x_0 is isolated, but can be degenerate). Then*

$$\mathrm{Ind}\,\mathrm{grad}\,f \leqslant 1.$$

The proof of the theorem can be found, for example, in [H] and we omit it here. To prove the corollary, it suffices to notice for the gradient field that $E \geqslant I$. The corollary has applications in the description of the singularities of smectic crystals [Pol].

The degree of a map and index of a vector field admit various non-trivial multidimensional generalizations in subsequent sections.

Now, we turn to one of the key examples in the topology of manifolds, illustrating the force and variety of topological applications, i. e., the Hopf fibration.

2.6 HOPF INVARIANT

Here, we study invariants of maps

$$S^{2n-1} \to S^n \tag{2.35}$$

of spheres.

2.6.1 Hopf Fibration

The first non-trivial example is the map

$$S^3 \xrightarrow{S^1} S^2. \tag{2.36}$$

The proof of nontriviality of the group $\pi_3(S^2)$ due to Hopf has been the first great success of the homotopy theory. Numerous generalizations and extensions of the discovery became afterwards a basis for many modern achievements in topology.

Taking into account the extreme importance of (2.36), we consider fiber bundles from different viewpoints, observing how the technique developed in the preceding chapters works here. As a physical example, we consider bundles appearing in several problems of mechanics (see applications to the field theory and condensed matter in Chs. 4 and 5).

We first show that $\pi_3(S^2) = Z$.

The proof is immediate from the exactness of the sequence of homotopy groups

$$\pi_i(S^1) \to \pi_i(S^3) \to \pi_i(S^2) \to \pi_{i-1}(S^1).$$

For $i \geqslant 3$, we have

$$0 \to \pi_i(S^3) \to \pi_i(S^2) \to 0,$$

in particular, $\pi_3(S^2) = Z$.

Constructing the Hopf fibration. The homotopy classes of maps $S^3 \to S^2$ are characterized by the group of integers Z. What is their geometric meaning?

We show that all fiber maps $S^3 \to S^2$ can be obtained by the composition of the maps f: $S^3 \to S^3$ and a special map p: $S^3 \to S^2$, which is called a *Hopf fibration.*

We represent S^3 as a pair of complex numbers (z_1, z_2) so that $|z_1|^2 + |z_2|^2 = 1$. We associate (z_1, z_2) with the point $z_1/z_2 = w$ (w is a point in the complex plane). The map $(z_1, z_2) \to w$ is extended to the completion of C with the point $z_2 = 0$ at infinity. We thus obtain a map $S^3 \to S^2$ into the Riemann sphere. It is obvious that under this map the points of the form $\exp(i\varphi)z_1$ and $\exp(i\varphi)z_2$ are sent into the same point w. Therefore, the fiber of $S^3 \to S^2$ is the set of points $\lambda = \exp(i\varphi)$, or the unit circle S^1.

We obtained a fiber map p: $S^3 \to S^2$ with the fiber S^1. It is easy to see that p is not trivial, i.e., not equivalent to the direct product

$$S^3 \neq S^2 \times S^1.$$

The simplest way to see this is to calculate the homotopy group (e.g., π_1 or π_2) of both sides, namely,

$$\pi_2(S^3) = 0, \quad \pi_2(S^2 \times S^1) = Z.$$

The second proof illustrates the calculation of $\pi_1(S^3)$ and $\pi_1(S^2 \times S^1)$ geometrically.

Any circle in S^3 can be contracted to a point in S^3, but not in $S^1 \times S^2$.

Proposition 2.22 *The set of classes of homotopy maps $S^3 \to S^2$ is the composition of mapping f: $S^3 \to S^3$ and of the Hopf fibration p: $S^3 \to S^2$.*

The proof follows from Theorem 2.1 applied to the fiber map $S^3 \to S^2$ (see Subsec. 2.1.2).

We now investigate different properties of the Hopf fibration.

Linking numbers. Let f be a smooth map $S^3 \to S^2$, y_0 and y_1 two regular points in S^2, and M_0 and M_1 the preimages of y_0 and y_1, equal to $f^{-1}(y_0)$ and $f^{-1}(y_1)$, respectively. We define $H(f) = \{M_0, M_1\}$ to be the linking number for the inverse images. It follows from the definition of regularity that $f^{-1}(y_i) \sim S^1$.

Definition 2.27 The *linking number* (or *coefficient*) of two disjoint curves $\gamma_i(t)$, $i = 1, 2$, lying in the Euclidean space R^3 ($\gamma_i(t) = r_i(t)$, $0 \leqslant t \leqslant 2\pi$, r is the radius-vector of a point in R^3) is the number

$$\{\gamma_1, \gamma_2\} = \frac{1}{4\pi} \int_{\gamma_1} \int_{\gamma_2} \frac{\langle [dr_1, dr_2], r_1 - r_2 \rangle}{|r_1 - r_2|^3}. \tag{2.37}$$

The following property formulated as a theorem yields an equivalent definition of a linking number in terms of homology.

Theorem 2.25 (a) *A linking number is an integer remaining unaltered under disjoint deformations of the closed curves γ_1 and γ_2.*

(b) *Let F: $D^2 \to R^3$ be a map that coincides with γ_1 at the boundary $\partial D^2 = S^1$ and is in general position (i. e., possesses t-regularity) on γ_2. Then* Ind $(F(D^2) \cap \gamma_2)$ *is equal to $\{\gamma_1, \gamma_2\}$.*

Outline of the proof. The closed curves $\gamma_1(t)$ and $\gamma_2(t)$ define a two-dimensional oriented surface $\gamma_1 \times \gamma_2$: $(t_1, t_2) = (r_1(t_1), r_2(t_2))$ in R^6. Let γ_1 and γ_2 be disjoint. Then the map

$$\varphi(t_1, t_2) \to S^2 = \frac{r_1(t_1) - r_2(t_2)}{|r_1(t_1) - r_2(t_2)|} \tag{2.38}$$

is defined, with the degree given by integral (2.37); therefore, deg φ is an integer. Since the integer is unaltered under homotopies, (a) is thus proved. To prove (b), it suffices to calculate (2.37) when γ_1 and γ_2 are two orthogonally linked circles (γ_1 is in the (x, y)-plane and γ_2 in the (y, z)-plane) (see detailed data in [DFN]).

It follows from Theorem 2.25 that, by analogy with the linking number for two curves, we can define a linking number for two manifolds, M^k and N^l.

Let M^k and N^l be two closed smooth manifolds ($\dim M^k = k$ and $\dim N^l = l$), while f and g are their continuous maps into the Euclidean space R^{k+l+1}, and let the sets $f(M^k)$ and $g(N^l)$ be disjoint. Let S^{k+l} be the unit sphere in R^{k+l+1}, center at the origin, with orientation as of the ball's boundary. We also give orientation on $M^k \times N^l$. Then we can define the linking number for $f(M^k)$ and $g(N^l)$.

Definition 2.28 The *linking number* $k(f(M^k), g(N^l))$ of two manifolds M^k and N^l is the degree of the Gauss map $\chi: M^k \times N^l \to S^{k+l}$.

To construct the map $\chi(x, y)$, $x \in M^k$, $y \in N^l$, as in formula (2.38), we have to take the segment from the point $f(x)$ to $g(y)$, where $f(x), g(y) \in R^{k+l+1}$, translate it into the origin, and take the point where it meets S^{k+l}.

If M^k and N^l are submanifolds in R^{k+l+1} and the maps f and g are the identity, then the linking number is denoted by $k(M, N)$. The following is obvious.

1. If the manifolds $f(M)$ and $g(N)$ are disjoint under the homotopies f_t and g_t, then the linking number is unaltered.
2. The anticommutativity of the linking number

$$k(g(N^l), f(M^k)) = (-1)^{(k+1)(l+1)} k(f(M^k), g(N^l)). \tag{2.39}$$

To prove this, we consider the map $\eta: M \times N \to N \times M$ such that $(x, y) \to (y, x)$. It is easy to see that the map $\chi': N \times M \to S^{k+l}$ is the composition

$$\chi' = \varepsilon\chi\eta \tag{2.40}$$

of η, the antipodal map ε of the sphere S^{k+l} into itself (each point is sent into the diametrically opposite one), and the Gauss map $\chi: M \times N \to S^{k+l}$.

$\deg \eta = (-1)^{kl}$ and $\deg \varepsilon = (-1)^{k+l+1}$ thus entailing (2.39).

Exercise 2.8 Show that the orientability of RP^n is determined by the degree of the antipodal map $\varepsilon: S^n \to S^n$.

Linking numbers play an important role in studying defects in liquid crystals and superfluid liquids. No less interesting applications arise in the physics of polymers and DNA theory [FV]. The corresponding results are mentioned in Chs. 5 and 6.

We now turn to the Hopf invariant, which is intimately related to the concept of link.

2.6.2 Hopf Invariant

Given a Hopf fibration $f: S^3 \to S^2$, consider two regular points, a and b, on the sphere S^2 and take their preimages $l_a^1 = f^{-1}(a)$ and $l_b^1 = f^{-1}(b)$.

The manifolds l_a^1 and l_b^1 are two closed curves in S^3. Consider the linking number $k(l_a^1, l_b^1)$.

Definition 2.29 The *Hopf invariant* $h(f)$ of f is the linking number $k(l_a^1, l_b^1)$.

Theorem 2.26 *$h(f)$ is the homotopic invariant of f and is independent of the choice of a, b.*

The theorem is valid in a considerably more general situation of a $(2n-1)$-dimensional Hopf fibration $f: S^{2n-1} \to S^n$, to be considered below.

Remark 2.13 The linking number $k(M^k, N^l)$ has been given in the preceding subsection for manifolds in Euclidean space R^{k+l+1}. It is easy to see that a similar definition is possible for manifolds in S^{k+l+1}, for which it suffices to consider the stereographic projection

$$S^{k+l+1} \setminus s_0 \to R^{k+l+1}, \quad s_0 \in S^{k+l+1}.$$

It is obvious that the point s_0 can be selected so that $f^{-1}(s_0)$ does not belong to the preimages $f^{-1}(a)$ and $f^{-1}(b)$.

Remark 2.14 The Hopf invariant $h(f)$ is defined for a $(2n-1)$-dimensional Hopf fibration similarly to the fiber bundle $S^3 \to S^2$. It should only be noticed that the inverse preimages of two points in an n-dimensional sphere are $(n-1)$-dimensional closed submanifolds M_1^{n-1} and M_2^{n-1}.

Outline of the proof of Theorem 2.26 We show that the Hopf invariant is unaltered under a homotopy of the map f.

1. Let f_0 and f_1 be two mutually homotopic maps $S^{2n-1} \to S^n$ and f_t the connecting homotopy. To prove that $h(f_0) = h(f_1)$, it suffices to show that the deformation $f_t: S^{2n-1} \times I \to S^n$ connecting f_0 to f_1 and not passing through the points a and b can be constructed. Then the submanifolds $f_t^{-1}(a)$ and $f_t^{-1}(b)$ are disjoint under the homotopy; therefore, the linking number is unaltered.

2. The independence of $h(f)$ from the choice of regular values a and b in S^n is proved quite simply. Let a_1 and b_1 be two other regular points in S^n. There exists a map $\gamma: S^n \to S^n$ of the sphere onto itself, homotopic to the idendity (since $\pi_1(S^n) = 0$) and such that $\gamma(a) = a_1$ and $\gamma(b) = b_1$. Then the maps f and γf are homotopic; therefore, $h(f) = h(\gamma f)$.

We now formulate the following general result, a consequence of the anticommutativity of the linking number and Theorem 2.26.

Corollary 2.3 *Let $S^{2n-1} \to S^n$ be a Hopf fibration. If n is odd, then $h(f) = 0$.*

Since $h(f)$ does not depend on the choice of points a, $b \in S^n$, we interchange them. Then $k(M_0^{n-1}, M_1^{n-1}) = k(M_1^{n-1}, M_0^{n-1})$. However, the linking number is $k(M_0^{n-1}, M_1^{n-1}) = (-1)^{n^2} k(M_1^{n-1}, M_0^{n-1}) = 0$ due to (2.39).

Let us see how the Hopf invariant behaves under a map g of the sphere S^{2n-1} into itself with degree σ and under a map j of the sphere S^n into itself with degree τ. Let f be a map $S^{2n-1} \to S^n$, $f' = jfg$.

Proposition 2.23

$$h(f') = \sigma\tau^2 h(f)$$

(see the proof in [Mil7]).

2.6.3 Integral Representation of the Hopf Invariant

The Hopf invariant $h(f)$ admits an interesting integral representation due to Whitehead [Wh, Whi]. The result has applications in magnetohydrodynamics, chiral fields, liquid crystals, and other branches of physics.

The Hopf invariant acts as the topological conservation law (see Chs. 4, 5 and the literature cited).

First, we formulate the Whitehead result for the classical Hopf fibration.

Theorem 2.27 (Whitehead) *Let ω^2 be a normalized 2-form on S^2, i.e., $\int_{S^2} \omega^2 = 1$ and $f: S^3 \to S^2$ is a smooth map. Consider the 2-form $\Omega^2 = f_*(\omega^2)$ induced by f on S^3. The form is exact on S^3, i.e., $f_*(\omega^2) = d\omega^1$, where ω^1 is a 1-form. Then*

$$\int f_*(\omega^2) \wedge \omega^1 = h(f). \tag{2.41}$$

The multidimensional generalization of the Whitehead formula for the Hopf fibration $S^{2n-1} \to S^n$ is the content of the following.

Theorem 2.28 (Whitehead) *Let ω^n be a normed n-form on S^n, i.e., $\int_{S^n} \omega^n = 1$. There exists the n-form $f_*\omega^n$ induced by f and exact, i.e., $f_*\omega^n = d\xi^{n-1}$, where ξ^{n-1} is a form on S^{2n-1}. Then*

$$q(f) = \int (f_*\omega^n) \wedge \xi^{n-1} = h(f).$$

We give the central ideas of the proof, following Whitney while not proving some of the justifications [Whi].

The *proof* is separated into several steps.

1. That $f_*\omega^n$ is closed is obvious, since $df_*\omega^n = f_*d\omega^n = 0$ (ω^n is a form on S^n). The exactness of the n-form $f_*\omega^n$ follows from the triviality of the cohomology groups $H^i(S^{2n-1}, R)$ for $1 \leqslant i \leqslant 2n-2$.

2. We show that $q(f)$ is an integer.

(a) We first prove that $q(f)$ does not depend on the choice of ξ^{n-1}. Let there exist $\tilde{\xi}^{n-1}$ such that $d\tilde{\xi}^{n-1} = f_*\omega^n$. Then $d(\tilde{\xi}^{n-1} - \xi^{n-1}) = 0$ and $\tilde{\xi}^{n-1} - \xi^{n-1} = d\eta^{n-2}$, where η^{n-2} is an $(n-2)$-form. Since $\partial S^{2n-1} = 0$ and $d(f_*\omega^n) = 0$, making use of the formula for $d(\eta^{n-2} \wedge f_*\omega^n)$, we obtain

$$\int_{S^{2n-1}} \tilde{\xi}^{n-1} \wedge f_*\omega^n - \int \xi^{n-1} \wedge f_*\omega^n = \int d(\eta^{n-2} \wedge f_*\omega^n) - \eta^{n-2} \wedge d(f_*\omega^n)$$

$$= \int_{\partial S^{2n-1}} \eta^{n-2} \wedge f_*\omega^n = 0.$$

(b) We prove that $q(f)$ does not depend on ω^n.

Let $\tilde{\omega}^n$ be cohomologous to ω^n, i.e., $\int_{S^n} (\tilde{\omega}^n - \omega^n) = 0$, $\tilde{\omega}^n - \omega^n = d\nu^{n-1}$, where ν^{n-1} is a form on S^n, e.g., let $d\xi^{n-1} = f_*\omega^n$. Since $\nu^{n-1} \wedge \tilde{\omega}^n = 0$ and $\nu^{n-1} \wedge \omega^n = 0$

on S^n $(2n-1>n)$, $d\xi^{n-1}\wedge f_*\nu^{n-1}=f_*(\omega^n\wedge\nu^{n-1})=0$. We have

$$\begin{aligned}&(\xi^{n-1}+f_*\nu^{n-1})\wedge f_*\tilde{\omega}^n-\xi^{n-1}\wedge f_*\omega^n\\&=\xi^{n-1}\wedge(f_*(\tilde{\omega}^n-\omega^n)+f_*\nu^{n-1}\wedge f_*\tilde{\omega}^n)\\&=\xi^{n-1}\wedge f_*(d\nu^{n-1})=\pm\, d(\xi^{n-1}\wedge f_*\nu^{n-1})\\&\pm\, d\xi^{n-1}\wedge f_*\nu=\pm\, d(\xi^{n-1}\wedge f_*\nu^{n-1}).\end{aligned}$$

Therefore,

$$\begin{aligned}\int_{S^{2n-1}}(\xi^{n-1}+f_*\nu^{n-1})\wedge f_*\tilde{\omega}^n&=\int_{S^{2n-1}}\xi^{n-1}\wedge f_*\omega^n\\&=\int_{S^{2n-1}}\xi^{n-1}\wedge f_*\tilde{\omega}^n+\int_{S^{2n-1}}f_*\nu^{n-1}\wedge f_*\tilde{\omega}^n.\end{aligned}$$

However, $\int_{S^{2n-1}}f_*\nu^{n-1}\wedge f_*\tilde{\omega}^n=0$; hence

$$\int_{S^{2n-1}}\bar{\xi}^{n-1}\wedge f_*\omega^n=\int_{S^{2n-1}}\xi^{n-1}\wedge f_*\tilde{\omega}^n. \tag{2.42}$$

Since $d(\xi^{n-1}+f_*\nu^{n-1})=d\xi^{n-1}+f_*(\tilde{\omega}^n-\omega^n)=f_*(\tilde{\omega}^n)$, the integrals in (2.42) determine $q(f)$ through ω^n and $\tilde{\omega}^n$, respectively, and the two definitions coincide.

(c) Prove that $q(f_0)=q(f_1)$ if f_0 and f_1 are homotopic.

Proof. Let I be the unit interval $[0, 1]$. There exists a continuous map F of the direct product $I\times S^{2n-1}$ (of a manifold with boundary) into the sphere S, so that $F(0, x)=f_0(x)$, $F(1, x)=f_1(x)$.

Assume that F is smooth. Pick a form ω^n on S^n so that $\int\omega^n=1$.

Since $dF_*\omega^n=0$, we can find a form ξ^{n-1} on $I\times S^{2n-1}$ such that $d\xi^{n-1}=F_*\omega^n$. (The existence of ξ^{n-1} follows from the triviality of $H^i(S^{2n-1})$.)

Since $\partial(I\times S^{2n-1})=1\times S^{2n-1}-0\times S^{2n-1}$, we have

$$\begin{aligned}&\int_{1\times S^{2n-1}}\xi^{n-1}\wedge F_*\omega^n-\int_{0\times S^{2n-1}}\xi^{n-1}\wedge F_*\omega^n\\&=\int_{I\times S^{2n-1}}d(\xi^{n-1}\wedge F_*\omega^n)=\int_{I\times S^{2n-1}}d\xi^{n-1}\wedge F_*\omega^n\\&=\int F_*\omega^n\wedge F_*\omega^n=\int F_*(\omega^n\wedge\omega^n)=0.\end{aligned}$$

Considered only on $0\times S^{2n-1}$, the form $F_*\omega^n$ coincides with $f_{0*}\omega^n$ on S^{2n-1}, while ξ^{n-1} on $0\times S^{2n-1}$ yields a form ξ_0^{n-1} such that $d\xi_0^{n-1}=f_{0*}\omega^n$. Similarly,

$d\xi_1^{n-1} = f_{1*}\omega^n$. Therefore,

$$q(f_0) = \int_{S^{2n-1}} \xi_0^{n-1} \wedge f_{0*}\omega^n = \int_{0\times S^{2n-1}} \xi^{n-1} \wedge F_*\omega^n = \int_{1\times S^{2n-1}} \xi^{n-1} \wedge F_*\omega^n = q(f_1). \quad ■$$

The case where f is continuous follows from the lemma below.

Lemma 2.7 *If two smooth maps f_0 and f_1 of a smooth manifold M to a smooth manifold N are homotopic, then they are smoothly homotopic* (see the proof in [Whi]).

We now show that the quantity determined by integral (2.41) is the same as the Hopf invariant introduced above in terms of the linking number. We need some preliminary information about smooth mappings of manifolds.

Definition of a normal map. A *map $f: N^n \to M^m$ is said to be normal* if, for each point $a \in N$ and $b = f(a)$, the map of tangent spaces $T_aN \to T_bM$ is that onto the whole of T_bM. For any point $x \in M$, f is normal at x if f is normal at each point of the preimage $f^{-1}(x) = y$. Points x at which f is normal are said to be *regular.*

Lemma 2.8 *Given a smooth map $f: N \to M$ and $x_0 \in M$* (dim $N \geqslant$ dim M), *there exists a smooth map $\tilde{f}$ arbitrarily close to f, homotopic to f, and normal at x_0*. The proof makes use of Sard's lemma or Morse's theorem [Mil4, GG].

Lemma 2.9 *Let f be a smooth map $N^n \to M^m$ of manifolds. Assume that f is normal at a point $x_0 \in M$. Then $N_{x_0}^{\perp} = f^{-1}(x_0)$ is a smooth $(n - m)$-dimensional manifold in N^n.*

The *proof* can be obtained locally in neighborhoods of points $z_0 \in R^n$ and $f(z_0) \in R^m$, followed by the application of the implicit-function theorem [Whi].

We retrace our steps to the proof that the two definitions of the Hopf invariant are equivalent. We select a regular point b in the base space S^n of the Hopf fibration $f: S^{2n-1} \to S^n$. Let f be normal over b. By Lemma 2.9, the manifold $N_b^{\perp} = f^{-1}(b)$ is a smooth $(n - 1)$-dimensional (not necessarily connected) submanifold in S^{2n-1}.

We orient $N_b^{\perp}$ as follows: we select a basis $e_1', \ldots, e_{2n-1}'$ determining the positive orientation of the sphere S^{2n-1} contained by space R^{2n} in a neighborhood of the point a, $b = f(a)$, so that grad $f(a, e_i') = 0$ for $i \leqslant n - 1$ and the vectors $e_1 = \text{grad}\, f(a, e_n'), \ldots, e_n = \text{grad} f(a, e_{2n-1}')$ define the positive orientation of the sphere S^n. Then $e_1', \ldots, e_{n-1}'$ determine the positive orientation of $N_b^{\perp}$.

We show that if ω^n is an n-form on S^n so that $\int \omega^n = 1$ and if $d\xi^{n-1} = f_*\omega^n$ in S^{2n-1} and the map f is normal over b, then

$$q(f) = \int_{S^{2n-1}} \xi^{n-1} \wedge d\xi^{n-1} = \int_{N_b^{\perp}} \xi^{n-1}. \tag{2.43}$$

The existence of a neighborhood $\mathscr{U}$ of b so that f is normal over each point $b' \in \mathscr{U}$ follows from the normality. Then all $N_{b'}^{\perp}$ ($b' \in \mathscr{U}$) are smooth manifolds which form the "fiber bundle" of $f^{-1}(\mathscr{U})$.

We assume the neighborbood $\mathscr{U}$ to be connected.

First, we assume that $\omega^n = 0$ outside of $\mathscr{U}$. Given a point $b' \in \mathscr{U}$, let A be a smooth arc in $\mathscr{U}$, joining b to b'. The set $f^{-1}(A)$ is a smooth n-dimensional manifold N with boundary $N_{b'} - N_b$.

If we put $f_A = f|_{N_A}$, then $f_A: N_A \to A$. For $n > 1$, the Jacobian of the map f_A vanishes at all N_A. Therefore, $f_{A*}\omega^n = 0$ on N_A and

$$\int_{N_{b'}^{\perp}} \xi^{n-1} - \int_{N_b^{\perp}} \xi^{n-1} = \int_{\partial N_A} \xi^{n-1} = \int_{N_A} d\xi^{n-1} = \int_{N_A} f_*\omega^n = 0. \tag{2.44}$$

Given a point $a \in N_b$, we select two bases, $e_1', e_2', \ldots, e_{2n-1}'$ and $e_1, \ldots, e_n$ as above. Since grad $f(a, e_i') = 0$ for $i < n$, $f_*\omega^n(a) \cdot e_n' \ldots e_{2n-1}' = \omega^n(f(a))e_{1\ldots n}$ is a unique nonzero component of the form $f_*\omega^n(a)$. Hence, by the familiar relation $[\omega_1 \wedge \omega_2] \cdot e_{1\ldots n} = \langle \omega_1 | e_{1\ldots m} \rangle \cdot \langle \omega_2 | e_{m+1\ldots n} \rangle$, we obtain

$$[\xi^{n-1}(a) \wedge f_*\omega^n(a)] \cdot e_{1\ldots 2n-1}' = \langle \xi^{n-1}(a) | e_{1\ldots n-1}' \rangle \cdot \langle \omega^n(f(a)) | e_{1\ldots n} \rangle.$$

Therefore, by representing integral (2.42) as iterated one and taking into account (2.43), we find that

$$\int_{S^{2n-1}} \xi^{n-1} \wedge f_*\omega^n \quad \int_{f^{-1}(\mathscr{U})} \xi^{n-1} \wedge f_*\omega^n = \int_{\mathscr{U}} \omega(b') \int_{N_{b'}^{\perp}} \xi(a)da\, db'$$

$$= \int_{N_b^{\perp}} \xi^{n-1} \cdot \int_{\mathscr{U}} \omega^n = \int_{N_b^{\perp}} \xi^{n-1},$$

which proves the formula for $q(f)$ in the case of $\omega = 0$ on $S^{2n-1} \setminus \bar{\mathscr{U}}$. The general case is reduced to the above if we notice that with the help of an appropriate partition of unity (see [Rh]) the form ω^n can be made cohomologous to $\tilde{\omega}^n$, which vanishes outside of $\mathscr{U}$.

To complete the proof, it remains for us to show that the integral $\int_{N_b^{\perp}} \xi^{n-1} = q$ is the degree of the map f; therefore, q is an integer. We choose f to be normal over b. Let C^n be the differentiable chain bounded by N_q, which means that there exists a cell complex K, an n-dimensional chain C_0^n in K and a smooth cellular map $\varphi: K \to S^{2n-1}$ such that $\varphi(C_0^n) = C^n$, $N_b^{\perp} = \varphi(\partial C_0^n)$, the latter always existing for mappings to a sphere. Then

$$\int_{N_b^{\perp}} \xi^{n-1} = \int_{\varphi(\partial C_0^n)} \xi = \int_{(C^n)} f_*\omega^n = \int_{f(C^n)} \omega^n \tag{2.45}$$

(by duality). Since f maps $\varphi(\partial C_0^n) = N_b^{\perp}$ into a point, $f(C^n)$ is an n-cycle on S^n, equal to the degree of the map f given on S^n. It then follows from (2.45) that

$$q(f) = \int_{\deg S^n} \omega^n = \deg f \int_{S^n} \omega^n = \deg f.$$

It follows from Lemma 2.9 and the properties of the degree of a map that this is valid for any continuous map f. ∎

2.6.4 Application of the Hopf Invariant. New Examples

Two equivalent classical problems in mathematics lead to the Hopf invariant: construction of division algebras and finding all parallelizable spheres S^n. The latter is a particular case of the problem of the maximal number of linearly independent fields on S^n. It turns out that their solution is related to the following property.

Proposition 2.24 *Spheres S^{2n-1} are parallelizable if and only if there exists a Hopf bundle $p: S^{2n-1} \to S^n$ with Hopf invariant $h(f) = 1$.*

Adams has proved that this only holds in the three classical cases $n = 2, 4, 8$ (and of course $n = 1$).

The *proof* of this exceptionally complicated theorem uses the whole power of modern algebraic topology (K-theory, cohomology operations, etc.). The best treatment is in [At1], [Sc].

Adams has also solved the problem of the maximum number of linearly independent fields on an arbitrary odd-dimensional sphere [Ad], [At1].

Mostly confining ourselves to the sphere S^3, we clarify what distinguishes the manifolds S^1, S^3, S^7, and S^{15} from the other spheres of the set.

Existence of the Hopf map $h(f) = 1$ for $S^3 \to S^2$. The Hopf fibration realized by a pair of complex numbers (z_1, z_2) and $|z_1|^2 + |z_2|^2 = 1$ has been constructed in Subsec. 2.6.1. We see that the preimages of points in the sphere S^2 are the linked unit circles S_1^1 and S_2^1. We can prove that $k(S_1^1, S_2^1) = 1$.

We show that the existence of a Hopf fibration with $h(f) = 1$ is equivalent to that of three linearly independent fields on S^3; hence, S^3 is parallelizable.

We select a point $z = (z_1, z_2) \in S^3$ and construct a field $v_h = i(z_1, z_2)$ on S^3 at z. A simple calculation shows that v_h is the vector tangent at z to S^3. The field can be called Hopf, since the integral curves $z(t) = (z_1 e^{it}, z_2 e^{it})$ are closed periodic and form fibers of the Hopf fibration. We obtain the 3-frame on S^3 with the help of two fields orthogonal to v_h.

Note that the parallelizability of S^3 also follows from the isomorphism $S^3 \simeq SU(2)$, since all Lie groups are parallelizable (see Subsec. 1.2.2).

Remark 2.15 All orbits of v_h are closed. The Seifert conjecture has long been of interest. It states that each non-singular field of class $C^r \geqslant 1$ on S^3 has a closed (periodic) orbit. In 1972 Schweitzer [Sz], [Tam] constructed on S^3 a vector C^1-field on which there is not a single periodic orbit. The problem remains open for C^r-fields, $r > 1$.

Remark 2.16 A vector field v_h arises in the following problem in mechanics. Consider the motion of a spatial pendulum with two degrees of freedom.

The Hamiltonian function is

$$H = (1/2)(p_1^2 + p_2^2 + q_1^2 + q_2^2)$$

and the Hamiltonian equation is

$$\dot{q}_i = p_i, \quad \dot{p}_i = -q_i.$$

It follows from the law of conservation of energy $H(p, q) = C = \text{const}$ that the motion occurs on the sphere S^3 in the phase space $(p_1 q_1, p_2 q_2)$.

The Hamiltonian equations determine the tangent field v (velocity in the phase space) on S^3 which is naturally non-singular, since $c \neq 0$. The field v coincides with the Hopf field v constructed above.

Quaternion Hopf fibration. By analogy to the fiber bundle $S^3 \to S^2$, we can construct the Hopf fibration $S^7 \to S^4$ related to the quaternion field.

Quaternion algebra (field) is the only associative division algebra along with the fields of real and complex numbers (the classical result due to Frobenius).

Quaternions only arise in physics episodically. However, they can be regarded as coefficient domains along with real and complex numbers. Other applications of quaternions are related to their introduction into a Clifford algebra as a subalgebra [Ze]. Quaternions appear in this book when calculating the fundamental group of liquid crystals of two types, cholesterics and biaxial nematics (see Sec. 5.1).

We recall the basic operations on quaternions.

Definition 2.30 *Quaternions* are elements of a set H, representable as $q = \alpha_0 \cdot 1 + \alpha_1 i + \alpha_2 j + \alpha_3 k$, where α_i are real numbers and the generators 1, i, j, k are units of H satisfying the relations

$$1 \cdot i = i,\ 1 \cdot j = j,\ 1 \cdot k = k,\ ij = -ji = k,\ ik = -ki = -j,$$
$$jk = -kj = i,\ i^2 = j^2 = k^2 = -1. \tag{2.46}$$

The notation of quaternion units is due to the British mathematician Hamilton, who discovered quaternions in 1843. The letter H is employed in his honor. In modern symbols, (2.46) can be written in more compact form. Let e_0, e_1, e_2, e_3 be the quaternion units. Then (2.46) is

$$e_0^2 = 1, \quad e_i^2 = -1\ (i = 1, 2, 3), \quad e_i e_j = \varepsilon_{ijk} e_k \tag{2.47}$$

(where ε_{ijk} is the unit skew-symmetric tensor).

Multiplication of a quaternion by a scalar α and addition of quaternions are defined similarly to usual vectors. We can define the product of two quaternions $q = \alpha_i e_i$, $q' = \beta_j e_j$ by the formula

$$qq' = \alpha_i \beta_j e_i e_j. \tag{2.48}$$

The set H thereby turns into the algebra of quaternions. It follows from (2.46) that H is not a commutative but an associative algebra containing the field of real numbers $R = \{\alpha_0 e_0\}$ and that of complex numbers $C = \{\alpha_0 e_0 + \alpha_1 e_1\}$ as two subalgebras.

H admits the isomorphic matrix representation

$$e_0 = \begin{pmatrix} 1 & 0 \\ 0 & 1 \end{pmatrix},\ e_1 = i\begin{pmatrix} 0 & 1 \\ 1 & 0 \end{pmatrix} = i\sigma_x,$$
$$e_2 = i\begin{pmatrix} 0 & i \\ -i & 0 \end{pmatrix} = i\sigma_y,\ e_3 = i\begin{pmatrix} 1 & 0 \\ 0 & -1 \end{pmatrix} = i\sigma_z \tag{2.49}$$

in terms of the Pauli matrices σ_x, σ_y, σ_z, where $i = \sqrt{-1}$.

For each quaternion $q = \alpha_0 e_0 + \alpha_1 e_1 + \alpha_2 e_2 + \alpha_3 e_3$, the conjugate is $\bar{q} = \alpha_0 e_0 - \alpha_1 e_1 - \alpha_2 e_2 - \alpha_3 e_3$ and the norm

$$q = N(q) = q\bar{q} = |q|^2 = \alpha_0^2 + \alpha_1^2 + \alpha_2^2 + \alpha_3^2 \tag{2.50}$$

is defined.

The inverse quaternion of q is $q^{-1} = \bar{q}/N(q)$. It follows from formula (2.50) that the set of quaternions with norm $N(q) = 1$ is isomorphic to S^3. We denote the set by ${}_1H$.

Each nonzero quaternion has its inverse. Algebra with such a property is called a *division algebra.*

Along with the fields R and C, the algebra H is the only associative division algebra (*Frobenius' theorem*). It follows from Frobenius' theorem that, together with real and complex numbers, quaternions can be regarded as coefficient domains in various problems of the representation theory of groups, topology, and physics.

We now return to the quaternion Hopf fibration $p: S^7 \to S^4$. Consider the two-dimensional quaternion space H^2 (dim $H = 8$). Realize the unit sphere $S^7 \subset H^2$ as the space of quaternions with the unit norm. The Hopf fibration is realized as

$$p: S^7 \to HP^1, \tag{2.51}$$

where HP^1 is the quaternion projective line, an analog of the complex projective line CP^1 equivalent to S^2 (the Riemann sphere).

The quaternionic projective space HP^n corresponding to the space H^{n+1} is defined in a similar manner to the real, RP^n, or complex, CP^n, projective spaces. We introduce an equivalence relation for the elements of H^{n+1}. Let h_1 and h_2 be two nonzero elements of H^{n+1}; we say that $h_1 \sim h_2$ if there exists a scalar $\lambda \in H$ such that $h_2 = \lambda h_1$. The relation is reflexive, symmetric, and transitive and therefore separates the set of nonzero elements of H^{n+1} into equivalence classes, i.e., the points of the new space HP^n. Consider the map

$$f: S^{4n-1} \to HP^n, \tag{2.52}$$

which associates each of the points $s_0 \in S^{4n-1}$ with its equivalence class. Since any nonzero vector can be normalized to unity, f is an epimorphism (a map onto HP^n). It is obvious that the pre-image of a point $h_i \in HP^n$ is the set of quaternions with norm $N = 1$. Since the set is isomorphic to the three-dimensional sphere S^3, we obtain the fiber map $S^{4n-1} \to HP^n$ with fiber S^3. It is easy to show (this can be done by the reader independently; see also [Stee]) that (2.52) is a fiber bundle with base space HP^n and fiber ${}_1H$. The group G acting transitively on S^{4n-1} is the symplectic group $\mathrm{Sp}(n)$. The stability subgroup G_0 at a point $h_i \in HP^n$ is $\mathrm{Sp}(1) \times \mathrm{Sp}(n-1)$. Fiber bundle (2.52) can then be represented as

$$S^{4n-1} \xrightarrow{\mathrm{Sp}(1)} \mathrm{Sp}(n)/(\mathrm{Sp}(1) \times \mathrm{Sp}(n-1)), \tag{2.53}$$

or

$$\mathrm{Sp}(n)/\mathrm{Sp}(n-1) \xrightarrow{\mathrm{Sp}(1)} \mathrm{Sp}(n)/\mathrm{Sp}(1) \times \mathrm{Sp}(n-1). \tag{2.54}$$

The realization shows that HP^n can be regarded as a special case of a quaternionic Grassmannian

$$G_{n,k}^{H} = \mathrm{Sp}(n)/\mathrm{Sp}(k) \times \mathrm{Sp}(n-k).$$

It is seen from (2.53) that the fiber bundle $(S^{4n-1}, f, \mathrm{Sp}(1), HP^n)$ is principal. We retrace our steps to quaternionic Hopf fibration (2.51). The space HP^1 can be identified with the sphere S^4 by completing the quaternionic space H with the point at infinity. We thereby find that the Hopf fibration

$$p\colon S^7 \to S^4$$

is equivalent to (2.51).

The Hopf fibration for $n = 8$, which associated with the Cayley algebra, can be realized in a more complicated way. The algebra is not associative and analogs of the projective space over Cayley numbers cannot be formed any more. The sphere S^{15} can nevertheless be embedded into the space of pairs (a, b) of Cayley numbers, forming the completion of the eight-dimensional space of Cayley numbers Ca, when turning the space into the sphere S^8 (for the detail see [Stee]).

Hopf fibrations are characterized by another important property: universality.

1. The fiber bundle $S^3 \xrightarrow{S^1} S^2$. A Hopf fibration with the fiber group $G = S^1$ is principal. Since $\pi_i(S^3) = 0$ for $0 \leqslant i < 3$, $S^3 \to S^2$ is 3-universal for the bundle with group $U(1) \sim S^1$.

2. Similarly, the fiber bundle $S^7 \to S^4$ is 7-universal for a fiber bundle with group $SU(2)$.

The universal fiber bundle is an extremely important object since it reduces the study of fiber bundles over a fixed base space B and the structure group G to one universal fiber bundle U. The other bundles are obtained by a factorization of U under the action of a subgroup H of G. The simplest example is the universal covering space $\tilde{M}$ of the manifold M, i.e., a 2-universal fiber bundle.

We now give precise definitions.

Definition 2.31 A principal bundle $E \xrightarrow{G} B$ is said to be *n-universal* if and only if $\pi_i(E) = 0$ for all $0 \leqslant i < n$. The base space B is said to be *n-classifying* and is denoted by BG. If we put $n = \infty$, then we will speak simply of *universal* and *classifying* spaces. The universality property has already been encountered in Subsec. 2.1.7 defining the n-connection of the space E. Examples of universal and classifying spaces are real, complex, and quaternionic Stiefel and Grassmannian ones, respectively. We restrict ourselves to real and complex spaces.

1. The real case is

$$V_{n,k} \xrightarrow{SO(k)} G_{n,k}.$$

The bundle is $(n-k-1)$-universal.

2. The complex case is

$$V_{n,k}^{C} \xrightarrow{U(k)} G_{n,k}^{C}.$$

The bundle is $(2n - 2k + 1)$-universal. It is obvious that the spaces $V_{\infty,k}$ and $V^C_{\infty,k}$ are universal.

We now state fundamental classification.

Theorem 2.29 *For any (e.g., compact) manifold M admitting a finite cellular decomposition the set* Map $[M, BG]$ *of the homotopy classes of maps $M \to BG$ is in a one-to-one correspondence with the equivalence classes of principal bundles over M with group G.*

In particular, it follows from the theorem that the classical Hopf bundle is 3-universal for fiberings over the sphere S^2 with fiber S^1. Since Map $[S^2, S^2]$ is characterized by one parameter (the group Z), all fiberings of S^2 with fiber S^1 are characterized by integers called *characteristic numbers.* If $n_1 \neq n_2$ and $n_1, n_2 \in Z$, then the corresponding fiber bundles are not equivalent.

We shall show that the total space for a characteristic number n is the lens space $L^2(S^3, n)$.

Consider the principal bundle

$$S^3 \xrightarrow{Z_n} S^3/Z_n,$$

where the action is given by the formula

$$(w^0, w^1, w^2, w^3) \xrightarrow{\varphi} \left(e^{\frac{2\pi i}{n}} w^0, e^{\frac{2\pi i}{n}} w^1, e^{\frac{2\pi i}{n}} w^2, e^{\frac{2\pi i}{n}} w^3\right),$$

and w^i are the coordinates of a point $w \in S^3$. It is obvious that the bundle is principal with fiber Z_n.

$L^2(S^3)$ is universal for $G = Z_n$. Therefore, $L^2(S^3, n)$ can be regarded as classifying for the group Z_n. On the other hand, the space can be obtained from the Hopf fibration. In fact, consider the map $\varphi_0 \colon u \to u^n$ of the two-dimensional sphere S^2 into itself, realized as the Riemann sphere. Let p be a Hopf map. If each point $u \in S^2$ is assigned $\tilde{u} = p^{-1}\varphi$, then we obtain a lens space $L^2(S^3, n)$. Since two maps $u \to u^{n_1}$ and $u \to u^{n_2}$ are not homotopic for $n_1 \neq n_2$ (they have different degrees), the spaces $L^2(S^3, n_1)$ and $L^2(S^3, n_2)$ are not isomorphic. The reasoning can be summarized as the statement that the diagram

$$\begin{array}{ccc} S^3 & \xrightarrow{\varphi} & S^3/Z_n \\ p \downarrow & & p\downarrow \\ S^2 & \xrightarrow{\varphi_0} & S^2 \end{array}$$

is commutative.

Lens spaces arise when investigating the geometric properties of a Dirac monopole (see Subsec. 4.1.1).

If principal bundles with group G over M are known, then we can also construct the associated ones. The construction is as follows:

Let

$$E \to M \tag{2.55}$$

be a bundle with fiber F and G a fiber structure group. The bundle is associated with a principal bundle $P \xrightarrow{G} M$ via the action of G in the fiber

$$G \times F \to F. \tag{2.56}$$

The principal bundle is induced by a map $f: M \to BG$ and $P = f'EG$.
We now construct the fiber bundle

$$E'G \to BG \tag{2.57}$$

with fiber F and structure group G, associated with the universal principal bundle, via action (2.56). Fiber bundle (2.55) is then induced by f and by $E'G \to BG$.

We have thereby obtained a correspondence between the homotopy classes of $M \to BG$ and equivalence classes for fiber bundles over M, with fiber F and group G acting in a fixed way in F as in (2.56).

In the general case, this correspondence is not one-to-one, since non-homotopic map of M into BG can be associated with equivalent fiber bundles (e. g., if F consists of one point). However, this correspondence is one-to-one if the action of G into F is free (according to (2.56)), e. g., for vector bundles with fiber R^n if the structure group G is $GL(n, k)$ or some of its subgroups act with respect to the identity representation.

The importance of introducing universal spaces becomes obvious when the problem of classification of all fiber bundles over a certain base space and fiber G is solved. Another important application of universal bundles is given in the next subsection, where the study of singularities of vector fields and other structures in fiber bundles is shown to be reducible to the study of series of universal fiberings.

In conclusion, consider the problem of the classification of fiberings of the sphere S^4, which has important applications (see Sec. 4.2). The example is quite typical and, therefore, deserves consideration from different viewpoints. Meanwhile we adduce certain facts characteristic of fiber bundles over an arbitrary base space.

1. We notice that for an arbitrary universal n-bundle

$$EG \xrightarrow{G} BG, \tag{2.58}$$

where $\pi_1(BG) = \pi_{i-1}(G)$, $i \leqslant n - 1$, which is immediately apparent from the exactness of the sequence of homotopy groups.

By Theorem 2.29, the problem of classification of principal bundles of spheres S^k is reduced to calculating the set Map (S^k, BG), i.e., $\pi_k(BG)$. As we know, for each series of classical Lie groups $O(n)$, $SO(n)$, $U(n)$, $\mathrm{Sp}(n)$, the corresponding universal and classifying spaces are the Stiefel (EG) and Grassmann (BG) manifolds.

Hence, the corresponding fiber bundles of S^4 are classified by the group $\pi_3(G)$, and in the general case of S^n, by $\pi_{n-1}(G)$, e. g., the universal Hopf fibration $S^7 \to S^4$ is realized as the universal bundle $G = \mathrm{Sp}(1) \sim SU(2)$, $EG = S^7 \subset H^2$, and $BG = HP^1 \sim S^4$. The bundle is characterized by the integer $\pi_3(SU(2)) = Z$. Since

TABLE 4

G	EG	BG
$O(n)$	$V_{\infty,n}$	$G_{\infty,n}$
$SO(n)$	$V_{\infty,n}$	$\hat{G}_{\infty,n}$
$U(n)$	$V^{C}_{\infty,n}$	$G^{C}_{\infty,n}$
$\mathrm{Sp}(n)$	$V^{H}_{\infty,n}$	$G^{H}_{\infty,n}$

Here $V_{\infty,n}$, $G_{\infty,n}$, ... were introduced in Subsec. 2.1.7.

the homotopy groups of classical Lie groups are known (the corresponding results given in Table 1), we can classify all the principal bundles over spheres. The corresponding results for fiber bundles of S^4 are used in Ch. 4.

It is noteworthy that classification of principal bundles over S^n can be obtained directly without the construction of universal bundles.

Lemma 2.10 *A principal bundle P over a disk D^n and its associated bundles are equivalent to the direct product.*

Proof. Let x_0 be the center of D^n which can be contracted to x_0 along the radii. Pick an arbitrary point x and consider the interval $[x_0, x]$. Let G_x be the fiber of P. The mapping of the direct product $\varphi: D^n \times G_x \to P$, defined by the formula

$$\varphi(x, g) = \varphi_{[x_0, x]}g$$

for $g \in G_x$, introduces the structure of a direct product in P. ∎

Remark 2.17 The general fact is valid that if the base space B is contractible to a point, then the fiber bundle over B is trivial, since the triviality is equivalent to the existence of a section s, and, given the contraction of the base, s is deformed into a section s' over the point. The fiber bundle over the point is trivial.

Since the sphere S^n is covered by two disks, D^n_+ and D^n_-, intersecting in the equator S^{n-1}, the fiber bundle over S^n is determined by two coordinate neighborhoods of D^n_+ and D^n_- with the gluing function $\lambda_{12}: D^n_+ \cap D^n_- \to G$. The function λ_{12} defines the map $\varphi_{12}: S^{n-1} \to G$ by the formula

$$\lambda_{12}(x, y) = (x, \varphi_{12}(x)y), \quad x \in S^{n-1}, \ y \in G.$$

Lemma 2.11 *During the homotopy for the function φ_{12}, the fiber bundle equivalence class does not change.*

Proof. We construct the cylindrical strip $S^{n-1} \times [-\varepsilon, \varepsilon]$ by extending the disks D^n_+ and D^n_- below and above the equator, respectively. Consider the homotopy

$$\psi_t: S^{n-1} \times [-\varepsilon, \varepsilon].$$

It is obvious that the fiber bundle constructed by composition of the maps ψ_t and φ_{12} is equivalent to the original one. ∎

The theorem follows from Lemmas 2.10 and 2.11. All principal bundles of the sphere S^n with group G are classified by the group $\pi_{n-1}(G)$.

We now study the topological invariants related to the finer properties of manifolds.

2.7 CHARACTERISTIC CLASSES

The existence of complex and spin structures on manifolds, the generalization of theorems on singularities of vector fields, the embeddability of one manifold in another, various relations between differential-geometrical and topological properties of manifolds are in some way or another connected to the important topological invariants, i. e., characteristic classes. Here, we give a brief analytic treatment of certain facts from this basic topological theory.

We begin with the classical example of the Gauss-Bonnet theorem (1848).

2.7.1 Gauss-Bonnet Theorem

Let M^2 be a closed orientable two-dimensional manifold and K its Gaussian curvature.

Theorem 2.30 (Gauss-Bonnet)

$$\frac{1}{2\pi}\int_{M^2} K\,dS = \chi(M^2). \tag{2.59}$$

The proof can be obtained by using the formula for the degree of a map of a surface and applying the Morse index theorem according to the following scheme (e. g., for detail see [DFN]).

(a) We embed M^2, a pretzel with g holes, in R^3 (Fig. 1) and consider the height function f. The indices of critical points of f are: $\operatorname{Ind} f|x_0 = -2$ for one minimum x_0, $\operatorname{Ind} f|x_{2g+1} = 0$ for one maximum x_{2g+1}, and $\operatorname{Ind} f|x_i = -1$ for $2g$ saddles. For any Morse function, the formula

$$\deg f = (1/2)\sum(-1)^{\alpha(x_j)} \tag{2.60}$$

holds, where $\alpha(x_j) = 0$ for the minimum and maximum of f and $\alpha(x_j) = 1$ for the saddles. For f, $\deg f = 2 - 2g$. On the other hand, the degree of the map given by f is $1/(4\pi)\int K\,ds$ (see Proposition 2.19), where $1/(4\pi)$ arises from normalizing to unity the volume of the sphere S^2 and (2.59) follows.

Formula (2.60) holds for any Morse function and the Gaussian curvature does not depend on the embedding of M^2 in R^3. It remains to note that for any manifold M^2 there exists a Morse function and to use the Morse equality $\deg f = \chi(M)$. ∎

We now give another proof of the theorem that admits a multidimensional generalization and takes us closer to the concept of characteristic class [MSt].

We specify on M^2 the bundle ξ^2 of unit vector with the structure group $SO(2)$. Let θ_1 and θ_2 be a local orthonormal basis. The matrix of a connection ω is given by 1-forms which are defined by the condition

$$D\theta_i = \omega_i^j \otimes \theta_j, \ \omega = \begin{pmatrix} 0 & \omega_2^1 \\ -\omega_2^1 & 0 \end{pmatrix}. \tag{2.61}$$

The curvature matrix

$$\Omega = \begin{pmatrix} 0 & \Omega_{12} \\ -\Omega_{12} & 0 \end{pmatrix} \tag{2.62}$$

defined as $D\omega$ consists of 2-forms Ω_{12}, where $d\omega_{12} = \Omega_{12}$.

Recall that the connection ω and curvature Ω matrices are vector-valued forms. For principal bundles with structure Lie group and the associated bundles, matrices ω and Ω take values in the Lie algebra. Here, the group G is $SO(2)$ and we could simply consider real-valued forms. However, with subsequent generalizations in mind, we represent the forms ω and Ω as 2×2 matrices.

It is obvious that (2.62) is invariant under the action $\Omega \to g\Omega g^{-1}$, where $g \in SO(2)$.

Therefore, Ω_{12} does not depend on the choice of an oriented orthonormal basis and, therefore, determines the global 2-form on M^2.

We can give the form $dS = \theta_1 \wedge \theta_2$, the element of area of the surface. Ω_{12} must be proportional to dS, i.e., $\Omega_{12} = K\, dS$, where K is a scalar function shown in (2.30) and (2.31) to be the Gaussian curvature of the surface, and $\Omega_{12} = g_{xx} dx \wedge dy = -\, g_{xx}/g\, dS$.

We now turn to the proof of the Gauss-Bonnet theorem

$$\int \Omega_{12} = \int K\, dS = 2\pi\chi(M^2). \tag{2.63}$$

Proof ([MSt]). Consider an arbitrary real two-dimensional vector bundle on M^2 with the Euclidean metric. Note that ξ^2 can be turned into a one-dimensional complex bundle ζ by introducing the canonical complex structure in the fiber R^2 in the standard way. Let $s_1(x)$, $s_2(x)$ be an oriented local orthonormal basis. The complex structure operator $J: R^2 \to R^2$ is defined as $Js_1 = s_2$ and the vector $e \in C$ is $s_1 + Js_2$, where $J^2 = -1$, J rotates each vector through $\pi/2$ counterclockwise.

We pick the connection ∇ corresponding to the metric on ξ^2 so that $\nabla s_1 = \omega_{12} \otimes s_2$, $\nabla s_2 = -\,\omega_{12} \otimes s_1$. ∇ naturally determines the connection also on ζ as $\nabla s_1 = \omega_{12} \otimes is_1 = i\omega_{12} \otimes s_1$; therefore,

$$\nabla (is_1) = i\nabla (s_1) = -\,\omega_{12} \otimes s_1.$$

The matrix of this complex connection is a 1×1 matrix $[i\omega_{12}]$, while the curvature matrix $[i\Omega_{12}]$. Since $\mathrm{Tr}[i\Omega_{12}] = i\Omega_{12}$, the form $i\Omega_{12}$ is invariant and, therefore, closed. By de Rham's theorem, each closed 2-form determines a cohomology class in $H^2(M^2, C)$. Here, the cohomology 2-class is given on a two-dimensional manifold, i.e., a cycle. The class is said to be *characteristic*.

It is clear from the subsequent treatment of the theory of characteristic classes that a complex one-dimensional manifold has only one characteristic class

$c_1(\zeta) \in H^2(M, C)$ and that all characteristic classes are determined by $c_1(\zeta)$ or its multiples.

The class $c_1(\zeta)$ determines the class $e(\xi_R^2)$ of the real bundle and is the complexification of $c_1(\zeta)$. Therefore,

$$i\Omega_{12} = \alpha c_1(\zeta) = \alpha e(\xi_R^2), \tag{2.64}$$

where α is a complex number.

To find α, it suffices to calculate both sides of the equation for a concrete fiber bundle. We pick the bundle $T^*(M^2)$ cotangent to the two-dimensional manifold M^2. We have

$$\int i\Omega_{12} = \alpha e(T^* M^2) = \alpha e(M^2), \tag{2.65}$$

or

$$i\int K\, ds = \alpha e(M^2). \tag{2.66}$$

The value of the class $e(M^2)$ equals the Euler characteristic $\chi(M^2)$. We take this statement without proof (see [MSt]). We calculate both sides of (2.66) for the unit sphere S^2. Since $K_{S^2} = 1$, $V(S^2) = 4\pi$, $\chi(S^2) = 2$, we find that $\alpha = 2\pi i$.

2.7.2 General Construction of Characteristic Classes

Characteristic classes are the main topological invariants of fiber bundles directly related to differential geometric and analytic structures of fiber bundles.

Essentially, characteristic classes are far-reaching generalizations of two classical results in topology, i.e., the theorems of Gauss-Bonnet and of Poincaré on the index of a vector field on a surface.

An example of a characteristic class is the Euler characteristic of a manifold.

Let ξ be a smooth vector field on M^n with isolated zeros. The multiplicity of a zero in ξ is Ind ξ. Below is the multidimensional generalization of Poincaré's theorem due to Hopf.

Theorem 2.31 $\chi(M^n) = \text{Ind}\ \xi$.

If $x(M) = 0$, then there exists a field ξ on M^n without singularities. The condition $\chi(M) \neq 0$ is thus an obstruction to constructing ξ. On the other hand, the Gauss-Bonnet formula expresses $\chi(M)$ in terms of Gaussian curvature. Multidimensional generalizations of these results lead to the theory of characteristic classes.

Instead of ξ alone, consider k vector fields $\xi_1, \ldots, \xi_k$. In general position, the set of points at which the exterior product $\xi_1 \wedge \xi_2 \wedge \ldots \wedge \xi_k$ is zero, i.e., the vectors are linearly dependent, forms a $(k-1)$-dimensional manifold N^{k-1}. Depending on whether $n-k$ is odd or even, N^{k-1} either defines an integral $(k-1)$-cycle or a cycle with coefficients in Z_2. We can prove (see [Stee], [MSt] that the homology class for the cycle is independent of the choice of vector fields. Since the linear dependence of vector fields can be dually defined in terms of forms, we obtain conditions of cohomology nature*; in particular, we can define characteristic classes

$$w^i \in H^i(M, Z_2), \quad 1 \leqslant i < n - 1, \quad i = n - k,$$

Stiefel-Whitney classes or, e.g., the nth Euler class $e(M^n) \subset H^n(M^n, Z)$, which is related to the Euler characteristic

$$\int_{M^n} e(M^n) = \chi(M^n). \tag{2.67}$$

An analytic approach to the theory of characteristc classes is based on two key ideas of using universal bundles for classical Lie groups and of constructing cohomology classes via connection forms.

The realization technique is the *Weil homomorphism,* which enables us to represent characteristic classes in terms of curvature forms.

It has been shown in Subsec. 2.6.4 that for each series of classical Lie groups $O(n)$, $U(n)$, $\mathrm{Sp}(n)$, there exist universal bundles: real, complex, and quaternionic Stiefel spaces, respectively. The importance of universal bundles is determined by the Classification Theorem 2.24 due to Whitney-Pontrjagin. In particular, the following consequence is essential.

Corollary 2.4 *Let $u \in H^i(G_{N,q}, A)$ be a cohomology class with coefficients in a group A. Then the element $f_* u \in H^i(M, A)$ only depends on the fiber bundle and is called the characteristic class corresponding to the universal one.*

A certain characteristic class can be related to each series of universal bundles. To carry the operation out, we need explicit expressions of cohomology classes in terms of differential forms and the description of the map f_*.

2.7.3 Cohomology Classes and Connection Forms

The principal goal of the subsection is to prove that curvature forms in a principal bundle, associated with different connections, are cohomologous. The corresponding scalar closed 2-form Ω_1 defines a 2-cocycle on the base space and the cohomology classes of the space $H^i(M, R)$ are given by the exterior products of forms Ω_1.

We now turn to precise formulations.

Let (P, G, M^n) be the principal bundle over the base space M^n and Ω a $\mathscr{G}$-valued local curvature form in P for a connection ω.

A connection θ is defined in each coordinate neighborhood $\mathscr{U} \subset M^n$ with respect to ω so that at the intersection of two neighborhoods $\mathscr{U}$ and $\mathscr{V}$ and in transition to another set of coordinates, θ is transformed according to formula (1.55) as

$$\theta_{\mathscr{U}} = g_{\mathscr{U}\mathscr{V}} \theta_{\mathscr{V}} g_{\mathscr{U}\mathscr{V}}^{-1} + dg_{\mathscr{U}\mathscr{V}} g_{\mathscr{U}\mathscr{V}}^{-1} \tag{2.68}$$

* A definition directly in terms of cohomology groups is more general, since certain characteristic classes (e.g., in $H^i(M, Z_2)$) cannot be expressed in terms of differential forms directly.

(assuming the right action of the group G in the fiber). The matrix of curvature forms Ω_4 is meanwhile transformed as

$$\Omega_{\mathscr{U}} = g_{\mathscr{U}\mathscr{V}} \Omega_{\mathscr{V}} g_{\mathscr{U}\mathscr{V}}^{-1}. \tag{2.69}$$

Consider the scalar form $\Omega_1 = \operatorname{Tr} \Omega$, where the trace is taken in the Lie algebra $\mathscr{G}$ of G. Under "gauge transformations" $\operatorname{Tr}(g_{(x)} \Omega g_x^{-1}) = \operatorname{Tr}(\Omega) = \Omega_1$, $x \in M^n$ from G, Ω_1 is unaltered; therefore, the form is defined globally on the whole of M^n.

We now formulate the properties of Ω_1 as follows.

Theorem 2.32

(1) Ω_1 *is closed.*

(2) *In changing the connection* $\theta_\mu \to \theta_\mu'$, Ω_1 *is replaced by an exact form.*

Proof. (1) The form Ω satisfies the Bianchi identity $D\Omega = d\Omega + [\theta, \Omega] = 0$, i.e., $d\Omega = -[\theta, \Omega]$ and $d \operatorname{Tr} \Omega = \operatorname{Tr}(d\Omega) = -\operatorname{Tr}[\theta, \Omega] = 0$, since the trace of the commutator of two matrices is zero.

(2) Let ω and ω' be two connection 1-forms on P. We write the difference $\omega - \omega'$ by means of a structural equation (1.45). Taking into account (1.58), we have $d(\omega - \omega') = d(\omega) - d(\omega') = \Phi - \Phi' - (1/2)[\omega, \omega] - [\omega', \omega']$.

Locally,

$$d\theta - d\theta' = \Omega - \Omega' - (1/2)[\theta, \theta] - (1/2)[\theta', \theta'], \tag{2.70}$$

where Φ and Φ' are the fiber bundle curvature forms associated with the connections ω and ω', respectively. Considering the traces of both sides of equality (2.70) and taking $\operatorname{Tr}[\theta, \theta] = \operatorname{Tr}[\theta', \theta'] = 0$ into account, we obtain

$$d(\operatorname{Tr}\theta - \operatorname{Tr}\theta') = \operatorname{Tr}\Omega - \operatorname{Tr}\Omega' = p_*(\operatorname{Tr}\Phi - \operatorname{Tr}\Phi'), \tag{2.71}$$

where p_* is the map of forms, induced by the projection $p: P \to M$. Finally, we obtain

$$\operatorname{Tr}\Omega - \operatorname{Tr}\Omega' = du \quad (\text{on } M^n). \tag{2.72}$$

Thus, scalar 2-forms are constructed for curvature ones.

It follows from (2.72) and the Stokes formula that the integrals of $\operatorname{Tr}\Omega$ on two-dimensional cycles in M are topological invariants.

We define the ith characteristic classes

$$c_i = \operatorname{Tr}(\Omega \wedge \Omega \wedge \ldots \wedge \Omega) = \operatorname{Tr}(\Omega^i), \ i \geqslant 1, \tag{2.73}$$

where the exterior product of forms with the values in a Lie algebra was defined in Subsec. 1.2.4. Recall that the product $\omega = \omega^p \wedge \omega^q$ is a form with coefficients of the matrix product of ω^p and ω^q in the matrix realization of a Lie algebra.

Theorem 2.33 (1) *The forms* c_i *are closed.*

(2) *The forms* c_i *are replaced by exact forms if the connection is replaced.*

Proof. (1) The closedness of $c_i = \mathrm{Tr}(\Omega^i)$ follows from the Bianchi identity

$$d(\mathrm{Tr}\,\Omega^i) = \mathrm{Tr}(d\Omega^i) = \sum_{j=1}^{i} \mathrm{Tr}(\Omega^{j-1}(d\Omega) \wedge \Omega^{i-1}) = 0,$$

since $d\Omega = -\ [\theta, \Omega]$ and $\mathrm{Tr}(\Omega^{j-1}[\theta, \Omega]\Omega^{i-1}) = \mathrm{Tr}[\theta, \Omega] = 0$.

(2) If a connection form ω is replaced by another connection $\hat{\omega}$, c_i is replaced by an exact form du_i, where the preimage of u_i under the map p_* induced by the projection $p: P \to M$ is defined as

$$p_* u_i = \sum_{j=1}^{i} (-1)^j \mathrm{Tr}(\Omega^{j-1} \wedge (\theta - \hat{\theta}) \wedge \Omega^{i-j}),$$

and θ and $\hat{\theta}$ are the connections associated with the connection forms ω and $\hat{\omega}$, respectively.

This is carried out in the same way as for formula (2.71) and is left to the reader.

The integrals $\int_{N^{2i}} c_i$ along closed oriented $2i$-dimensional submanifolds $N^{2i} \subset M$ are thus topological invariants.

Before considering concrete characteristic classes related to a special choice of structure groups G and fiber bundles, we discuss an approach due to Weil and Chern, enabling us to describe completely characteristic classes for compact groups.

2.7.4 Weil Homomorphism

Condition (2.69) shows that when passing from one neighborhood on the base space to another and for any polynomial of degree k, which is invariant under the adjoint representation, we can define a global form $f(\Omega)$ of degree $2k$ on M.

The goal of the subsection is to show that the algebra AdG of invariant polynomials on $\mathscr{G}$ describes characteristic classes of M^n.

Consider the ring of n-linear (real or complex) functions $F(X_1, X_2, \ldots, X_n)$, where $X_i \in \mathscr{G}$ and $\mathscr{G}$ is the Lie algebra of the group G satisfying the conditions:

(1) F is n-linear and unaltered under permutations of arguments.

(2) F is G-invariant, i.e.,

$$F(\mathrm{Ad}(g)X_1, \ldots, \mathrm{Ad}(g)X_n) = F(X_1, \ldots, X_n)$$

for all $g \in G$.

Each such F is associated with a polynomial $F\,(\underbrace{X, X, \ldots, X}_{n})$ of degree n for which $F(X_1, \ldots, X_n)$ is the complete polarization*.

The polynomial F is said to be *invariant* under G. All invariant polynomials form a ring $I(G)$.

We now adduce another property of polynomials and functions F following from (2):

---

* In algebra, *polarization* is the reconstruction of a multilinear function in n arguments from its value on the coincident ones, e.g., a bilinear form can be uniquely determined from a quadratic form $F(x, x)$.

(2′) Infinitesimal invariance

$$\sum_{1 \le i \le n} F(X_1, \ldots, [Y, X_i], X_n) = 0, \quad Y, X_i \in \mathscr{G}.$$

The condition (2′) is easily generalizable. Let Y be a $\mathscr{G}$-valued 1-form and X_i be a $\mathscr{G}$-valued form of degree $m_i\,(1 \leqslant i \leqslant n)$. Then

$$\sum(-1)^{m_1 + \cdots + m_{i-1}} F(X_1, \ldots, [Y, X_i], \ldots, X_n) = 0. \tag{2.74}$$

As mentioned at the beginning of the subsection, if F is an invariant polynomial of degree n, then the global form

$$F(\Phi) = F(\Omega_{\mathscr{U}}), \tag{2.75}$$

induced by the connection form on M is defined, where Φ is $\mathscr{G}$-valued 2-form on the principal bundle P, which can be locally represented as $\Phi = \mathrm{Ad}(s_{\mathscr{U}})\Omega_{\mathscr{U}}$.

It follows from the Bianchi identity and (2.74) that

$$dF(\Phi) = nF([\omega, \Phi], \Phi, \ldots, \Phi) = 0, \tag{2.76}$$

where ω is a connection 1-form.

Therefore, the form $F(\Phi)$ is closed. By de Rham's theorem, its cohomology class $\{F(\Phi)\}$ determines an element in $H^{2n}(M, R)$ or in $H^{2n}(M, C)$.

It remains to show that the corresponding cohomology class does not depend on the choice of a connection.

Lemma 2.12 (Chern) *Let ω_0 and ω_1 be two $\mathscr{G}$-valued connection forms and $F \in I(G)$ an invariant polynomial of degree n. We put*

$$\omega_t = \omega_0 + t\alpha, \ \alpha = \omega_1 - \omega_0, \tag{2.77}$$

$$\Phi_t = d\omega_t - (1/2)[\omega_t, \omega_t]. \tag{2.78}$$

Then

$$F(\Phi_1) - F(\Phi_0) = nd\int_0^1 F(\alpha, \Phi_t, \ldots, \Phi_t)dt. \tag{2.79}$$

Proof. It follows from structural equation (2.78) that

$$\Phi_t = \Phi_0 + t(d\alpha - [\omega_0, \alpha]) - (1/2)t^2[\alpha, \alpha]. \tag{2.80}$$

We substitute Φ_t in F and differentiate with respect to t. We obtain

$$\frac{1}{n}\frac{d}{dt}F(\Phi_t) = F(d\alpha - [\varphi_t, \alpha], \Phi_t, \ldots, \Phi_t).$$

On the other hand,

$$dF(\alpha, \Phi_t, \ldots, \Phi_t) = dF(d\alpha, \Phi_t, \ldots, \Phi_t)$$
$$- (n-1)F(\alpha, [\omega_t, \Phi_t], \Phi_t, \ldots, \Phi_t).$$

It follows from the invariance of F in (2.74) that

$$F([\omega_t, \alpha], \Phi_t, \ldots, \Phi_t) - (n-1)F(\alpha, [\omega_t, \Phi_t], \Phi_t, \ldots, \Phi_t) = 0,$$

i.e.,

$$\frac{d}{dt}F(\Phi_t) = n\,dF(\alpha, \Phi_t, \ldots, \Phi_t). \tag{2.81}$$

Integrating the expression with respect to t, we obtain (2.79).

As a consequence of Chern's lemma, we immediately obtain the following.

Corollary 2.5 *The forms $F(\Phi_0)$ and $F(\Phi_1)$ corresponding to the connections ω_0 and ω_1 are closed and cohomologous to each other on M^n.*

Corollary 2.6 *Let ω be a connection in a fiber bundle $P \to M$ and $F \in I(G)$. Then $F(\Phi)$ is a coboundary in P.*

We put $\omega_t = t\omega$. Then

$$\Phi_t = t\,d\omega - (1/2)t^2[\omega, \omega] = t\Phi + (1/2)(t - t^2)[\omega, \omega] \tag{2.82}$$

and

$$F(\Phi) = nd\int_0^1 F(\omega, \Phi_t, \ldots, \Phi_t), \tag{2.83}$$

where $F(\omega, \Phi, \ldots, \Phi)$ is a form of degree $2n - 1$.

It follows from Corollary 2.6 that $F(\Phi)$ can be projected into a unique form on M.

We denote the graded cohomology ring for a manifold M^n by

$$H^*(M^n, R) = \bigotimes_{i=1}^{n} H^i(M^n, R)$$

and specify a mapping of rings

$$W: I(G) \to H^*(M^n, R) \tag{2.84}$$

by the formula

$$W(F) \to F(\Omega). \tag{2.85}$$

It is obvious that W is a homomorphism of rings.

Definition 2.32 W is called a *Weil homomorphism.*

If G is a compact connected Lie group, then the Weil homomorphism admits the following interpretation (see the proof in [Ch]). Let

$$EG \to BG$$

be a universal principal bundle with a group G and a classifying space as the base space BG. Then the ring $I(G)$ is isomorphic to the cohomology ring of the classifying space $H^*(BG, R)$.

The isomorphism W_0 between $I(G)$ and $H^*(BG, R)$ is determined by W. Relations between the cohomologies of the principal bundle $P \to M$, $I(G)$, and $H^*(BG, R)$ are characterized by the commutative diagram below:

$$\begin{array}{ccc} I(G) & \xrightarrow{W} & H^*(M^n, R) \\ & {\scriptstyle W_0}\searrow \quad \nearrow{\scriptstyle f_*} & \\ & H^*(BG, R) & \end{array} \tag{2.86}$$

The map f_* is induced by $f: M \to BG$, which is involved in the definition of the universal bundle $EG \to BG$ for $P \xrightarrow{G} M$ (2.57). We now turn to the construction of characteristic classes of bundles related to classical Lie groups.

2.7.5 Chern Characteristic Classes

Let us consider the result of the general construction of characteristic classes for principal bundles of compact complex manifolds with the structure group $U(n)$.

We begin with a description of the ring of invariant polynomials $I(G)$ for $G = U(n)$ consisting of unitary matrices $A^{(n)}$ (n is the order) and for the Lie algebra $u(n)$ of skew-Hermitian matrices $B^{(n)}$.*

Let E_j^k be a matrix with unity at (j, k) and zeros at other places. Then the connection ω is written as $E_j^k \omega_k^j$, where ω_k^j are linear forms defined in M^n. It is clear that

$$\omega \wedge \omega = E_j^k \omega_p^j \wedge \omega_k^p. \tag{2.87}$$

Making use of the identity

$$[E_j^k, E_s^r] = \delta_r^k E_j^s - \delta_j^s E_r^k, \tag{2.88}$$

we obtain

$$\begin{aligned} [\omega, \omega] &= [E_j^k, E_s^r] \omega_k^j \wedge \omega_r^s = E_j^s \omega_k^j \wedge \omega_s^k - E_r^k \omega_k^j \wedge \omega_j^r \\ &= 2E_j^k \omega_p^j \wedge \omega_k^p = 2\omega \wedge \omega. \end{aligned} \tag{2.89}$$

* Since the order of matrices is fixed, it is omitted below.

For the curvature form Φ, we have

$$\Phi = d\omega + (1/2)[\omega, \omega] = d\omega + \omega \wedge \omega. \tag{2.90}$$

If $B \in u(n)$, then $(\mathrm{Ad}\, A)B = ABA^{-1}$.

We select a characteristic polynomial $F = \det(\lambda I + B)$. The coefficients in

$$\det(\lambda I + B) = \lambda^n + b_1(B)\lambda^{n-1} + \ldots + b_n(B) \tag{2.91}$$

of the powers of λ are invariant multilinear functions of B.

If the matrix B is replaced by the curvature form $\Omega/(2\pi i)$, then we obtain the Weil homomorphism $I(G) \to H^*(M^n, R)$ of $I(G)$. The factor $1/(2\pi i)$ is selected by the normalization condition

$$\frac{1}{2\pi i} \int_{CP^1} \Omega = -1 \tag{2.92}$$

and real-valuedity of the forms c_i.

If we consider the universal bundle $V^C_{n,q} \to G^C_{n,q}$ for the group $U(n)$ and compute explicitly the curvature forms Ω for $G^C_{n,q}$, then the Weil homomorphism determines the isomorphism of the ring $I(U(n))$ and of the cohomology ring $H^*(G^C_{n,q}, R)$.

In defining the ring $I(G)$ for arbitrary compact groups G, the following assertion is useful. Let T be the maximal torus of a group G. The set of elements $g \in G$ such that $gTg^{-1} = T$ is called the *normalizer* of the torus and denoted by $N(T)$. It is proved in the theory of Lie groups (e.g., see [Sem]) that the group $\hat{W} = N(T)/T$ is finite. The group is said to be *Weyl*. The Weyl group belongs to automorphisms preserving the Lie algebra of the group T, or a Cartan subalgebra $t \subset \mathcal{G}$. It is obvious that the algebra $I(G)$ can be restricted to the Cartan subalgebra of $I(N(T))$. It is essential that a stronger converse statement holds, i.e., any polynomial F invariant under Ad G is uniquely determined by its value on the Cartan subalgebra t. In other words, we have the following.

Proposition 2.25 *$I(G)$ is isomorphic to $I(N(T))$.*

We stress that G is assumed to be compact (e.g., see the proof in [KN2]).

Proposition 2.25 considerably simplifies finding $T(G)$ in the concrete cases of classical Lie groups of interest and thereby gives an explicit description of the characteristic classes.

Group $U(n)$. A Cartan subalgebra $t \subset u(n)$ consists in this case of diagonal skew-Hermitian matrices. If we select a basis for t as $t_0^1, \ldots, t_0^n$, where $(t_0^j)_{jj} = i$ $(i = \sqrt{-1})$ (the only nonzero element of the diagonal is i), then it is easy to see that the group $\hat{W}$ acts in the same way as that of all permutations of this basis, namely, $\hat{W} \simeq S(n)$. Define basis linear forms $l_j \in \mathcal{G}^*$ by putting $l_j(t_0^k) = \delta_{kj}$. Then the restriction of symmetric polynomials F invariant under Ad G to a Cartan subalgebra t consists of symmetric polynomials of l_j. A basis of the algebra $I(N(T))$ can be selected in the form of elementary symmetric functions

$$F_k = \mathrm{Tr}\,[l_j^k] \tag{2.93}$$

of l_j. It follows from Proposition 2.25 that $I(G)$ is generated by classes (2.93). Taking into account formulas (2.91) and extending the forms F_k to the whole Lie algebra $\mathscr{G}$ according to (2.74), we obtain an explicit description of the characteristic classes for the group $U(n)$.

In physical applications, $U(n)$ (and, more often, $SU(n)$) arise as the internal symmetry groups of gauge field theories.

In the subsequent chapters the characteristic classes arise as topological conserved quantities of field configurations, or topological charges. Theories of essential interest are already those with simplest symmetry groups, in particular, with $SU(2)$.

Example 2.17 Group $SU(2)$.

Let us see how the characteristic classes c_i look for this group.

The Lie algebra $su(2)$ is the space of 2×2 skew-Hermitian matrices A with $\operatorname{Tr} A = 0$. The characteristic polynomial is

$$\det |\lambda I + A| = \lambda^2 + \lambda \operatorname{Tr} A + \det A = \lambda^2 + c_1(A)\lambda + c_2(A).$$

Since $\operatorname{Tr} A = 0$, the only nonzero class is c_2. We find the explicit expression for c_2 in terms of the curvature forms

$$\Omega = \|\Omega^{ij}\| = \|(1/2)F^{ij}_{\mu\nu}\, dx^\mu \wedge dx^\nu\|, \tag{2.94}$$

where the upper indices label elements of $su(2)$, while the subscripts the spatial (base) coordinates. We calculate

$$c_2(A) = \det A = \begin{Vmatrix} \Omega^{11}, & \Omega^{12} \\ \Omega^{21}, & \Omega^{22} \end{Vmatrix} = \Omega^{11}\Omega^{22} - \Omega^{21}\Omega^{12}$$

$$\begin{aligned} &= (1/4)\{F^{11}_{\mu\nu}\, dx^\mu \wedge dx^\nu \wedge F^{22}_{\varrho\gamma} dx^\varrho \wedge dx^\gamma - F^{21}_{\mu\nu} dx^\mu \wedge dx^\nu \wedge F^{12}_{\varrho\gamma} dx^\varrho \wedge dx^\gamma\} \\ &= (1/4)\{F^{11}_{\mu\nu}\, F^{22}_{\varrho\gamma} - F^{21}_{\mu\nu} F^{12}_{\varrho\gamma}\}\, dx^\mu \wedge dx^\nu \wedge dx^\varrho \wedge dx^\gamma \\ &= (1/4)\{-(1/2)(F^{11}_{\mu\nu}\, F^{11}_{\varrho\gamma} + F^{22}_{\mu\nu} F^{22}_{\varrho\gamma}) - (1/2)(F^{21}_{\mu\nu} \bar{F}^{21}_{\varrho\gamma} + \bar{F}^{12}_{\mu\nu} F^{12}_{\varrho\gamma})\} \\ &\quad \times dx^\mu \wedge dx^\nu \wedge dx^\varrho \wedge dx^\gamma. \end{aligned}$$

We recall that

$$F^{11}_{\mu\nu} + F^{22}_{\mu\nu} = 0, \quad F^{12}_{\mu\nu} = -\bar{F}^{21}_{\mu\nu} \tag{2.95}$$

and obtain

$$c_2(\Omega) = -(1/2) \operatorname{Tr} (\Omega \wedge \Omega). \tag{2.96}$$

By integrating $c_2(\Omega)$ with respect to the 4-cycle, we obtain the number

$$C = \int_{M^4} c_2(\Omega) dV. \tag{2.97}$$

We can show that when normalizing $c_2(\Omega)$

$$\frac{1}{8\pi^2}\int_{CP^2} c_2(\Omega) = 1,$$

C is an integer.

We postpone discussing applications of formulas of type (2.96) to the field theory until Sec. 4.2 and retrace our steps to the theory of characteristic classes for fiber bundles with group $U(n)$.

Each principal bundle P over M with group G is associated with the vector bundle E^C over M.

Consider the complex vector bundle E^C over M. The principal bundle P^C over M, which is associated with E^C, has the structure group $GL(n, C)$. We show that P^C with $G = GL(n, C)$ is reduced to the fiber bundle P with $G = U(n)$ which is the compact form of $GL(n, C)$. It is thereby proved that the characteristic classes c_i of a principal bundle determine those of E^C.

In fact, contracting $GL(n, C)$ to $U(n)$, we reduce P^C to the principal bundle P (with $G = U(n)$). We pick a connection form ω on P and the corresponding curvature form Ω. Let ω^C be the connection form on P^C extending ω and Ω^C the curvature form for ω^C. If f^C is a multilinear function invariantly Ad $GL(n, C)$ on $\mathscr{G}L(n, C)$, then its restriction to $u(n)f$ is obviously invariant under $U(n)$. Since the restriction of $f^C(\Omega^C)$ to P is $f(\Omega)$, then the characteristic class for P^C defined by f^C coincides with that for P defined by f. Since we have described all polynomials invariant under Ad $U(n)$ on $u(n)$, we thereby have all the characteristic classes of the fiber bundle E^C.

The characteristic classes of a fiber bundle with a non-compact structure group G always coincide with those of a fiber bundle with maximal compact subgroup $K \subset G$.

Definition 2.33 The *Chern characteristic classes* $c_i(M^n)$ of a complex n-dimensional manifold M^n are those of the tangent bundle $c_i(T^C M^n)$.

It follows from the construction of characteristic classes that

$$c_i(M) \in H^{2i}(M, R).$$

If normalization condition (2.92) is taken into account, then

$$c_i(M) \in H^{2i}(M, Z). \tag{2.98}$$

We define the complete Chern class $C(E) = \sum_{i=1}^{n} c_i(E)$ as satisfying the relation

$$C(E_1 \oplus_W E_2) = C(E_1)\cdot C(E_2),$$

where $E_1 \oplus_W E_2$ is the Whitney sum of the complex fiber bundles over M.

2.7.6 Pontrjagin Characteristic Classes

Two types of characteristic classes are related to real bundles. We now discuss the theory of Pontrjagin characteristic classes defined for vector bundles over $4n$-dimensional manifolds.

Let E be a real vector bundle over M with fiber R^n. Complexify R^n, i.e. construct the fiber bundle E^C over M by complexifying each fiber. Also, we define the principal bundles P and P^C associated with the vector bundles E and E^C, respectively. P is a subbundle of P^C and the structure groups for P and P^C are $GL(n, R)$ and $GL(n, C)$.

Definition 2.34 The Chern class

$$p_i(E) = (-1)^i c_{2i}(E^C) \in H^{4i}(M, R) \tag{2.99}$$

is called the *Pontrjagin class* $p_i(E)$.

The *complete Pontrjagin class* is accordingly

$$p(E) = \sum_{i=0}^{n} p_i(E). \tag{2.100}$$

The *Pontrjagin classes for a manifold* M^{4n} are $p_i(TM^{4n})$.

Pontrjagin classes like Chern ones can be described by taking into account the obvious modifications which arise from replacing the structure group $GL(n, C)$ with $GL(n, R)$ and the maximal compact subgroup $U(n)$ with $O(n)$, respectively. We give the corresponding results.

Algebra $I(O(n))$. The algebra Ad $O(n)$ of invariant polynomials on $o(n)$ is determined by the expansion

$$\det\,[\lambda I^{(n)} - (1/(2\pi))X] = \sum f_k(X)\lambda^{n-k}, \tag{2.101}$$

where $x \in o(n)$, $f_0, \ldots, f_n$ is the basis of $I(O(n))$ and $I^{(n)}$ the unit matrix.

Remark 2.18 The polynomials $f_k(X)$ vanish if k in (2.101) is odd. Indeed, since C is a skew-symmetric matrix,

$$\det\,[\lambda I^{(n)} - (1/(2\pi))X] = \det\,[\lambda I^{(n)} - X/(2\pi)]^t = \det\,[\lambda I^{(n)} + X/(2\pi)]. \tag{2.102}$$

Hence,

$$\sum f_k(X)\lambda^{n-k} = \sum(-1)^k f_k(X)\lambda^{n-k} \quad \text{for} \quad X \in O(n) \tag{2.103}$$

and $f_{2p+1} = 0$.

By analogy with the construction of Chern classes, we construct p_i from curvature forms. Let ω be a connection on P and Ω the curvature form of ω. There exists a unique closed $4k$-form β_k on M such that

$$p_*(\beta_k) = f_{2k}(\Omega). \tag{2.104}$$

β_k represents the kth Pontrjagin class for the vector bundle.

Indeed, we extend the connection form ω to ω^C on P^C and do the same with the curvature forms Ω. The Chern classes c_i are constructed from Ω^C. Like (2.102),

$$\det [\lambda I^{(n)} - (1/(2\pi i))\tilde{X}] = \sum f_k(\tilde{X})\lambda^{n-k}, \quad \tilde{X} \in u(n).$$

Hence, by replacing $\tilde{X}$ with X, we have

$$f_{2k}(X) = (-1)^k \tilde{f}_{2k}(X), \quad X \in O(n). \tag{2.105}$$

Therefore, the restriction of $(-1)^k \cdot f_{2k}(\Omega^C)$ coincides on P with $f_k(\Omega)$. Our assertion follows from the representation $f_{2k}(\Omega^C)$ for a Chern class.

Remark 2.19 If a form Ω is given in matrix form $\Omega = \Omega^i_j$, then the form $f_{2k}(\Omega)$ has the explicit representation

$$f_{2k}(\Omega) = \frac{1}{(2\pi)^{2k}(2k)!} \delta^{j_1 \ldots j_{2k}}_{i_1 \ldots i_{2k}} \Omega^{i_1}_{j_1} \wedge \ldots \wedge \Omega^{i_{2k}}_{j_{2k}},$$

where $\delta^{j_1 \ldots j_k}_{i_1 \ldots i_k}$ is determined by the parity of the permutation $(i_1, \ldots, i_k) \to (j_1, \ldots, j_k)$. The summation is over all ordered subsets $(i_1, \ldots, i_{2k})$ consisting of $2k$ elements from $1, \ldots, n$, in particular, $p_1(E) = c_2(E)$ and once again we find formula (2.96) on M^4, i.e.,

$$p_1(E) = \frac{1}{8\pi^2} \delta^{j_1 j_2}_{i_1 i_2} \Omega^{i_1}_{j_1} \wedge \Omega^{i_2}_{j_2} = -(1/2) \operatorname{Tr} (\Omega \wedge \Omega).$$

2.7.7 Euler Class

Another type of characteristic class, i.e., the Euler class $e(E)$, can be related to a real bundle E over M^n.

The Euler class is defined for orientable fiber bundles. Poincaré's theorem in the two-dimensional case on vector fields and the Gauss-Bonnet theorem can be generalized with the help of $e(E)$.

Recall that the real vector bundle E over M^n with fiber R^n is said to be *orientable* if the structure group $G = GL(n, R)$ acting in the fiber can be reduced to the subgroup $GL^+(n, R)$ of $GL(n, R)$ consisting of matrices with a positive determinant.

Definition 2.35 A fiber bundle with the fiberwise Riemannian metric g_x on R^n, i.e., with scalar product g_x in the fiber R^n over each point x and with the family of metrics g_x which depend on x differentiably, is said to be *Riemannian*.

Given the orientable vector bundle E over M^n, then giving the fiberwise metric reduces the principal bundle associated with E to the fiber bundle with structure group $SO(n)$.

Note that a complex bundle E^C with fiber C^n, which is a real bundle, is always orientable since C^n has the structure of R^{2n} and the fiber group $GL(n, C) \subset GL^+(2n, R)$.

We define the Euler class $e(E) \in H^n(M^n, R)$ using the Weil homomorphism W: $I(G) \to H^*(BG, R)$, for which we need the description of the ring $I(SO(n))$.

We now give the corresponding results. The reader can reconstruct or find (e.g., in [KN]) the proof of the omitted statements independently.

Theorem 2.34 (1) *The algebra of* Ad $SO(n)$ *invariant polynomials coincides with that of* Ad $O(n)$*-invariant ones for odd* $n = 2m + 1$. *The basis for* $I(SO(2m + 1))$ *consists of the functions* $f_1, \ldots, f_m$ *determined by the expansion*

$$\det (\lambda I^{(n)} - X) = \lambda^n + f_1(X)\lambda^{n-2} + f_2(X)\lambda^{n-4} + \ldots + f_{[n/2]}(X) \qquad (2.106)$$

for $X \in o(n)$.

(2) *If* $n = 2m$, *then there exists a polynomial function* h, *unique apart from the sign and such that* $f_m = h^2$ *and* $f_1, \ldots, f_{m-1}, h$ *are algebraically independent, generating the algebra of* Ad $SO(n)$*-invariant functions on* $o(n)$.

The proof of statement (1) is similar to the theorem on the structure of the ring $I(U(n))$. It follows from the antisymmetry of matrices in the algebra $o(n)$ that functions f_i vanish for odd i. The corresponding justification is given by formulas (2.103) and (2.104). It is essential that $f_n = 0$.

We now turn to even n and make use of the following auxiliary construction.

Pfaffian (Pf). There exists one and (apart from the sign) only one polynomial with integer coefficients associating each skew-symmetric $2n \times 2n$ matrix A over a commutative ring Q with its element Pf (A) whose square is the determinant of A. In this case

$$\text{Pf}\,(BAB^t) = \text{Pf}\,(A) \det B \qquad (2.106')$$

for any $2n \times 2n$ matrix B (see the proof in [MSt]).

Remark 2.20 In topological applications, Q is either the field R, or the field C, or, e.g., the ring Z of integers.

Example 2.18

$$\text{Pf}\begin{pmatrix} 0 & a \\ -a & 0 \end{pmatrix} = a.$$

It follows from (2.106′) that Pf is invariant under Ad B, where $B \in SO(n)$. Therefore, Pf (A) is in the ring $I(SO(n))$. Since the polynomial f_m in (2.106) is det X, it follows from the definition of Pf (A) that $f_m(X) = [\text{Pf}\,(X)]^2 = h^2(X)$. We have thus defined the invariant polynomial $h^2(X)$, which is equal to the polynomial of the greatest power on the Lie algebra $o(n)$. If we recall that polynomials invariant under Ad $SO(n)$ are uniquely determined by their values on the Cartan subalgebra $t \subset SO(n)$ and are representable as

$$\begin{matrix} 1\left\{ \begin{matrix} \\ \\ \end{matrix} \right. \\ \vdots \\ \\ m\left\{ \begin{matrix} \\ \\ \end{matrix} \right. \end{matrix}
\begin{pmatrix} 0 & a_1 & & & & & \\ -a_1 & 0 & & & & & \\ & & 0 & a_2 & & & \\ & & -a_2 & 0 & & & \\ & & & & \ddots & & \\ & & & & & 0 & a_m \\ & & & & & -a_m & 0 \end{pmatrix} \qquad (2.107)$$

where

$$t = \mathscr{G}(\underbrace{SO(2) \times \ldots \times SO(2)}_{m})$$

with $h(X) = a_1 \cdot \ldots \cdot a_m$ on t. If $X \in SO(n)$, i.e., $a_{ij} = -a_{ji}$, then we put

$$h(X) = \frac{1}{2m \cdot m!} \varepsilon_{i_1 i_2 \ldots i_{2m-1} i_{2m}} a_{i_1 i_2} \cdot \ldots \cdot a_{i_{2m-1} i_{2m}}, \tag{2.108}$$

where the summation is over all permutations of numbers 1, ..., $2m$, and the antisymmetric tensor $\varepsilon_{i_1 \ldots i_{2m}} = 1$ if the permutation is even and $\varepsilon_{j_1 \ldots j_{2m}} = -1$ if odd, $\varepsilon_{1 \ldots 2m} = 1$.

Direct verification of (2.108) shows that $h(x)$ is Pfaffian.

Like the Chern and Pontrjagin classes, we can now express the Euler class in terms of the closed $2n$-form on M^{2n}. We pick a fiberwise metric in E and construct the principal bundle P with fiber $SO(2n)$, which is associated with E.

Let ω_j^i be the connection forms corresponding to the metric and Ω_j^i the curvature forms. We define a closed $2n$-form by substituting the curvature forms

$$e(\Omega) = \frac{(-1)^n}{2^{2n} \pi^n n!} \varepsilon_{i_1 \ldots i_{2n}} \Omega_{i_2}^{i_1} \wedge \ldots \wedge \Omega_{i_{2n}}^{i_{2n-1}} \tag{2.109}$$

in the Pfaffian.

Proposition 2.26 *The Euler class $e(E)$ of an oriented real bundle E over M^{2n} with fiber R^{2n} is represented by the closed $2n$-form $e(\Omega)$.*

It is actually the independence of the Euler class from the choice of a fiberwise Riemannian metric that should be proved.

Let the metric g_x in the fiber be Euclidean in R^{2n}. Any other Riemannian metric in the fiber is obtained by an orthogonal transformation A with $\det A = 1$. Since $\mathrm{Pf}\,(A\Omega A^t) = \mathrm{Pf}\,(A\Omega A^{-1}) = \mathrm{Pf}\,(\Omega)$, the Euler class does not depend on the choice of a Riemannian metric in E.

Remark 2.21 When calculating the Chern and Pontrjagin classes in terms of closed forms, we can select any connection in the fiber bundle. For the Euler class, the connection should be consistent with the fiberwise metric on E. The case $n = 1$ is the only exception and the Euler class $e(E^2)$ then coincides with $c_1(E^2)$. An example due to Milnor (1958) of an oriented vector bundle with flat connection (i.e., vanishing curvature tensor) for which the Euler class with real coefficients does not vanish is given in [MSt].

Remark 2.22 The Euler class $e(M)$ with real coefficients is usually defined axiomatically.

Axiom I *For each orientable fiber bundle E over M^n with fiber R^n, the Euler class $e(M) \in H^n(M, R)$, $e(M) = 0$, is defined for odd n.*

Axiom II (naturality) *If E is an oriented fiber bundle over M^n and f map $M' \to M$, then*

$$e(f^{-1}(E)) = f_*(e(E)) \in H^n(M', R),$$

where $f^{-1}(E)$ is the vector bundle over M' induced by f.

Axiom III *Let $E_1, \ldots, E_r$ be r fiber bundles over M with fiber R^2. Then*

$$e(E_1 \oplus_W E_2 \oplus_W \ldots \oplus_W E_r) = e(E_1)\cdot e(E_2) \ldots e(E_r).$$

Axiom IV *If E^1 is a one-dimensional complex bundle over CP^1, then $e(E^1) = c_1(E^1)$.*

It is easy to verify that if $e(E)$ is defined in terms of the $2n$-form, then all the axioms hold.

In conclusion, we formulate a multidimensional analog of the Gauss-Bonnet formula due to Chern.

Theorem 2.35 *Let M be an orientable compact Riemannian manifold of dimension 2n and TM the tangent bundle over M. We put $e(M) = e(TM)$, i.e., the closed 2n-form defined in* (2.109). *Then*

$$\int_M e(M)\,dV = \chi(M),$$

where $\chi(M)$ is the Euler characteristic of M.

2.7.8 Stiefel-Whitney Characteristic Classes

At the beginning of this section, the Stiefel-Whitney characteristic classes w_i of real bundles were mentioned as elements of the cohomology groups $H^i(M, Z_2)$. We now discuss the classes in more detail.

Unfortunately, Stiefel-Whitney classes cannot be constructed explicitly in terms of curvature forms as Chern classes can, since w_i are in the cyclic group Z_2. However, w_i can be defined axiomatically,

The uniqueness and existence of cohomology classes satisfying the axioms was proved in [MSt].

The axioms used to define the classes are similar to the corresponding assertions for the Chern classes. Let ξ be the vector bundle over M^n and $H^i(M, Z_2)$ the cohomology group with coefficients in Z_2.

Axiom 1 *For each vector bundle ξ, there exists a sequence of cohomology classes*

$$w_i(\xi) \in H^i(M, Z_2)$$

said to be Stiefel-Whitney.

The class w_0 equals the identity element of the group $H^0(M, Z_2)$ and the classes $w_i(\xi)$ vanish for $i > n$ if ξ is an n-dimensional fiber bundle.

Axiom 2 (naturality) *If a map $f\colon M \to M_1$ is covered by a fiberwise map $\xi \to \eta$, then*

$$w_i(\xi) = f_* w_i(\eta). \tag{2.110}$$

Axiom 3 *If ξ and η are two vector bundles over the same base space M, then*

$$w_i(\xi \oplus_W \eta) = \sum w_i(\xi) \cup w_{k-i}(\eta). \tag{2.111}$$

Here the operation $\cup$ is cup product, e.g.,

$$w_1(\xi \oplus_W \eta) = w_1(\xi) \oplus w_1(\eta).$$

It is convenient to define the complete Stiefel-Whitney class $W(\xi) = \sum_0^n w_i(\xi)$. In a similar manner as for a complete Chern class, Axiom 3 can be written as

$$W(\xi \oplus \eta) = W(\xi) \cdot W(\eta). \tag{2.112}$$

Axiom 4 (normalization) *For the linear bundle ξ^1 over $S^1 \simeq RP^1$,*

$$w_1(\xi^1) = H^1(RP^1, Z_2) \neq 0. \tag{2.113}$$

Definition 2.36 The classes $w_i(TM)$, where TM is the tangent bundle to M^n, are said to be *Stiefel-Whitney* of M^n.

The Stiefel-Whitney classes w_i are defined for real bundles with structure group $GL(n, R)$ and are reducible to the group $O(n)$. The base space is not assumed to be orientable. It would be interesting to formulate the condition for fiber bundles to be orientable in terms of the Stiefel-Whitney classes.

Theorem 2.36 *A fiber bundle ξ is orientable if and only if*

$$w_1(\xi) = 0.$$

Proof. Consider the exact sequence of groups

$$1 \to SO(p) \to O(p) \to Z_2 \to 1$$

associated with that of the cohomology

$$H^1(M, SO(p)) \to H^1(M, O(p)) \to H^1(M, Z_2) \to \dots .$$

The orientability condition is equivalent to the condition for the isomorphism

$$H^1(M, SO(p)) \to H^1(M, O(p))$$

to exist. Therefore, the image

$$\varphi: H^1(M, O(p)) \to H^1(M, Z_2) = w_1$$

vanishes. ∎

We now formulate a number of simple, but quite useful, properties of w_i immediately following from the axioms.

Proposition 2.27 *If a fiber bundle ξ is isomorphic to η, then $w_i(\xi) = w_i(\eta)$.*

Proposition 2.28 *If ε is the trivial bundle, then $w_i(\varepsilon) = 0$ for all i.*

In fact, if ε is trivial, then it can be fiberwise mapped into the fiber bundle over a point. Combining this result with Axiom 3, we obtain

Proposition 2.29 *If a fiber bundle ξ is trivial, then $w_i(\varepsilon \oplus_y \eta) = w_i(\eta)$.*

Proposition 2.30 *If ξ is an n-dimensional fiber bundle with the Euclidean metric, with a section nonvanishing anywhere, then $w_n(\xi) = 0$.*

Proof. We notice that if ζ is a subbundle of ξ and ξ is equipped with the Euclidean metric, then we can define in ξ the orthogonal complement $\zeta^{\perp}$ such that $\xi = \zeta \oplus \zeta^{\perp}$, which is obvious since the metric existence enables us to introduce the orthogonal complement to fibers of ξ. We now use Proposition 2.29. Since the existence of a section is equivalent to the distinction of the one-dimensional trivial subbundle ε, we obtain

$$w_n(\varepsilon \oplus \zeta_{n-1}) = w_n(\zeta_{n-1}) = 0. \quad \blacksquare$$

Proposition 2.30 is immediately generalized.

Proposition 2.31 *If ξ possesses k everywhere linearly independent sections, then*

$$w_{n-k+1}(\xi) = w_{n-k+2}(\xi) = \ldots = 0.$$

Note that the condition $w_n(\xi) = 0$ is necessary, but not sufficient. Such is the vanishing of the Euler class $e(\xi)$. Formulate the relation between $e(\xi)$ and the corresponding Stiefel-Whitney class $w_n(\xi)$.

Proposition 2.32 *The natural homomorphism $H^n(M, Z) \to H^n(M, Z_2)$ sends $e(\xi)$ into the higher class $w_n(\xi)$.*

In physical applications, the class $w_2 = H^2(M, Z_2)$ is of special interest. That w_2 is zero supplies a necessary condition for the existence of a spin structure on the manifold M. This is duscussed in Subsec. 5.2.8.2.

Calculating Stiefel-Whitney classes for concrete manifolds is rather complicated. Even for the simplest S^n or RP^n type manifold, this requires nontrivial considerations.

We introduce the following general concept and consider the set of infinite series $\{W\} = (1 + w_1 + \ldots + w_n + \ldots) \in H^*(M, Z_2)$, beginning with unity.

Lemma 2.13 The set $\{W\}$ *forms a commutative ring.*

Proof. We show that each element of W can have an inverse. The other statements are obvious. We calculate the inverse element W^{-1} by expanding it into a series

$$\begin{aligned} W^{-1} &= (1 + w_1 + \ldots + w_n + \ldots)^{-1} = 1 - (w_1 + w_2 + \ldots) \\ &\quad + (w_1 + w_2 + \ldots)^2 - (w_1 + w_2 + \ldots)^3 + \ldots \\ &= 1 - w_1 + (w_1^2 - w_2) + (-w_1^3 + 2w_1 w_2 - w_3) + \ldots \end{aligned} \tag{2.114}$$

(the signs being unimportant). The coefficient of $w_1^{i_1} \ldots w_k^{i_k}$ is $(i_1 + \ldots + i_k)!/(i_1! \ldots i_k!)$ in the expansion W^{-1}.

Specify two fiber bundles, ξ and η, over the same base space M. It follows from Lemma 2.13 that the equation

$$W(\xi \oplus_{\mathrm{W}} \eta) = W(\xi) \cdot W(\eta)$$

is uniquely solvable and $W(\eta) = W(\xi \oplus_{\mathrm{W}} \eta) \cdot W(\xi)^{-1}$.

In particular, if the fiber bundle $(\xi \oplus_{W} \eta)$ is trivial, then

$$W(\eta) = W(\xi)^{-1}, \tag{2.115}$$

e.g., if ξ is the tangent bundle TM and η the normal bundle $N(M)$, we obtain the result

$$w_i(TM) = w_i^{-1}(N(M)) \tag{2.116}$$

due to Whitney.

Example 2.19 We consider $w_i(S^n)$. We can now calculate the Stiefel-Whitney classes for the sphere S^n which we embed standardly in R^{n+1}. The normal bundle is trivial; therefore,

$$W(TS^n) = W(N(S^n))^{-1} = W(N(S^n)) = 1,$$

i.e., $w_i(S^n) = 0$.

This shows that the Stiefel-Whitney classes do not enable us to distinguish the trivial bundle over S^n from the tangent one.

A calculation of the Stiefel-Whitney classes of fiber bundles over RP^n requires knowledge of the cohomology ring of RP^n.

Proposition 2.33 (1) *The group $H^i(RP^n, Z_2)$ is Z_2 for $0 \leqslant i \leqslant n$ and $H^i(RP^n, Z_2) = 0$ for $i > n$.*

(2) *If $a \neq 0$, $a \in H^1(RP^n, Z_2)$, then each $H^i(RP^n, Z_2)$ is generated by the $\cup$-product of a taken i times.*

Thus, the ring $H^*(RP^n, Z_2)$ is the algebra with identity over the field Z_2 with one generator a and one defining relation $a^{n+1} = 0$ (see the proof in [MSt]).

Example 2.20 The complete class $W(\gamma_n^1(RP^n))$, where γ_n^1 is the linear bundle over RP^n, is specified by the formula $W(\gamma_n^1(RP^n)) = 1 + a$.

Proof. The standard embedding j: $RP^1 \subset RP^n$ is covered fiberwise by the fiber bundle γ_1^1 in γ_n^1. Therefore,

$$j_* w_1(\gamma_n^1) = w_1(\gamma_1^1) \neq 0.$$

Since $w_1(\gamma_n^1) \neq 0$, the class must be a and the remaining classes $w_i(\gamma_n^1) = 0$ by Axiom 1.

Example 2.21 Consider $W(RP^n)$. We give, without proof, the results of calculations for the tangent bundle $T(RP^n)$, namely,

$$W(RP^n) = (1 + a)^{n+1} = 1 + C_{n+1}^1 a + \ldots + C_{n+1}^n a^n, \quad a^{n+1} = 0 \tag{2.117}$$

(see the proof in [MSt]). The binomial coefficients are considered mod 2, e.g.,

$$W(RP^2) = 1 + a + a^2, \quad W(RP^3) = 1.$$

We draw the reader's attention to an interesting corollary. The necessary condition for the manifold RP^n to be parallelizable is $W(TRP^n) = 1$. It follows from formula (2.117) that $W(RP^n) = 1$ only for $n - 1 = 2^p$, which was proved by Stiefel.

However, a considerably stronger result due to Adams is valid. The only parallelizable projective manifolds are $RP^1 \sim S^1$, $RP^3 \sim SO(3)$, RP^7, RP^{15}, which follows from the parallelizability of the spheres S^1, S^3, S^7, S^{15}.

We introduce another important concept enabling us to compare the Stiefel-Whitney classes of two manifolds. We use the corresponding results in the very next subsection.

Stiefel-Whitney numbers. Let M^n be a closed (possibly, disconnected) n-dimensional manifold. There exists a unique fundamental homology class $\mu_M \in H_n(M, Z_2)$. If M^n is compact, then we can regard the whole manifold as the fundamental class.

For any n-cohomology class $v \in H^n(M, Z_2)$, the Kronecker index

$$\delta(M) = \langle v | \mu \rangle,$$

the value of v on μ, is defined.

Let $r_1, \ldots, r_n$ be n non-negative integers such that $r_1 + 2r_2 + \ldots + nr_n = n$. For any vector bundle $\xi \to M$, we can form the monomial

$$w_1^{r_1}(\xi) \ldots w_n^{r_n}(\xi)$$

in the cohomology group $H^n(M, Z_2)$. Consider the tangent bundle TM to M.

Definition 2.37 The *Kronecker index*, i.e., the residue mod 2

$$\langle w_1^{r_1}(TM) w_2^{r_2}(TM) \ldots w_n^{r_n}(TM) | \mu \rangle,$$

or, in abbreviated form, $w_1^{r_1} \cdot w_2^{r_2} \cdot \ldots \cdot w_n^{r_n} [M]$, is called the *Stiefel-Whitney number* of the manifold M and is associated with the class $w_1^{r_1}, \ldots, w_n^{r_n}$.

Two manifolds M and M_1 have the same Stiefel-Whitney numbers if $w_1^{r_1} \ldots w_n^{r_n}([M] = w_1^{r_1} \ldots w_n^{r_n}[M_1]$ for any class $w_1^{r_1} \ldots w_n^{r_n}$ of total dimension n.

Example 2.22 Stiefel-Whitney numbers of the projective space RP^n. (1) First, consider even-dimensional projective spaces RP^{2n}. The classes $w_{2n} = (2n + 1)a^{2n}$ and $w_1 = (2n + 1)a$ do not vanish and therefore neither do the Stiefel-Whitney numbers $w_{2n} = w_{2n}[RP^{2n}]$ and $w_1^{2n}[RP^{2n}]$. If $2n$ is a power of two, then the complete Stiefel-Whitney class is $1 + a + a^{2n}$; therefore, the other w_i are zero. In the general case, w_i are calculated in terms of the binomial coefficients.

(2) RP^{2n-1}. $W(TRP^{2n-1}) = (1 + a)^{2n} = (1 + a^2)^n$; therefore, $w_i(TRP^{2n-1}) = 0$ for any odd i. Since each monomial of total dimension $2n - 1$ should contain a factor w_i of odd dimension i, all Stiefel-Whitney numbers $w_i[RP^{2n-1}] = 0$.

2.7.9 Cobordism Theory

We start with two examples.

Example 2.23 Consider two manifolds: $M_1 = S_1^1$ and the disjoint union M_2 of two circles S_2^1 and S_3^1. Can we construct a manifold W^2 with boundary $\partial W^2 = M_1 \cup M_2$? It is obvious that the answer is yes and W^2 is a so-called "pants" (Fig. 7).

The following example is less obvious.

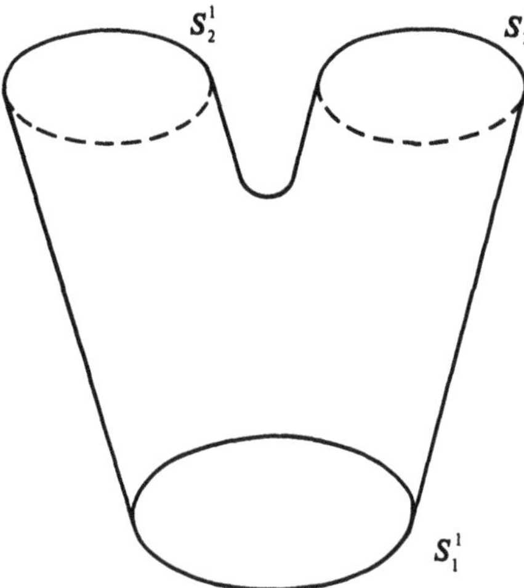

Fig. 7. Cobordant manifold constructed from three circles (" pants")

Example 2.24 Does a three-dimensional manifold W^3 with two boundaries ∂W_1^3 and ∂W_2^3 equal to the sphere S^2 and the torus $T^2 = S^1 \times S^1$ respectively exist? Yes, it does. We pick in R^3 two disjoint two-dimensional surfaces, S^2 and T^2, which bound the ball D^3 and solid torus B^3. We cut these domains out and compactify the remaining domain with the point at infinity. We obtain the compact oriented manifold W^3 with boundaries T^2 and S^2. It is obvious that the result is valid for any two-dimensional closed orientable surfaces.

We now formulate the generalization of these problems, leading to the principal definitions of the cobordism theory. Let M_1^n and M_2^n be two compact manifolds. Under what conditions for the manifolds does an $(n + 1)$-dimensional manifold W^{n+1} called a *film* with boundaries M_1^n and M_2^n exist?

We turn to exact formulations.

Definition 2.38 Let M^n be a closed manifold. M^n is said to be *null-cobordant* if there exists a compact manifold W^{n+1} with boundary diffeomorphic to M^n.

For the present, we assume that M^n is orientable.

Definition 2.39 Two orientable manifolds, M_1^n and M_2^n are said to be *cobordant* if a manifold W^{n+1} with boundary ∂W^{n+1} equal to the disjoint union $M_1^n \cup -M_2^n$ (where $-M_2^n$ is M_2^n with opposite orientation) exists.

It is easy to verify that the set of all orientable compact n-dimensional manifolds separates into equivalence classes relative to the cobordance relation. Let the equivalence class of a manifold M be $[M]$. The addition operation can be defined for the classes as $[M] + [M_1] = [M \cup M_1]$, relative to which the equivalence classes form a commutative group.

The inverse of a class $[M]$ is $[-M]$. In fact, the space $M \cup -M$ is the boundary of $M \times I$ and therefore null-cobordant, namely, $[M] + [-M] = 0$. The group is called the *cobordism group of dimension n* and is denoted by Ω^n.

If a manifold M is cobordant to M_1, then for any oriented manifold V the product $M \times V$ is cobordant to $M_1 \times V$. Therefore, multiplication can be defined for $[M]$. The multiplication is anticommutative and distributive with respect to addition. The direct sum $\Omega = \oplus\Omega^n$ is thereby a ring.

Remark 2.23 Cobordism of non-orientable manifolds. Since giving an orientation to a manifold separates a generator out of $H_n(M, Z)$, turning to the group $H_n(M, Z_2)$ we can skip mentioning orientation and consider the ring N of cobordant manifolds mod 2.

The condition that two manifolds are cobordant is equivalent to the null-cobordance condition for one manifold M^n. Therefore, finding necessary and sufficient tests for null-cobordance of M^n is of great interest. We thereby construct the topological invariants of the cobordism classes.

Theorem 2.37 (Pontrjagin-Thom) *A compact manifold M^n is the boundary of W^{n+1}, i.e., null-cobordant, if and only if all Stiefel-Whitney numbers are zeros.*

The necessity was proved by Pontrjagin (see the proof in [MSt]). The proof of sufficiency is considerably more complicated. It was obtained by Thom [Th].

Example 2.25 As was shown in Example 2.22, all Stiefel-Whitney numbers of RP^{2n+1} are zeros. Therefore, RP^{2n+1} is the boundary of a $(2n + 2)$-dimensional compact manifold.

We now give an example of a simpler condition imposed on the topological characteristic of a null-cobordant manifold M^n.

Proposition 2.34 *If $[M^n] \sim 0$, then the Euler characteristic $\chi(M^n)$ is even.*

Proof. Let M^n be the boundary of a manifold W^{n+1}. We consider the "double" of W^{n+1}, i.e., the manifold $\tilde{W}^{n+1}$ without boundary obtained from W^{n+1} by "gluing" the second copy of W^{n+1} along the boundary of M^n (see Example 2.16 for detailed discussion of "doubles"). Then, using the cellular decomposition of $\tilde{W}^{n+1}$ and the definition of the Euler characteristic, we obtain

$$\chi(\tilde{W}^{n+1}) = \chi(W^{n+1}) - \chi(M^n) + \chi(W^{n+1}). \tag{2.118}$$

If $n = 2k + 1$, then $\chi(M^n) = 0$. Therefore, the case $n = 2k$ only remains. Consequently, $\chi(\tilde{W}^{n+1}) = 0$ and $\chi(M^n) = 2\chi(W^{n+1})$. ■

Finding the structure of rings Ω^n for different classes of n-dimensional manifolds seems to be quite a complicated problem. Actually, a number of classical problems in topology can be formulated in terms of cobordism theory. In physical applications, manifolds of dimension less than or equal to four are now of great interest. We now give the results of calculating Ω^4 for them.

1. Manifold Ω^0 consisting of finitely many points. It is obvious that two collections of point are cobordant if the number of the points is the same. If they are given signs, then the signs sum determines the only invariant of Ω^0. Obviously, $\Omega^0 \simeq Z$.

2. $\Omega^1 \sim \Omega^2 = 0$. It is obvious that each compact one-dimensional or orientable two-dimensional manifold is a boundary.

2′. N^2. For non-orientable manifolds, this is already not so. It follows from Theorem 2.37 that RP^2 is not the boundary of a three-dimensional manifold. On the other hand, the Klein bottle K is null-cobordant.

Since any non-orientable surface is the connected sum of k-projective planes RP^2 [Mas2], $\#_k RP^2 \sim M^2_{(k-1)/2} \# RP^2$ (where $M^2_{(k-1)/2}$ is an orientable surface of genus

$(k-1)/2$ if k is odd and $\#_k RP^2 = M^2_{(k-2)/2} \# K$ (where K is the Klein bottle) if k is even . Hence, $N^2 \sim Z_2$.

3. The ring $\Omega^3 \sim 0$, i.e., any three-dimensional orientable manifold is the boundary of a four-dimensional one. This important result was first proved by Rokhlin by geometric methods developed in the Pontrjagin theory of framed manifolds. Later, it was obtained by Thom as a corrollary of his remarkable cobordism theory [Th].

4. Ring $\Omega^4 \sim Z$. This was also proved by Rokhlin. We consider this case in more detail. To find cobordism invariants of four-dimensional manifolds, similarly to the Stiefel-Whitney characteristic classes, it is useful to consider the Pontrjagin numbers defined in terms of the Pontrjagin characteristic classes of four-dimensional manifolds.

Pontrjagin numbers. Let M^{4n} be a smooth compact oriented manifold. Any disordered sequence $I = i_1, \ldots, i_r$ of positive numbers with sum k is called a *partition* of the non-negative whole number k.

Definition 2.40 *For each partition $I = i_1, \ldots, i_r$ of an integer k, the Ith Pontrjagin number is the integer*

$$p_I[M^{4n}] = p_{i_1 \ldots i_r}[M^{4n}]$$

equal to the Kronecker index

$$\langle p_{i_1}(TM^{4n}) \ \ldots \ p_{i_r}(TM^{4n}) | \mu_{4n} \rangle .$$

Here TM^{4n} is the tangent bundle to M^{4n} and μ_{4n} the fundamental homology class for M^{4n}.

We give without proof the formula

$$p_{i_1 \ldots i_r}[CP^{2n}] = C^{i_1}_{2n+1} \ \ldots \ C^{i_r}_{2n+1} \tag{2.119}$$

for the Pontrjagin numbers of the manifold CP^{2n} (e.g., see the proof in [MSt]).

To clarify the question of null-cobordance of a four-dimensional manifold M^{4n}, we conclude that the Pontrjagin numbers play the same role as the Stiefel-Whitney numbers. However, taking into account the constructive calculation of Pontrjagin classes, the Pontrjagin numbers are more convenient.

Proposition 2.35 *If M^{4n} is a boundary, then all the Pontrjagin numbers p_i are zeros.*

Example 2.26 Manifold CP^2. There exists only the Pontrjagin number $p_1(CP^2) = 3$. Therefore, CP^2 cannot be a boundary. The same holds for any manifold CP^{2n}.

By definition (as well as the Pontrjagin classes), the Pontrjagin numbers are zeros for manifolds M^n with $n \neq 4k$. Such manifolds can be boundaries.

Example 2.27 The complex odd-dimensional manifold CP^{2n+1} is the boundary of the manifold W^{4n+2}_R.

The stronger theorem due to Wall holds [Sto].

Theorem 2.38 *A manifold M^n is an orientable boundary if and only if all its Pontrjagin and Stiefel-Whitney numbers are zeros.*

We retrace our steps to the ring Ω^4. It turns out that, for the ring of four-dimensional manifolds, the space CP^2 plays the role of the generator. We formulate the exact result as the theorem due to Thom and Rokhlin.

Theorem 2.39 *The product of complex projective spaces* $CP^{2i_1} \times \ldots \times CP^{2i_r}$, *where* $i_1, \ldots, i_r$ *run over the set of all partitions of a number* k, *determines the maximum number of linearly independent elements in the group* Ω^{4k}.

Corollary 2.7 *The ring* Ω^4 *is isomorphic to the group* Z *and is generated by the class* CP^2.

It thus follows from the Wall and Rokhlin-Thom theorems that the cobordism group Ω^n is always the direct sum of several specimens of Z_2 and, given $n = 0$ (mod 4), it is the direct sum of the corresponding number of specimens of the group Z.

Signature of manifold. Another important topological characterization of a manifold is its signature, or index $\tau(M)$, and is related to the Pontrjagin numbers. I will use the more modern term *signature.*

Definition 2.41 The *signature* τ of a compact orientable manifold M^n is assumed to be zero if $n \neq 4k$. Let $n = 4k$. Consider the cohomology space of "average" dimension $H^{2k}(M^{4k}, Q)$ (Q the field of rational numbers). We select a basis $a_1, \ldots, a_r$ in the vector space $H^{2k}(M^{4k}, Q)$, in which the matrix $A = \langle a_i \cup a_j | \mu \rangle$ is diagonal (μ the fundamental cycle of M^{4k}). The *signature* $\tau(M^{4k})$ is the difference of the number of positive and negative eigenvalues of A.

The definition coincides with that of the signature $\langle a \cup a | \mu \rangle$ of a quadratic form $a \in H^{2k}(M^{4k}, Q)$.

Note that, as it follows from the definition, $\tau(M^{4k})$ is an integer. That the form A exists follows from Poincaré's duality theorem. It also follows that A can be given by the dual intersection operation for the cycles $\tilde{a}_i$ associated with the cocycles a_i. We have

$$\langle \tilde{a}_i \cap \tilde{a}_j \rangle = \text{ind}\,(\tilde{a}_i, \tilde{a}_j),$$

or the intersection number for $\tilde{a}_i$, $\tilde{a}_j$.

The signature $\tau(M)$ possesses the following properties.

Theorem 2.40 (Thom)

(1) $\tau([M] + [M']) = \tau(M) + \tau(M')$.

(2) $\tau(M \times M') = \tau(M) \cdot \tau(M')$.

(3) *If the manifold* M *is an oriented boundary, then* $\tau(M) = 0$.

Property (1) is obvious. The proof of (2) and (3) is in [MSt, Hi].

It follows from (1) and (3) that $\tau(M)$ can be represented as a polynomial $L(p_1, \ldots, p_k)$ of the Pontrjagin numbers $p_i(M)$. The exact form of the polynomials was indicated by Hirzebruch [Hi]. Knowing the Hirzebruch polynomials and Pontrjagin classes, we can find $\tau(M)$. The polynomials are

$$(1)\ L_1(M^4) = p_1/3, \tag{2.120}$$

$$(2)\ L_2(M^8) = (1/45)(7p_2 - p_1^2) \tag{2.121}$$

for manifolds M^{4n}, $n = 1, 2$.

It follows from the Hirzebruch signature theorem that the value of polynomials $L_k(p_1, \ldots, p_k)$ on M^{4k} is $\tau(M^{4k})$.

Consider the simplest example of calculating the signature $\tau(CP^2)$ explicitly. The vector space $H^2(CP^2, Q)$ has a unique generator a such that

$$\langle a \cup a | \mu \rangle = 1,$$

which follows from the structure of the ring $H^*(CP^2, Q)$ (see [MS]). Therefore, $\tau(CP^2) = 1$. We find from formula (2.120) that $p_1(CP^2) = 3$, which coincides with the calculation in Example 2.26.

Similarly, the signature $\tau(CP^{2n}) = 1$ is calculated.

The Hirzebruch signature formulas of type (2.120) and (2.121) possess one remarkable property. Since the signature $\tau(M)$ of a manifold is an integer, the coefficients L_k must satisfy strong divisibility conditions, e.g., it follows from (2.120) and (2.121) that $p_1(M^4) \vdots 3$ and $7p_2(M^8) - p_1^2(M^8) \vdots 45$.

Chapter 3
Physical Principles and Structures

In this chapter, the physical foundations for modern ideas in quantum field theory and statistical mechanics are discussed.

Such are the gauge invariance of the Yang-Mills field equations, spontaneous symmetry breakdown in continuous and discrete systems, of mass appearance via the Higgs mechanism, etc. By giving simple examples, we seek to explain what is behind these fundamental concepts, which form the foundation of modern physics unifying ideas.

The efficiency of topological methods for field and condensed media theories is just based on these ideas. The proximity of certain concepts of topology to field theory is seen when the same objects are described in two different languages.

3.1 GAUGE INVARIANCE

3.1.1 Maxwell and Yang-Mills Fields

The modern theories of elementary particles are based on several main propositions.

The interaction is carried by certain particles related to the fields. The Lagrangians are invariant under the transformation group associated with the conserved quantities of the model.

By way of example, consider Maxwell's classical electrodynamics as a free field.

The equations of a free electromagnetic field are

$$\partial^{\mu} F_{\mu\nu} = 0, \quad \partial_{\mu} F_{\nu\varrho} - \partial_{\nu} F_{\mu\varrho} + \partial_{\varrho} F_{\mu\nu} = 0, \tag{3.1}$$

where $F_{\mu\nu}$ is the electromagnetic field's strength tensor.

The first equation in (3.1) requires a metric.

We select the pseudo-Euclidean metric $g_{\mu\nu}$ $(+ + + -)$ for the Minkowski space $\check{M}^4$. The geometric meaning of the Maxwell equations is clarified in Subsec. 3.1.2. They possess the fundamental property of invariance under the transformations

$$A_\mu \to A_\mu + \frac{\partial f}{\partial x^\mu} \tag{3.2}$$

arbitrarily depending on the (world) point x_μ and are said to be *local gauge.* They form a gauge group G.

Conserved quantities in the theory are always related to the existence of a symmetry group, e.g., conserving the isospin in nuclear interactions means the invariance property of the corresponding Lagrangian under the isotopic space's rotation group $SU(2)$ which does not depend on the space-time coordinates. Such symmetry groups are said to be *global.*

From this standpoint consider the Lagrangian in quantum electrodynamics describing the interaction between electrons and photons.

The interaction between electrons is carried out via photon radiation and absorption.

The Lagrangian* is of the form

$$L = \bar{\psi}(i\hat{\partial} + e\hat{A} - m)\psi - (1/4)F_{\mu\nu}F^{\mu\nu}, \tag{3.3}$$

where $\hat{\partial} = \gamma_\mu \partial_\mu$, $\hat{A} = \gamma_\mu \partial_\mu$, γ_μ are Dirac matrices (see [IZ]), and is invariant under the relative local group $U(1)$, i.e.,

$$\psi \to \exp(i\alpha(x))\psi = S\psi,$$

$$A_\mu \to A'_\mu = A_\mu + (1/e)\frac{\partial \alpha(x)}{\partial x_\mu} = A_\mu + (1/e)\partial_\mu \alpha(x), \tag{3.4}$$

where $\alpha(x)$ is a parameter depending on the (world) point x, the interaction constant e is an electric charge and the selected system of units is $\hbar = c = 1$.

The strength of the field $F_{\mu\nu}$ is invariant under these transformations.

The requirement for local invariance of Lagrangian (3.3) leads to the following fundamental physical consequences:

(a) conservation of the electric charge,

(b) interaction being given by the massless photon field A_μ.

In fact, if the photon had mass, then the Lagrangian with the additional term $m^2 A_\mu A^\mu$ would not be invariant under the gauge transformations of the group $U(1)$. It is thus necessary for local invariance that the conserved charge be the source of a massless vector field.

Modern theories of strong and weak interactions are based on the non-Abelian generalization of gauge theory, i.e., the Yang-Mills theory.

The theory was originally designed to describe the forces relating nucleons, i.e., protons and neutrons, possessing the remarkable property that they can be regarded as the same particles in strong interactions. This was noticed even within the framework of quantum mechanics and was called *isotopic invariance.*

---

* More precisely, the Lagrangian density. The real Lagrangian is $\mathscr{L} = \int L(\mathbf{x}, x^0)\,d\mathbf{x}$. However, L is also often called a Lagrangian in field theory.

The invariance is a global symmetry acting in the space of internal degrees of freedom of a nucleon and not affecting the space-time continuum.

Let ψ be the nucleon's wave function and

$$\psi = \begin{pmatrix} \psi_1 \\ \psi_2 \end{pmatrix}, \quad \psi \in C^2,$$

be the isodoublet. Isotopic invariance means the invariance of the nuclear forces Lagrangian under the transformations

$$\psi \to \psi' = S\psi, \quad \bar{\psi} \to \bar{\psi}' = S^+ \bar{\psi}, \tag{3.5}$$

where S, $S^+ = S^{-1} \in SU(2)$, $S = \exp(\alpha_i T^i)$, α_i are parameters independent of x, $T^i = \frac{1}{2\sqrt{-1}} \tau^i$, and τ^i are the Pauli matrices* $\tau^1 = \sigma_x$, $\tau^2 = \sigma_y$, $\tau^3 = \sigma_z$ in the notation of (2.49). Since the τ^i do not commute, the symmetry is not Abelian.

Yang and Mills assumed the local character of certain internal symmetries, and not only global, postulating in [YM] that the properties of nuclear forces, connected to the isotopic symmetry, and, therefore, the indistinguishability between a neutron and proton, are also preserved under considerably more general transformations. Transformations of an isospin vector can be carried out independently at each point in spacetime, i.e., α is of the form $\alpha(x)$.

Models of strong interactions are constructed on the basis of Yang-Mills fields by modelling the ideas of quantum electrodynamics.

That the gauge group G ($SU(2)$ in our case) is non-Abelian leads to a number of new physical consequences.

An analog of the field A_μ is the Yang-Mills field W_μ transformed under the action of representation of $SU(2)$.

An analog of the conserved "charge" e is the isospin T operator with three components.

We now write the Lagrangian of a Yang-Mills field (in vacuum**, without fields ψ), for which we have to derive an analog of the covariant derivative of W_μ of the form

$$\nabla_\mu = \partial_\mu - \Gamma(W_\mu), \tag{3.6}$$

where $\Gamma(W_\mu)$ is the representation of the matrix W_μ, associated with the given representation of the group G.

We indicate the form of transformations of W_μ and the field strength tensor $F_{\mu\nu}$ by requiring that the Lagrangian be gauge invariant and quadratic with respect to $F_{\mu\nu}$.

A field transformation $\psi(x)$ similar to the local phase one in quantum electrodynamics is of the form

$$\psi(x) \to \psi^g(x) = \Gamma[g(x)]\psi(x), \tag{3.7}$$

---

* In physical literature, Pauli isospin matrices are denoted by τ^i and spin ones by σ^i.

** Such fields are said to be purely *Yang-Mills*.

where $g(x)$ is a function of x with values in G and can be regarded as a matrix in the adjoint representation of G. Derivative (3.6) is covariant with respect to (3.7) if W_μ is transformed as

$$W_\mu \to W'_\mu = gW_\mu g^{-1} + \partial_\mu g g^{-1}.$$

We put

$$F_{\mu\nu} = \partial_\mu W_\nu - \partial_\nu W_\mu + [W_\mu, W_\nu]. \tag{3.8}$$

It is easy to see that $F_{\mu\nu}$ is transformed as

$$g(x) F_{\mu\nu} g^{-1}(x). \tag{3.9}$$

Therefore, the expression $\mathrm{Tr}\,(F_{\mu\nu}F^{\mu\nu})$ is invariant under gauge transformations. The trace is taken in the space of the representation of G (of internal degrees of freedom).

The expression with $\mathrm{Tr}\, F_{\mu\nu}F^{\mu\nu}$ is the natural generalization of the Lagrangian of the electromagnetic field to the case of a non-Abelian gauge group, which is what the Lagrangian of a purely Yang-Mills field is.

The choice of the constant c, which is proportional to the interaction constant, depends on the normalization of the fields W_μ and on the basis for the space of the representation of G.

Let two fields W_μ and $F_{\mu\nu}$ take their values in the Lie algebra $\mathscr{G}$ of a compact semi-simple group G, or $U(1)$. We denote by T^a the generators $\mathscr{G}$ ($a = 1, \ldots, N$, $N = \dim \mathscr{G}$). The invariant metric on G is the Killing form B on $\mathscr{G}$. We normalize $B(X, Y)$ by the condition $\mathrm{Tr}\,(T^a T^b) = -(1/2)\delta^{ab}$. This normalization for L_{YM} is the most frequent, e.g., for a group $SU(2)$, it corresponds to the choice of generators as $T^j = \frac{1}{2i}\tau^j$, in which case*

$$L_{YM} = \frac{1}{2g^2}\,\mathrm{Tr}\, F_{\mu\nu}F^{\mu\nu} = -\frac{1}{4g^2} F^a_{\mu\nu}F^{\mu\nu a}.$$

For $U(1)$, the Lagrangian of Yang-Mills coincides with that of the electromagnetic field. The coefficient $-1/4$ is associated with the choice of the Heaviside system of units in the electromagnetism theory [LL1]. In writing the tensor $F_{\mu\nu}$ in the form of (3.8), the interaction (coupling) constant g is only involved in the coefficient $1/(2g^2)$ and can be eliminated by redefining $W_\mu \to gW_\mu$ the fields. However, g will then be used in the definition of the covariant derivative $\nabla_\mu = \nabla'_\mu = \partial_\mu - gW_\mu$. The Lagrangian describing the interaction of the Yang-Mills field with matter fields is of the form

$$L_{YM} + L_M(\psi, \nabla_\mu \psi), \tag{3.10}$$

where L_M describes the gauge-invariant interaction of the fields W_μ and $\psi(x)$ and

* The coupling constant in Yang-Mills theory is traditionally denoted by g (not to be mixed up with an element g of a group).

can be obtained from the Lagrangian for the fields ψ by replacing the usual derivatives with the covariant ones.

3.1.2 Geometric Interpretation of Maxwell and Yang-Mills Fields

In the previous subsection, the physical ideas were outlined bringing us to the concept of a non-Abelian gauge field. We now consider its geometric interpretation from the point of view of fibration theory.

1. Maxwell field. Let $\check{M}^4$ be Minkowski space-time. Consider the principal bundle P over $\check{M}^4$ with the structure group $G = U(1)$. The connection forms ω and θ take values in the Lie algebra $u(1) \sim R$. Since the group $U(1)$ is commutative, the structural equation (1.45) for the curvature form Ω (1.58) is turned into

$$\Omega = d\theta, \tag{3.11}$$

or, in coordinates*,

$$\theta = A_\mu dx^\mu, \quad \Omega = (1/2)(\partial_\mu A_\nu - \partial_\nu A_\mu)dx^\mu \wedge dx^\nu, \quad \mu, \ \nu = 1, \ \ldots, \ 4. \tag{3.12}$$

The Bianchi identity turns into the relation

$$d\Omega = 0 \tag{3.13}$$

and the Maxwell equations (3.1) into

$$d\Omega = 0, \quad d^*\Omega = 0. \tag{3.14}$$

The equation $d^*\Omega = 0$ is written in terms of the tensor $F_{\mu\nu}$ in the usual form $(\partial^\mu F_{\mu\nu} = 0)$, while $d\Omega = 0$ in terms of the dual tensor ${}^*F_{\mu\nu}$ as $\partial_\mu {}^*F_{\mu\nu} = 0$, where ${}^*\Omega$ is the form dual to the operation $*$ and ${}^*F_{\mu\nu} = (1/2)\varepsilon_{\mu\nu\varrho\gamma}F^{\varrho\gamma}$ the dual tensor. The curvature form is invariant, i.e., $g\Omega g^{-1} = \Omega$, $g \in U(1)$. The connection transformation is

$$\theta' = g\theta g^{-1} + dgg^{-1} = \theta + dgg^{-1}, \quad A_\mu' = gA_\mu g^{-1} + \partial_\mu gg^{-1} = A_\mu + \partial_\mu gg^{-1},$$

where dgg^{-1} is an element of the Lie algebra of the group $U(1)$, i.e., $\mathscr{G} = u(1) = \{i\lambda\} \sim R$. If we put $g = \exp(i\lambda(x))$, then $\partial_\mu gg^{-1} = i\partial_\mu\lambda$. We thus obtain a formula for gauge transformations equivalent to (3.2).

The Maxwell field A_μ is a connection (a gauge field) on $\check{M}^4$. Since $\check{M}^4$ is contractible, P is trivial.

However, there are important examples of the Maxwell equations with sources (the Dirac magnetic monopoles) where we have to consider (stationary) solutions given on the manifold $R^3 \setminus x_0$. A magnetic charge is placed at the point x_0. The manifold is contractible to the two-dimensional sphere S^2 and the fiber bundle aris-

* We replace θ_μ by A_μ in conformity with the universally accepted designation for vector potential in physics.

ing with the fiber $U(1)$ is non-trivial. We consider the Dirac magnetic monopoles in Subsec. 4.1.1.

2. Yang-Mills field. The geometric interpretation of free Yang-Mills fields is essentially the same as that of the Maxwell ones; however, the non-Abelianness of the gauge group has far-reaching consequences.

Let the gauge group G be a semisimple Lie one. The base space is a four-dimensional manifold M^4 with nonsingular metric $g_{\mu\nu}$.

Consider the principal bundle P over M^4 with G and a local connection form $\theta = W_\mu dx^\mu$, the values being in the Lie algebra $\mathscr{G}$ of G. If we define the curvature form $\Omega = D\theta$, then, according to (3.8), we obtain

$$\Omega = (1/2)F_{\mu\nu}dx^\mu \wedge dx^\nu, \quad \mu, \nu = 1, \ldots, 4. \tag{3.15}$$

Since a metric is given on the space M^4, we can define the operator $*: \Omega \to *\Omega$ on the curvature forms.

The Bianchi identity

$$D\Omega = 0 \tag{3.16}$$

holds for Ω. If we turn to the form $*\Omega$, then applying D yields

$$D^*\Omega = 0. \tag{3.17}$$

If we write (3.16) and (3.17) in terms of the covariant derivatives ∇_μ

$$\nabla_\mu F = \partial_\mu F - [W_\mu, F],$$

then

$$[\nabla_\mu[\nabla_\nu, \nabla_\sigma]] + [\nabla_\nu[\nabla_\sigma, \nabla_\mu]] + [\nabla_\sigma[\nabla_\mu, \nabla_\nu]], \tag{3.18}$$

or

$$\nabla_\mu F_{\nu\sigma} + \nabla_\nu F_{\sigma\mu} + \nabla_\sigma F_{\mu\nu} = 0$$

and

$$[\nabla_\mu[\nabla_\mu, \nabla_\nu]] = 0, \tag{3.19}$$

or

$$\nabla_\mu F_{\mu\nu} = 0. \tag{3.20}$$

The latter is what we usually call the Yang-Mills equation.

We now discuss the important question of the association of the Yang-Mills equations with the metric on the bundle P. The Bianchi identity does not require a metric (only a connection). However, to turn to dual forms, or to specify the operator $*$, the metric $g_{\mu\nu}$ must exist. Yang-Mills equations can be considered on any manifold M with a metric, however, only the Minkowski $\check{M}^4$ and Euclidean R^4 spaces and their compactifications are of physical interest.

Equation (3.20) is derived by a variational method from the action

$$S = -\frac{1}{2g^2}\int_{M^4} \mathrm{Tr}\,(F_{\mu\nu}F^{\mu\nu})dV. \tag{3.21}$$

The equation was obtained by integrating the Lagrangian density L with respect to the space M^4.

Raising the indices in $F^{\mu\nu}$ is carried out by means of $g_{\mu\nu}$, namely, $F^{\mu\nu} = g^{\mu\alpha}g^{\nu\beta}F_{\alpha\beta}$. The trace of the matrix $F \subset \mathscr{G}$ is determined in the sense of the Killing form B given in the Lie algebra. Recall that the algebra is assumed to be semi-simple, therefore, B is non-singular.

For the present, we have assumed that the following two requirements for G and M are fulfilled: semisimplicity of the group G, i.e., nondegeneracy of the Killing form and the existence of a non-singular metric $g_{\mu\nu}$. Reasonable physical requirements strongly restrict the classes of the groups. For example, if we regard internal symmetry groups as compact—and we only encounter such situations now—then it is natural to confine ourselves to compact semi-simple Lie groups.

In real physical theories only certain series of classical Lie groups are encountered: $O(n)$, $U(n)$, $SU(n)$, and their products. The symplectic group $\mathrm{Sp}(n)$ appears extremely rarely.*

If we confine ourselves to compact simple Lie groups, then the Killing form is negative definite. The sign in (3.21) is chosen using the energy positivity condition.

We now indicate the differences in writing the Yang-Mills equation in Minkowski $\check{M}^4$ and Euclidean R^4 spaces. Since the difference is determined by the metrics $g_{\check{M}} = (+++-)$** and $g_E = (++++)$, the corresponding changes arise under the action of the operator $*$ on the forms. Let Ω_E be associated with the tensor $F_{\mu\nu}$ in R^4, while $\Omega_{\check{M}}$ with $F_{\mu\nu}$ in $\check{M}^4$. Then

$$**\Omega_E = \Omega_E, \quad **\Omega_{\check{M}} = -\Omega_{\check{M}}. \tag{3.22}$$

We can turn from $\check{M}^4$ to R^4 with replacing the coordinates $x^1, x^2, x^3, t \in \check{M}^4$ with $x^1, \ldots, x^4 \in R^4$, where $x^4 = it$. The transformation, called a *Wick rotation* in physics, enables us to turn from calculations in the Minkowski space to Euclidean. The exact meaning of the assertion is in what follows. If the real time t in the coordinate system in $\check{M}^4$ is extended to imaginary values, then $x_E^4 = it$ can be regarded as the fourth coordinate of R^4, leaving the others unaltered. After such an analytic continuation with respect to t, $g_{\check{M}}$ and g_E coincide. It turns out that many calculations can be carried out in the Euclidean domain to obtain finite expressions and analytically continue them into the "physical" region or the Minkowski space.

On the other hand, there exist direct physical interpretations of motions along "trajectories" in a Euclidean domain, which are well-known quantum mechanics analogs of particles' tunnelling. The classical solutions to the Yang-Mills equation are sought for in Euclidean space. The corresponding considerations are in Sec. 4.2.

Pure Lagrangians of Yang-Mills fields and those interacting with other fields are the basis for the theory of strong interactions and the unified theories of weak and electromagnetic interactions. The theory of weak and electromagnetic interactions now seems to be considerably more complete. The brilliant confirmations of

* Lately [GSW], the exceptional Cartan subalgebras have been used, e.g., the group E_8 in field theories based on models of strings.

** The metric $g'_{\check{M}} = (---+)$, $g'_{\check{M}} = -g_{\check{M}}$, is also often used on $\check{M}^4$.

Weinberg-Salam-Glashow model (the discoveries of the *neutral currents* and *W*- and *Z*-bosons predicted in the theory) suggest that the theory is basically true. The situation with strong interactions is not so convincing, though quantum chromodynamics is thought by many to be the basis for a future theory.

In addition to the Yang-Mills fields, interaction carriers, a method for introducing particle masses forms the basis of modern unified field theories. The origins are in the mechanism of *spontaneous symmetry breaking*, well-known in phase-transition theory. The appearance of spontaneous magnetization in a ferromagnet is possibly the first example of this effect.

Statistical mechanics ideas turn out to be quite fruitful in the theory of elementary particles.

3.2 SYSTEMS WITH SPONTANEOUS SYMMETRY BREAKING

Liquid crystals, magnets, and superfluid helium are just a few media, different in their properties, which can be included into a wide class of systems with spontaneous symmetry breaking.

The essence of the phenomenon can be clarified by considering a concrete example from the theory of magnetism. A number of crystals are known to possess magnetism in the absence of an external magnetic field and low temperatures. The phenomenon is called *ferromagnetism* and due to the existence of special exchange interaction between crystal lattice atoms. The magnetism is said to be *spontaneous* (formed without exterior field application) and characterized by the vector of magnetization **M**, or the *magnetic moment* of the ferromagnet.

We do not go deep into ferromagnetism theory, but consider the familiar model of an isotropic Heisenberg ferromagnet with spontaneous symmetry breaking. Imagine a crystal lattice with particles of half-integer spin, e.g., electrons, at the sites. The interaction between electrons at adjacent lattice points is determined via the spin vectors $\mathbf{S}(x)$. For definiteness, we give the Hamiltonian (energy operator)

$$\hat{H} = \sum_{x,\, x'} I(x - x')\mathbf{S}(x)\cdot\mathbf{S}(x'). \tag{3.23}$$

The function $I(x)$ vanishes if $x \neq \mathbf{a}$, where $\mathbf{a}$ is the lattice's basis, $I(a) = \lambda$, $\lambda < 0$. The quantity $I(x)$ is called an *exchange integral* and the interaction itself the *nearest-neighbor,* since only two neighboring points contribute to (3.23). The operator-valued vector of magnetization **M** is

$$\mathbf{M} = \sum_{x} \mathbf{S}(x). \tag{3.24}$$

We can show that (3.23) is invariant under rotation in a spin space, i.e., the symmetry group H is $SO(3)$.

However, the lowest energy state, the ground state of the system, is associated with the greatest $\langle \mathbf{M}^2 \rangle$. Since $I < 0$, all spins **S** are similarly oriented along a fixed axis **n** in this state and possess maximum projection of **M** onto **n**. It follows that the ground state is not invariant under the full rotation group for the spin space.

Since the symmetry group must preserve the fixed direction of **M**, the group coincides here with the rotation group $SO(2)$ about **M**.

We can finally define a spontaneous symmetry breaking.

Systems in which the symmetry of the ground state does not coincide with that of the Hamiltonian are said *to have spontaneous symmetry breaking*, however, it would be more correct to call them *systems with hidden symmetry.* In fact, Hamiltonian symmetry is only hidden rather than broken. A higher symmetry of the system cannot be observed in the ground state.

Examples of spontaneous symmetry breaking are encountered in various problems of physics. For instance, nuclear forces are known to be rotation-invariant, while the ground state of a nucleus with nonzero spin is not invariant under the rotation group.

The spontaneous symmetry breaking is one of the mechanisms responsible for phase transitions in matter.

The same substance is known to be in different states, or phases, depending on environmental conditions (such as temperature or pressure). *Transition* from one phase into another is said to be *phase*. Phase transitions arise in various materials, e.g., the transition of a metal from the normal superconducting state arises at ultralow temperatures.* Superfluidity of helium under temperatures of about 2 K is another example. The phase transition into the superfluid state of another helium isotope ^{3}He was recently discovered. It occurs at the fantastically low temperature of 0.0026 K.

The fruitful approach underlying the modern theory of phase transition is based on the study of system symmetry in various phases. There are two kinds of phase transitions. The classical examples of the first-kind transition are liquid-vapor or solid-liquid, such as ice melting under high pressure. The second kind of transitions are those of liquid helium from the normal to superfluid state and of a metal to the superconducting one.

The first- and second-kind transitions are essentially different in nature. The difference can be clarified if we turn to the classical work by Landau, 1937, underlying the modern phase-transition theory. Landau considered phase transitions from one state into another as changes in the substance's symmetry. To describe a transition quantitatively, he introduced the *degree*, or *parameter, of order.* The order parameter is defined in a different way for different systems, however, it possesses an important property in common, namely, it vanishes in the "disordered" phase and is nonzero in the "ordered" one.

Phase transitions of the first and second kind differ in the behavior of the order parameter. For the first kind, the parameter varies jumpwise, while for the second kind it varies continuously.

Consider the ice-water phase transition (melting process) from the above viewpoint. If we select the order parameter η as the ratio of the number of atoms at the vertices of the crystal lattice to that of all atoms, then in the "ordered" phase η is nonzero, in the "disordered" one, i.e., in water, η vanishes. Meanwhile, η changes jumpwise, therefore, melting is a phase transition of the first kind.

* The recent discovery of superconductors with $T_c \sim 90$ K revised our understanding of superconductivity radically.

We now sharpen what "ordered" and "disordered" mean. Consider the symmetry of two states, ice and water. It is natural to regard the crystal lattice's symmetry group Γ as that of ice. Γ is discrete with the lattice translations, discrete spatial rotations, and reflections as its elements. However, the symmetry group G_{H_2O} of water, which is that of all transformations preserving the invariance of the fluid dynamics equations for an incompressible fluid, is considerably wider than Γ, continuous of infinite dimensional.

In our example, G_{H_2O} of the less ordered phase, water, contains the ice symmetry group as a subgroup. However, the situation holds mostly in the second-kind phase transitions. In phase transitions of the first kind, it is feasible that the symmetries of both phases are not related at all.

Let us discuss phase transitions for the Heisenberg ferromagnet. In the ground state ($T = 0$), all spins are codirectional. We specified this direction above by a vector $\mathbf{n}$ corresponding to the maximal value of $\langle \mathbf{M} \rangle^2$. When heating ($T > 0$), the correlation between the spins is weakened, their orientation chaotic, and the average of $\langle \mathbf{M}^2 \rangle$* decreases accordingly. At a certain temperature $T = T_C$, the *Curie point*, $\langle \mathbf{M}^2 \rangle$ becomes zero. The ferromagnet then loses its properties, i.e., changes into the paramagnetic phase. The order parameter defining the phase transition in the system can be taken as $\mathbf{M}$.

It is appropriate to stress the symmetry relation to the properties of ordered and disordered phases. In the ordered or ferromagnetic state the system possesses a ground state symmetry, and the symmetry group coincides with the rotation one leaving invariant the magnetism direction along which the spins were aligned. When passing to the paramagnetic state, a system has formed with a wider group of all rotations in three-dimensional space, since there is no distinguished direction in the paramagnetic phase when the Hamiltonian's full symmetry group coincides with the symmetry group of the ground state, or as the physicist would say, symmetry has been restored.

We can observe all the features of the general scheme: the symmetry group associated with the ordered phase is smaller than that with the unordered phase. Meanwhile, the unordered phase's symmetry group contains that of the ordered one as a subgroup.

Landau's macroscopic theory of phase transitions enabled us to describe a very large number of phenomena. At the same time, the difficulty of the detailed description of phase transitions near the critical points remains. Essential progress in recent years has occurred after applying field theory ideas. Still, the problem of constructing a microscopic theory of phase transitions is not solved completely.

We now turn to spontaneous symmetry breaking in field theory. The effect occurs in field theory in the same manner as in statistical physics. Certain systems in solid-state theory actually model the field theory on a lattice.

Let us see how internal symmetry breaking occurs in the simplest case of a scalar field in two-dimensional space-time. The field theory to be studied is mostly assumed to be classical. The transition to quantized fields is a separate serious problem. Nevertheless, quantum mechanics terminology can be employed.

---

* The exact definition of phase transition involves a concept of local magnetization or appropriate correlators.

3.2.1 Discrete Spontaneous Symmetry Breaking

The Lagrangian of a scalar real field is

$$L = (1/2)(\partial_t \varphi)^2 + (1/2)(\partial_x \varphi)^2 - U(\varphi),$$

while the Hamiltonian and potential are, respectively,

$$\hat{H} = (1/2)(\partial_t \varphi)^2 + (1/2)(\partial_x \varphi)^2 + U(\varphi) \tag{3.25}$$

and

$$U(\varphi) = (\mu^2/2)\varphi^2 + (\lambda/4)\varphi^4,$$

where $\lambda > 0$ and μ^2 can be either negative or positive.

The lowest energy state is called the *vacuum state* and denoted by $\langle \varphi \rangle$. The quantity $\langle \varphi \rangle$ is governed by the minimum condition for $U(\varphi)$. The Hamiltonian (3.25) is invariant under transformations $\varphi \to -\varphi$.

Let us see how the vacuum states are arranged in the model. The graphs of $U(\varphi)$ are in Fig. 8 as a function of the sign of μ^2. If (a) $\mu^2 > 0$, then $\langle \varphi \rangle = 0$. Suppose (b) $\mu^2 < 0$. The potential minima are at the points

$$\langle \varphi \rangle = \pm\sqrt{-\mu^2/\lambda}$$

and the vacuum is degenerate because its energy is the same at two different points.

In quantum language, the "masses" of particles are defined as a small oscillation spectrum in the neighborhood of the equilibrium point, or vacuum. The mass of the scalar particle generated by the field φ (normally called a *scalar meson*) is determined by the coefficient of φ^2 in the neighborhood of the vacuum state.

Thus, in the case of discrete symmetry breaking, the transition to a nonzero vacuum state only leads to mass change.

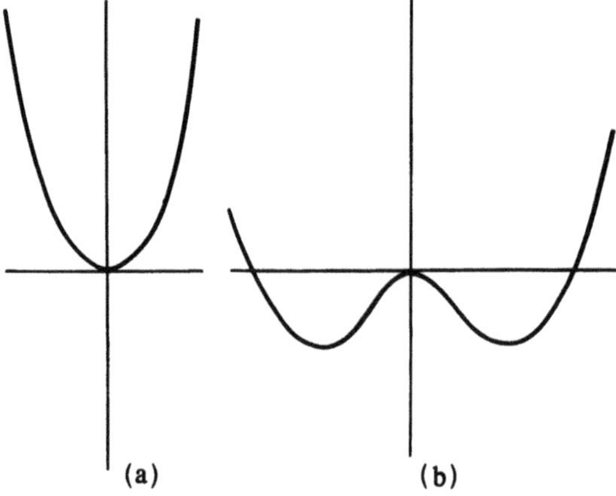

Fig. 8. Graph of potential $U(\varphi)$, (a) $\mu^2 > 0$, (b) $\mu^2 < 0$

The new effect shows up when the field equations have a continuous symmetry group.

3.2.2 Spontaneous Breaking of a Continuous Symmetry Group

Consider a theory with two scalar fields A and B and the potential

$$U(A, B) = \lambda(A^2 + B^2 - a^2)^2, \tag{3.26}$$

where λ is a fixed constant. $U(A, B)$ is invariant under the group $SO(2)$ of rotations

$$A \to A' = A \cos \omega + B \sin \omega,$$
$$B \to B' = -A \sin \omega + B \cos \omega$$

of the field $\varphi(A, B)$.

The minima of $U(A, B)$ are on the circle

$$A^2 + B^2 = a^2.$$

As in the above case, it is immaterial which minimum is taken as a vacuum state. As soon as the vacuum state is fixed, the internal symmetry is broken spontaneously.

We select

$$\langle A \rangle = a, \quad \langle B \rangle = 0.$$

Expanding $U(A, B)$ in the neighborhood of the vacuum in terms of the variables $A' = A - \langle A \rangle$, $B' = B$, we obtain

$$U = \lambda(A'^2 + 2aA' + B'^2)^2. \tag{3.27}$$

Hence, the A-meson acquires mass and the B-meson becomes massless (the terms quadratic in B' absent). This result does not depend on the concrete form of U and is only related to the symmetry group $SO(2)$. It becomes especially obvious if we turn to the complex field $\psi = \varrho \exp(i\varphi)$ and the Lagrangian invariant under the group $U(1) \sim SO(2)$ being the fields of the form

$$A = \varrho \cos \varphi, \quad B = \varrho \sin \varphi.$$

The Lagrangian turns into

$$L = (1/2)(\partial_\mu \varrho)^2 + (1/2)\varrho^2(\partial_\mu \varphi)^2 - U(\varrho). \tag{3.28}$$

The invariance of U under $U(1)$ means that the potential $U(\varrho)$ does not depend on φ.

If we introduce, as before, the translated fields $\varrho' = \varrho - \langle a \rangle$ and $\varphi' = \varphi$ in the neighborhood of the vacuum $\langle \varrho \rangle = a$, $\langle \varphi \rangle = 0$, then (3.28) is of the form

$$L = (1/2)(\partial_\mu \varrho')^2 + (1/2)(\varrho' + a)^2(\partial_\mu \varphi')^2 - U(\varrho' + a)$$

in terms of the new variables. Hence it is clear that φ is the massless meson because the Lagrangian only involves derivatives of the φ field.

The massless particles appearing in continuous symmetry's spontaneous breaking are called *Goldstone bosons* in honor of the American physicist J. Goldstone.

3.2.3 Group-Theoretic Interpretation of Goldstone Bosons

If we consider a Lagrangian of general form with potential $U(\varphi)$ invariant under a global symmetry group G (assumed to be compact), then we can give the following description of the Goldstone bosons.

Given a collection of n real vector fields $\varphi = \{\varphi_i\}$ such that the corresponding potential U is invariant under the transformation group G, namely, $U(g\cdot\varphi) = U(\varphi)$, $g \in G$.

We take a point $\langle\varphi\rangle$ belonging to the set of minima of $U(\varphi)$ and consider the subgroup H of G which leaves $\langle\varphi\rangle$ fixed. H is the stability subgroup of G. Depending on the structure of U, H can be any subgroup of G (from the trivial one to the whole of G).

Let $\dim G = N$, $\dim H = m$. If we embed H in G so that the first m generators of G are the same as those of H, then the corresponding equations for the vacuum state are

$$T_i\langle\varphi\rangle = 0, \quad i \leqslant m.$$

The set of vacua $\{\langle\varphi\rangle\}$ is an orbit of G in the space of fields. The orbit is the surface of constancy of $U(\varphi)$ passing through the point $\langle\varphi\rangle$ and has the dimension $N - m$.

If we introduce coordinates on the $(N - m)$-dimension orbit Ω by means of one-parameter groups $g_i(t)$, $i = 1, \ldots, N - m$, then their generators are spontaneously broken infinitesimal symmetries of the Lagrangian, each of which is associated with massless particles (generated by shifts along the corresponding orbits). It is obvious that there are as many of them as $\dim \Omega$.

The appearance of massless particles, Goldstone bosons, is characteristic of local field theories with symmetry breaking, obeying the Wightman axioms (Lorentz-invariance, locality, existence of the Hilbert space of states with positive norm, spectrality) [SW].

At the time, because the Goldstone result was proved so rigorously as to be a theorem it caused serious concern. The massless particles were not observed in the real world and so they cast doubt on all field theories with a mechanism for spontaneous breaking.

However, as often happens to rigorous theorems in physics, the more serious the conclusions from the assertions, the more attention should be paid to the starting points. The same happened to Goldstone's theorem. As the saying goes, "every cloud has a silver lining". Almost at the same time another theory (the Yang-Mills) of gauge fields encountered similar difficulties. The Yang-Mills gauge field had to generate massless gauge vectorial particles. Coleman [Col] in his lecture characterized the situation as follows:

"It is pleasant to remember that, at the time of their inventions, both the theory of non-Abelian gauge field and the theory of spontaneous symmetry breaking were thought to be theoretically amusing but physically untenable, because both predicted unobserved massless particles, the gauge mesons and the Goldstone bosons. It was only later that it was discovered that each of these diseases was the other's cure."

The "panacea" turned out to be the Higgs mechanism.

3.2.4 Higgs Mechanism

The mechanism's action is most conspicuous in the technically simplest case of interaction between a complex field φ (with the internal symmetry group $SO(2) \sim U(1)$) and a gauge electromagnetic field A_μ.

The Lagrangian invariant under the relative local group $U(1)$ is of the form

$$L = [(\partial - ieA)\varphi]\cdot[(\partial + ieA)\bar{\varphi}] - (1/4)F^2 - U, \tag{3.29}$$

where

$$U(\varphi) = \mu^2(\varphi\bar{\varphi}) + \lambda(\varphi\bar{\varphi})^2, \quad \mu^2 < 0, \quad F_{\mu\nu} = \partial_\mu A_\nu - \partial_\nu A_\mu.$$

We write φ in polar coordinates as

$$\varphi(x) = (1/\sqrt{2})\varrho(x) \exp(i\theta(x)). \tag{3.30}$$

The invariance of the potential $U(\varphi)$ under $U(1)$ means the independence of U from θ.

Vacua, i.e., minima of U, are on the circle $\varrho = a$. Fixing a circle point, e.g., $\langle\varrho\rangle = a$, we fix the vacuum. It is not symmetric relative to $U(1)$, due to which a massless particle associated with the component θ of the field appears in the purely scalar model.

As φ interacts with the gauge field, a totally new effect arises.

It follows from the invariance of (3.29) under $U(1)$ that the field A must be transformed as in

$$A_\mu \to A'_\mu = A_\mu + (1/e)\partial_\mu\theta(x), \tag{3.31}$$

substituting which together with (3.30) in the Lagrangian, we obtain

$$L = (1/2)[(\partial_\mu - ieA')\varrho][(\partial_\mu + ieA')\varrho] - U(\varrho^2) - (1/4)F^2, \tag{3.32}$$

or

$$L = (1/2)(\partial_\mu\varrho)^2 + \frac{e^2}{2}\varrho^2 A'^2 - U(\varrho^2) - (1/4)F^2. \tag{3.33}$$

In the interaction of φ with A, θ is eliminated from the Lagrangian and the massless θ-meson disappears under the transformation from $U(1)$.

The physical meaning of the above transformations becomes clear if we find the spectrum of the obtained Lagrangian's masses. According to the standard recipe we expand (3.33) with respect to the deviation of ϱ from the vacuum expectation value $\langle \varrho \rangle$ and only retain the terms quadratic with respect to fields. Let $\varrho = a + \varrho'$. Then

$$L = (1/2)(\partial\varrho')^2 - (\mu^2/2)\varrho'^2 + (1/2)e^2a^2A'^2 - (1/4)F^2 + \ldots + \text{const.} \quad (3.34)$$

The Lagrangian describes the free field ϱ' with particles of mass μ and the free field A' with particles of mass $m_A = ea$, which can be seen from the equation of motion following from the Lagrangian, namely,

$$\partial^2\varrho' + \mu^2\varrho' = 0, \quad (3.35)$$

$$\partial_\mu F_{\mu\nu} = m_A^2 A'_\nu. \quad (3.36)$$

The equations are said to be *Proca* and describe particles with unit spin and mass m_A.

We now discuss the physical meaning of the miraculous transformations. If the Goldstone field φ interacts with a gauge one A_μ, the massless boson θ vanishes and a vectorial particle with a mass appears. The massive meson associated with the scalar field also remains.

Originally, there is a two-component field φ and a Maxwell vector field A_μ. If only quadratic terms are retained in (3.29) for $\mu^2 > 0$, then the Lagrangian splits into the sum of the Lagrangians of two free fields described by the (positive and negative) charged particles (with spin zero) and photons of zero mass and two states of polarization; altogether four types of particles. Now, in breaking $U(1)$-symmetry, the number of particle types does not change, but the particles themselves become different. There are no charged ones any more, the field ϱ' being real. On the other hand, the particles of the field A acquire mass, the number of their polarizations is three (of spin unity), the Maxwell field turns into the Proca one, and there are no Goldstone particles now.

In the case of global $U(1)$-symmetry, Goldstone bosons arise, compensating thereby for the decrease in the number of degrees of freedom of the complex field φ in symmetry breaking.

In local symmetry breaking, the extra degree of freedom is taken by the vector field with gauge $U(1)$-symmetry breaking.

This mechanism of acquiring mass by the originally massless particle is called the *Higgs effect**.

Quite similar reasons account for interesting phenomena in superconductivity. In particular, gauge symmetry breaking in the macroscopic equation for superconductors in the presence of a magnetic field shows up in the unusual Meissner effect in which magnetic flux is excluded from superconductors.

---

* The honor of the discovery does not belong to Higgs only, but also to a number of other researchers, including Brout, Englert, Kibble *et al.* See in [Col] and [O'R].

3.2.5 Meissner Effect

To explain most observed phenomena in superconductors, we can operate within the framework of the phenomenological Ginzburg-Landau (G-L) equation describing changes in the macroscopic wave function Ψ (system's order parameter) of the superconductor. The reader may find the detailed description of the Meissner effect in [LL3, TT]. We write the Helmholtz free-energy potential

$$F = (1/2)B^2 + (1/m^*)(\overline{\nabla}\overline{\Psi})(\nabla\Psi) + U(\Psi\overline{\Psi}), \tag{3.37}$$

where Ψ is the complex-valued function, order parameter, $\overline{\Psi}$ the complex-conjugate function, $B = \text{rot}\, A$ the proper magnetic field of the superconductor, $\nabla = \partial_\mu - ie^*A$ the covariant derivative, and $\overline{\nabla} = \partial_\mu + ie^*A$ the conjugate operator. The constants $m^* = 2m$ and $e^* = 2e$ are, respectively, the effective mass and the effective charge of the Cooper pair, and potential $U(\Psi\overline{\Psi})$ is represented in the form

$$U(\Psi\overline{\Psi}) = U_0 + \frac{\mu^2}{2}|\Psi|^2 + (\lambda/4)|\Psi|^4 \tag{3.38}$$

already familiar to us, where U_0 is the free-energy density in the normal state and with null magnetic field, μ^2, λ are real parameters:

$$\mu^2 = \mu_*(T - T_C), \quad \lambda = \text{const}, \quad \mu^2 < 0.$$

Since $\mu^2 < 0$, the minimum of $U(\Psi)$ is attained for $\Psi \neq 0$ and $\langle\Psi\rangle = -\mu/\sqrt{\lambda} = \delta$ is the spontaneous symmetry breaking.

If an external magnetic field H is switched on, then in order to find the stable state we should minimize the Gibbs free energy $G(T, H) = F - B\cdot H$, where G depends on Ψ and A.

The corresponding variation with respect to Ψ yields

$$\delta G/\delta\overline{\Psi} = 0. \tag{3.39}$$

The complex conjugate equation (since G is real)

$$\delta G/\delta\Psi = 0$$

is equivalent to (3.39).

Equation (3.39) is written as

$$(1/m^*)(\partial_\mu - ie^*A_\mu)^2\Psi + \mu^2\Psi + \lambda|\Psi|^2\Psi = 0 \tag{3.40}$$

(under gauge $\partial_\mu A_\mu = 0$).

The boundary condition for (3.40), making the surface terms vanish in variation of the functional, is

$$(\mathbf{n}\cdot\nabla\Psi) = 0, \tag{3.41}$$

where $\mathbf{n}$ is the normal to the surface.

The variation of the functional G with respect to A yields

$$I_e = \frac{ie^*}{m^*}(\bar{\Psi}\nabla\Psi - \Psi\overline{\nabla\Psi}) - \frac{2e^{*2}}{m^*}|\Psi|^2 A, \tag{3.42}$$

where $I_e = \text{rot } B$ is the superconducting current defined by the Maxwell equations.

For reasons similar to (3.41), the boundary condition is

$$\mathbf{n} \times (\mathbf{B} - \mathbf{H}) = 0. \tag{3.43}$$

Completed by the requirement that rot $\mathbf{H} = 0$, together with (3.40)-(3.42), (3.43) forms the G-L system for superconductors.

The G-L equations can be used to explain the Meissner effect.

We consider their solution with $\Psi = \text{const}$, $\langle\Psi\rangle \neq 0$, minimizing the Gibbs potential due to the existence of gradient terms. The equation for I_e turns into

$$I_e = \frac{-2e^{*2}}{m^{*2}}|\Psi|^2 A = \frac{-2e^{*2}}{m^*}\delta^2 A \tag{3.44}$$

said to be *Londons.*

By means of the Londons equation, it is easy to see that the magnetic flux is displaced out of a one-connected conductor.

We take the rotor of both sides of (3.44) and obtain

$$\text{rot } I_e = \text{rot rot } B = -\frac{2e^{*2}}{m^*}\delta^2 \text{ rot } A = -c^2 B. \tag{3.45}$$

Taking the Maxwell equations for B into account, div $B = 0$, we obtain

$$\Delta B = -c^2 B.$$

The solution for a plane surface in a parallel field is

$$B = B_0 \exp(-\chi/c)$$

where B_0 is the value of B on the surface and χ the distance from the surface. Magnetic induction B vanishes in a superconductor. The constant c is called the *penetration depth.*

Thus, the Meissner effect consists in the displacement of the magnetic flux B out by the "quantum" current I.

Note that the exterior field H has not influenced the effect of displacing B. Hence, B is driven out irrespective of whether the specimen is in the exterior field under the temperature below the critical point T_c or H is applied after the transition into the superconducting state, which means that the equilibrium thermodynamics is applicable to superconductors.

If we compare the above example with interaction between a scalar and gauge field A_μ, then far-reaching analogies can be seen.

TABLE 5

Meissner effect for superconductor	Higgs mechanism
Macroscopic wave function Ψ	Scalar field φ
Vector potential A_μ field	Gauge field A_μ
Penetration depth c	Higgs boson mass m

3.2.6 Higgs Effect and a Yang-Mills Field

We now consider the interaction between a Yang-Mills field and the Goldstone-Higgs spontaneously symmetry breaking. We take the technically simplest case. The Yang-Mills field W_μ belongs to the adjoint representation of the group $SU(2)$ and the Goldstone field φ is with isospin.

We define $\varphi = \begin{pmatrix}\varphi_1\\ \varphi_2\end{pmatrix}$ as the two-dimensional spin representation of $SU(2)$ in C^2, $\bar{\varphi}^+(\bar{\varphi}_1, \bar{\varphi}_2)$. The action of $SU(2)$: $\varphi \to g\varphi$, $g \in SU(2)$ is isospin rotation

$$\begin{pmatrix}\alpha, & \beta\\ -\bar{\beta}, & \bar{\alpha}\end{pmatrix}\begin{pmatrix}\varphi_1\\ \varphi_2\end{pmatrix} = \begin{pmatrix}\varphi_1'\\ \varphi_2'\end{pmatrix}, \quad \begin{pmatrix}\alpha, & \beta\\ -\bar{\beta}, & \bar{\alpha}\end{pmatrix} \in SU(2).$$

The gauge-invariant Lagrangian L is

$$L = L_{\mathrm{YM}} + (\nabla_\mu \varphi)^+ (\nabla^\mu \varphi) - [\lambda^2(\varphi^+\varphi - a^2)]^2, \tag{3.46}$$

where $L_{\mathrm{YM}} = -(1/4)\,\mathrm{Tr}\,(F_{\mu\nu}F^{\mu\nu})$ and $\nabla_\mu\varphi = \partial_\mu\varphi + (i/2)g\tau^a W_\mu^a \varphi$.

The gauge transformation is given by the formula

$$\delta\varphi(x) = (i/2)g\tau^a\alpha^a(x)\varphi(x), \tag{3.47}$$

where we consider the infinitesimal transformation from $SU(2)$, therefore, elements $g(x) \in SU(2)$ are representable as $g(x) = 1 + \alpha^a(x)T^a = 1 + \alpha(x)$, T^a being generators of the Lie algebra $su(2)$, $\alpha(x) \in su(2)$.

It is easy to see that the vacua $U(\varphi)$ are on the three-dimensional sphere S^3 in C^2, namely,

$$(\varphi^+\varphi) = a^2. \tag{3.48}$$

We fix a minimum, e.g., $\varphi_0 = \begin{pmatrix}0\\ a\end{pmatrix}$, where a is a real positive number.

We have thereby broken the symmetry.

By the gauge group $SU(2)$, we can turn the spinor $\frac{1}{\sqrt{2}}\begin{pmatrix}0\\ \varrho(x)\end{pmatrix}$, where $\varrho(x)$ is a real function $\langle\varrho\rangle = a$, into an arbitrary spinor $\begin{pmatrix}\varphi_1\\ \varphi_2\end{pmatrix}$, which means that the gauge $\varphi_1(x) = 0$, Im $\varphi_2(x) = 0$, can be selected.

With this gauge, we have one scalar field $(1/\sqrt{2})\varrho(x)$. Since (3.46) is invariant under transformations from $SU(2)$,

$$L = (1/2)(\nabla\varrho)^+(\nabla\varrho) - U(\varphi) + L_{\mathrm{YM}}. \tag{3.49}$$

All variables of the field but ϱ are eliminated by the gauge transformation. Lagrangian (3.49) can therefore be written as

$$L = (1/2)(\partial_\mu\varrho)(\partial^\mu\varrho) + (g^2/2)\varrho^2 W^a_\mu W^a_\mu - U(\varphi) - (1/4)F_{\mu\nu}F^{\mu\nu}, \tag{3.50}$$

where $U(\varphi) = (\varrho^2/2 - a^2)^2$.

For small deviations from the vacuum state, when expanding U in terms of $\varrho' = \varrho - a$ and leaving in Lagrangian (3.50) terms quadratic in fields, we obtain

$$L = \text{const} + (1/2)(\partial_\mu\varrho')(\partial_\mu\varrho') - (m^2/2)\varrho'^2 - (1/4)F_0^2 + (1/2)g^2a^2(W^a_\mu W^a_\mu),$$

where $F_0 = \partial_\mu W_\nu - \partial_\nu W_\mu$, $U = \text{const} + (m^2/2)\varrho'^2$.

The Lagrangian describes the free scalar field ϱ and the triplet of free vector ones W_μ. The scalar one is of mass m and the other particles associated with W_μ of mass $m_W = ga$.

Thus, thanks to the Higgs mechanism, all Yang-Mills field particles get a mass.

We examine the group-theoretic nature of the problem. We notice that the terms of type $W^a_\mu W^a_\mu$ completely violate Lagrangian's gauge invariance. From the physical viewpoint, this is what leads to the appearance of mass in all the fields W_μ. The mathematical explanation is as follows. The group $SU(2)$ acts freely (without fixed points) in the space S^3 of vacua, i.e., S^3 is a unique orbit of $SU(2)$. Therefore, the only invariant under the action of $SU(2)$ is the scalar ϱ. In the general case, the recipe for finding massive vector fields consists in combining the result of Subsec. 3.2.3 with the Higgs mechanism.

Another example of interacting scalar and Yang-Mills fields with the gauge group $SO(3)$ is considered in connection with the classical monopole solutions in Subsec. 4.1.2.

In conclusion of this introductory physical chapter we note that the ideas of gauge invariance and the Higgs mechanism play the key role in the unified theory of weak and electromagnetic interactions, i.e., the Weinberg-Salam model. The detailed discussion of the theory may be found in a number of books and surveys. Especially [Ok, Tay] can be recommended.

Chapter 4
Topology of Gauge Fields

We finally reap the topological harvest. Topological methods are applied to analyze the solutions of gauge field equations with internal symmetry groups. Different classical solutions are associated with real or hypothetical particles, excitations, vortices. Two classes of solutions are of greatest interest for physics, viz., 'tHooft-Polyakov monopoles and instantons, which we mostly consider in the chapter. Even though instanton analogs, two-dimensional pseudoparticles, exist in two-dimensional chiral fields, too, we consider this latter case separately.

4.1 MONOPOLES IN GAUGE FIELD THEORIES

Consider the special class of stationary solutions of gauge field equations, first found by 'tHooft [Ho1] and Polyakov [Pol1] and now called *'tHooft-Polyakov monopoles.*

The term "monopole" calls to mind an association with a Dirac monopole. This is not accidental and seems important. The modern treatment of the Dirac monopole due to Wu and Yang [WY3] is based on elementary ideas from the theory of fibrations.

4.1.1 Dirac Monopole

The magnetic monopole was introduced by Dirac in 1931 in his paper *Quantized singularities in an electromagnetic field* [Dir] where he shed light on fundamental physical problems in a totally different manner. Why are there single particles with an electric charge and no particles with a magnetic charge g? The Maxwell equations do admit the symmetry

$$H \to E, \quad E \to -H.$$

The second question is related to the least electric charge. The charges of all particles in appropriate units turn out to be integer multiples of that on the electron *e*. Dirac hypothesized the existence of separate magnetic poles and found that the conjecture led to a natural explanation of electric charge quantization. He showed by an elegant analysis that the existence of a magnetic pole did not lead to any contradiction in principle with modern physical ideas, in particular, with the Maxwell equations. In his unique style, he concluded: "Under these circumstances one would be surprised if Nature had made no use of it".

Dirac monopoles should have a number of fascinating properties. As for electrically charged particles, magnetic charge should be conserved, which means that, once generated, the monopole cannot vanish without colliding with another monopole with the opposite magnetic charge. The elementary magnetic charge of a monopole should be 137/2 times greater than the electron charge. Therefore, the force of interaction between two monopoles should exceed that between two electrons at the same distance by 4 692 times. "This very large force may perhaps account for why poles of opposite sign have never yet been separated," wrote Dirac in 1931.

Fifty years of effort to discover Dirac monopoles have not changed the situation. Since there are at present no convincing arguments why magnetic monopoles should not form, their appearance cannot be excluded.

From the history, we pass on to the meaning of Dirac construction.

Let a magnetic monopole, i.e., a particle with a magnetic charge $g \neq 0$, be at the point O in space R^3. We surround it with a sphere of fixed radius r (e.g., $r = 1$). Then the following holds.

Proposition 4.1 *The vector field generated by the vector potential A and defining the magnetic field of the monopole $H = \operatorname{rot} A$ has at least one singular point on S^2.*

For a proof of this, it suffices to notice that the field A is tangent to S^2.

Another and identical proof is based on Stokes' theorem.

We cover the sphere with two disks D_a and D_b overlapping on a strip around the equator and assume that A does not vanish. Consider the magnetic fluxes through D_a and D_b

$$\Phi_a = \int_{D_a} \operatorname{rot} A\, dS = \int_{S^1} A \cdot dl = \Phi_b = \int_{D_b} \operatorname{rot} A\, dS = \int_{S^1} A \cdot dl. \tag{4.1}$$

Then

$$\Phi_a - \Phi_b = 0.$$

On the other hand, $\int_{S^2} \operatorname{rot} A\, dS = 4\pi g \neq 0$ (Gauss' theorem), a contradiction, which proves Proposition 4.1. ■

In fact, we have obtained nothing new but another (physical) illustration of Poincaré's vector field theorem (see Subsec. 2.5.3).

Vector potential A can be given on S^2 in spherical coordinates in D_a as

$$(A_r)_a = (A_\theta)_a = 0, \quad (A_\varphi)_a = g\,\frac{(1 - \cos\theta)}{\sin\theta}, \quad r = 1, \tag{4.2}$$

where θ is the polar angle, φ the zenith angle, and in D_b as

$$(A_r)_b = (A_\theta)_b = 0, \quad (A_\varphi)_b = -\frac{g(1 + \cos\theta)}{\sin\theta}. \tag{4.3}$$

The vector potential $(A_\mu)_a$ has no singularities in D_a, nor the vector potential $(A_\mu)_b$ in D_b'. The rotor of each determines the monopole's magnetic field. In the region $D_a \cap D_b \neq \varnothing$, the difference is

$$(A_\mu)_a - (A_\mu)_b = \frac{2g}{\sin\theta} = \partial_\mu \alpha, \quad \alpha = 2g\varphi, \tag{4.4}$$

where $\mu = (r, \theta, \varphi)$, $\partial_\mu = \left(\frac{\partial}{\partial r}, \frac{1}{\sin\theta}\frac{\partial}{\partial\varphi}, \frac{\partial}{\partial\theta}\right)$, $r = 1$.

Now, consider the interaction of an electron with a monopole. The wave functions of the electron in the monopole field satisfy the Schrödinger equation

$$\left[\frac{1}{2m}(p - eA_a)^2 + U\right]\psi_a = E\psi_a \quad (\text{for } D_a),$$

$$\left[\frac{1}{2m}(p - eA_b)^2 + U\right]\psi_b = E\psi_b \quad (\text{for } D_b),$$

where ψ_a and ψ_b are the wave functions determined in the regions D_a and D_b, respectively. Since the functions determine the physical states if the phase factor is not considered, it follows from gauge invariance that

$$\psi_a = \exp(ie\alpha)\psi_b \tag{4.5}$$

and that α is determined by (4.4). Consider the wave function ψ for the electron in the region $D_a \cap D_b$. Let S_e^1 be the equatorial circle in the region. Obtained by restricting either ψ_a or ψ_b, ψ must be single-valued in traversing a closed contour l. We have

$$\psi_a = \exp(i2eg(\varphi + 2\pi))\psi_a = \exp(i2eg\varphi)\exp(i4\pi eg)\psi_a,$$

whence,

$$eg = n/2, \tag{4.6}$$

where n is an integer, which is the condition for the quantization of the wave function's phase and is called the *Dirac quantization condition.*

Assuming that a magnetic charge exists, (4.6) entails the quantization of the electric charge $e = n/2g$.

We now consider the Dirac monopole from the fiber space standpoint. It is perfectly clear that the definition of the monopole by vector potential A is equivalent to providing a connection in the fiber bundle over S^2. What then are the fiber bundles? The example of monopole field-electron interaction is interpreted as follows. Consider the principal bundle over S^2 with fiber $U(1) \sim S^1$ and the associated bundle E with group $U(1)$ and fiber C. By defining wave functions on D_a and D_b and their transformations (4.5), we specify the cross-sections explicitly. The assertion that global cross-sections do not exist (in this physical interpretation, that a global wave function does not exist) means that the respective fiber bundles are not trivial.

We now pass to principal bundles over S^2 with fiber S^1. The integer n involved in the Dirac quantization condition (4.6) characterizes the corresponding fiber bundle. We know (see Subsec. 2.6.4) that all principal bundles over S^2 with fiber S^1 are classified by the group $\pi_1(S^1) = Z$ coinciding with $H^2(S^2, Z)$ and characterized by the Chern class c_1. Thus, each quantum number n is associated with its monopole (fiber bundle). Constructing the fiber space characterizing the monopole with charge $g = n/2e$ is quite easy and illustrative.

We apply general results concerning fibration classification (see Subsec. 2.6.4) to the fiber bundle over S^2 with fiber S^1. Since $\pi_1(S^3) = \pi_2(S^3) = 0$, S^3 is 3-universal for the Hopf fibration $p: S^3 \to S^2$; by Theorem 2.29, all principal bundles over S^2 with fiber S^1 are in a one-to-one correspondence with $\pi_2(S^2)$. It follows from Hurewicz's theorem that $\pi_2(S^2)$ coincides with $H_2(S^2, Z)$ and $H^2(S^2, Z)$ and this is another interpretation of the Chern number. We exhibit the fiber spaces explicitly by using the simple observation already familiar to us that the diagram

$$\begin{array}{ccc} S^3 & \xrightarrow{Z_k} & S^3 \\ {\scriptstyle p}\downarrow & & \downarrow{\scriptstyle p} \\ S^2 & \xrightarrow{Z_k} & S^2 \end{array} \tag{4.7}$$

is commutative, with the assertion that S^3/Z_k, the lens space L^k, is the fiber bundle over the two-dimensional sphere with the characteristic class $c_1 = k$. A convenient explicit formulation of the result is suggested in the note [Sol]. Consider the action of the group Z_k on S^3 induced by the standard action in the complex space C^2:

$$\exp(2\pi i/k)z^1, \quad \exp(2\pi i/k)z^2, \tag{4.8}$$

where

$$S^3 = \{z^1, z^2\}, \ |z^1|^2 + |z^2|^2 = 1, \ S^3 \subset C^2,$$

where the coset space is $S^3/Z_k = L^k$ on which the gauge group $U(1) \sim S^1$ acts. We notice that the action of $U(1)$ is free on (4.8), therefore, L^k is the principal fiber bundle over S^2 with fiber S^1 and class $c_1 = k$ as seen from (4.7). The coordinate transformations for fiber bundles $L^k \to S^2$ are the functions $\exp(ik\varphi)$. We now formulate the final result.

Proposition 4.2 *The (Wu-Yang) bundle spaces for a monopole with quantum number k are lens spaces L^k.*

In the next subsection we show how ideas of modern field theory enable us to look at the problem of monopole existence from a new angle.

4.1.2 'tHooft-Polyakov Monopole (Gauge Monopole)

In 1974, 'tHooft [Ho1] and Polyakov [Pol1] offered a new class of solutions for one model of the unified theory of weak and electromagnetic interactions, i.e., Georgi-Glashow, regarded at the time as a strong opponent of the Weinberg-Salam one. However, experimental justification of the existence of neutral currents and the observation of the Z-Boson, which were absent in the Georgi-Glashow model, made the Weinberg-Salam model the undisputed leader. Still, the Georgi-Glashow model was undoubtedly important. One valuable theoretical result is the 'tHooft-Polyakov monopole.

4.1.2.1 'tHooft-Polyakov monopoles in the Georgi-Glashow model

Consider the collection of interacting Yang-Mills fields which are transformed relative to the adjoint representation of the group $SO(3)$ with Higgs isovector fields.

The Lagrangian L is representable as

$$L = -(1/4)F_{\mu\nu}F^{\mu\nu} + (1/2)(\nabla_\mu \varphi)(\nabla^\mu \varphi) - U(\varphi),$$
$$U(\varphi) = (\mu^2/2)(\varphi \cdot \varphi) + (\lambda/4)(\varphi \cdot \varphi)^2, \ \lambda > 0, \ \mu^2 < 0, \tag{4.9}$$

where

$$F_{\mu\nu} = \partial_\mu W_\nu - \partial_\nu W_\mu + \tilde{g}[W_\mu, W_\nu], \tag{4.10}$$

$$\nabla_\mu \varphi^a = \partial_\mu \varphi^a + \tilde{g}[W_\mu, \varphi] = \partial_\mu \varphi^a - \tilde{g}\varepsilon_{abc}W_\mu^b \varphi^c, \tag{4.11}$$

where $\tilde{g}$ is the Yang-Mills coupling constant.

The motion equations for system (4.9) are

$$(\nabla_\mu F_{\mu\nu})_a = \tilde{g}\varepsilon_{abc}\varphi^b(\nabla_\mu \varphi^c), \tag{4.12}$$

$$(\nabla^\mu \nabla_\mu \varphi)_a = -\partial U/\partial \varphi^a. \tag{4.13}$$

The classical vacuum solution of equations (4.12) and (4.13) is

$$\varphi(x) = \sqrt{-\mu^2/\lambda}\,\hat{e}_3, \ W_\mu = 0, \tag{4.14}$$

where $\hat{e}_3$ is the unit vector in isospin space, pointing along the expectation z-axis. The classical solution of (4.14) with nonvanishing vacuum value of the field operators defines the minimum of potential $U(\varphi)$.

However, there are other stationary solutions in (4.9), generated by local minima of $U(\varphi)$, i.e., local vacua of the following form provided that $|x| \to \infty$:

$$\varphi^i(x) \to (\mu/\sqrt{\lambda})(\hat{x})^i, \ W_\mu^i(x) \to -\varepsilon_{\mu ij}\frac{(\hat{x})^j}{\tilde{g}|x|}, \tag{4.15}$$

where $\hat{x}$ is the unit vector in coordinate space and $\varepsilon_{\mu ij}$ the antisymmetric tensor for $i, j, \mu = 1, 2, 3$, $\varepsilon_{0ij} = 0$. The classical energy of such locally vacuum solutions is $-E_{\text{cl}} \sim \sqrt{-\mu^2/\lambda}/\tilde{g}$ [Hol].

Following 'tHooft [Hol], [GO], we now show why they can be treated as monopoles.

We define the magnetic field

$$B_{\mu\nu} = \varphi_a F^a_{\mu\nu} - \frac{1}{\tilde{g}|\varphi|^3} \varepsilon_{abc} \varphi^a (\nabla_\mu \varphi^b)(\nabla_\mu \varphi^c). \tag{4.16}$$

The formula is obtained from the following argument. We align the Higgs field strictly along the third φ-axis, where $\varphi = |\varphi|$ (0, 0, 1). Then (4.16) turns into

$$B_{\mu\nu} = F^3_{\mu\nu}, \tag{4.17}$$

which satisfies the Maxwell equation everywhere but at the point $\varphi^a = 0$. A massless particle identified with the photon γ can be related to φ. The strength $B_{\mu\nu}$ in (4.16) is a gauge invariant quantity associated with the field B for an arbitrary direction of the Higgs vector.

For 'tHooft-Polyakov asymptotic behaviour (4.15) we have

$$\varphi_a F^a_{\mu\nu} = -\frac{1}{\tilde{g}r^3} \varepsilon_{\mu\nu a} x^a,$$

$$\nabla_\mu \varphi^a = \partial_\mu \varphi^a + \tilde{g} \varepsilon_{abc} W^b_\mu \varphi^c;$$

therefore,

$$B_{\mu\nu} = -\frac{1}{\tilde{g}r^3} \varepsilon_{\mu\nu a} r^a. \tag{4.18}$$

The corresponding radial field $B_a = \dfrac{r^a}{\tilde{g}r^3}$ has total flux $4\pi/\tilde{g}$.

Thus, the 'tHooft-Polyakov solution is associated with a monopole with magnetic charge $g = 1/\hat{g}$.

4.1.3 Topological Criterion for the Existence of Monopole-Like Solutions

We consider the Yang-Mills equation with an arbitrary compact group G and interacting with Higgs fields φ. The goal of this subsection is to find a topological criterion for monopole-like solutions.

Given a system of vector Yang-Mills fields W^j_μ (μ is the spatial index and j the isotopic one, $j = 1, \ldots, n = \dim G$) and the Higgs scalar fields $\varphi^a (a = 1, \ldots, N)$ interacting with them, which are transformed relative to the N-dimensional representation $T(g)$ of G, we assume, for simplicity, that $T(g)$ is irreducible. The Lagrangian describing the system is

$$L = -(1/4)F^j_{\mu\nu}F^{\mu\nu,j} - (1/2)(\nabla_\mu\varphi^a)(\nabla^\mu\varphi^a) - U(\varphi), \tag{4.19}$$

where

$$F^j_{\mu\nu} = \partial_\mu W^j_\nu - \partial_\nu W^j_\mu + g[W_\mu, W_\nu] = \partial_\mu W^j_\nu - \partial_\nu W^j_\mu + gC^j_{kl}W^k_\mu W^l_\nu,$$

$$\nabla_\mu\varphi^a = \partial_\mu\varphi^a + gW^j_\mu(T_j)^b_a\varphi^b. \tag{4.20}$$

Here C^j_{kl} are the structural constants of the Lie algebra of G and T_j are generators of $T(g)$.

Assume that potential $U(\varphi)$ satisfies the conditions:

(1) $U(\varphi)$ is invariant under the action of G and

(2) $U(\varphi)$ is Higgs, i.e., the absolute minimum is attained for finite $\varphi^a = \chi^a \neq 0$.

We write the motion equations by changing Lagrangian (4.18), varying W and φ as

$$\partial_\mu F_{\mu\nu} + g[F_{\mu\nu}, W] = g[\nabla_\mu\varphi, \varphi], \quad (\nabla^2_\mu\varphi) = -\frac{\partial U}{\partial\varphi}, \tag{4.21}$$

i.e., in invariant form

$$D^*\Omega = {}^*I, \quad I = g[\nabla\varphi, \varphi],$$

$$\Delta\varphi = -\frac{\partial U}{\partial\varphi}, \tag{4.22}$$

where D is the covariant differentiation of the field W in (4.20).

The relation of φ to W follows from the motion equations. The vacuum solutions of (4.21) and (4.22) are $\varphi^a = \chi^a$, $W_\mu \equiv 0$.

However, there exist other solutions supplying local minima. Such are monopole ones to be sought for as

$$\varphi^a(r) \to \varphi^a(\mathbf{n}), \; W^j_\mu(r) = (1/r)W^j_\mu(\mathbf{n}) \tag{4.23}$$

as $r \to \infty$ and $\varphi^a(\mathbf{n})$ depend on $\mathbf{n}$ explicitly.

Note that it follows from the motion equations that W^j_μ can be expressed in terms of φ^a similarly to the isovector $SO(3)$-monopole. We select the vectors

$$\varphi^a(\mathbf{n}) = \Omega^b_a\varphi^{(0)}_b, \quad \Omega^b_a(\mathbf{n}) = T^b_a(g(\mathbf{n})), \tag{4.24}$$

where φ^0_b is fixed in the space of minima of $U(\varphi)$ so that $U(\varphi^{(0)}) = U_0 \neq 0$.

As seen from Subsec. 4.2.4, vectors $\varphi(\mathbf{n}) = \{T(g)\varphi^{(0)}\}$ belong to the orbit of representation of G. By definition (see Subsec. 1.3.1), $\{T(g)\varphi^{(0)}\}$ is representable as G/H, where H is the stability subgroup of the vector $\varphi^{(0)}$. Recall that the functions $\varphi^a(\mathbf{n})$ according to boundary conditions (4.23) now define a mapping of the two-dimensional sphere S^2: $\{\mathbf{n}, (\mathbf{n}\cdot\mathbf{n}) = 1\}$ to the orbit G/H. Since the two maps φ

and φ' which can be sent into each other by a continuous gauge transformation Ω are equivalent physically, we should consider classes of maps $S^2 \to G/H$ taking into account the equivalence, easily seen to be homotopy. Thus, we obtain the following [MP], [TFS].

Theorem 4.1 *The necessary condition for monopole existence is the nontriviality of the group*

$$\pi_2(G/H) \neq 0. \tag{4.25}$$

If the group G is simply connected, then it follows from the exactness of the homotopy group sequence with the fiber map $G \to G/H$ that $\pi_2(G/H) = \pi_1(H)$. In physical applications, H is associated with massless particles and is the product of the group $U(1)$ taken n times. If $\pi_1(G) = 0$, then the number of monopoles is characterized by the rank r of the group H, namely, $\pi_1(H) = \underbrace{Z \oplus Z \oplus \ldots \oplus Z}_{r}$.

However, if G is not simply connected, then $\pi_2(G/H) \neq \pi_1(H)$ in the general case, and to find monopoles a straight-forward calculation of $\pi_2(G/H)$ is necessary, e.g., for the Weinberg-Salam model, the group G is $U(2)$, the subgroup H is $U(1)$, and there are no monopoles, as is shown in the next subsection. For the group $G = SO(3)$ and subgroup $H = U(1) \sim SO(2)$ (in the Georgi-Glashow model) there are monopoles.

4.1.4 Application of the Topological Criterion to Different Physical Models

The simple criterion from the foregoing subsection enables us to clarify the existence of monopoles in a number of physically important examples.

Monopoles in the model with isospin and with group $SU(2)$. Consider the gauge-invariant $SU(2)$ model with the multiplet of complex scalar fields given in the two-dimensional (spinor) representation

$$\varphi = \begin{pmatrix} \varphi_1 \\ \varphi_2 \end{pmatrix}, \quad \varphi^+ = (\bar{\varphi}_1, \bar{\varphi}_2).$$

The Lagrangian L is of form (3.49) and the gauge transformation of the fields φ as in formula (3.50). The stable extremum is associated with a constant φ such that $\varphi\varphi^+ = \chi^2$. The vacuum manifold is the three-dimensional sphere S^3. To show that there are no monopoles here, it suffices to notice that $\pi_2(S^3) = 0$.

We formulate the general criterion of monopole existence in $SU(2)$-models in terms of the representations $T(g)$, $g \in SU(2)$.

Recall from the theory of group representations that unitary representations of the group $SU(2)$ are given by index spin I with values from the set of integers and half-integers.

1. If the isospin is an integer, then there exists a vector with zero projection onto the t_3-axis with the stability subgroup $H \sim U(1)$ of rotations around the axis,

in which case

$$\pi_2(G/H) = \pi_2(SU(2)/U(1)) = \pi_1(U(1)) = Z,$$

and the monopoles are characterized by an integer.

Note that for integers I the representation of $SU(2)$ coincides with that of $SO(3)$.

2. If the isospin is half-integer, then the stability subgroup H of any vector is trivial and there are no monopoles.

Consider another interesting physical example showing that there are no monopoles in the Weinberg-Salam model.

From the mathematical viewpoint, we have the full gauge group $U(2) \simeq SU(2) \dot{\times} U(1)$ and the subgroup $H = U_e(1) \subset U(2)$ associated with the electric charge e and intersecting both $SU(2)$ and $U(1)$. Such an embedding is said to be *nonregular.*

The question of monopole existence is again reduced to the calculation of $\pi_2(G/H) = \pi_2(U(2)/U(1))$.

However, it is impossible to calculate π_2 only on the basis of the exactness of the homotopy group sequence. Therefore, we resort to some other argument.

H is representable as $\exp(it\tau^3 + Y)$ (see the notation in Subsec. 3.1.1). We prove that $\pi_2(G/H) = 0$. There exist several elementary proofs, e.g., by finding the intersections $H \cap SU(2)$, $H \cap U(1)$. Our proof applies to more general situations. Consider the action of the group $SU(2)$ on G/H, namely,

$$g_0 : x \to g_0 x, \tag{4.26}$$

where $x \in G/H$ and $g_0 \in SU(2)$. We denote by Ω the orbit of $SU(2)$ relative to (4.26).

We prove that $\Omega \simeq G/H$. We find the stability subgroup at x_0. Since the stability subgroups at two orbit points are conjugate, it suffices to find the stability group at the simplest point $x_0 = eH$ (e is the unit element of G). We have $x_0 = \{\exp(it\tau^3), \exp(itY)\}$, $g_0 x_0 = x_0 \Rightarrow \{g_0^{-1}\exp(it\tau^3), \exp(itY)\} = \{\exp(it\tau^3), \exp(itY)\}$ ($SU(2)$ acting on the left). The stability group at x_0 is trivial. Therefore, $\Omega \simeq SU(2)$ and is compact as the image of a compact set. Hence, along with dimension argument ($\dim \Omega = \dim G/H$), we have $\Omega \simeq G/H$ and $\pi_i(\Omega) = \pi_i(G/H) = \pi_i(SU(2))$. In particular, $\pi_2(G/H) = 0$. We have thus shown that there are no monopoles in the standard Weinberg-Salam model.

However, various possibilities in theories with gauge groups of rank $r \geqslant 2$ have physical and mathematical interest, e.g., in models of unified field theories, the gauge groups $SU(5)$, $SO(10)$, $SO(32)$, etc., appear. A large number of possibilities arise during the internal symmetry group's spontaneous breaking.

In the next subsection, we consider the simplest group $SU(3)$ of rank 2 containing the $SU(2) \times U(1)$ group.

4.1.5 *SU*(3)-Monopoles

$SU(3)$-monopole solutions are constructed according to the scheme similar to $SU(2)$, $SO(3)$-monopoles, but particular due to a large number of subgroups of $SU(3)$.

To classify $SU(3)$-monopoles, we need some information from the theory of the unitary representations of $SU(3)$ (see the detail in [We]).

Irreducible unitary representations of $SU(3)$ are characterized by two non-negative integers m_1 and m_2 (*weights* of the representations). The number of m_i equals the rank of the group G and is two in our case.

Adopting Weyl's formula for the dimension of an irreducible representation of the compact group G [We], we obtain

$$\dim T(g)^{(m_1, m_2)} = \frac{(m_1 + 1)(m_2 + 1)(m_1 + m_2 + 2)}{2}$$

if $G = SU(3)$, e.g., for $m_1 = 1$ and $m_2 = 0$, we obtain 3 for the minimal dimension of the irreducible representation $T(g)^{(1,0)}$. For $m_1 = 1$ and $m_2 = 1$, dimension of the adjoint representation $T(g)^{(1,1)}$ of $SU(3)$ is 8.

(1) The simplest three-dimensional representation of $SU(3)$ can be realized in a three-dimensional complex space C^3. $SU(3)$ acts by translations as follows:

$$T(g)f(z) = f(g^{-1}z), \quad g \in SU(3), \quad z \in C^3. \tag{4.27}$$

Given a G-invariant potential $U(\varphi)$, $\varphi \in C^3$, we fix a vector $|\varphi^{(0)}\rangle$ in C^3 belonging to the vacuum manifold and consider the orbit of $SU(3)$: $\Omega = T(g)|\varphi^0\rangle = |\varphi_g\rangle$. The stability subgroup of $|\varphi^0\rangle$ consists of unitary rotations in space C^2 orthogonal to $|\varphi^0\rangle$. This group obviously coincides with $SU(2)$. Thus, $\Omega = G/H = SU(3)/SU(2) \simeq S^5$, $\pi_2(S^5) = 0$.

If we consider the Higgs potential

$$U(\varphi) = -\frac{\mu^2}{2}(\varphi^i\bar{\varphi}^i) + (1/4)\lambda(\varphi^i \cdot \bar{\varphi}^i)^2 \quad (\mu^2, \lambda > 0),$$

then the necessary condition for the existence of a minimum yields

$$\frac{\partial U}{\partial \varphi^i} = [-\mu^2 + \lambda(\varphi^i\bar{\varphi}^i)]\bar{\varphi}^i = 0, \quad i = 1, 2, 3.$$

We obtain the equation of a five-dimensional sphere $(\varphi^i \cdot \bar{\varphi}^i) = \mu^2/\lambda$. $SU(3)$ acts on S^5 transitively. This follows from the general criterion of monopole solutions that there are no monopoles. This result holds for any $SU(3)$-invariant potential and representation (4.27).

(2) Of greatest interest is the adjoint representation of $SU(3)$, in which case the scalar Higgs fields φ and Yang-Mills fields are transformed via the adjoint representation.

Let the model's Lagrangian be representable in standard form

$$L = -(1/4)\,\mathrm{Tr}\,(F_{\mu\nu}F^{\mu\nu}) - (1/2)(\nabla_\mu\varphi)(\nabla^\mu\varphi) - U(\varphi), \tag{4.28}$$

where $U(\varphi)$ is an $SU(3)$-invariant function and $F_{\mu\nu}$ and $\nabla_\mu \varphi$ are defined in (4.20).

The matrix of the adjoint representation can be conveniently selected in the form $W \in \mathrm{Ad}\, G = \tilde{\mathcal{G}}$, W being a Hermitian 3×3 matrix with trace zero. The orbits of group $SU(3)$ acting in the space $\tilde{\mathcal{G}}$ can be classified relative to the eigenvalues λ_i.

There exist three types of orbit: (a) $\lambda_1 = \lambda_2 = \lambda_3 = 0$, the trivial orbit Ω_0 consisting of one point, (b) $\lambda_1 = \lambda_2 \neq \lambda_3$, the orbit Ω_1 isomorphic to the space $SU(3)/U(2) = CP^2$, (c) $\lambda_1 \neq \lambda_2 \neq \lambda_3$, the orbit $\Omega_2 \sim SU(3)/U(1) \times U(1)$ in general position.

If a G-invariant potential $U(\varphi)$ is a polynomial of degree n in invariants of the field φ, then the minima of $U(\varphi)$ can be both Ω_1 and Ω_2, i.e., monopoles with $\pi_1(\Omega_1) = Z$, $\pi_1(\Omega_2) = Z + Z$ can be realized.

In the former case, we can speak about existence of one massless particle, in the latter, of two (see p. 200).

It would be interesting to consider the situation arising for a polynomial $U(\varphi)$ under G-invariant at degree four. The condition that the polynomial should be of a degree not higher than four is determined by the requirement that the Lagrangian (4.28) should be renormalized.

The renormalizability condition is an important (and, at the present stage of theory development, fundamentally important) technical requirement for the Lagrangian. Since (4.28) and $U(\varphi)$ have an energy density dimension proportional to l^{-4}, φ is of dimension l^{-1}, therefore, μ^2 is proportional to l^{-2}, the mass squared, and, if φ is involved in degrees less than, or equal to, four, then the coefficient λ of φ^4 is dimensionless and can be expanded into series in λ perturbatively. In case when $U(\varphi)$ involves terms in φ of degree greater than four, then the dimensionless quantity is $\lambda_k(\sqrt{E})$ (k is the highest degree in the polynomial), i.e., increases with energy E. Therefore, an expansion into a series by perturbation theory is inadmissible. This argument is only an outline for a proof of renormalizability. Note that, in phase-transition theory, potentials of degree not higher than four are also used. With this condition, the absence of terms of third degree is necessary for transitions of the second kind. That there are no terms of third degree is in field theory related to the choice of symmetric vacua (i.e., invariance under $\varphi \to -\varphi$) and is of no essential importance.

Let $U(\varphi)$ be an $SU(3)$-invariant polynomial of fourth degree. Since the matrix is Hermitian and $\mathrm{Tr}\,\varphi = 0$, $U(\varphi)$ can be represented as

$$U(\varphi) = c_1 + c_2\,\mathrm{Tr}(\varphi^2) + c_3\,\mathrm{Tr}(\varphi^3) + c_4\,\mathrm{Tr}(\varphi^4) \tag{4.29}$$

(we have used the identity $(\mathrm{Tr}(\varphi^2))^2 = 2\mathrm{Tr}(\varphi^4)$).

The necessary extremum conditions yield

$$2c_2\lambda_i + 3c_3\lambda_i^2 + 4c_4\lambda_i^3 = 0, \quad \lambda_1 + \lambda_2 + \lambda_3 = 0. \tag{4.30}$$

There are *a priori* three types of orbits:

(1) $\lambda_1 = \lambda_2 = \lambda_3 = 0$, Ω_0 is the point 0 associated with the trivial orbit $H = SU(3)$ and $U(0) = U_{\min}$.

(2) $\lambda_1 = \lambda_2 \neq \lambda_3$; Ω_1.

Here, the stability subgroup $H \sim SU(2) \times U(1)$ and

$$\varphi^{(0)} = \begin{pmatrix} \lambda & 0 & 0 \\ 0 & \lambda & 0 \\ 0 & 0 & -2\lambda \end{pmatrix}.$$

(3) $\lambda_1 \neq \lambda_2 \neq \lambda_3$, Ω_2, in which case all eigenvalues λ_i satisfy the same cubic equation following from (4.29). Therefore, $\lambda_1 + \lambda_2 + \lambda_3 = 3c_3 = 0$, i.e., $c_3 = 0$.

$U(\varphi)$ can be represented as $c_1 + c_2 \mathrm{Tr}(\varphi^2) + c_4 \mathrm{Tr}(\varphi^4) = a[\mathrm{Tr}\varphi^2 - b^2)^2 \geqslant 0$. Hence, the minima of $U(\varphi)$ are on the sphere S^7: $\mathrm{Tr}(\varphi^2) = b^2$. The group $SU(3)$ does not act transitively on S^7 and we are only interested in the minima associated with the orbits of $SU(3)$ in the adjoint representation's space (the monopoles being associated only with solutions $\varphi(\mathbf{n})$ with values in the orbits of G). Contrasting the conditions on the coefficients $c_4 > 0$, $c_2 < 0$ with equations (4.29) shows that a minimum of $U(\varphi)$ can only be attained on the orbit Ω_1. The same result also follows from the general theorem of Michel [Mi].

Let $T(g)$ be a linear representation of a compact (or finite) group G in R^n with no invariant nonzero vectors (i.e., $Hx = x \Rightarrow x = 0$). Following Michel, we call a polynomial $P(x)$ of degree four a *Higgs-Landau polynomial* if $P(T(g)x) = P(x)$ is bounded from below and takes the least values at points $x \neq 0$.

Theorem 4.2 *If a representation $T(g)$ is irreducible, then the Higgs-Landau polynomial has no extrema on an open dense stratum.*

As a corollary, we derive that a Higgs-Landau polynomial cannot attain a minimum on orbits with minimal stability subgroup H. For $SU(3)$, $H = U(1) \times U(1)$.

For polynomials of arbitrary degree, there are no such restrictions. Each orbit relative to the action of G in the representation's space can be the manifold of minima of an arbitrary G-invariant polynomial $P_n(x)$ of degree higher than four.

Thus, in theories with gauge group $SU(3)$ and potentials $U(x) = P_n(x)$ there exist monopole solutions associated with the group $U(1) \times U(1)$. The solutions are classified by two charges, namely, $\pi_1(U(1) \times U(1)) = Z + Z$.

Consider the monopole solutions associated with the subgroup $H = U(2)$ from the topological viewpoint. By Theorem 4.1, monopole solutions are classified by elements of the group $\pi_2(G/H)$, in our case, $\pi_2(CP^2) = Z$. H is the stability subgroup of a vector $|\varphi^0\rangle$ in the adjoint representation's space, while monopole solutions are characterized by the asymptotic behaviour of the scalar fields $\varphi(x)$ at infinity, i.e., by the boundary values of the field $\varphi(x) = \varphi(\mathbf{n})$, where $\varphi(\mathbf{n})$ is the map of the two-dimensional unit sphere to space G/H.

In the terminology of fiber spaces, the problem of monopole classification is reduced to the following chain of fiber bundles. The original bundle is the principal one over $\check{M}^4$ with gauge group G as the fiber. By considering only static solutions we distinguish a submanifold R^3 in the base space $\check{M}^4$ (or R^4). Specifying the asymptotic conditions for Yang-Mills and Higgs fields determines a fiber bundle over the base space S^2. It is natural to consider the fiber bundles E associated with the principal ones $P \xrightarrow{G} M$ and $P \xrightarrow{G} S^2$ which are generated by fields φ taking values in the group G representation space.

The existence of a monopole solution enables us to reduce the original principal bundle with group G to the fiber bundle with group $H \subset G$. The reduction is based on the following general theorem in the theory of fiber spaces (see [KN1]).

Theorem 4.3 *The structure group G of a principal fiber bundle (P, G, M) is reducible to a closed subgroup H if and only if the associated bundle $(E, G/H, G, M)$ admits a section $s: M \to P/H = E$ (H acting on P on the right).*

This enables us to reduce the fiber bundle P with $G = SU(3)$ to a fiber bundle Q with $H = U(2)$. Note that the correspondence between sections s and reducible submanifolds Q is one-to-one. Hence, for fiber bundles $E \to M \sim S^2$, different reductions are determined by the group $\pi_2(G/H)$.

We use this result to clarify the role of the electromagnetic $U(1)$-subgroup, which is associated with the magnetic charge $\hat{g}$ [Ma].*

Let P be a principal bundle with group $G = SU(3)$ over S^2 and E the associated bundle over S^2 with fiber G/H, where $H = U(2)$. We denote an element of the orbit G/H by φ. Let $n = \varphi/|\varphi|$,** $\varphi = \varphi^j k_j$, k_j be the basis of the Lie algebra $su(3)$, $k_j = -i\lambda_j/2$, $i = \sqrt{-1}$, and λ_j be the Gell-Mann matrices:

$$\lambda_1 = \begin{pmatrix} 0 & 1 & 0 \\ 1 & 0 & 0 \\ 0 & 0 & 0 \end{pmatrix}, \quad \lambda_2 = \begin{pmatrix} 0 & -i & 0 \\ i & 0 & 0 \\ 0 & 0 & 0 \end{pmatrix}, \quad \lambda_3 = \begin{pmatrix} 1 & 0 & 0 \\ 0 & -1 & 0 \\ 0 & 0 & 0 \end{pmatrix},$$

$$\lambda_4 = \begin{pmatrix} 0 & 0 & 1 \\ 0 & 0 & 0 \\ 1 & 0 & 0 \end{pmatrix}, \quad \lambda_5 = \begin{pmatrix} 0 & 0 & -i \\ 0 & 0 & 0 \\ i & 0 & 0 \end{pmatrix}, \quad \lambda_6 = \begin{pmatrix} 0 & 0 & 0 \\ 0 & 0 & -i \\ i & 0 & 0 \end{pmatrix},$$

$$\lambda_7 = \begin{pmatrix} 0 & 0 & 0 \\ 0 & 0 & -i \\ 0 & i & 0 \end{pmatrix}, \quad \lambda_8 = (1/\sqrt{3}) \begin{pmatrix} 1 & 0 & 0 \\ 0 & 1 & 0 \\ 0 & 0 & -2 \end{pmatrix}.$$

The Gell-Mann matrices form a basis for the space of Hermitian 3×3 matrices with trace zero.

It is easy to see that the orbit G/H can be obtained by applying elements from G to a point $\lambda_8 \in \mathscr{G}$ ($\mathscr{G}$ is the adjoint representation's space). Such orbits are said to be λ_8-*similar.* Note also that the matrices $-i\sigma_j/2 = -i\lambda_j/2$ ($j = 1, 2, 3$) from a basis of the Lie algebra of the group $SU(2)$ diagonally embedded in $SU(3)$. Following [Ma1] we explicitly construct the reduction of P to Q.

We identify the orbit $G/H \sim CP^2$ with a four-dimensional submanifold of the unit sphere $S^7 \subset R^8$. The projection p of P onto P/H is given by the fiber homomorphism $p_r G \to G/H$, where,

$$p_r(g) = \mathbf{n}, \quad g k_8 g^{-1} = n = \mathbf{n} \cdot \mathbf{k}, \quad g \in SU(3),$$

$p(\mathscr{P}) = (x, \mathbf{n})$, where $\mathscr{P} \in P$, $\mathscr{P} = (x, g)$, $x \in M$. The kernel of the mapping is $U(2)$.

* To avoid confusion, we use the notation $\hat{g}$ for magnetic charge, $\tilde{g}$ for the Yang-Mills coupling constant, and g for an element of the group G.

** The norm is determined by the Killing metric in $su(3)$.

Let ω be a connection on Q with group H, i.e., that with values in the Lie algebra $\mathscr{H}$ of H. ω is the restriction of a connection $\tilde{\omega}$ on P under reduction $P \to Q$.*

We write ω in terms of the original connection $\tilde{\omega}$. Let s be a local section of Q. Any such section can be given as $s = b(x^\alpha)$, $x^\alpha \in S^2$, where b is the solution of the equation $n = bk_8b^{-1}$. Then, locally, $\tilde{\omega}/Q = g^{-1}\Gamma_\mu' g\, dx^\mu + g^{-1}\, dg$, $\Gamma_\mu' = b^{-1}\Gamma_\mu b + b^{-1}\partial_\mu b$. Here $g \in U(2)$ and Γ_μ is the connection in the fiber bundle P generated by the unit element of the group $G = SU(3)$.

The connection ω is defined as

$$\omega = g^{-1}\Gamma_\mu g\, dx^\mu + g^{-1}\, dg.$$

To find how the connection $\tilde{\omega}/Q$ is related to ω, we use the auxiliary associated bundle with fiber C^2.

The covariant derivative ∇_μ generated by the connection Γ_μ' acts in $C^3 = \begin{Bmatrix} \psi^1 \\ \psi^2 \\ \psi^3 \end{Bmatrix} = \{\psi\}$ by the formula

$$\nabla_\mu \psi^a = \partial_\mu \psi^a + \Gamma_\mu'^j \lambda_j \psi^a + \Gamma_\mu'^8 \lambda_8 \psi^a \quad (j = 1, \dots, 7), \tag{4.31}$$

while ∇_μ generated by Γ_μ by the formula

$$\nabla_\mu \psi = \partial_\mu \psi^a - \frac{i}{2}\Gamma_\mu^j \sigma_j \psi^a \quad (j = 1, 2, 3), \tag{4.32}$$

where $\begin{Bmatrix} \psi^1 \\ \psi^2 \end{Bmatrix}$ is a subspace of C^2 in C^3.

Comparing (4.31) with (4.32), we have

$$\Gamma_\mu = -\frac{i}{2}\left(\Gamma_\mu'^j \sigma_j + \frac{1}{\sqrt{3}}\Gamma'^8 \begin{pmatrix} 1 & 0 \\ 0 & 1 \end{pmatrix}\right). \tag{4.33}$$

We can obtain information about the magnetic charge by carrying out the other reduction of Q with group $U(2)$ to the subbundle R with group $U(1)$, which can be embedded in $U(2)$ as follows:

(1) $U(1) \cap SU(2) = e$,

(2) $U(1) \cap SU(2) = Z_2$,

(3) $U(1) \subset SU(2)$.

We consider all the three embeddings one by one.

(1) $U(1) \cap SU(2) = e$.

This is given in matrix form as

$$U(1) \to \begin{pmatrix} e^{i\alpha} & 0 \\ 0 & 1 \end{pmatrix}, \quad 0 \leqslant \alpha \leqslant 2\pi.$$

---

* The conditions under which the connection on P is reduced to that on the subbundle Q can be found in [KN1] and always hold in constructing the monopole solutions.

The corresponding homogeneous space $U(2)/U(1)$ is S^3. Reduction to the $U(1)$ bundle is determined by a section of the associated bundle E' with fiber S^3. It is convenient to select the section $s = (0, 0, 1) \in S^3$ which is associated with the unit element $e \in U(2)$. Since $\pi_2(U/(2)/U(1)) = 0$ here, all the sections (cross-sections) determine the same reduction to $U(1)$ bundle.

Let R be a $U(1)$ bundle and $\hat{\omega}$ the connection form projected onto R. Locally, it can be determined as

$$\hat{\omega} = i\hat{\Gamma}_\mu \, dx^\mu + i \, d\alpha, \quad \hat{\Gamma}_\mu = \tilde{e}A_\mu, \tag{4.34}$$

where $\tilde{e}$ is the "electric" charge and A_μ the electromagnetic field's connection.

The coefficients of the connection $\hat{\Gamma}_\mu$ can be expressed in terms of restrictions of Γ_μ given on the fiber bundle Q, which can also be done conveniently by considering the associated vector bundle $\tilde{E}$ with fiber C^2. If, for simplicity of calculation, we assume that the local expression for ω in Q is a section of R, then we derive from (4.34) that

$$\hat{\Gamma}_\mu = \Gamma_\mu = -1/(2\sqrt{3})\Gamma_\mu'^8, \quad \tilde{e} = \tilde{g}/(2\sqrt{3}).$$

The existence of such a reduction of the $U(2)$ bundle to the $U(1)$ bundle means that $\tilde{E}$ is decomposed into the Whitney sum of the R bundle and the trivial bundle ε. The group $U(1)$ acts on the fiber C as

$$\begin{pmatrix} e^{i\alpha} & 0 \\ 0 & 1 \end{pmatrix} \begin{matrix} \psi^1 \\ \psi^2 \end{matrix} = \begin{matrix} e^{i\alpha}\psi^1 \\ \psi^2 \end{matrix}. \tag{4.35}$$

The characteristic class of $\tilde{E}$, or the first Chern class c_1, is determined by the Whitney formula (see Subsec. 2.7.8) as

$$c_1(R \oplus_y \varepsilon) = c_1(Q) = c_1(R). \tag{4.36}$$

The characteristic class $c_1(R)$ of the $U(1)$ bundle R determines a magnetic charge by the formula

$$2r\hat{g} = m_1, \quad m_1 \in Z,$$

like the Dirac quantization condition in the Wu-Yang approach (see Subsec. 4.1.1).

The formula for the electromagnetic field $B_{\mu\nu}$ tensor (analogous to (4.16)) is obtained by straightforward calculation, namely,

$$\sqrt{3}e \, B_{\mu\nu} = \mathrm{Tr}\,(nF_{\mu\nu}) - (4/3)\,\mathrm{Tr}\,(n[\nabla_\mu n, \ \nabla_\mu n]), \tag{4.37}$$

where $F_{\mu\nu}$ is the curvature tensor of the connection ω.

(2) $U(1) \cap SU(2) = Z_2$.

The embedding of $U(1)$ in $U(2)$ is given by the matrix

$$e^{i\alpha} \in U(1) \to \begin{pmatrix} e^{i\alpha} & 0 \\ 0 & e^{i\alpha} \end{pmatrix}. \tag{4.38}$$

The space $U(2)/U(1) \sim SO(3)$, $\pi_2(SO(3)) = 0$, therefore, all sections of the associated bundle $\check{E}$ with fiber $SO(3)$ give equivalent reductions. A section s in $\check{E}$ can be given by the unit element in $SO(3)$. All other reductions obviously yield equivalent connections with the same characteristic class. The formulas for the reduced connection and curvature are the same as (4.34) and (4.37).

However, the action of the group $U(1)$ in the fiber bundle $\check{E}$ differs from corresponding action (4.35). $\check{E}$ is decomposed into two fiber bundles R_1 and R_2 isomorphic to $U(1)$. We obtain from the Whitney formula for the Chern class c_1 that

$$c_1(Q) = c_1(R_1 \oplus_w R_2) = 2c_1(R),$$

which takes us to another quantization rule

$$4e\hat{g} = m_1, \quad m_1 \in Z.$$

The electromagnetic field $B_{\mu\nu}$ tensor is also given by (4.37).

(3) $U(1)$ is embedded in $SU(2)$ as

$$e^{i\alpha} \in U(1) \to \begin{pmatrix} e^{i\alpha} & 0 \\ 0 & e^{-i\alpha} \end{pmatrix}. \tag{4.39}$$

The quotient space is $U(2)/U(1) = S^2 \times S^1$. The fiber bundle reduction is determined by the section of the associated bundle $\mathring{E} = (\mathring{E}, S^2 \times S^1, S^2)$. Since $\pi_2(S^2 \times S^1) = Z$, there exist infinitely many nonequivalent reductions to the $U(1)$ (R) bundle. However, two reductions determined by homotopic sections of $\mathring{E}$ determine the same electromagnetic field $B_{\mu\nu}$ and magnetic charge, respectively.

We proceed as above by selecting a section of the Q bundle so that the local representation for ω is also that for ω in R and then Γ_μ is locally given as

$$\hat{\Gamma}_\mu = (1/2)\Gamma_\mu^3 = (1/2)\Gamma_\mu'^3, \; e = \tilde{g}/2. \tag{4.40}$$

That Q can be reduced to R is determined by action (4.39). E is decomposed into two $U(1)$ bundles F_1, $F_2 = \bar{F}_1$.

The Whitney formula yields

$$c_1(Q) = c_1(F_1 \oplus_w \bar{F}_1) = 0, \tag{4.41}$$

i.e., Q is isomorphic to the trivial bundle. This case is associated with the 'tHooft-Polyakov $SU(2)$ monopole.

The quantization condition is $2e\hat{g} = m_2$, $m_2 \in Z$.

Direct reduction $SU(3) \to SU(2) \to U(1)$ monopole can be determined by the choice of the canonical section $n_3 = bk_3b^{-1}$ of E, where $b \in SU(3)$ is an element determining a local section of Q and R, in which case $m_2 = 0$ and $B_{\mu\nu}$ is given by

$$eB_{\mu\nu} = \mathrm{Tr}\,(n_3F_{\mu\nu}) - \mathrm{Tr}\,(n_3[\nabla_\mu n_3, \; \nabla_\nu n_3]) - \sqrt{3}\,\mathrm{Tr}\,(n[\nabla_\mu n_3, \; \nabla_\nu n_3]). \tag{4.42}$$

Like the above construction, we can consider the case of $G = SU(3)$, $H = U(1) \times U(1)$, where the monopole solutions are realized for potentials of degrees less than, or equal to, four. The topological charges are determined by the group $\pi_1(H)$ and characterized by two integers (m_1, m_2).

Another class of solution can be realized, for instance, in the six-dimensional space V of representation of $SU(3)$. The space V consists of real symmetric 3×3 matrices. The stability subgroup of the vector $|\varphi^{(0)}\rangle = \lambda I$ under the action

$$\varphi \to u\varphi u', \quad u \in SU(3),$$

is the group $SO(3)$.

It is natural to regard monopole solutions of this sort as spherically symmetric, characterized by the topological charge $\pi_1(SO(3)) = Z_2$.

The topological criterion enables us to supply the necessary conditions for monopoles. We can obtain sufficient conditions, too, mostly by variational methods (see [JT]). However, explicit solutions could only be found in the one singular case of the *Bogomolny-Prasad-Sommerfeld* (BPS) limit, which is obtained for λ, $\mu^2 \to 0$ and finite μ^2/λ, where μ^2 and λ are involved in the definition of the Higgs potential

$$U(\varphi) = (\lambda/8)(\mu^2/\lambda - \varphi^2)^2.$$

In this limit, $U(\varphi) \to 0$, and, as shown in [Bg], we can pass from the second-order motion equation to the first-order equation

$$F_{\mu\nu} = -\varepsilon_{\mu\nu\gamma} \nabla_\mu \varphi, \tag{4.43}$$

where $F_{\mu\nu}$ and $\nabla_\mu \varphi$ are determined as in (4.20).

At present, great progress has been achieved in the problem of finding monopole solutions with arbitrary topological charge q for an arbitrary compact gauge group G [N], [DoK].

Basically, the methods are related to the twistor approach, which was worked out earlier for the classical solutions (instantons) to the pure Yang-Mills equation. The approach to a considerable extent relies on the basis of algebraic geometry methods and, therefore, is beyond the framework of this book. To look into this promising topic, I would recommend [DoK], [AW].

Another interesting question relates to the dimension of the space of monopole solutions with a given topological charge. The work is also intimately related to the methods for finding the instanton solution's parameter space [ADHM], [Dok].

The end of this section is devoted to one physical problem where the homotopy group technique is applied to describe monopole behavior in phase transitions. I will follow [B].

4.1.6 Topology of Monopoles Crossing a Phase Boundary

Consider the following hypothetical situation.

Given a gauge group G and two phases V_1 and V_2 determined by two subgroups H_1 and H_2, associate a monopole φ_1 with H_1 and a monopole φ_2 with H_2. We are interested in the behavior of φ_1 as one phase turns into another. For simplicity, we regard the original group as connected and simply connected ($\pi_0(G) = \pi_1(G) = 0$).

From the point of view of the theory of phase transitions, we have three possible cases of location of H_1 and H_2:

$$\text{(i) } H_1 \supset H_2, \quad \text{(ii) } H_2 \supset H_1, \quad \text{(iii) } H_1 \cap H_2 = H_3.$$

We call the chain of mappings

$$G \to H_1 \to H_2$$

a *phase transition* if the transition from the group G to H_1 and H_2 is carried out by spontaneous symmetry breakdown.

(i) $H_1 \supset H_2$. This "phase transition" is typical for unified strong and weak interaction theories, where spontaneous breakdown goes from the larger group to the smaller one. Consider the exact sequence

$$0 \to \pi_2(H_1/H_2) \xrightarrow{\varphi} \pi_1(H_2) \xrightarrow{\psi} \pi_1(H_1) \to \pi_1(H_1/H_2) \to \pi_0(H_2) \to \pi_0(H_1). \quad (4.44)$$

Since $\pi_2(G/H_1) = \pi_1(H_1)$, the topological charges of the monopole in the phase H_1 are characterized by the elements of $\pi_1(H_1)$ and in the phase H_2 by the elements of $\pi_1(H_2)$. Since $\pi_2(H_1) = 0$, $\pi_2(H_1/H_2) = \text{Ker } [\pi_1(H_2) \xrightarrow{\psi} \pi_1(H_1)]$. The group $\pi_2(H_1/H_2)$ characterizes monopoles in H_2 which correspond to the trivial elements in H_1, or new monopoles in H_2. We can say in physical terminology that the elements of $\pi_1(H_2)$ not in Ker ψ are associated with the monopoles of H_2 surviving the transition into H_1. Besides, there exist monopoles from the group Im φ.

The fundamental group $\pi_1(H_1/H_2)$ determines a nontrivial tube of linear vortex solutions in the H_2-phase, not obtained from the linear vortices of the H_1-phase. This physical assertion follows by considering a segment of the exact sequence

$$\pi_1(H_1/H_2) \to \pi_1(G/H_2) \to \pi_1(G/H_1) \to \pi_0(H_1/H_2). \quad (4.45)$$

Without loss of generality, we can assume that H_1/H_2 is connected, e.g., both H_1 and H_2 are connected or consist of the same number of connected components. Then $\pi_0(H_1/H_2) = 0$ and our interpretation of the group $\pi_1(H_1/H_2)$ immediately follows from the exactness of (4.45).

The groups $\pi_0(H_1)$ and $\pi_0(H_2)$ determine the topological characterization (i.e., charges) of linear vortex solutions, namely,

$$\pi_1(G) \to \pi_1(G/H_i) \to \pi_0(H_i) \to \pi_0(G),$$

or

$$\pi_0(H_i) = \pi_1(G/H_i) \quad (i = 1, 2).$$

We track the fate of the H_1-monopole under the transition $H_1 \to H_2$. We obtain from (4.44) that

$$\operatorname{Im}\left\{\pi_1(H_2) \xrightarrow{\psi} \pi_1(H_1)\right\} = \operatorname{Ker}\left\{\pi_1(H_1) \to \pi_1(H_1/H_2)\right\}. \tag{4.46}$$

In physical terminology, this means that the monopoles generated by elements of $\pi_1(H_1)$, which are associated with the image of $\pi_1(H_2)$, do not form a vortex tube, since they correspond to the trivial element of $\pi_1(H_1/H_2)$. On the other hand, the elements of $\pi_1(H_1) \neq \operatorname{Im} \psi(\pi_1(H_2))$ determine the topological charge of the tube in the H_2-phase.

Thus, a monopole in the H_1-phase under the transition into the H_2-phase may ($\pi_1(H_1/H_2) \neq 0$) or may not ($\pi_1(H_1/H_2) = 0$) generate a vortex tube. A purely topological argument does not help clarify the behavior of an H_1 monopole in both cases. However, following Bias, we can assume that possible transformations of the H_1 monopole differ depending on the conditions (1) $\pi_1(H_1/H_2) \neq 0$, (2) $\pi_1(H_1/H_2) = 0$.

(1) The vortex filaments forming the tube can break (due to dynamic instability), forming monopole-antimonopole pairs (and conserving the topological charge) and either annihilating into a vacuum or forming new monopoles in the H_2-phase.

Since $\operatorname{Im}(\pi_1(H_2)) \neq \pi_1(H_1)$, there exist proper H_2 monopoles, too.

(2) Here, the following situation is possible: transition of the H_1 monopole into the H_2 monopole with the same topological charge. However, it follows from (4.46) that a monopole from the H_1-phase is possibly associated with several from the H_2-phase with the configuration determined by the energy minimum conditions. Meanwhile, quantum mechanical (tunnel) transitions are possible between different monopoles.

(ii) The group $H_2 \supset H_1$, in which case the transitions $G \to H_1 \to H_2$ are considerably simpler. Each element of $\pi_1(H_1)$ can be associated with the corresponding elements of $\pi_1(H_2)$ by lifting a closed path $\gamma_1 \in \pi_1(H_1)$ into $\gamma_2 \in \pi_1(H_2)$. The monopole in the H_1-phase is turned into the configuration determined by minimization of system energy and corresponding to a certain element of $\pi_1(H_2)$.

Vortex tubes cannot be formed.

(iii) $H_1 \cap H_2 = H_3$. This case covers all possibilities. It is natural to consider the chain of transitions

$$\begin{array}{ccccc} G & \to & H_1 & \to & H_3 \\ & \searrow & & & \\ & & H_2 & \to & H_3. \end{array}$$

The analysis of this case should include various assumptions on the set H_3.

The nontrivial example with which the effects under discussion can be illustrated is given by familiar $SU(3)$ monopoles.

4.2 INSTANTONS

Here we study the topological properties of one class of solutions of field theory equations and of analogs of the equations in statistical mechanics.

We mean the classical solutions minimizing the action for free Yang-Mills and chiral field equations. Solutions of the type are called *instantons,* or *pseudoparticles.* The first name due to 'tHooft is more usual in the theory of Yang-Mills fields; the second, due to Polyakov, is more often used in solid-state physics. We will use both.

Instantons are related to various physical phenomena. In field theory, they describe vacuum fluctuations, in the theory of ferromagnetism, fluctuations in spin waves. Wherever appropriate, other applications are mentioned.

Much literature is devoted to the physical aspects of instanton theory. For a first acquaintance, I would recommend a nice survey [VZNS] and also [R], [Ho2], [Co2]. [Co2].

The structure of instanton solutions is quite complicated and requires refined and modern topological methods to analyze.

4.2.1 Instantons in One-Dimensional Models

Here, we briefly discuss the physical meaning of an instanton by way of example (see the detailed discussion in [VZNS]).

Consider the model of a scalar field with interaction $\lambda\varphi^4$ we have already repeatedly used. We write the Lagrangian as

$$L = (\varphi_t)^2 - (\varphi_x)^2 + \frac{\mu^2}{2}\varphi^2 - \frac{\lambda}{4}\varphi^4 \quad (\mu^2, \lambda > 0). \tag{4.47}$$

The motion equation is accordingly

$$\varphi_{tt} - \varphi_{xx} = \mu^2\varphi - \lambda\varphi^3 \tag{4.48}$$

with constant solutions

$$\varphi = 0, \quad \varphi = \pm\mu/\sqrt{\lambda}. \tag{4.49}$$

The minima of potential

$$U(\varphi) = -(\mu^2/2)\varphi^2 + (\lambda/4)\varphi^4 \tag{4.50}$$

are the points $\varphi = \pm\mu/\sqrt{\lambda}$, viz., vacuum solutions.

Equation (4.48) has a constant solution, a so-called *kink,* or *domain wall,* with the vacuum boundary conditions:

$$\varphi_{\text{kink}} = \varphi(x - x_0) = \pm(\mu/\sqrt{\lambda})\tanh[(\mu/\sqrt{\lambda})(x - x_0)], \tag{4.51}$$

$$\varphi(\infty) = \mu/\sqrt{\lambda}, \ \varphi(-\infty) = -\mu/\sqrt{\lambda}$$

(here x_0 is an arbitrary point on the straight line R^1, called the *center of solution.* This can easily be seen to be independent of the choice of x_0). The solution is a local minimum of potential $\tilde{U}(\varphi) = (\varphi_x)^2 + U(\varphi)$.

Solution (4.51) is called an *instanton.*

We show that (4.51) possesses the important characteristic property of minimizing the action for Lagrangian (4.47).

We pass to the purely imaginary variable $\tau = it$ and consider the Lagrangian L in the space of (x, τ). The transition $t \to it$ to the Euclidean metric is essential in instanton theory. In Euclidean form, the action S is typically

$$iS(\varphi(t)) = -\int_{-\tau_0}^{\tau_0} [(1/2)(\varphi_\tau)^2 + \tilde{U}(\varphi)]\, d\tau \tag{4.52}$$

for (4.47) if the boundary condition is

$$\varphi(\tau_0) = \mu/\sqrt{\lambda}, \quad \varphi(-\tau_0) = -\mu/\sqrt{\lambda}.$$

The integral

$$S_E = \int_{-\tau_0}^{\tau_0} [(1/2)(\varphi_\tau)^2 + \tilde{U}(\varphi)]\, d\tau \tag{4.53}$$

is called *Euclidean action.* Below we use the term action to mean the action S defined on Minkowski space, or S_E defined on Euclidean space. The action functional is the main ingredient of the path-integral method along trajectories of Feynman integrals.

We recall the principal idea of the Feynman method, taking a simple mechanical example of motion of a spinless particle in potential (4.50).

Let the particle be at a point φ_0 at the initial moment $(-t_0)$ and at φ_1 at the final moment (t_0). Then the amplitude of transition of the particle from the point $(-t_0, \varphi_0)$ into (t_0, φ_1) is determined by functional integration along all trajectories joining the world points $(-t_0, \varphi_0)$ and (t_0, φ_1) with weight $\exp(iS)$ and equals

$$\langle \varphi_1 | \exp(-i\hat{H}t_0) | \varphi_0 \rangle = N \int \exp(iS(\varphi(t))D[\varphi], \tag{4.54}$$

where $\hat{H}$ is the Hamiltonian of the system, N the normalizing factor, and D a functional measure. In Euclidean formulation, (4.54) assumes the form

$$\langle \varphi_1 | \exp(-\hat{H}\tau_0) | \varphi_0 \rangle = N \int \exp(-S)D[\varphi]. \tag{4.55}$$

Without going into the detail of the definition of the measure $D[\varphi]$ (what does "integration along all trajectories" mean?), note that from the standpoint of calculating integral (4.55), e.g., by the steepest descent method, a knowledge of extremals of the action S is essential. In particular, if an extremal trajectory (or, which is the same, a stationary point) of the action S is unique, then $S([\varphi(\tau)]) = S_0$ and

$$N\int \exp(-S)\,D[\varphi] \sim N_1 \exp(-S_0). \tag{4.56}$$

In the general case, summation over all stationary points is necessary.

Formulas (4.54)-(4.56) show that in order to calculate the complete transition amplitude, including the preexponential factor in (4.56), we should know all the extremals of the action functional. Generally speaking, there may be solutions associated with the saddle points in solution's function space. However, it is especially important to know the minima of the action functional.

Thus, we retrace our steps to example (4.47), where the Lagrangian can be treated in two ways. On the one hand, it describes the motion of the domian wall (the stationary solution determining the wall's shape) and, on the other hand, it defines the motion of a spinless particle in the potential $\tilde{U}(\varphi)$.

Consider the second interpretation and the problem of describing the amplitude of transitions from one vacuum $\varphi_0 = -\mu/\sqrt{\lambda}$ into another $\varphi_1 = \mu/\sqrt{\lambda}$ which is the *tunnelling* phenomenon familiar from quantum mechanics [LL4]. This is conveniently written in Euclidean terms.

It turns out that during the transition to imaginary time the classical trajectory (absent in physical two-dimensional Minkowski space) responsible for the transition arises. The trajectory $\varphi(\tau)$ relates two classical vacuum solutions φ_0 and φ_1. $\varphi(\tau)$ is already familiar as the kink (4.51) (if we replace x by τ in the formula and put $\varphi_x = 0$).

Euclidean action is finite on this solution and has a local minimum ($S_E \neq 0$). The solution is the instanton in the model $\lambda\varphi_2^4$.

Note that (4.51) is an exact solution of equation (4.48) only if the boundary values on $\pm\infty$ are selected. For finite τ, the solution is more complicated. Together with a solution (4.51), it is obvious that another one exists, starting at the point $\tilde{\varphi}_0 = \mu/\sqrt{\lambda}$ and terminating at $\tilde{\varphi}_1 = -\mu/\sqrt{\lambda}$. It is natural to call such a solution an *anti-instanton.* The reader who would like a detailed analysis of one-dimensional instantons from the point of view of general instanton problems can turn to the survey [VZNS]. However, we confine ourselves to the topological side of the problem.

Topological charge of a kink. Consider a solution of domian wall type from the topological viewpoint. We have a mapping $\hat{\varphi}$ of the straight line R^1 with two different points at infinity, $\tau_0 = -\infty$ and $\tau_1 = \infty$, to the solutions of kink type. It is easy to see that, for a solution (4.51) with fixed boundary conditions φ_0 and φ_1, we can define the conserved quantity $Q = (1/2)(\sqrt{\lambda}/\mu)\,[\hat{\varphi}(\infty) - \hat{\varphi}(-\infty)]$, $Q = 1$.

For anti-instanton solution, a domain wall with interchanged boundary conditions, Q is equal to -1.

However, this solution is invariant under the gauge transformation $\varphi \to -\varphi$. Therefore, both solutions should be identified. On the other hand, there are also simply vacuum solutions $\hat{\varphi} = 0$ among those of equation (4.48), for which $Q = 0$.

By analogy with the topological charges of monopoles, the topological charge Q can be treated as an element of the group $\pi_0(V)$, where the vacuum manifold V is zero-dimensional sphere S^0 consisting of the boundary points $(-\mu/\sqrt{\lambda}, +\mu/\sqrt{\lambda})$.

For Lagrangian (4.47), $\pi_0(V)$ is Z_2.

The map $\hat{\varphi}\begin{pmatrix}-\infty\\+\infty\end{pmatrix} = \begin{pmatrix}a\\b\end{pmatrix}$ determines two classes of solutions: (1) $\varphi(-\infty) = \varphi(+\infty)$, $Q = 0$ and (2) $\varphi(-\infty) \neq \varphi(+\infty)$, $Q = 1$ (mod 2) which are not homotopy equivalent.

Note that Q is of the same nature as the topological charge of a monopole. In fact, we define Q as an element of the group of homotopy nontrivial maps of an asymptotic manifold (consisting of the points $x_0 = -\infty$, $x_1 = \infty$) to the orbit space of the gauge group G of Lagrangian (4.47), $G = Z_2$, H is trivial.

In scalar theory, Q is a particular case of the general structure of topological charges, based on cobordism theory. The structure is discussed in Subsec. 5.2.8.3 after we study various examples. Q can be introduced in scalar two-dimensional field theory on the basis of currents conservation laws (in a homological approach).

Consider the current $I_\mu = \varepsilon_{\mu\nu}\partial_\nu\varphi$ in two-dimensional space-time M^2 ($\mu = 0, 1$ and $\varepsilon_{\mu\nu}$ is an antisymmetric tensor). I_μ is conserved, namely,

$$\partial_\mu I_\mu = 0.$$

Therefore, we can determine the conserved quantity

$$\frac{d}{dt}\int_{-\infty}^{+\infty} I_0\,dx = 0, \text{ i.e., } \int_{-\infty}^{+\infty} \frac{\partial\varphi}{\partial x}\,dx = \varphi(\infty, t) - \varphi(-\infty, t) = Q. \tag{4.57}$$

The value of Q is determined by the spatial derivative. Q is not trivial in the case of nontrivial boundary conditions.

Consider another example of scalar field theory with the topological charge Q, namely,

The sine-Gordon equation

$$(\varphi_t)^2 - (\varphi_x)^2 = -\sin\varphi. \tag{4.58}$$

This fascinating relation is encountered in a vast number of physical and mathematical problems, e.g., it determines the metric for a space of constant negative curvature, is the basis for two-dimensional relativistically invariant field theory (due to the Thirring model [R], [FT]), arises in plasma theory [Soli], etc.

The most important property of (4.58) is that the equation is in a wide class of totally integrable evolution systems. A powerful mathematical method for the solution of such equations is that of the inverse scattering problem [ZMNP]. The sine-Gordon equation and its ilk have infinitely many conservation laws (of no topopogical character), soliton solutions, and other remarkable properties (see [ZMNP], [FT]). I restrict myself to the purely topological aspects of (4.58).

The equation is generated by the Lagrangian

$$L = (1/2)[(\varphi_t)^2 - (\varphi_x)^2] + \cos\varphi - 1 \tag{4.59}$$

admitting the discrete symmetry gauge group $D_\infty = Z_2 \dot{\times} Z_\infty$, i.e., the dihedral group consisting of transformations $\varphi \to -\varphi$, $\varphi \to \varphi + 2\pi n$.

Equation (4.58) has the static solutions

$$\varphi_{xx} = \sin\varphi, \; x - x_0 = \pm \int \frac{d\varphi}{\sqrt{2(k - \cos\varphi)}}, \quad x_0,\, k = \text{const.} \tag{4.60}$$

In addition to the constant solutions $\varphi = 2\pi n$ associated with degenerate vacua, there also exists the soliton solution

$$\varphi_{\text{sol}} = 4\tanh^{-1}[\exp(x - x_0)] \tag{4.61}$$

(where $k = 1$ for the plus sign either) with the same property as domain wall solutions, viz., that the soliton's energy is

$$E(\varphi_{\text{sol}}) - E(\varphi_{\text{vac}}) = \int_{-\infty}^{+\infty} \left[(1/2)\left(\frac{\partial\varphi_{\text{sol}}}{\partial x}\right)^2 + (1 - \cos\varphi_{\text{sol}})\right] dx = 8.$$

A soliton solution cannot be obtained by perturbation theory over a vacuum solution.

We shall now discuss the topological properties of soliton solutions. The homotopy approach immediately yields a description of topologically nontrivial solutions. In fact, the vacuum manifold of solutions is isomorphic to the orbit D_∞/Z_2 of the group D_∞. The stability group of the vacuum H is Z_2. Thus, the topological charge Q is classified by elements of the group $\pi_0(D_\infty/Z_2) = \pi_0(Z) = Z$. Consider the matter from the point of view of topological currents.

Expressed in terms of the field φ in (4.57), Q also obviously exists for the sine-Gordon model and assumes discrete values $\pm 1, \pm 2, \ldots$. We have

$$Q = [1/(2\pi)] \int_{-\infty}^{+\infty} \frac{d\varphi}{dx}\, dx = [1/(2\pi)][\varphi(+\infty) - \varphi(-\infty)]. \tag{4.62}$$

In the soliton solution (4.61), $Q = 1$. Accordingly, the solution with $Q = -1$ (antisoliton) is

$$\varphi_{\text{antisol}} = -4\tanh^{-1}\exp(x - x_0)$$

and obviously topologically inequivalent to a soliton (having different boundary conditions). It would be interesting to note that solutions with $Q > 1$ cannot be obtained via a linear combination of static solutions. Nevertheless, multisoliton solutions in the sine-Gordon model are known.

The topological charges introduced in this subsection admit different modifications and can be generalized to multidimensional theories; they play a key role in classifying the solutions.

Before turning to the topology of classical solutions in concrete physical models, we make a brief mathematical diversion. In the next section, a special class of maps of manifolds related to action functional minimization is considered. Those maps said to be *harmonic* were introduced in the mid-fifties [Fu], [ES] with no immediate application to physics and then they turned out to be extremely useful in the theory of chiral and gauge fields.

4.2.2 Harmonic Maps

Harmonic maps are defined as the critical points of the action functional, naturally generalizing harmonic functions. We now turn to exact formulations.

Let $\varphi: M \to N$ be a map of Riemannian manifolds. The action functional is defined as

$$S(\varphi) = (1/2) \int_M |d\varphi(x)|^2 \, dv(g), \tag{4.63}$$

where $d\varphi$ is the differential of φ at a point $x \in M$ and $dV(g)$ the volume element induced by the metric g. The norm $|d\varphi(x)|$ is defined as the Hilbert-Schmidt norm of the linear operator $d\varphi(x)$. In local coordinates $x = (x^i) \in U$ and $u = (u^\alpha) \in V(\varphi(x))$ we have

$$|d\varphi(x)|^2 = g^{ij} h_{\alpha\beta} \frac{\partial \varphi^\alpha}{\partial x^i} \frac{\partial \varphi^\beta}{\partial x^j} \, . \tag{4.64}$$

In the orthonormal basis $e_1, \ldots, e_m$ for TM_x^m we have

$$|d\varphi|_x^2 = \sum_{i=1}^m \frac{|d\varphi(e_i)|_{TN}^2}{|e_i|_{TM}^2} \, . \tag{4.65}$$

The Euler-Lagrange equation for S is

$$\tau(\varphi) = \operatorname{div}(d\varphi) = 0. \tag{4.66}$$

Definition 4.1 A map φ is said to be *harmonic* if the field $\tau(\varphi)$ associated with φ is identically zero.

$\tau(\varphi)$ is called the *tension field.* "Physically" the condition of harmonicity for φ means that the surface $\varphi(M)$ is spanned by N extremally, i.e., is in elastic equilibrium.

We describe the operator $\tau(\varphi)$ in local coordinates. Let the local coordinates on C^∞ manifolds M^m and N^n be $(x^1, \ldots, x^m) \in M^m$, $(y^1, \ldots, y^n) \in N^n$ and the Riemannian metrics be g_{ij} on M^m and $h_{\alpha\beta}$ on N^n. We also introduce the (*Levi-Cività*) connection coefficients Γ_{ij}^k on M^m and $L_{\alpha\beta}^\gamma$ on N^n associated with g_{ij} and $h_{\alpha\beta}$, respectively. Then the operator $\tau(\varphi)$ is representable for the map $\varphi: M^m \to N^n$ as follows:

$$\tau^\gamma(\varphi)\Big|_{x_0} = g^{ij} \left\{ \frac{\partial^2 \varphi^\gamma}{\partial x^i \partial x^j} - \Gamma_{ij}^k \frac{\partial \varphi^\gamma}{\partial x^k} + L_{\alpha\beta}^\gamma \frac{\partial \varphi^\alpha}{\partial x^i} \frac{\partial \varphi^\beta}{\partial x^j} \right\}. \tag{4.67}$$

It is especially simple in form in normal coordinates, namely,

$$g_{ij} = \delta_{ij}, \quad \Gamma^k_{ij}(x_0) = 0, \quad L_{\alpha\beta}(\varphi(x_0)) = 0. \tag{4.68}$$

The operator $\tau(\varphi)$ can be written as a collection of scalar Laplacians $\Delta\varphi^\gamma$ in R^n, namely,

$$\tau^\gamma(\varphi) = \frac{\partial^2\varphi^\gamma(x_0)}{\partial x^i\,\partial x^i}, \; 1 \leqslant \gamma \leqslant n, \; 1 \leqslant i \leqslant m. \tag{4.69}$$

The immediate consequence is that $M^m \to R^n$ is harmonic if and only if $\Delta\varphi = (\Delta\varphi^1, \ldots, \Delta\varphi^n) = 0$, i.e., the coordinates of φ are harmonic functions, in particular, if M^m is compact, then a harmonic map is constant.

$\tau(\varphi)$ can be given invariantly by the theory of harmonic forms with values in vector bundles [EL1]. Since a detailed treatment of harmonic maps is not my purpose (there are detailed discussions in [EL1, 2]), I do not give the corresponding formulations.

We consider several examples of harmonic maps.

Example 4.1 **1.** If dim $M = 1$, then harmonic maps $M \to N$ are geodesics in N. For proof, it suffices to notice that

$$\tau(\varphi) = \nabla_t\left(\frac{d\varphi}{dt}\right) = \nabla_{tt}\varphi,$$

where ∇ is the covariant derivative of the field along the curve $\varphi(t)$. The vector field $\nabla_{tt}\varphi$ is called the *acceleration field.* A statement now follows from the definition of a geodesic line (see Subsec. 1.5.9).

2. If M as a Riemannian submanifold embedded in N is of minimum volume, then the embedding map φ: $M \to N$ is harmonic [EL1].

3. If M and N are two Kähler manifolds, then holomorphic maps φ: $M \to N$ are harmonic relative to any compatible metric.

In physical applications the fact that φ supplies an extremum for the action functional is the most essential property.

We now formulate another result from the theory of harmonic maps, in particular, enabling us to describe all instanton solutions in the two-dimensional chiral σ-model (see the next subsection for the result obtained by explicit formulas) [Wod], [EL1].

Theorem 4.4 *Let $M = S^2$ and $N = S^2$ be two two-dimensional spheres. Then each harmonic map φ: $S^2 \to S^2$ is either holomorphic or antiholomorphic.*

In the general case, harmonic maps of a closed surface M^2 of genus g to a surface N^2 of genus p are arranged as follows [Wod], [EL1].

Theorem 4.5 *Let M^2 and N^2 be compact and orientable. Then harmonic maps φ are classified as follows:*

(a) *If $p = 0$, φ is either holomorphic or antiholomorphic.*

(b) *If $p > 0$, there only exist constant maps.*

To prove Theorems 4.4 and 4.5, we need certain auxiliary relations between harmonic and holomorphic maps of two-dimensional manifolds. We consider two-dimensional closed orientable surfaces as Riemann ones equipped with a complex structure and a Riemannian metric.

Remark 4.1 If we confine ourselves to the case of a Riemann surface M^2 (and of an arbitrary Riemann manifold N^n), then it is not hard to see that the action functional in (4.63) is unaltered under a conformal change $g \to \lambda g$ ($\lambda(x)$, $x \in M^2$) of the metric. Therefore, the harmonicity property only depends on the conformal equivalence's class of g. It is known that there exist (isothermal) coordinates on M^2, in which g is conformally equivalent to the flat one $g = \varrho^2((dx^1)^2 + (dx^2)^2)$. The operator τ is represented in the latter metric as

$$\tau(\varphi) = \varrho^{-2}\left[\frac{\partial^2\varphi^\gamma}{\partial x^1\,\partial x^1} + \frac{\partial^2\varphi}{\partial x^2\,\partial x^2} + \delta^{ij}L^\gamma_{\alpha\beta}\frac{\partial\varphi^\alpha}{\partial x^i}\frac{\partial\varphi^\beta}{\partial x^j}\right]. \tag{4.70}$$

A conformal metric g generates a complex structure on M^2. Therefore, to study harmonic maps $\varphi: M^2 \to N^n$, it suffices to restrict ourselves to specifying M^2 as a Riemann surface.

Definition 4.2 Let M^2 be a Riemann surface with a local coordinate z. A symmetric quadratic tensor of rank (2, 0), representable locally as $\eta(z) = a(z)\,dz^2$, where $dz^2 = dz \otimes dz$, is called a *quadratic differential* η *on* M^2. Under coordinate change $z \to z'$, $\eta = a'(z')dz'^2$, where

$$a'(z') = a(z(z'))\left(\frac{dz}{dz'}\right)^2. \tag{4.71}$$

A differential η is said to be *holomorphic* if the function $a(z)$ is.

Let us see how harmonic maps and holomorphic differentials are related in the two-dimensional case. First, we consider a local problem.

Let z be a local coordinate in a domain $U \subset M^2$, $z = x + iy$, and φ a harmonic map of U onto $V \subset N^2$.

φ can be regarded as a harmonic function, i.e.,

$$\Delta\varphi = \frac{\partial^2\varphi}{\partial x^2} + \frac{\partial^2\varphi}{\partial y^2} = 0.$$

Lemma 4.1 (1) *If a map* $\varphi: M^2 \to N^2$ *is harmonic, then the differential* $\eta = \left\langle\frac{\partial\varphi}{\partial z}\,\middle|\,\frac{\partial\varphi}{\partial z}\right\rangle dz^2 = a(z)\,dz^2$ *is holomorphic.*

(2) *If* η *is holomorphic and a map* φ *is of rank* 2 *at all points* $z \in M^2$, *then* φ *is harmonic.*

Here $\langle\varphi|\psi\rangle$ is the scalar product $\langle\varphi|\psi\rangle_{TN^c} = h_{\alpha\beta}\varphi^\alpha\psi^\beta$ defined in the standard way for any smooth map. Note that as usual the scalar product $\langle e_i|e_j\rangle$ defined on vector of real Euclidean space R^n is extended to a complex bilinear form on $C^n = R^n \otimes C$, the complexification of R^n. The role of R^n is here played by the tangent space to N^2. The form η can be defined in another way.

We carry the metric $h_{\alpha\beta}$ given on N^2 over to M^2 by a smooth map φ. We obtain a symmetric covariant tensor $(\varphi_* h)$ of rank 2 on M^2. We take its component (2, 0), which is η by definition. In real coordinates (for $z = x + iy$),

$$a(z) = \frac{1}{4}\left\{\left|\frac{\partial\varphi}{\partial x}\right|^2_{TN} - \left|\frac{\partial\varphi}{\partial y}\right|^2_{TN} - 2i\left\langle\frac{\partial\varphi}{\partial x}\,\middle|\,\frac{\partial\varphi}{\partial y}\right\rangle_{TN}\right\}. \tag{4.72}$$

We now *prove Lemma 4.1.*

(1) We calculate $\frac{\partial}{\partial\bar{z}}a(z)$ in local coordinates $a(z) = \left(\frac{\partial\varphi}{\partial z}\right)^2$, therefore,

$$\frac{\partial a}{\partial\bar{z}} = 2\frac{\partial^2\varphi}{\partial\bar{z}\,\partial z}\frac{\partial\varphi}{\partial z} = 2\left\langle\frac{\partial^2\varphi}{\partial z\,\partial\bar{z}}\,\middle|\,\frac{\partial\varphi}{\partial z}\right\rangle, \tag{4.73}$$

or, in real coordinates,

$$\frac{\partial a}{\partial\bar{z}} = \frac{1}{4}\left\langle\frac{\partial^2\varphi}{\partial x^2} + \frac{\partial^2\varphi}{\partial y^2}\,\middle|\,\frac{\partial\varphi}{\partial x}\right\rangle_{TN} - \frac{i}{4}\left\langle\frac{\partial^2\varphi}{\partial x^2} + \frac{\partial^2\varphi}{\partial y^2}\,\middle|\,\frac{\partial\varphi}{\partial y}\right\rangle_{TN}. \tag{4.74}$$

Since $\Delta\varphi = \partial^2\varphi/\partial z\partial\bar{z} = \frac{\partial^2\varphi}{\partial x^2} + \frac{\partial^2\varphi}{\partial y^2}$, we immediately derive from (4.73) and (4.74) that the holomorphy of the differential η follows from the harmonicity of φ. It is obvious that this is also true for any complex analytic map $\varphi\colon M^2 \to N^n$.

(2) Let $\varphi\colon M^2 \to N^2$ be a smooth map. We take a complex connection $\nabla\partial/\partial z$ on M^2 generated by the field $\partial/\partial z$. The operator $\tau(\varphi)$ in (4.66) can be written as $\tau(\varphi) = \nabla_{\partial_{\bar{z}}}\partial_z\varphi$ and the condition for harmonicity as

$$\nabla_{\partial_{\bar{z}}}\partial_z\varphi = 0, \text{ or } \nabla_{\partial_z}\partial_{\bar{z}}\varphi = 0. \tag{4.75}$$

This formula is proved by straightforward calculation in local coordinates. We now write the holomorphy condition for $\partial a/\partial\bar{z}$ in terms of the operator $\nabla_{\partial_{\bar{z}}}$. We have

$$\frac{\partial a}{\partial\bar{z}} = \frac{\partial}{\partial\bar{z}}\left\langle\frac{\partial\varphi}{\partial z}\,\middle|\,\frac{\partial\varphi}{\partial z}\right\rangle = \left\langle\nabla_{\bar{z}}\frac{\partial\varphi}{\partial z}\,\middle|\,\frac{\partial\varphi}{\partial z}\right\rangle$$

$$+ \left\langle\frac{\partial\varphi}{\partial z}\,\middle|\,\nabla_{\partial\bar{z}}\frac{\partial\varphi}{\partial z}\right\rangle = 2\left\langle\nabla_{\partial\bar{z}}\frac{\partial\varphi}{\partial z}\,\middle|\,\frac{\partial\varphi}{\partial z}\right\rangle. \tag{4.76}$$

It follows from the conditions for harmonicity and formula (4.75) that if $(\partial/\partial\bar{z})a(z) = 0$, then $\tau(\varphi) \in TN_{\varphi(z_0)}$ is orthogonal to the differential $d\varphi(TM_{z_0}) \subset TN_{\varphi(z_0)}$. Therefore, if the rank of the map $d\varphi\colon TM^2 \to TN^2$ is two, then the harmonicity of φ follows from the holomorphy of η.

Remark 4.2 It follows from the Sard theorem that it suffices to require that the condition for the map's rank should be fulfilled for an everywhere dense set in M^2.

We now retrace our steps to the *proof of Theorem 4.4*. It follows from (4.75) that holomorphic and antiholomorphic maps are harmonic. Let φ be a harmonic map other than holomorphic or antiholomorphic. It follows from Lemma 4.1 that φ is associated with a holomorphic differential η. It follows from the classical Riemann-Roch theorem [Sp] that there are no holomorphic quadratic differentials on S^2. Theorem 4.4 is thereby proved.

The *proof of Theorem 4.5* is considerably more complicated, being based on a study of the zeros of holomorphic differentials and the classical Hurwitz and Riemann-Roch theorems. We formulate the Eells and Wood theorem from which the statement of Theorem 4.5 can now be obtained easily.

Theorem 4.6 *Let M^2 and N^2 be two compact Riemann surfaces, $\chi(M)$ and $\chi(N)$ their Euler characteristics, φ a smooth map $M^2 \to N^2$ and harmonic relative to the Hermitian metric on N^2,* $\deg \varphi = n$. *Then if*

$$|n| > -\chi(M)/|\chi(N)|, \tag{4.77}$$

φ is holomorphic or antiholomorphic.

See the *proof* of the theorem in [EL1].

The first part of Theorem 4.5 immediately follows from Theorem 4.6.

If the genus p of N^2 is greater than zero, then to prove the second part of Theorem 4.5 it suffices to use the following Riemann-Hurwitz relation.

Let φ be a holomorphic map of a compact Riemann surface M^2 of genus g to a compact Riemann surface N^2 of genus p, φ a branched covering of N^2, n the degree of φ, and q the number of branch points with multiplicity. The equality

$$q = n\chi(N^2) - \chi(M^2) \tag{4.78}$$

holds (Riemann-Hurwitz relation).

Remark 4.3 Formula (4.78) is a particular case of the general Riemann-Hurwitz relation for punctured manifolds M^2 and N^2 with a boundary and r, S points removed, respectively.

It follows from (4.78) that there exists no holomorphic map of degree $n > \chi(M^2)/\chi(N^2)$ (for $p > 1$). Therefore, neither do holomorphic nor antiholomorphic maps for $|n| > \chi(M^2)/\chi(N^2)$. Theorem 4.5 is thereby proved completely.

We can immediately derive the following corollary due to Lemaire and Wood from Theorem 4.6 [EL1].

Corollary 4.1 *There exists no harmonic map of degree ± 1 of a torus onto a sphere (for any given metric).*

From the point of view of physical applications, it would also be interesting to study the following classes of harmonic maps.

1. Map $S^2 \to RP^2$. Here, harmonic maps can be interpreted as two-dimensional textures in nematic liquid crystals (see Subsec. 5.1.4). For arbitrary boundary conditions, this problem is transformed into description of harmonic maps $\varphi: M^2 \to RP^2$. I give some results in this respect in Subsec. 5.1.4.

2. Map $S^2 \to CP^n$. A description of the harmonic maps is obtained in [Z], [EW2]. In contrast to the case of CP^1, instanton and anti-instanton solutions which in the

general case are holomorphic and antiholomorphic maps $\varphi: CP^1 \to CP^n$ do not generate the whole class of harmonic maps. Exact formulations are given in Subsec. 4.2.4.

3. A study of harmonic maps of three- and four-dimensional manifolds are of great interest for the theories of chiral fields, Yang-Mills and Einstein equations. Only partial results and many unsolved problems exist. Some can be discussed with the help of the surveys in [Gi], [EGH].

We now pass to the explicit description of special classes of harmonic maps, of instantons and anti-instantons in two-dimensional models.

4.2.3 Instantons in Two-Dimensional Ferromagnets

Here, we describe the class of two-dimensional solutions in the two-dimensional chiral model of double interest. In solid-state physics, the model describes a two-dimensional ferromagnet. In particular, classical solutions are related to longwave length fluctuations of spin waves and affect the nature of phase transition in the Heisenberg model (see the physical details in [BP]).

On the other hand, this is the simplest two-dimensional analog of realistic four-dimensional theories. Chiral models have a nontrivial topological charge and admit an explicit description of instanton solutions.

The action is

$$S = (1/2) \int (\partial_\mu n^a)(\partial_\mu n^a)\, d^2x, \tag{4.79}$$

where $(n^a)^2 = 1$, $a = 1, 2, 3$, $\mu = 1, 2$, and d^2x is the element of area in R^2. The action extremals are determined by the equation

$$\delta S = 0 \Rightarrow \partial_\mu \partial_\mu n + (\partial_\mu n \cdot \partial_\mu n) n = 0, \tag{4.80}$$

or

$$\partial_\mu [n, \partial_\mu n] = 0. \tag{4.81}$$

We look for solutions of (4.81) with finite action and consider the class with the boundary behavior

$$n(x) \to n_0 \tag{4.82}$$

as $|x| \to \infty$, in which case the field variable assumes values on the extended plane $R^2 \cup \infty = S^2$. Since the metric on $R^2 \cup \infty$ is conformally equivalent to that on S^2, all solutions of (4.81) with condition (4.82) can be extended uniquely on S^2.

The map

$$S^2_{\text{Sp}} \to S^2_{\text{F}} \tag{4.83}$$

arises due to the field variable, where the subscript Sp means "space" and F "field".

The set of classes of homotopy-equivalent maps $S^2_{\text{Sp}} \to S^2_{\text{F}}$ is characterized by the two-dimensional homotopy group $\pi_2(S^2_{\text{F}}) = Z$, i.e., different field configurations which are not gauge invariant are specified by an integer. Since the sphere S^2 is simply connected, the corresponding topological invariant admits an integral representation by Hurewicz's theorem ($\pi_2(S^2) = H^2(S^2)$). Henceforth the topological charge

$$Q = \frac{1}{8\pi^2} \int \varepsilon_{\mu\nu}(n[\partial_\mu n, \partial_\nu n])\, d^2x \tag{4.84}$$

is an integer

That the right-hand side is an integer follows from the equality of Q to the degree of the map $S^2 \to S^2$.

For proof, we calculate (4.84) by introducing a convenient parametrization of S_F in terms of spherical coordinates:

$$n = (\sin\theta\cos\varphi,\ \sin\theta\sin\varphi,\ \cos\theta),\ Q = \frac{1}{4\pi}\int \sin\theta\, d\theta(x)\, d\varphi(x). \tag{4.85}$$

Since integral density equals the Jacobian of $S^2 \to S^2$, that Q takes an integral value follows from definition of the degree of a map (see Subsec. 2.5.1).

The inequality

$$(\partial_\mu n \pm \varepsilon_{\mu\nu}[n, \partial_\nu n])^2 \geqslant 0$$

enables us to estimate the action S from below as

$$S \geqslant 4\pi Q,$$

the equality holding if

$$\partial_\mu n = \pm\varepsilon_{\mu\nu}[n, \partial_\nu n]. \tag{4.86}$$

The equations are said to be *self-dual* (*anti-self-dual* if the sign is minus). Their solutions are called *instantons* (*anti-instantons*).

Equations (4.86) admit an explicit solution if we pass to complex coordinates, first mapping the Riemann sphere S^2 the z-plane) by the stereographic projection onto the w-plane, where

$$w = w_1 + iw_2 = \frac{n^1 + in^2}{1 - n^3} = \cot(\theta/2)\exp(i\varphi).$$

In these variables, the expression for action (4.79) is

$$S = 2\int \frac{d^2x}{(1 + |w|^2)^2}\partial_\mu w \partial_\mu \overline{w}, \tag{4.87}$$

or

$$S = 4\int \frac{d^2x}{(1 + |w|^2)^2}\left(\frac{\partial w}{\partial z}\frac{\partial \overline{w}}{\partial \overline{z}} + \frac{\partial w}{\partial \overline{z}}\frac{\partial \overline{w}}{\partial z}\right),$$

and the topological charge is

$$Q = \frac{1}{\pi} \int \frac{d^2x}{(1 + |w|^2)^2} \left(\frac{\partial w}{\partial z} \frac{\partial \bar{w}}{\partial \bar{z}} - \frac{\partial w}{\partial \bar{z}} \frac{\partial \bar{w}}{\partial z} \right).$$

Hence, the self-dual (anti-self-dual) equations are reduced to Cauchy-Riemann equations $\partial w/\partial \bar{z} = 0$ (self-dual) or $\partial w/\partial z = 0$ (anti-self-dual).

The general solution of (4.86) together with the boundary conditions $n(\infty) = n_0$ is given by the rational function

$$w(z) = \prod_{j=1}^{n} \frac{(z - a_j)}{(z - b_j)} \tag{4.88}$$

on the sphere S^2, where the boundary value $n_0 \equiv 1$ is chosen. It is obvious that such a choice does not diminish the generality of (4.88), since, due to the invariance of S relative to the groups $SO(3)$, any boundary value can be turned into unity.

The topological charge Q is n for an instanton and $-n$ for an anti-instanton.

The n-instanton solution is characterized by $4n - 3$ real parameters. $4n$ parameters are determined by $2n$ complex numbers a_j and b_j; however, we have to take into account that the group $SO(3)$ acts on the sphere globally. Therefore, the total number of independent parameters is $4n - 3$.

Remark 4.4 In Subsec. 5.1.4, we describe the space of instanton solutions associated with the chiral RP^2 model.

Since each extremal solution of equation (4.81) with finite action is harmonic, it follows from Theorem 4.4 that all solutions are either instanton or anti-instanton. The result is obtained by direct argument in [Wo].

The chiral σ-model with isotopic space S^2 admits different equivalent representations, which are convenient for possible generalizations. The corresponding realizations are given in [Z].

4.2.4 Instantons in Two-Dimensional Chiral Models

Chiral models are a first example of a whole class of two-dimensional models admitting instanton solutions.

From the physical point of view, most chiral models have no direct interpretation, however, they are of interest as four-dimensional field prototype models. Besides, chiral models have an important integrability property, e.g., there exist an infinite series of conservation laws, the Lax representation, etc. (see [Z]).

4.2.4.1 Topologically nontrivial chiral models

An immediate generalization of the S^2 model is a σ model with the field variable $\varphi(x)$, $x = (x^1, x^2)$ taking values in a manifold M^n. If we define the same asymptotic behavior

$$\varphi(x) \to \varphi_0, \quad |x| \to \infty, \tag{4.89}$$

then the condition for a nontrivial topological charge Q to exist is

$$\pi_2(M^n) \neq 0.$$

An immediate generalization of action (4.79) is

$$S = (1/2) \int g_{\alpha\beta}(\partial_\mu \varphi^\alpha \cdot \partial_\mu \varphi^\beta) d^2x, \tag{4.90}$$

where $g_{\alpha\beta}$ is the metric on M^n, $\mu = 1, 2$.

Let M^n be the homogeneous space relative to the action on the group G.

(1) Since $\pi_2(G) = 0$ for any compact Lie group G (by Theorem 2.17), there are topologically nontrivial solutions only for those models where the field variable takes values in the homogeneous space $M = G/H$ (H being a nontrivial subgroup of G), provided

$$\pi_2(G/H) \neq 0. \tag{4.91}$$

(2) Another condition restricting the class of chiral models is the following. We want the solutions on R^2, with fixed asymptotic behavior $\varphi(\infty) = \varphi_0$, to coincide with those on the sphere. It is then necessary that the action should be conformally invariant.

We consider the so-called chiral *CP^n-model.*

A chiral manifold is one to which the values of the chiral field φ belong. CP^n is isomorphic to

$$S^{2n+1}/S^1 = SU(n+1)/SU(n) \times U(1), \quad \pi_2(CP^n) = \pi_1(S^1) = Z. \tag{4.92}$$

Therefore, there exist topologically nontrivial solutions in the model.

The field variable $\varphi \in CP^n$ can be conveniently represented as an element of coset space (4.92) in the form of a Hermitian $(n+1) \times (n+1)$ matrix with trace zero and eigenvalues

$$\lambda_1, \ \ldots, \ \lambda_n = \frac{\lambda}{n+1}, \quad \lambda_{n+1} = -\frac{n\lambda}{n+1}, \tag{4.93}$$

where λ can be chosen to be unity.

The field φ can be represented as

$$\varphi(x) = \frac{1}{n+1} I - u \otimes \bar{u}, \tag{4.94}$$

where I is the identity matrix and $u \otimes \bar{u} = u^\alpha \cdot \bar{u}^\beta$, $\alpha, \beta = 1, \ldots, n+1$. φ is determined by the unit complex $(n+1)$-dimensional vector u (if we ignore the multiplication by the phase factor $\exp(i\alpha(x))$, which is another notation of the statement that CP^n is the quotient $S^{2n+1}/U(1)$.

We obtain from (4.94) that

$$(1/2)\ \mathrm{Tr}\ (\partial_\mu\varphi,\ \partial_\mu\varphi) = (\partial_\mu\bar{u},\ \partial_\mu u) - (\bar{u},\ \partial_\mu u)(u,\ \partial_\mu\bar{u}) \tag{4.95}$$

and

$$\mathrm{Tr}\ ([\varphi,\ \partial_\mu\varphi],\ [\varphi,\ \partial_\mu\varphi]) = \mathrm{Tr}\ (\partial_\mu\varphi,\ \partial_\mu\varphi). \tag{4.96}$$

It follows from (4.95) that

$$S = (1/2)\int \mathrm{Tr}\ (\partial_\mu\varphi,\ \partial_\mu\varphi)\ d^2x = \int \{\partial_\mu\bar{u},\ \partial_\mu u - (\bar{u},\ \partial_\mu u)(u,\ \partial_\mu\bar{u})\}\ d^2x. \tag{4.97}$$

Topological charge density is represented as

$$\mathrm{Tr}\ (\varphi[\partial_1\varphi,\ \ \partial_2\varphi]) = (\partial_1 u,\ \ \partial_2\bar{u}) - (\partial_2 u,\ \partial_1\bar{u}),$$

therefore,

$$Q = \frac{1}{c}\int \varepsilon_{\mu\nu}(\partial_\mu\bar{u},\ \partial_\nu u)\ d^2x. \tag{4.98}$$

To obtain the dual equation, as in the S^2 model, we introduce the matrices

$$\Psi_\mu^\pm = \partial_\mu\varphi \mp i\varepsilon_{\mu\nu}[\varphi,\ \partial_\nu\varphi]. \tag{4.99}$$

The relation

$$(1/2)\ \mathrm{Tr}\ (\Psi_\mu^\pm,\ \Psi_\mu^\pm) \geqslant 0 \tag{4.100}$$

is true, the equality holding for fields satisfying the dual equations

$$\partial_\mu\varphi = \pm\ i\varepsilon_{\mu\nu}[\varphi,\ \partial_\nu\varphi]. \tag{4.101}$$

In terms of variables u, (4.101) is represented as

$$\partial u = \pm(u,\ \partial\bar{u})u, \quad \bar{\partial} u = \pm(\bar{u},\ \bar{\partial} u)u, \tag{4.102}$$

where

$$\partial = \partial/\partial z = (1/2)(\partial_1 - i\partial_2), \quad \bar{\partial} = \partial/\partial\bar{z} = (1/2)(\partial_1 + i\partial_2).$$

Equations (4.101) can be reduced to Cauchy-Riemann ones on CP^n if we pass to complex projective coordinates

$$(u^{1,}\ ...,\ u^{n+1}) \to (w^1,\ ...,\ w^n,\ u^{n+1}),$$

$$(u^1,\ ...,\ u^{n+1}) = (u_{n+1}w^1,\ ...,\ u^{n+1}w^n,\ u^{n+1}).$$

Equations (4.102) are given as

$$\bar{\partial} w^j = 0 \quad \text{or} \quad \partial w^j = 0, \quad j = 1, \ldots, n,$$

i.e., the dual equations are reduced to Cauchy-Riemann ones. Selecting the boundary condition

$$\varphi(x) \to 1, \quad |x| \to \infty,$$

we obtain a representation for solutions $w^j = f^j(z)$, where $f^j(z)$ are rational functions in z.

The important difference of general CP^n models ($n > 1$) from the case of CP^1 is the existence of harmonic maps φ: $M^2 \to CP^n$ other than holomorphic or antiholomorphic. The corresponding solutions of the chiral field's equations like (4.81) do not coincide either with instantons or anti-instantons. General solutions with finite action (not instanton ones) were investigated in a number of papers (see [Z], [CCR]). The most complete and mathematically rigorous results are in [EW2]. I shall discuss several results from [EW2] on the structure of harmonic maps.

Theorem 4.7 *Let M_g^2 be a two-dimensional compact Riemannian surface of genus g and the manifold CP^n equipped with the Fubini-Studi metric* (see [GH] or [Wu]). *Then*:

(1) *There exist nonholomorphic harmonic maps* φ: $M_g^2 \to CP^n$, $\deg \varphi > g$.

(2) *There exist nonholomorphic harmonic maps of $M_0^2 = S^2$ and of $M_1^2 = T^2$ to CP^n, of any degree greater than, or equal to, zero.*

It follows from the Eells-Wood theorem [EW2] that these mappings are not minima of the action functional.

Explicit solutions of necessary type see in [Z].

Example 4.2 Two-dimensional chiral fields can be constructed if we pass from spaces CP^n to complex Grassmannians. Examples of noninstanton solutions are also known [Z]. The most general example of two-dimensional chiral fields with instanton solutions is given by a system with "isotopic" Kählerian space N for which holomorphic maps $CP^1 \to N$ exist. The statement that instanton (anti-instanton) solutions are given by holomorphic maps $f: CP^1 \to N$ follows from the Eells-Wood theorem [EW1].

Theorem 4.8 *Let M_g^2 be a compact Riemannian surface and N a simply connected Kähler manifold, the group $\pi_2(N)$ being generated by a holomorphic map $CP^1 \to N$. Then each map φ: $M_g^2 \to N$ with minimal action in its homotopy class is holomorphic.*

Remark 4.5 The condition for $\pi_2(N)$ is essential, e.g., for the familiar two-dimensional complex algebraic (simply connected Kähler) surface K_3, the holomorphic map f: $M^2 \to K_3$ is constant [GH]. However, it follows from the results of [SU] that $\pi_2(N)$ is generated by harmonic mappings minimizing the action.

Remark 4.6 We can consider chiral models on an arbitrary compact manifold M. The condition for topologically nontrivial solutions to exist is

$$\text{Map}\,(M, V) \neq 0,$$

where V is the chiral manifold. We shall consider the interesting physical example of a nematic liquid crystal in Sec. 5.1, where a three-dimensional chiral model with the chiral manifold $V = RP^2$ arises in a particular case. The topological invariant is then coincident with the Hopf invariant h.

We now pass to the most important solutions from the physical standpoint of pure Yang-Mills equations with finite action.

4.2.5 Instantons in the Yang-Mills Equations

Here, we study the topological properties of classical solutions of pure Yang-Mills equations.

Consider the Lagrangian of a pure Yang-Mills equation with gauge group $G = SU(2)$ given in the space R^4

$$L_{\mathrm{YM}} = \frac{1}{2g^2} \operatorname{Tr} (F_{\mu\nu})^2, \tag{4.103}$$

where $F_{\mu\nu} = \partial_\mu W_\nu - \partial_\nu W_\mu + [W_\mu, W_\nu]$.

We retain the notation of Ch. 3.

The action functional is

$$S = \frac{1}{2g^2} \int d^4x \langle F_{\mu\nu} | F_{\mu\nu} \rangle = -\frac{1}{4g^2} \int d^4x F^a_{\mu\nu} F^a_{\mu\nu}, \tag{4.104}$$

$x^\mu = (x^1, \dots, x^4)$, $a = 1, 2, 3$ ($\dim SU(2) = 3$).

We seek solutions of the equations $\delta S = 0$ for the extremals, or, in explicit form,

$$\partial_\mu F_{\mu\nu} + [W_\mu, F_{\mu\nu}] = 0 \tag{4.105}$$

with finite action S, which assumes that

$$F_{\mu\nu} \to 0, \text{ or } W_\mu \sim g^{-1} \partial_\mu g, \; g \in SU(2) \tag{4.106}$$

as $|x| \to \infty$.

The four-dimensional Euclidean space R^4 completed with the point ∞ at infinity is turned into S^4. The functional S is then conformally invariant. Uhlenbeck [Uhl] has proved that all solutions of (4.105) with condition (4.106) are carried over to the sphere S^4.* Thus, we have a connection W_μ given on the fiber bundle associated with the principal one with group $SU(2)$ over S^4.

We know from the classification of principal bundles with group G over S^n that they and all the associated bundles are characterized by the topological invariant $\pi_{n-1}(G)$. Here $\pi_3(SU(2)) = Z$, therefore, there exists one topological invariant,

* The deep results of Uhlenbeck are discussed in more detail below in Remark 4.8.

which takes integer values and separates the classes of gauge invariant fields according to the topological charge.

The topological charge Q can be represented as

$$Q = \frac{1}{8\pi^2} \int \mathrm{Tr}\ (F_{\mu\nu} {}^*F_{\mu\nu}) d^4x, \tag{4.107}$$

where

$${}^*F_{\mu\nu} = (1/2)\varepsilon_{\mu\nu\varrho\sigma} F^{\varrho\sigma}, \tag{4.108}$$

or, in invariant form,

$$Q = \frac{1}{16\pi^2} \int \mathrm{Tr}\ (\Omega \wedge \Omega). \tag{4.109}$$

The invariant (apart from the sign) coincides with the second Chern class c_2 or with the first Pontrjagin class p_1 (see Subsec. 2.7.5).

Finding all solutions of equation (4.105) with finite action and selecting local minima is a complicated (possibly unsolvable) problem if we take into account the nonintegrability of the complete Yang-Mills equation. However, there exists a subsystem of (4.105) for which the class of solutions providing for a minimum of the functional S has been found.

By analogy with chiral models, we consider the inequality

$$\frac{1}{2g^2} \int d^4x \left\{ -\mathrm{Tr}\ (F_{\mu\nu} - {}^*F_{\mu\nu})(F_{\mu\nu} + {}^*F_{\mu\nu}) \right\} \geqslant 0. \tag{4.110}$$

Hence,

$$S = \frac{1}{2g^2} \int d^4x \left\{ -\mathrm{Tr}\ (F_{\mu\nu} F_{\mu\nu}) \right\} \geqslant \frac{1}{2g^2} \int d^4x \left\{ -\mathrm{Tr}\ ({}^*F_{\mu\nu} {}^*F_{\mu\nu}) \right\}, \tag{4.111}$$

for

$$F_{\mu\nu} F_{\mu\nu} = {}^*F_{\mu\nu} {}^*F_{\mu\nu}, \tag{4.112}$$

the equality in (4.110) and (4.111) holding only if

$$F_{\mu\nu} = {}^*F_{\mu\nu} \tag{4.113}$$

(self-dual equation) or if

$$F_{\mu\nu} = -{}^*F_{\mu\nu} \tag{4.114}$$

(anti-self-dual equation).

We have

$$\frac{1}{2g^2} \int d^4x \left\{ -\mathrm{Tr}\ ({}^*F_{\mu\nu} {}^*F_{\mu\nu}) \right\} = \frac{8\pi^2 Q}{g^2},$$

hence,

$$S \geqslant \frac{8\pi^2 Q}{g^2} \tag{4.115}$$

and the equality holding only for self-dual or anti-self-dual equations.

Solutions of self-dual (anti-self-dual) equations are called *instantons* (*anti-instantons*) and the topological charge Q is called an *instanton number.*

Finding explicit solutions of a dual equation with given topological number Q and the problem of describing instanton solution's space are of interest.

The solution with $Q = 1$ was obtained in the basic paper [BPST] introducing the concept of instanton (calling it a *pseudoparticle*).

Later, different classes of instanton solutions were found [A], [Wit1], [R].

A complete solution of the problem, depending on the total number of parameters, i.e., finding n-instanton solutions, was given in [ADHM]. However, there are no formulas to describe effectively a complete n-instanton solution. The method is reduction to a complicated problem in algebraic geometry. The latter admits an ineffective solution within the framework of linear algebra (see the detailed discussion in [DoK]).

Below, we give the original solution of the BPST instanton (see [BPST]) with $Q = 1$ and two special classes of n-instanton solutions due to 'tHooft and Witten, admitting an explicit description.

4.2.5.1 The BPST instanton

There are two ways of representing an instanton solution with $Q = 1$. We describe both, since they are related to two equivalent specifications of a Yang-Mills field.

First method [BPST]. We consider a connection W_μ on Euclidean space R^4 with asymptotic behavior at infinity $W_\mu \sim g^{-1}\partial_\mu g$.

Since we would like to find the solution associated with $Q = 1$, it is natural to look for it in the class of spherically $SO(4)$-symmetric solutions (invariant under $SO(3)$-spatial and $SO(3)$-isotopic rotations), taking into account that, locally, $SU(2) \sim SO(3)$. The gauge $SO(4)$ field $W_\mu{}^{\alpha\beta}$ antisymmetric with respect to the upper indices is related to the gauge $SU(2)$ field

$$W_\mu^i = (1/2)(W_\mu^{0i} \pm \varepsilon_{ikl} W_\mu^{kl}). \tag{4.116}$$

Then the two equations for the strength

$$\pm F_{\mu\nu}^i = \pm(1/2)\varepsilon_{\mu\nu\lambda\gamma} F_{\lambda\gamma}^i \tag{4.117}$$

are equivalent, i.e.,

$$\varepsilon_{\alpha\beta\gamma\delta} F_{\mu\nu}^{\gamma\delta} = \varepsilon_{\mu\nu\lambda\gamma} F_{\lambda\gamma}^{\alpha\beta} . \tag{4.118}$$

The spherically symmetric solution of self-dual equation (4.118) can be uniquely represented as

$$W_\mu^{\alpha\beta} = f(\tau)(x^\alpha \delta_\mu^\beta - x^\beta \delta_\mu^\alpha), \tag{4.119}$$

where

$$\tau^2 = (x^4)^2 + (x^1)^2 + (x^2)^2 + (x^3)^2.$$

Then

$$F^{\alpha\beta}_{\mu\nu} = (2f(\tau) - \tau^2 f^2(\tau)), \tag{4.120}$$

where $f(\tau) = \dfrac{2}{\tau^2 + \lambda^2}$ is a solution of the equation $f'/\tau + f^2 = 0$, f' the derivative with respect to τ and λ an arbitrary scalar.

Calculating the action S for the solution, we obtain

$$S = \frac{1}{2g^2} \int \mathrm{Tr}\,(\pm F^2_{\mu\nu}) d^4x = 8\pi^2, \tag{4.121}$$

i.e., derive from (4.115) that the solution of (4.119) supplies the absolute minimum for $Q = 1$.

Second method. We use the conformal equivalence of the Yang-Mills equations on R^4 and S^4 and write

$$W_\mu = \frac{\tau^2}{\tau^2 + \lambda^2} g^{-1} \partial_\mu g(x) = -\eta^\alpha_{\mu\nu} \frac{2x^\nu}{x^2 + \lambda^2}, \tag{4.122}$$

where $g(x) = (x^4 + i\mathbf{x}\sigma)/((x^4)^2 + (\mathbf{x})^2)^{1/2}$, $\sigma = (\sigma^1, \sigma^2, \sigma^3)$, σ^i are Pauli matrices and $\eta^\alpha_{\mu\nu}$ 'tHooft tensors.

We dwell on the properties of the latter at length [Ho2].

'tHooft tensors. Consider two four-dimensional vectors

$$\tilde{\sigma}_\mu = \begin{cases} I^{(2)}, & \mu = 4, \\ i\sigma^\alpha, & \mu = \alpha = 1, 2, 3, \end{cases} \qquad \tilde{\sigma}^+_\mu = \begin{cases} I^{(2)}, & \mu = 4, \\ -i\sigma^\alpha, & \mu = \alpha = 1, 2, 3, \end{cases}$$

where $I^{(2)}$ is the identity 2×2 matrix and σ^α Pauli matrices.

Definition 4.3 *'tHooft tensors* η and $\bar{\eta}$ are defined by the Pauli matrices

$$\begin{aligned} \eta_{\mu\nu} &= -(1/4)(\tilde{\sigma}^+_\mu \tilde{\sigma}_\nu - \tilde{\sigma}^+_\mu \tilde{\sigma}_\nu), \\ \bar{\eta}_{\mu\nu} &= -(1/4)(\tilde{\sigma}_\mu \tilde{\sigma}^+_\nu - \tilde{\sigma}_\nu \tilde{\sigma}^+_\mu). \end{aligned} \tag{4.123}$$

The 'tHooft tensors are quite convenient for calculations in instanton theory. The main property of η and $\bar{\eta}$, following from the commutation relations for the Pauli matrices, is that the tensors are self-dual and anti-self-dual, respectively, namely,

$$\begin{aligned} \eta_{\mu\nu} &= (1/2)\varepsilon_{\mu\nu\lambda\varrho}\eta_{\lambda\varrho} = {}^*\eta_{\mu\nu}, \\ \bar{\eta}_{\mu\nu} &= -(1/2)\varepsilon_{\mu\nu\lambda\varrho}\bar{\eta}_{\lambda\varrho} = {}^*\bar{\eta}_{\mu\nu}. \end{aligned} \tag{4.124}$$

The tensors $\eta_{\mu\nu}$ and $\bar{\eta}_{\mu\nu}$ are anti-Hermitian traceless matrices of order two and can be written relative to the basis for the Pauli matrices as

$$\eta_{\mu\nu} = \eta^{\alpha}_{\mu\nu} \frac{\sigma^{\alpha}}{2i}, \quad \bar{\eta}_{\mu\nu} = \bar{\eta}^{\alpha}_{\mu\nu} \frac{\sigma^{\alpha}}{2i}, \tag{4.125}$$

where the tensors $\eta^{\alpha}_{\mu\nu}$ and $\bar{\eta}^{\alpha}_{\mu\nu}$ only have real components. The main identities for 'tHooft tensors were obtained in [Ho2] and listed in a number of surveys (e.g., see [A]).

By means of the 'tHooft tensors, the field strength tensor $F_{\mu\nu}$ for the one-instanton solution is written simply as

$$F^{\alpha}_{\mu\nu} = -\eta^{\alpha}_{\mu\nu} \frac{4}{g} \frac{\lambda^2}{(x - x_0)^2 + \lambda^2}, \tag{4.126}$$

being self-dual due to (4.124).

The field W_μ for the one-instanton solution can be reduced to

$$W_\mu = \eta^{\alpha}_{\mu\nu} \partial_\mu \ln \left(1 + \frac{\lambda^2}{x^2}\right) \tag{4.127}$$

by a gauge transformation. The above solution admits generalization due to 'tHooft. We write the field W_μ as

$$W^{\alpha}_{\mu} = \eta^{\alpha}_{\mu\nu} \partial^{\nu} \ln \varrho, \tag{4.128}$$

where

$$\varrho = 1 + \sum_i \frac{\lambda_i^2}{(x - x_i)}, \quad \frac{\Box \varrho}{\varrho} = 0, \quad \Box = \partial_\mu \partial_\mu. \tag{4.129}$$

The solution possesses a number of interesting properties. First, note that the appearance of "singularities" in equation (4.128) is not physical in nature. We can make $F_{\mu\nu}$ vanish at the points $x = x_i$ by a pure gauge transformation of the form $g_i^{-1} \partial_\mu g_i$ (where g is a matrix from the group G).

The action functional's density can be represented as

$$\frac{1}{4} F^{\alpha}_{\mu\nu} F^{\alpha}_{\mu\nu} = -\frac{1}{2g^2} \Box \Box \ln \varrho. \tag{4.130}$$

After integrating, we get

$$S = \frac{8\pi^2 n}{g^2}, \tag{4.131}$$

i.e., an instanton solution whose topological charge $Q = n$ is obtained, depending, as can be seen from formula (4.129), on $5n$ parameters: $4n$ positions of x_i and n scale parameters λ_i. The number of parameters characterizing the general n-instanton solution is $8n - 3$. For this reason, a 'tHooft solution is not general n-instanton.

The solution is of some interest, being in a small number of n-instanton ones determined by explicit expressions.

4.2.5.2 Witten instanton solution

A Witten instanton solution is of interest from several points of view [Wit1]. First, as a 'tHooft solution, it is given explicitly with topological charge $Q > 1$. Second, finding n-instanton solutions is reduced to the solution of the Liouville equation which is important in modern field theory (in particular, in the theory of strings) [Pol2]. Third, Witten's *ansatz* will be used in the domain structure problem in the B-phase of superfluid ^{3}He (see Subsec. 5.2.6).

We dwell on the aspects of Witten's solution of interest. For a detailed discussion of this subject see [Wit1] and [A]. We seek the solution of self-dual equation (4.113) in R^4 invariant under three-dimensional spatial rotations. Following Witten, we call such solutions *cylindrically symmetric.* Witten's *ansatz* is

$$\begin{gathered} -gW_0^a = \frac{x^2}{r}A_0, \\ -gW_i^a = \varepsilon_{ian}\frac{x^n}{r^2}(1+\varphi_2) + \frac{x^a x^i}{r^2}A_1 + \left(\delta_{ai} - \frac{x^a x^i}{r^2}\right)\frac{\varphi_1}{r}, \end{gathered} \tag{4.132}$$

where i and n are spatial indices, a the isotopic index, functions A_0, A_1, φ_1, φ_2 depend on r and t and are subject to definition. Our goal is to find the general solution of equation (4.113) representable as (4.132).

Fields W^a are covariant relative to the subgroup $U(1)$ of the full gauge group $SU(2)$. Elements of the subgroup are representable as

$$u(x,\ t) = \exp\left[\frac{1}{2i}f(t,\ r)(\mathbf{x}\cdot\boldsymbol{\tau})\right],$$

where $\boldsymbol{\tau}$ is the vector made up of the Pauli matrices.

This gauge subgroup leads to the following form of transformations of A_0, A_1, φ_1, φ_2:

$$\begin{gathered} A_\mu \to A'_\mu = A_\mu - \partial_\mu f, \quad \mu = 0,\ 1, \\ \begin{pmatrix}\varphi_1\\ \varphi_2\end{pmatrix} \to \begin{pmatrix}\varphi'_1\\ \varphi'_2\end{pmatrix} = \begin{pmatrix}\cos f, & \sin f\\ -\sin f, & \cos f\end{pmatrix}\begin{pmatrix}\varphi_1\\ \varphi_2\end{pmatrix}. \end{gathered} \tag{4.133}$$

Hence, the remarkable reduction of the Yang-Mills fields in *ansatz* (4.132) to Abelian two-dimensional field theory.

We introduce the notation $\partial_0 = \partial/\partial t$, $\partial_1 = \partial/\partial r$. Straightforward calculation is left to the reader and shows that the Yang-Mills field stress tensor $F_{\mu\nu} = \partial_\mu W_\nu - \partial_\nu W_\mu + [W_\mu,\ W_\nu]$ can be rewritten as

$$F_{0i}^a = (\partial_0\varphi_2 - A_0\varphi_1)\varepsilon_{iak}x^k + (\partial_0\varphi_1 + A_0\varphi_2)\frac{(\delta_{ai}r^2 - x^a x^i)}{r^3} + (\partial_0 A_1 - \partial_1 A_0)\frac{x^a x^i}{r^2},$$

$$(1/2)\varepsilon_{ijk}\, F^a_{jk} = -\frac{\varepsilon_{ias}x^s}{r^2}(\partial_1\varphi_1 + A_0\varphi_2) + \frac{(\delta_{ai}r^2 - x^a x^i)}{r^3}(\partial_1\varphi_2 - A_1\varphi_1)$$
$$+\frac{x^a x^i}{r^4}(1 - \varphi_1^2 - \varphi_2^2).$$

We assume A_0 and A_1 to be components of two-dimensional field $A_\mu = (A_0, A_1)$ and (φ_1, φ_2) those of the Higgs (complex) field φ. We introduce the Abelian strength $\tilde{F}_{\mu\nu} = \partial_\mu A_\nu - \partial_\nu A_\mu$ (where μ, $\nu = 0, 1$) and the covariant differentiation of the field

$$\nabla_\mu\varphi_i = \partial_\mu\varphi_i + \varepsilon_{ij}A_\mu\varphi_j \quad (i, j = 1, 2). \tag{4.134}$$

The equivalence of the Yang-Mills theory in *ansatz* (4.132) and the Abelian Higgs two-dimensional model is meant in the following sense. We write the action S for the Lagrangian L_{YM}. Taking into account the invariance under $SO(3)$-rotations, we have

$$S = \frac{1}{4}\int d^3x \int dt F^a_{\mu\nu}F^a_{\mu\nu} = 8\pi \int\limits_{-\infty}^{+\infty} dt \int\limits_0^r \left[\frac{1}{2}(\nabla_\mu\varphi_i)^2 + \frac{1}{8}r^2\tilde{F}_{\mu\nu}\tilde{F}_{\mu\nu}\right.$$
$$\left. + \frac{1}{4r^2}(1 - \varphi_1^2 - \varphi_2^2)^2\right] dr. \tag{4.135}$$

Compare this action with the usual one of the Higgs two-dimensional model in the space with metric $g_{\mu\nu}$, namely,

$$\int d^2x\sqrt{g}\left[\frac{1}{2}g^{\mu\nu}(\nabla_\mu\varphi_i)(\nabla_\nu\varphi_i) + \frac{1}{8}g^{\mu\alpha}g^{\nu\beta}\tilde{F}_{\mu\nu}\tilde{F}_{\alpha\beta} + \frac{1}{4}(1 - \varphi_1^2 - \varphi_2^2)\right]. \tag{4.136}$$

It is easy to see that (4.135) coincides with (4.136) if the metric on the two-dimensional surface $R^2(r, t)$ is $\tilde{g}_{\mu\nu} = r^{-2}\delta_{\mu\nu}$.

The surface R^2 in this metric has constant negative scalar curvature K. This is the metric on the Lobachevski plane. The scalar curvature of $R^2(r, t)$ with the conformal metric $\tilde{g}(r, t)(dr^2 + dt^2)$ is found by the formula

$$K = -\frac{1}{2\tilde{g}}\Delta \ln \tilde{g}, \tag{4.137}$$

where $\Delta = \partial^2/\partial r^2 + \partial^2/\partial t^2$ is the Laplacian in Euclidean coordinates. The dual equation $F = {}^*F$ or the equivalent one $F^a_{0i} = (1/2)\varepsilon_{ijk}F^a_{jk}$ is, in coordinates (r, t), reduced to the system of equations

$$\begin{aligned} \partial_0\varphi_1 + A_0\varphi_2 &= \partial_1\varphi_1 - A_1\varphi_1, \\ \partial_1\varphi_1 + A_1\varphi_1 &= -(\partial_0\varphi_2 - A_0\varphi_1), \\ r^2(\partial_0 A_1 - \partial_1 A_0) &= (1 - \varphi_1^2 - \varphi_2^2). \end{aligned} \tag{4.138}$$

We note that the field $A_\mu = (A_0, A_1)$ also admits gauge symmetry. The Lorentz gauge $\partial_\mu A_\mu = 0$ is convenient. Hence,

$$A_0 = \partial_1 \psi, \; A_1 = -\partial_0 \psi. \tag{4.139}$$

The first two equations in (4.138) are transformed into

$$[\partial_0 - (\partial_0 \psi)]\varphi_1 = [\partial_1 - (\partial_1 \psi)]\varphi_2, \quad [\partial_1 - (\partial_1 \psi)]\varphi_1 = -[\partial_0 - (\partial_0 \psi)]\varphi_2. \tag{4.140}$$

After substituting

$$\varphi_1 = \exp(\psi)\chi_1, \quad \varphi_2 = \exp(\psi)\chi_2, \tag{4.141}$$

system (4.140) is transformed into

$$\partial_0 \chi_1 = \partial_1 \chi_2, \quad \partial_1 \chi_1 = -\partial_0 \chi_2. \tag{4.142}$$

In the variables $z = r + it$, equations (4.142) for $f = \chi_1 - i\chi_2$ turn into the Cauchy-Riemann equations $\partial f/\partial \bar{z} = 0$.

It remains for us to find the solution of the last equation in (4.138). After substitution (4.141) the equation becomes

$$-r^2 \Delta \psi = 1 - f\bar{f} \exp(2\psi). \tag{4.143}$$

It is solved in several steps.

Note that (4.143) also admits transformations

$$f \to fh, \quad \psi \to \psi - (1/2) \ln (h\bar{h}), \tag{4.144}$$

where $f(z)$ is analytic in the right-hand halfplane ($r > 0$) and admits a meromorphic continuation to the whole complex plane.

The invariance is related to the residual invariance in the conditions $\partial_\mu A_\mu = 0$, since A_μ admits the transformation $A_\mu \to A_\mu + \partial_\mu \lambda$, where λ satisfies the condition $\Delta\lambda = 0$.

The problem of singularities in f and ψ can be neglected in the first step of solving (4.143) due to the transformations in (4.144).

Let f have no isolated singularities (and h no zeros).

We substitute $\psi = \ln r - (1/2) \ln (f\bar{f}) + \varrho$, where ϱ is an unknown function. Equation (4.143) is transformed to

$$\Delta \varrho = \exp(2\varrho), \tag{4.145}$$

where we have used the condition $\Delta \ln|f^2| = 0$ for an analytic function f without zeros.

Equation (4.145) is *Liouville's*. It has the following remarkable geometric interpretation. If we specify the metric $g_{\mu\nu} = \exp(2\varphi)\delta_{\mu\nu}$ on the two-dimensional surface $R^2(u, v)$, then the scalar curvature K of $g_{\mu\nu}$ as calculated by formula (4.137) is

$$K = -\exp(-2\varphi)\Delta\varphi. \tag{4.146}$$

The scalar curvature of the surface is constant if the Liouville equation $\Delta\varphi = -K\exp(2\varphi)$ holds. All solutions of (4.145) can be found if we use the conformal invariance of the metric on R^2 (u, v).

Let $\varrho_1(z)$ be a particular solution of the Liouville equation, e.g., $\varrho_1(z) = -\ln[(1 - z\bar{z})/2]$. We take an arbitrary analytic function $\omega(z)$. The Laplacian relative to ω is $\Delta_\omega = \left|\frac{dz}{d\omega}\right|^2 \Delta_z$. Regarded as a function of ω, ϱ_1 satisfies the relation

$$\Delta_\omega \varrho_1 = \left|\frac{dz}{d\omega}\right|^2 \exp(2\varrho_1(\omega)) \tag{4.147}$$

which coincides with the Liouville equation except for the factor $\left|\frac{dz}{d\omega}\right|^2$. Putting $\varrho(\omega) = \varrho_1(\omega) - \frac{1}{2}\ln\left|\frac{dz}{d\omega}\right|^2$ and taking into account that $\Delta^2 \ln\left|\frac{dz}{d\omega}\right|^2 = 0$, we have

$$\Delta_\omega \varrho = \exp(2\varrho).$$

Hence, if $g(z)$ is an arbitrary analytic function,

$$\varrho(z) = -\ln\left[\frac{1 - g\bar{g}}{2}\right] + \frac{1}{2}\ln\left|\frac{dg}{dz}\right|^2 \tag{4.148}$$

satisfies the Liouville equation.

It is not hard to see that $\varrho(z)$ is a general solution of the Liouville equation, since, to find $g(z)$, it is necessary to specify two conditions, i.e., as many as are required to determine the general solution of a quadratic equation.

Comparing the solution (4.148) with (4.143), we see that the condition of no singularities in the latter equation means (1) neither zeros nor poles of $\frac{1}{f}\frac{dg}{dz}$ exist in the right-hand halfplane $r > 0$ and (2) its meromorphic continuation to the whole complex plane. If $\frac{1}{f}\frac{dg}{dz} <$ const everywhere on C, then, by the Liouville theorem, $\frac{1}{f}\frac{dg}{dz} \equiv \text{const} = k$. Given the possibility of gauge transformations (4.144), the general solution of (4.145) is thereby

$$\psi = -\ln\left[\frac{1 - g\bar{g}}{2r}\right], \quad f = \frac{dg}{dz}. \tag{4.149}$$

It follows from the theory of analytic functions that the two conditions, $|g| = 1$ for $r = 0$ and $|g| < 1$ for $r > 0$, should hold for the nonsingularity of ψ in the right-hand halfplane $(r > 0)$.

The general form of the function then is

$$g(z) = \prod_{i=1}^{n} \left(\frac{a_i - z}{\bar{a}_i + z} \right), \tag{4.150}$$

where a_i is the collection of complex numbers such that Re $a_i > 0$. Thus, the solution of (4.149) with the function determined by (4.150) is the general instanton solution of the Yang-Mills equation with the group $SU(2)$ and cylindrical symmetry.

It would be interesting to know the relation of (4.149) and (4.150) with respect to the formula for a two-dimensional chiral field. Other reductions of Yang-Mills fields are known and coincide with models of chiral fields [A].

It would be useful for the reader to analyze (4.150), e.g., for $n = 1$, f has no zeros. The corresponding solution determines $F_{\mu\nu} = 0$, or the vacuum solution. If $n = 2$, then we obtain the one-instanton solution of the type found by Belavin *et al.*, but in another gauge.

The points in the complex plane a_i ($i > 1$) have a simple physical interpretation. Re a_i determines the size of the ith instanton with $Q = 1$ and Im a_i its position on the t-axis. The exact meaning of the statement is clarified below. We formulate the basic result of the subsection as follows.

Proposition 4.3 *Solution* (4.149) *describes an* $(n - 1)$*-instanton solution whose real and imaginary parts are determined by the zeros of the function* f *(with multiple zeros).*

First, we discuss the zeros' relation to the instanton's position. The instanton solution in (4.149) is expressed in terms of the function $g(z)$ representable as (4.150) and depending on $2n$ real parameters, while the number of parameters of an $(n - 1)$-instanton is really $2(n - 1)$. This "paradox" can be explained simply. There exists a complex gauge transformation

$$g \to \tilde{g} = \frac{c + g}{\bar{c}g + 1}, \quad f \to \tilde{f} = \frac{d\tilde{g}}{dz}, \tag{4.151}$$

where c is a complex number, $|c| < 1$, not altering the form of solutions and belonging to the class of (4.144). Hence, the number of physical parameters in solution (4.149) is $2n - 2$. Meanwhile, the functions f and $\tilde{f}$ have the same set of zeros. Thus, the position of an $(n - 1)$-instanton solution is determined by the $(n - 1)$ zeros $\alpha_1, \ldots, \alpha_{n-1}$, $\alpha_i = \xi(a_1, \ldots, a_n)$ of f, where ξ depends on all the zeros $a_1, \ldots, a_n$.

It is in this sense that we have to understand the representation of a cylindrically symmetric $(n - 1)$-instanton solution as a linear composition of the $n - 1$ instantons on the t-axis. Assuming that a_i of $g(z)$ are far from each other, this solution can be decomposed into the sum of one-instanton solutions of the type found by Belavin *et al.*

The idea of proof is as follows: although there is no simple relationship between the zeros of f and g, we can select the gauge so that one of zeros of f, e.g., α_0, is also that of $g(z)$. Assuming that α_0 is a simple zero of f, we can represent $g(z)$ as

$$g(z) = \left[\frac{\alpha_0 - z}{\bar{\alpha}_0 + z} \right]^2 \prod \left(\frac{\beta_i - z}{\bar{\beta}_i + z} \right),$$

where β_i are functions of α_i, $i \geqslant 1$. If $|\alpha_i - \alpha_0|$ is large, then the distance $|\beta_i - \beta_0|$ also becomes large. In the neighborhood of α_0, the factor $\prod \left(\frac{\beta_i - z}{\bar{\beta}_i + z}\right)$ can be replaced by the constant $\prod \frac{\beta_i}{\bar{\beta}_i}$. Neglecting this immaterial factor, we see that, for large $|\alpha_i - \alpha_0|$, we have $g(z) \sim \left(\frac{\alpha_0 - z}{\bar{\alpha}_0 + z}\right)^2$, which is associated with (4.149) for $n = 2$. The solution localized at α_0 is a BPST instanton. We have thereby separated one instanton, from the remaining $n - 2$. This argument is qualitative in nature and not a rigorous proof, which can be found in [Taul].

We now turn to calculating the topological charge Q.

According to (4.107), Q for the Yang-Mills equation is given as

$$Q = \frac{1}{8\pi^2} \int F_{\mu\nu}{}^*F_{\mu\nu}\, d^4x.$$

Taking *ansatz* (4.132) for a cylindrically symmetric field into account, we obtain

$$Q = \frac{1}{2\pi} \int d^2x[\varepsilon_{\mu\nu}\varepsilon_{ij}\nabla_\mu\varphi_i\nabla_\nu\varphi_j + (1/2)\varepsilon_{\mu\nu}\tilde{F}_{\mu\nu}(1 - \varphi_1^2 - \varphi_2^2)], \qquad (4.152)$$

where μ, $\nu = 0, 1$, $i, j = 1, 2$.

Using the motion equation $\nabla_\mu\tilde{F}_{\mu\nu} = 0$, equation (4.152) can be rewritten conveniently as

$$Q = \frac{1}{2\pi} \int d^2x[\partial_\mu(\varepsilon_{ij}\varepsilon_{\mu\nu}\varphi_i\nabla_\nu\varphi_j) + (1/2)\varepsilon_{\mu\nu}\tilde{F}_{\mu\nu}], \qquad (4.153)$$

where $\tilde{F}_{\mu\nu}$ is two-dimensional Abelian field strength.

The first term in the integrand of (4.152) is the total derivative and vanishes since $\nabla_\mu\varphi_i = 0$ for solutions on the boundary. The second one is the topological charge in the two-dimensional Abelian-Higgs model.

A proof of the latter statement can be obtained directly taking into account the condition

$$\nabla_\mu\varphi_i = \partial_\mu\varphi_i + \varepsilon_{ij}A_\mu\varphi_j = 0 \quad \text{as } |x| \to \infty,\ |\varphi| = 1, \qquad (4.154)$$

fulfilled at infinity. These requirements are necessary conditions for solutions with finite action to exist in the Higgs model.

By replacing $\varphi = \varphi_1 - i\varphi_2$ with $f \exp \psi$, we represent (using the Stokes formula) the formula for topological charge as

$$Q = \frac{1}{4\pi} \int \varepsilon_{\mu\nu}\tilde{F}_{\mu\nu} = \frac{1}{2\pi} \int (\partial_0 A_1 - \partial_1 A_0) d^2x$$

$$= \frac{1}{2\pi i} \int_l ds \frac{d}{ds}(\ln f \exp(\psi)) = \frac{1}{2\pi i} \int_l d \ln f + \int_l d\psi = \frac{1}{2\pi i} \int_l \frac{df}{f} = \operatorname{Ind} f \quad (4.155)$$

($\int d\psi = 0$, since ψ is a single-valued function).

By Cauchy's theorem, the integral in (4.155) equals the number of zeros of the function f in the domain surrounded by the contour l, naturally regarded as the circle S^1_∞ of infinite radius.

Remark 4.7 That Q is an integer in the Higgs model, is from the topological viewpoint a consequence of the relation $\pi_1(S^1) = H^1(S^1) = c_1(\xi^1(S^2))$, where $\xi^1(S^2)$ is a one-dimensional complex fibering over the sphere S^2 with connection A_μ and curvature $\tilde{F}_{\mu\nu}$.

In fact, the conditions in (4.154) determine the homotopy class of mappings $\Phi: S^1_\infty \to S^1 = U(1)$. Formula (4.155) yields a topological charge, i.e., the Chern 1-class $c_1(\xi^1(S^2))$, where S^2 is obtained by the compactification of the plane R^2.

Therefore, the analytical verification of index integrality is an independent proof of the equality $H^1(S^1) = c_1(\xi^1(S^2)$. The equality $\pi_1(S^1) = H^1(S^1)$ which is simply one-dimensional Hurewicz's theorem follows from the coincidence of two definitions of the degree of $S^1 \to U(1) \sim S^1$.

Remark 4.8 Defining topological charges in Abelian Yang-Mills-Higgs field theory with the compact group $U(1)$ on R^2 and in Yang-Mills theory with compact group G on R^4 as

$$c_1 = \frac{1}{4\pi} \int_{R^2} F_{\mu\nu} d^2x \qquad (\text{YMH})_2, \tag{4.156}$$

$$c_2 = \frac{1}{8\pi^2} \int_{R^4} F_{\mu\nu} {}^*F_{\mu\nu} d^4x \qquad (\text{YM})_4 \tag{4.157}$$

is based on several nontrivial properties.

In topology, a characteristic class (in particular, a Chern one) is defined for fiber bundles over a compact manifold. Formulas (4.156)-(4.157) are meaningful for certain behavior at infinity of the fields (forms) $F_{\mu\nu}(\Omega_1)$ which enable us to compactify space R^n and thereby define topological charge correctly.

We confine ourselves to the case of the one-point compactification $S^n = R^n \cup \infty$. We note that Yang-Mills fields are distinguished in dimension four since Yang-Mills action is conformally invariant only in R^4, i.e.,

$$S = \int F_{\mu\nu} F_{\mu\nu} d^4x = \int F'_{\mu\nu} F'_{\mu\nu} d^4 g(x)$$

as $x \to g(x)$, where $g \in C(R^4)$ is the conformal group of transformations of R^4.

All smooth solutions of the Yang-Mills equation with finite action then satisfy the same equation on the sphere. However, a more general question can be posed. What restrictions can be placed on the form Ω for which (not necessarily smooth) solutions of the Yang-Mills equation with finite action on R^4 could be extended to S^n? The most complete results are due to Uhlenbeck [Uhl, 2]. We formulate two.

Theorem 4.9 *Given the integral*

$$\int_{R^4} |\Omega|^2 < \infty, \quad \Omega \in L^2_{1,\ \mathrm{loc}}, \tag{4.158}$$

then the form Ω *can be extended to* S^4 *and the integral in* (4.157) *is an integer.*

In modified form, this also holds for spaces $R^n(n = 2k, n \neq 2)$ if the condition in (4.158) is replaced by

$$\int_{R^n} |\Omega|^{n/2} < \infty.$$

The connection W is in a Sobolev space of type $L^{n/2}_{1,\ \mathrm{loc}}$ [JT].

An earlier result due to Uhlenbeck [Uhl] can be derived from Theorem 4.9.

Theorem 4.10 *A solution of the Yang-Mills equations with finite action on* R^4 *has no nonremovable isolated singularities.*

We actually prove that a Yang-Mills field with isolated singularities can be reduced to a smooth connection of S^4 with a gauge transformation.

In the general case ($n \neq 4$), the space S^n can also be conformally mapped onto R^n by a stereographic projection. The action of $\Omega\wedge{}^*\Omega$ on R^n is meanwhile not conformally equivalent to that on S^n. Hence, solutions of the Yang-Mills equation on R^n differ from those on S^n. However, it follows from Uhlenbeck's results that the invariant $C_2 = \int_{R^n} \Omega\wedge\Omega$ is integer on R^n too.

4.2.5.3 Space of instanton solutions

Instanton solutions for the group $SU(2)$ are parametrized by the values of one topological invariant, the Pontrjagin number p_1. According to the structure of a gauge group, there can be more topological charges characterizing an instanton solution. General analysis of topological charges which characterize classical solutions is given in Subsec. 4.2.6. I dwell here on another very interesting problem. How is the space of instanton solutions with given topological charge Q arranged? For two-dimensional chiral fields, the question is given a sufficiently simple answer. A two-dimensional chiral model with nonorientable isotopic space (i.e., the model of a nematic liquid crystal; see Subsec. 5.1.4) is interesting. However, the problem of describing instanton space is not solved even in four-dimensional space, i.e., the connection of solution space is unknown. The problem of describing instanton solution space is in spirit close to the *problem of moduli* in the theory of fibrations on algebraic curves, which is familiar from algebraic geometry [AW], [At2], [DoK].

Quite an important result relative to the structure of the instanton solution space ε for Yang-Mills fields on S^4 with gauge group $SU(2)$ is calculating the dimension of ε, which can be estimated if we proceed from the following intuitive argument.

First, any instanton is characterized by its position and scale (by five parameters). Second, the instanton can be rotated with the help of the three-parameter gauge group $SU(2)$.

We thus obtain $8n$ parameters for an n-instanton solution, from which the number of parameters determining the general global transformation of $SU(2)$ should be subtracted. Totally, we obtain $8n - 3$ parameters. The argument can in no way be a proof since n-instanton solutions are not a linear combination of n copies of one-instanton ones.* Nevertheless, the result turns out to be true and can be proved by the fundamental Atiyah-Singer index theorem [At2], [DoK], [Ma2].

4.2.6 Topological Charges in Multidimensional Field Theories

Classical vortex, monopole, and instanton type solutions are associated with the choice of physical space-time of dimension two, three or four, depending on the situation.

Here, we discuss the topological aspects of multidimensional solutions.

Multidimensional instantons are solutions of the free Yang-Mills equation with gauge group G, action S, and Euclidean physical space R^n. The fields W_μ have purely gauge form

$$W_\mu \to g^{-1}\partial_\mu g \tag{4.159}$$

at infinity as $|x| \to \infty$.

Condition (4.159) enables us to carry the connections W_μ from a fiber bundle on R^n with asymptotic behavior (4.159) over the compactification of R^n, or the n-dimensional sphere S^n. The transition from the fiber bundle on R^n to a fiber bundle over a compact manifold (and a sphere in this case) is favourable from the topological viewpoint. However, the transition is not as harmless from the analytic viewpoint as it seems from Remark 4.5. We assume that Yang-Mills fields are given on S^n. The classification of instanton solutions then only depends on the topological structure of the gauge group.

The classification of fiberings of S^n yields the following.

Proposition 4.4 *Instanton solutions for S^n are classified by elements of the group $\pi_{n-1}(G)$.*

Corollary 4.2 *The number of different vacuum states over S^4 in theories with group $SU(N)$, $N \geqslant 3$, is defined by $SU(2)$-theory.*

Proof.

$$\pi_3(SU(2)) = \pi_3(SU(N)),\ N \geqslant 3. \quad \blacksquare$$

Hence, all the physical corollaries for the theory with group $SU(2)$, e.g., chiral charge nonconservation or tunnelling between different instanton solutions, are

---

*In physical applications, the superposition of one-instanton solutions is often used. Such an approximation is called an *instanton gas*.

carried over to $SU(N)$-theories [Ho2]. However, the number of internal parameters characterizing the general instanton solution, or the dimension of instanton space $\varepsilon_{SU(N)}$, differs from dim ε for $SU(2)$, which follows from the general formula for ε_G which is obtained in [At2], [DoK].

Example 4.3 The gauge group G is $SO(4)$ in gravitational theory and four-dimensional instantons over S^4 are characterized by elements from

$$\pi_3(SO(4)) = \pi_3(SO(3) + SO(3)) = Z + Z. \tag{4.160}$$

It was shown in Subsecs. 2.7.6 and 2.7.7 that a pair of integers (m, n), $m \in Z$, $n \in \tilde{Z}$ ($\tilde{Z} = Z$, or the second summand in (4.160)) can be associated with the two topological charges, the Euler and Pontrjagin class, respectively,

$$e = \int_{S^4} R^{\gamma\delta}_{\alpha\beta} R^{\tau\nu}_{\varrho\sigma} \varepsilon^{\alpha\beta\varrho\sigma} \varepsilon_{\gamma\delta\tau\mu} \sqrt{g} dV, \tag{4.161}$$

$$p_1 = \int_{S^4} R_{\mu\nu\alpha\beta} R^{\mu\nu}_{\gamma\delta} \varepsilon^{\alpha\beta\gamma\delta} \sqrt{g} dV, \tag{4.162}$$

where $R_{\mu\nu\alpha\beta}$ is the curvature tensor and $g = \det g_{ij}$.

We recall that integral formula (4.161) only holds for the Euler class if the connections correspond to the Riemannian metric g on the manifold M^4.

Wilczek posed the following problem [Wi]. Is there any relation between e and p_1 for gravitational instantons? For example, is a solution with $e = 0$, $p_1 = 1$ possible?

It follows from the results of [No1] that, for fiber bundles ξ with fiber S^3 and group $SO(4)$ over S^4 ($SO(4)$ is the bundle of spheres S^3), the following restriction is placed on the numbers $e(\xi)$ and $p_1(\xi)$: p_1 is always even; if $e = 0$, then $p_1 = 4k$, otherwise $2e - p_1 = 4n$.

In the theory of gravitation, the Euler class e and the Pontrjagin p_1 are the characteristic classes of the tangent bundle of physical space M^4 and the relations between the classes impose restrictions on the topological structure of the manifold M^4, e.g., if $p_1(M^4) \neq 0$, then $H_2(M^4) \neq 0$.

A solution with a finite action for the Einstein equations is called a *gravitational instanton*. The modern state of the problem of gravitational instanton classification is surveyed in [Gi].

We now pass to an analysis of instanton solutions in space $S^n (n > 4)$. For definiteness, we pick the gauge group $SU(N)$. We know in the four-dimensional case that the instantons are characterized by the group $\pi_3(SU(N))$ or by the unique topological charge $p_1 = c_2$, the second Chern class. If we consider $S^n (n > 4)$, then other topological invariants appear. The first nontrivial example is S^6 with a new topological charge $c_3 \in H^6(S^n, Z)$ in the Yang-Mills bundle. The third Chern class is determined by the expression for a characteristic class (see Subsec. 2.7.5)

$$\begin{aligned} \det \|\Omega^{ij}\| &= \det \| (1/2) F^{ij}_{\mu\nu} dx^\mu \wedge dx^\nu \| = c_3 \\ &= \varepsilon^{i_1, i_2, i_3}_{j_1, j_2, j_3} F^{i_1 j_1}_{\mu_1 \nu_1} F^{i_2 j_2}_{\mu_2 \nu_2} F^{i_3 j_3}_{\mu_3 \nu_3} dx^{\mu_1} \wedge dx^{\nu_1} \wedge \ldots \wedge dx^{\mu_3} \wedge dx^{\nu_3}. \end{aligned} \tag{4.163}$$

Accordingly, the Chern number C_3 is the integral

$$C_3 = \int_{R^n} c_3 \tag{4.164}$$

of (4.163).

If we want c_3 to be nontrivial, then it is necessary that $n = 6$ (for space R^n). This follows from (4.163) since c_3 is a symmetric polynomial of the third degree in the components $F^{ij}_{\mu\nu}$. Note that the topological charge in (4.164) does not yield a lower estimate for energy, since energy density is a symmetric polynomial of degree two in $F^{ij}_{\mu\nu}$.

Multidimensional topological charges arise naturally in a number of models of modern field theory. Nonvacuous theories of this kind are in the Kaluza-Klein formalism in gravitational theory. Fiber bundles over multidimensional base spaces arise in field theories if internal degrees of freedom are incorporated into the base space. Base spaces of the form $M^N = R^n \times M^k$ arise.

Following [GM2] we consider an example of field theory with nontrivial topological charge c_3 and seek static solutions (instantons) of the pure Yang-Mills equation with gauge group $SU(3)$ defined over six-dimensional space R^6.

To find solutions of the corresponding equations for gauge fields, it is necessary to consider the symmetry property. By analogy with BPST instantons, it is natural to require that the gauge group should be embeddable in the symmetry group of "space-time". Space-time symmetries then interact with the internal symmetries of the gauge group. Assuming "space-time" to be six-dimensional, we should select six-dimensional representation of $SU(3)$ of the form $D(0, 2)$ and of dimension 6 as symmetric matrices.* The real part of the space of such matrices is of real dimension six. Considering only static solutions, we omit the time component. Therefore, the gauge field is represented by the tensor

$$W^a_k, \; a = 1, \ldots, 8, \quad k = 1, \ldots, 6 \tag{4.165}$$

in the tensor product $8 \times \bar{6}$, where 8 is the dimension of the adjoint representation of $SU(3)$.

Decomposing the tensor representation into irreducible ones by the familiar formula [BDFL], we obtain

$$8 \times \bar{6} = D(1, 0) \oplus D(0, 2) \oplus D(2, 1) \oplus D(3, 1). \tag{4.166}$$

It is best to write spatial coordinates as elements of the symmetric 3×3 matrix

$$z_{ij} = z_{ji}, \quad i, j = 1, 2, 3. \tag{4.167}$$

Taking (4.166) and (4.167) into account, we consider the *ansatz*

$$W^a_{ij} = (1/2)(\lambda^a_{in} z_{nj} + \lambda^a_{jn} z_{ni}) h(r), \quad r = \sqrt{z^2_{ij}}, \tag{4.168}$$

*We use the standard notation from representation theory (see [BDFL]).

for the field W_k^q, where λ_{in}^a and λ_{jn}^q are elements of Gell-Mann λ-matrices (see Subsec. 4.1.5), and the double subscripts are associated with spatial coordinates, e.g., an *ansatz* of the kind was considered in [WW, MP]. The concrete form of (4.168) is in our case determined by the $SU(3)$-symmetry of the problem.

Substituting expression (4.168) in the equation

$$F_{ij,kl}^q = W_{ij,kl}^q - W_{kl,ij}^q + gc_{bd}^a W_{ij}^b W_{kl}^d \tag{4.169}$$

for the field strength tensor, where $W_{i,k}^q = \partial_k W_i^q$, we obtain

$$F_{ij,kl}^q = U_{ij,kl}^q \dot{h} + V_{ij,kl}^q \frac{h}{2r} + X_{ij,kl}^q h^2 \tag{4.170}$$

with

$$U_{ij,kl}^q = \frac{z_{kl}}{2r^2}(\lambda_{ip}^a z_{pj} + \lambda_{jp}^a z_{pi}) - \frac{z_{ij}}{2r}(\lambda_{kp}^a z_{pl} + \lambda_{lp} z_{pk}),$$

$$V_{ij,kl}^q = (\lambda_{ip}^q \delta_{pj}^{kl} + \lambda_{jp}^q \delta_{pi}^{kl}) - (\lambda_{kp}^q \delta_{pl}^{ij} + \lambda_{lp}^q \delta_{pk}^{ij})$$
$$- \left[\frac{z_{kl}}{2r^2}(\lambda_{ip}^a z_{pj} + \lambda_{jp}^a z_{pi}) - \frac{z_{ij}}{2r^2}(\lambda_{kp}^a z_{pl} + \lambda_{lp}^a z_{pk}) \right],$$

$$X_{ij,kl}^a = \frac{1}{(2r)^2} c_{bd}^q (\lambda_{ip}^b z_{pj} + \lambda_{jp}^b z_{pi})(\lambda_{kp}^d z_{pl} + \lambda_{lp}^d z_{pk}).$$

After the corresponding calculations, we obtain for the Lagrangian

$$L = a_1 r^5 \dot{h}^2 + a_2 r^4 h\dot{h} + a_3 r^5 h^2 \dot{h} + a_4 r^3 h^2 + a_5 r^4 h^3 + a_6 r^5 h^4$$

satisfying the *ansatz* the representation

$$L = \int L \prod_{m,n} z_{mn} dr = \int_0^\infty r^5 \, dr L \int_{S^5} \Pi \, d\theta, \tag{4.171}$$

where $\Pi d\theta$ is the element of volume of S^5 with $r = 1$ (the dot indicating differentiation with respect to r), where

$$\begin{aligned} a_1 &= \int_{S^5} |U|^2 (\Pi \, d\theta), & a_4 &= (1/4) \int_{S^5} |V|^2 (\Pi \, d\theta), \\ a_2 &= \int_{S^5} \operatorname{Re}(UV^+)(\Pi d\theta), & a_5 &= \int_{S^5} (VX^+)(\Pi d\theta), \\ a_3 &= 2 \int_{S^5} \operatorname{Re}(UX^+)(\Pi d\theta), & a_6 &= \int_{S^5} |X|^2 (\Pi d\theta). \end{aligned} \tag{4.172}$$

The Euler equation for the functional in (4.171) can be written as

$$\ddot{h} + \frac{5}{r}\dot{h} + \frac{\alpha_1}{r^2}h + \frac{\alpha_2}{r}h^2 - \alpha_3 h^3 = 0, \tag{4.173}$$

$$\alpha_1 = \frac{2a_2 - a_4}{a_1}, \quad \alpha_2 = \frac{5a_3 - 3a_5}{2a_1}, \quad \alpha_3 = \frac{2a_6}{a_1},$$

where $\alpha_3 > 0$, since $a_1, a_6 > 0$.

Bounded solutions of (4.173) are found from the equation

$$\ddot{h} - \alpha^3 h^2 = 0 \tag{4.174}$$

as $r \to \infty$ (see the phase diagram of integral curves in Fig. 9).

To find solutions decreasing as $r \to \infty$, we seek them in the form of a series

$$h = \frac{b_1}{r} + \frac{b_2}{r^2} + \dots . \tag{4.175}$$

We can confine ourselves to the first term with coefficient

$$b_1 = \frac{5a_3 - 3a_5 \pm \sqrt{(5a_3 - 3a_5)^2 + 32a_6(3a_1 - 2a_2 + a_4)}}{4a_6}.$$

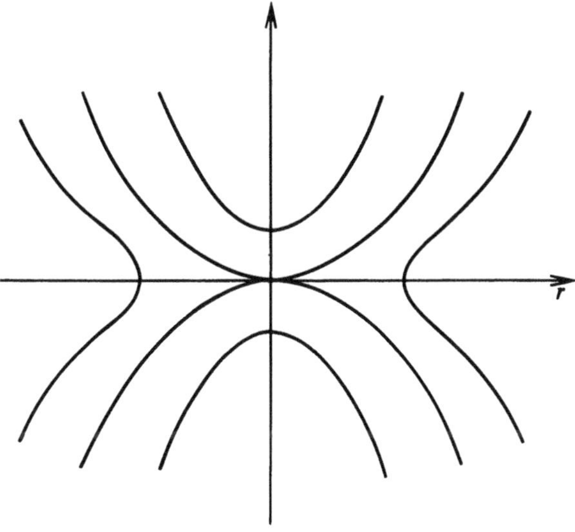

Fig. 9. Phase diagram of trajectories in Eq. (4.174)

Hence, the solution decreases as $r \to \infty$ as r^{-1}. The energy integral diverges as r^2 and the topological charge as ln r. We note that strictly speaking representations as in (4.164) only hold for compact manifolds.

In field theory, similar formulas are also written for noncompact manifolds, assuming that the solutions decrease rapidly at infinity. We can introduce a topological charge with the following construction.

We send S^6 onto R^6 by a conformal map f. Exterior differential forms on S^6 are the images of differential forms given on R^6. The corresponding third Chern class over S^6 is defined as the image $f_* c_3 = \tilde{c}_3$ of c_3. We have

$$\tilde{c}_3 = \int_{S^6} f_* c_3.$$

It is the value of integral that determines the topological charge in this case.

Remark 4.9 (on the origin of topological charges in field theory).

1. The existence of solutions with finite action with gauge group $SU(N)$ has topological foundations in Yang-Mills theory and depends on the dimension of physical space-time.

2. It would be interesting to consider the conditions at infinity in Minkowski space directly. To find its boundary or realization as a bounded domain we select the cone

$$V = \{ Y | (y^0)^2 - (y^1)^2 - (y^2)^2 - (y^3)^2 > 0, \quad y_0 > 0 \}$$

and consider an analog of the upper half-plane, viz., the domain

$$D = \{ Z | Z = X + iY, \quad X \in R^4, \quad Y \in V \},$$

which can be realized as a bounded domain $\tilde{D} = SO(4, 2)/SO(4) \times SO(2)$ in C^4. The conformal group $SO(4, 2)$ acts transitively on $\tilde{D} \simeq D$ (where $\simeq$ denotes analytic equivalence).

The Bergman-Šilov boundary of $\tilde{D}$ is the set

$$\check{S} = SO(4) \times SO(2)/SO(3) \sim S^3 \times S^1$$

[P.-S.]. The boundary is attainable along geodesics starting in the interior of $\tilde{D}$ and given in the Riemannian metric invariant under analytic automorphisms of $\tilde{D}$ [P.-S.].

Using the conformal invariance of transformations $D \to \tilde{D}$, it is not hard to see that the asymptotic conditions at infinity define a topological charge in $\check{M}^4$ as the set Map $(S^1 \times S^3, G)$ of homotopy classes of mappings (G is a gauge group).

Using a more refined topological technique, we can show that the group Map $(S^1 \times S^3, G)$ is determined by the exact sequence

$$0 \to \pi_4(G) \to \text{Map}\,(S^1 \times S^3, G) \to \pi_1(G) + \pi_3(G) \to 0.$$

3. Additional topological charges arise when considering physical space of dimension greater than four. Such situations occur when the coordinates describing internal degrees of freedom are involved in the coordinates of physical space. The

corresponding analogs of space $\breve{M}^4$ or R^4 are the manifolds $M^N = R^4 \times M^n$ or $\breve{M}^4 \times M^n$, respectively, which have rather complicated topologies. Physical examples of such systems are given in gravitational theory by the Kaluza-Klein formalism and by various grand unification models in unified field theories based on the ideas of spontaneous compactification [Wuy].

Topological charges naturally arise in a number of problems in condensed-matter physics, and not only in field and gravitational theories. The variety of topological charge forms can be described on the basis of cobordism theory. The corresponding construction is given in Subsec. 5.2.8.3.

There are many applications of topology to condensed-matter physics involving branches of topology which have not yet proved useful in field theory.

Chapter 5
Topology of Condensed Matter

The branches once called *solid-state physics* and now called condensed-state physics cover, for example, the crystals, para- and ferromagnets. However, materials can, given a change of conditions (pressure, temperature), transform into a completely different state of matter (e.g., liquid, plasma, etc.). Therefore, another terminology which more accurately reflects their properties is now being adopted. A more detailed classification separates physical systems into ordered and disordered media. The former includes liquid crystals, superfluid liquids, and other media naturally described by an order parameter while disordered systems constitute a comparatively new class. They include spin glasses, amorphous bodies, polymers, etc.

From the theoretical viewpoint, the investigation of disordered systems is only beginning. There are undoubtedly nontrivial topological applications to disordered systems, but it remains virgin territory.

If the reader is interested in disordered media, then [LGP] is recommended, in which the state of the subject is discussed brilliantly. We shall consider, in this chapter, ordered media which have a large field of topological applications and discuss the applications to two physical systems, namely, liquid crystals and superfluid ^{3}He. Other examples are also mentioned.

5.1 LIQUID CRYSTALS

The liquid crystal is an exceptionally interesting object for physical studies and a vast domain of topological applications, from elementary homotopy theory to refined applications of the theory of foliations combined with differential-geometric methods. On the other hand, liquid crystals can be investigated in quite a detailed manner experimentally.

5.1.1 Classification of Liquid Crystals

Here, we list the main physical characteristics of liquid crystals and the principles on which their classification is based.

The name itself implies a certain duality. In fact, they are in a state of matter intermediate between a solid (or crystalline) and a usual isotropic fluid. Liquid crystals are therefore called *mesomorphic (intermediate) phases*, or *mesophases*.

Mechanically liquid crystals are closer to isotropic fluids (their fluidity), while optically they are closer to normal crystals with Bragg reflections characteristic of crystal lattices. According on which property prevails liquid crystals are divided into several groups with essentially different qualities. The classification is nowadays relative to mesophase symmetry.

We first classify them qualitatively.

Smectics. Liquid crystals (type A and C* smectics) have periodic structure (or spatial order) in only one dimension. The system is represented as a stack of equally spaced two-dimensional liquid layers. The mechanical properties of these crystals are similar to soap, hence the name of "smectic" (from the Greek $\sigma\mu\eta\chi\mu\alpha$ which means soap) was introduced by the French physicist Georges Friedel [Fri]. A foliation is a naturally description and this interpretation was offered by the French mathematician Poénaru [Po1]. Although this is quite a promising domain only very preliminary results have been obtained [L, Po1].

Nematics. (a) Uniaxial nematics. Liquid crystals with orientational ordering but no spatial order are closest to normal isotropic fluids, but they differ in optical properties.

It would be natural if we were to represent a nematic as a collection of spatially elongated rods made up of organic molecules. The rods' centers of gravity are scattered at random, making the crystals similar to ordinary liquids, which have no correlation between the molecules' centers of gravity. However, there is an order in the molecules' orientation. Some molecules are aligned parallel to an axis $\mathbf{n}$ (of the nematic), defined as the average direction of the molecules' axes. By means of $\mathbf{n}$, the order parameter can be introduced for the nematic phase. In contrast with an ordinary liquid, nematics behave as a doubly refracting medium.

Nematic phases are only encountered in liquid crystals indistinguishable by mirror reflection. Each molecule in a crystal is reflected either into itself or into another molecule, the form itself of the crystal remaining unaltered.

An important property of a nematic liquid crystal, which has far reaching physical consequences (on which, as can be seen below, the defects structure depends) is the invariance of the system's states under transitions of $\mathbf{n}$ into $-\mathbf{n}$. Taking the property into account, $\mathbf{n}$ is called the *director*.

(b) Biaxial nematic. It follows from an analysis of the order parameter (see Subsec. 5.1.2) that a biaxial nematic given by two optical axes is possible in principle. This rare type has been discovered experimentally comparatively recently [MLLG]. The existence of biaxial nematics in another class of lyotropic liquid crystals was established earlier [Ge].**

---

* Type B smectics have a two-dimensional periodic structure and are close in properties to two-dimensional crystals.

** Relative to their phase transitions, liquid crystals are classified as *thermotropic* and *lyotropic*. In thermotropic crystals, transitions mostly occur due to temperature changes, while in lyotropic ones rod concentration changes are important. Later, we shall deal with thermotropic crystals.

(c) Cholesterics. If (chiral) molecules, which do not have mirror symmetry, are dissolved in a nematic liquid, then spiral distortion occurs. More precisely, it is essential that the number of left-chiral molecules should not equal that of right-chiral ones. This distortion was first observed in cholesterol ester, and so the phase is termed *cholesteric*.

For a topological analysis of defect (singularity) structure, it is necessary to describe precisely the structure of the space whithin which the order parameter ranges (the *space of internal states* according to [TK]) in the corresponding class of liquid crystals. The application of homotopy methods to classify linear and point defects is the same for any ordered matter. I therefore start with the general construction. Applications to concrete types of liquid crystals and superfluid liquids are given in subsequent sections.

5.1.1.1 Classification of singularities in systems with spontaneous symmetry breakdown

Let $\mathcal{U}$ be a region in physical space R^3 in which the material is concentrated and let the other parameter $P(x)$ be given at each point $x \in \mathcal{U}$. Depending on the system under consideration, $P(x)$ can be either a vector, or matrix, etc. The range of $P(x)$ is called the *space of internal states*, or, by analogy with field theory, the *vacuum manifold* M^n. Given a path γ in $\mathcal{U}$, $P(\gamma)$ maps it into the space M^n. We now give the mathematical definition of a linear defect in an ordered system.

Definition 5.1 Given a curve $\gamma \subset \mathcal{U}$ and a continuous map

$$P: \mathcal{U} \setminus \gamma \to M^n, \tag{5.1}$$

the curve $\gamma \subset \mathcal{U}$ is called a *linear defect* (or *linear disclination*) of P if P cannot be continued by continuity to any point in γ.

Consider a closed contour l encircling γ. We denote by $\tilde{l}$ the image of l under P and the class of curves in M^n homotopic to $\tilde{l}$ by $\{\tilde{l}\}$. It is obvious that if $\{\tilde{l}\}$ can be contracted to a point, then the preimages of $\{\tilde{l}\}$ in $\mathcal{U}$ can be continuously contracted onto γ, in which case γ is not a defect. However, if $\{\tilde{l}\}$ is not contractible to a point, then the corresponding curves l are not contractible to γ, therefore, P cannot be continued to γ. It is natural to assume that l cannot be contracted to a point by traversing γ, i.e., either γ is infinitely long or its boundary points are fixed on the boundary of $\mathcal{U}$.

The following criterion for topologically nontrivial defects to exist is a consequence of the above.

Theorem 5.1 *There exists a line defect in a domain $\mathcal{U}$ if $\pi_1(M^n) \neq 0$.*

The existence principle of topologically nontrivial point defects in $\mathcal{U}$ can also be proved in a simple manner. Let x_0 be a singular point of a field $P(x)$. We surround x_0 with the sphere S^2 and consider the classes of homotopic map f: $S^2 \to M^n$. Modifying the reasoning on line defects, we obtain the following.

Theorem 5.2 *There exists a point defect in a domain $\mathcal{U}$ if $\pi_2(M^n) \neq 0$.*

When applying a topological argument to real physical systems, we should always remember that topological criteria for a defect existence only give necessary conditions.

To destroy topologically nontrivial defects, energy is required (because the energy barrier should be overcome). On the other hand, metastable defects in the system are always possible and cannot be detected by topological methods.

Example 5.1 We consider superfluid ^{4}He. A system of vortex lines is known to exist in the superfluid state. Let us see how vortices can be found in ^{4}He by a topological argument. We use the familiar macroscopic description of ^{4}He (or, the more used term He II) by means of the complex order parameter $\psi = |\psi| \exp(i\varphi)$. We only need the basics of superfluidity theory. The detail and justification can be found in any book on superfluid He II, e.g., [TT]. Thus, the superfluid state is described by the macroscopic condensate function (macroscopically the number of particles filling one quantum state) $\psi(r) = |\psi| \exp(i\varphi(r))$, where the phase $\varphi(r)$ is a real function of radius-vector r and the amplitude $|\psi|$ is independent of r. The free-energy potential U_{cond} in London's theory is similar to (3.38) and has the form

$$U_{\text{cond}} = \alpha|\psi|^2 + \beta|\psi|^4 \tag{5.2}$$

and by assuming that $|\psi| = \text{const}$, the minima of U_{cond} are on the circle $|\psi| = \text{const}$. Thus, the vacuum manifold (5.2), which is the space of internal states, is the circle S^1.

Let γ be a vortex line in a container U. We encircle γ by a contour $l \sim S^1$. The function ψ maps l into the circle. A continuous map $l \to S^1$ is thereby defined. Since the mappings associated with a different number of traverses of S^1 are not homotopic ($\pi_1(S^1) = Z$), the vortex lines with nontrivial degrees of the mapping $l \to S^1$ are topologically stable. The contour $\tilde{l} = \psi(l)$ can be contracted to a point, only allowing deformations of $\tilde{l}$ in the whole complex plane. This is equivalent to the destruction of superfluidity in a large container and requires an energy threshold to be overcome.

Note that we have even obtained a stronger statement.

Line vortices in He II are characterized by the group $\pi_1(S^1)$. Since $\pi_1(S^1) = H_1(S^1)$, each vortex has an integral topological charge

$$Q = \frac{1}{2\pi} \int_l \nabla\varphi \, dr, \tag{5.3}$$

with the physical meaning of the vortex's circulation quantum, where Q is the number of circulation quanta.

Example 5.2 Ferromagnet. The states of an isotropic ferromagnet are specified by a three-dimensional magnetization vector $\mathbf{M}$ (3.24). The space of internal states $M^2 \sim S^2$. It is then obvious that there are no topologically stable line singularities and $\pi_1(S^2) = 0$. However, there exist point singularities with $\pi_2(S^2) = Z$.

We now turn to the principal goal of the section, namely, the classification of singularities in liquid crystals.

5.1.2 Linear and Point Defects in a Nematic

We begin with the choice of an order parameter. If we consider a nematic as a collection of molecules in the shape of elongated rigid rods, then we can introduce the unit vector ν^i pointing along the orientation axis of the ith molecule. The vector cannot be identified with the director **n** describing the molecule's averaged directions. Since the nematic has a center of symmetry, the averaged sum $\langle \nu^i \rangle$ is zero. Therefore, the natural order parameter characterizing the degree of order of the long axes in the nematic phase and vanishing under the transition into the isotropic phase is the tensor of rank two

$$S_{\alpha\beta}(r) = (1/N) \sum_i \nu_\alpha^{(i)} \nu_\beta^{(i)} - (1/3)\delta_{\alpha\beta}, \tag{5.4}$$

where summation is over all N molecules in certain macroscopic volume U in a neighborhood of a point r (ν_α are the components of the vector ν in fixed coordinates). $S_{\alpha\beta}$ is a symmetric traceless 3×3 matrix.

To describe the macroscopic properties of liquid crystals, it is better to introduce a macroscopic order parameter depending directly on the director **n**. Following de Gennes [Ge], it is convenient to select the anisotropic part

$$Q_{\alpha\beta} = G[\chi_{\alpha\beta} - (1/3)\delta_{\alpha\beta}\chi_{\gamma\gamma}] \tag{5.5}$$

of magnetic susceptibility as the order parameter, where G is a normalization constant. Usually, G is fixed by putting $Q_{zz} = 1$. As well as $S_{\alpha\beta}$, the tensor $Q_{\alpha\beta}$ is symmetric and has zero trace. For systems of rigid rods with uniaxial symmetry, $Q_{\alpha\beta}$ is related to $S_{\alpha\beta}$ as in

$$Q_{\alpha\beta} = N\chi S_{\alpha\beta}, \tag{5.6}$$

where $\chi = \chi_\parallel - \chi_\perp$ is the susceptibility's anisotropy, $\chi_\parallel$ and $\chi_\perp$ the susceptibility along and perpendicular to the orientation axis, respectively, and N is the number of molecules per unit volume. Note that (5.6) holds with the natural assumption that $\chi_{\alpha\beta}$ is the total susceptibility of the molecules.

If nematics are uniaxial, (5.5) and (5.6) can be written in terms of the components of the director **n**, assuming that the optical axis points along **n**, namely,

$$S_{\alpha\beta} = S[n^\alpha n^\beta - (1/3)\delta_{\alpha\beta}] \tag{5.7}$$

(n^α are the components of **n** in the laboratory's reference frame) and

$$Q_{\alpha\beta} = \chi[n^\alpha n^\beta - (1/3)\delta_{\alpha\beta}]. \tag{5.8}$$

The scalar factor S measures the order of molecules. In the isotropic phase $S = 0$ and in the nematic one $0 < S < 1$. The explicit formula for S is given in [Ge].

However, in the general case, the symmetric tensor of order two with $\mathrm{Tr}\, Q_{\alpha\beta} = Q_{\alpha\alpha} = 0$ has five components. If the matrix $Q_{\alpha\beta}$ is reduced to diagonal form, then the possible phases of a nematic liquid crystal are associated with

$$\text{(a) uniaxial crystal, where } Q = \begin{pmatrix} \lambda & & 0 \\ & \lambda & \\ 0 & & -2\lambda \end{pmatrix},$$

$$\text{(b) biaxial crystal, where } Q = \begin{pmatrix} \lambda_1 & & 0 \\ & \lambda_2 & \\ 0 & & -(\lambda_1 + \lambda_2) \end{pmatrix}. \tag{5.9}$$

Whether the nematic is uniaxial or biaxial depends on the structure of the free energy as a function of $Q_{\alpha\beta}$. The free-energy potential U_{cond}, in the Landau theory, has the form

$$U_{\text{cond}} = \alpha_1 \operatorname{Tr} [Q^2] + \alpha_2 \operatorname{Tr} [Q^3] + \alpha_3 \operatorname{Tr} [Q^4]. \tag{5.10}$$

From the point of view of the theory of spontaneous symmetry breaking, uniaxial and biaxial crystals are associated with different extrema of the free-energy potential. We assume that the extrema are realized in the five-dimensional order parameter space M^5 of matrices Q.

It follows from the general theory of Subsec. 1.3.4 that the extrema of U_{cond} are on the orbits of the symmetry group G. It is obvious that G coincides with the rotation group $SO(3)$. The classification of orbits in M^5 is given in Subsec. 1.3.4 with the consequence that the orbit associated with the uniaxial nematic (a) in (5.9) is isomorphic to $SO(3)/SO(2) \times Z_2$ (also seen from (5.7) and (5.8)) and the biaxial nematic (b) is isomorphic to the orbit $SO(3)/Z_2 \times Z_2 \sim SO(3)/D_2$.* The orbits coincide with the space of internal states of a uniaxial and biaxial nematic crystal.

Following the general scheme of Subsec. 5.1.1.1 (Theorems 5.1 and 5.2), it is easy to describe line and point singularities in a nematic.

Uniaxial nematic. The space of internal states $M^2 = SO(3)/SO(2) \times Z_2$ is isomorphic to the projective plane RP^2. By Theorem 5.1, line singularities are classified by the group $\pi_1(RP^2) = Z_2$ and the point ones are characterized by $\pi_2(RP^2) = Z$.

The interaction between two point defects in a nematic can be easily analyzed topologically [VM].

If $\pi_1(M) = 0$, then the coalescence of two singular points is associated with addition of two elements of $\pi_2(M)$. We recall that the latter group is commutative. However, if $\pi_1(M) \neq 0$, then we have to take into account its action on $\pi_2(M)$, i.e., the dependence of the interaction between two defects on the path. Let us consider this effect in a nematic. It was shown in Subsec. 2.1.5 that a nontrivial element $g_* \in Z_2 = \pi_1(RP^2)$ acts on $h_* \in Z = \pi_2(RP^2)$ by the formula $g_*(h_*) = -h_*$, where

* The group $Z_2 \times Z_2$ is isomorphic to the dihedral group D_2 of order 2. The dihedral, or n-gonal symmetry group $D_n \sim Z_n \dot{\times} Z_2$, is noncommutative for $n > 2$.

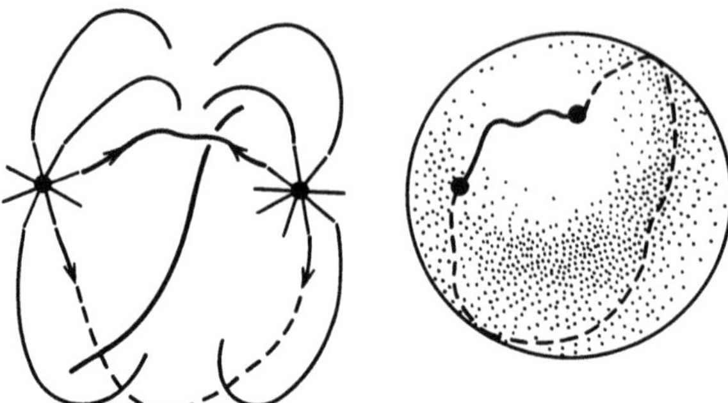

Fig. 10. Coalescence of two point singularities m_1 and m_2 in the presence of disclination line γ; l_1 and l_2 are two contours to be traversed; $\tilde{l}_1$ and $\tilde{l}_2$ are images of l_1 and l_2

the map g_* is induced by the element g: $S^2 \to S^2$ with $gx = -x$, $x \in S^2$ reversing the orientation. With a linear disclination γ, the coalescence of two point singularities m_1 and m_2 ($m_3 = m_1 \leftrightarrow m_2$) can proceed along a path l_1 or l_2 (see Fig. 10). If the path $\tilde{l}_1 = P(l_1)$ is contracted to a point, then, when two point defects coalesce, their indices N_1 and N_2 are added. However, if the path $\tilde{l}_2$ is not null-homotopic in RP^2, then the index of the point $m_3 = m_1 + m_2$ equals $N_1 - N_2$ (Fig. 10). It is obvious that the interaction of points with index $N_2 - N_1$ is also possible.

The problem of interaction between two disclinations is considerably more complicated and it is discussed in Subsec. 5.1.6.

Biaxial nematic. Its internal state space is the manifold $M^3 = SO(3)/D_2$. The fundamental group $\pi_1(M^3)$ can be found by the following construction.

We represent the group $SO(3)$ as the quotient group $SU(2)/Z_2$. This identification can be done in many ways, e.g., if we represent $SU(2)$ as S^3, then the identification of two diametrically opposite points yields the required factorization $SO(3) \sim RP^3$. To find $\pi_1(M^3)$, it is better to identify $SU(2)$ with the symplectic group Sp (1) or equivalently with the group of quaternion units ${}_1H$. All the definitions relative to the algebra H are in Subsec. 2.6.4. We reprise of this isomorphism's construction.

Definition 5.2 The group ${}_1H \sim$ Sp (1) is a *subgroup of the quaternion group* H *with norm* $N(q) = 1$.

We show that Sp (1) is isomorphic to $SU(2)$. We associate the elements e_k with matrices $i\sigma^k$ ($k = 1, 2, 3$) and the element e_0 with $-i\sigma^0 = I$ (I is the 2×2 matrix unit and the matrices σ^k are Pauli, see (2.49)).

Each quaternion q representable as $a_0 + a_ie_i$ can be written as

$$q = i(a_0\sigma^0 + a_i\sigma^i) = u = \begin{pmatrix} a_0 + ia_3, & a_1 + ia_2 \\ -a_1 + ia_2, & a_0 - ia_3 \end{pmatrix}, \tag{5.11}$$

$$N(q) = \det u.$$

Any complex 2×2 matrix can be written in form (5.11). If $N(q) = 1$, then the matrix u is representable as $u = \varrho v$, where v is a unitary matrix with det $v = 1$ and

$\varrho = \det u$. If we put $N(q) = \varrho = 1$, then we obtain the required isomorphism Sp (1) ~ $Su(2)$.

The subgroup $\mathscr{H} \sim Z_2 \subset$ Sp (1) consisting of two elements e_0 and $-e_0$ is a normal subgroup of Sp (1), therefore $SO(3) \sim$ Sp $(1)/Z_2$. Let $R^3 = \{a_1 t^1 + \ldots + a_3 t^3\} \subset H$ for any element $q \in$ Sp (1), $qR^3q^{-1} \subset R^3$. If we map Sp (1) into the group $GL(3, R)$ of all linear transformations of R^3, then Sp (1) is mapped into the group $SO(3)$ with kernel $\mathscr{H} \sim Z_2$. Let us consider the commutative diagram

$$\begin{array}{ccc} \text{Sp}(1) & \rightarrow & \text{Sp}(1)/Q \\ \downarrow & & \downarrow \\ SO(3) & \rightarrow & SO(3)/D_2 \end{array} \tag{5.12}$$

(the arrows denote the covering maps). It is easy to see that the mappings $e_i \rightarrow \pm e_i$ cover $t^i \rightarrow \pm t^i$, where t^i are basis vectors in R^3, $t^i t^j = \varepsilon_{ijk} t^k$, recalling that $e_i e_j = \varepsilon_{ijk} e_k$. We assume that the basis t^i $(i = 1, 2, 3)$ forms a right-hand coordinate system.

Using an exact sequence of homotopy groups for the fiber bundles

$$\text{Sp}(1) \rightarrow \text{Sp}(1)/Q, \tag{5.13}$$

we obtain from diagram (5.12) the following theorem.

Theorem 5.3 *The fundamental group $\pi_1(M^3)$ of a biaxial nematic is isomorphic to the group of unit elements of the quaternion algebra Q. The group Q is noncommutative and consists of the eight elements*: $\pm e_0$, $\pm e_i$ $(i = 1, 2, 3)$.

We retrace our steps to the classification of line and point singularities.

There are no point singularities in a nematic and $\pi_2(M^3) = 0$.

We pass to line singularities. When analyzing them in a medium with a noncommutative fundamental group $\pi_1(M)$ for the space of internal states, we encounter a new situation.

First, we begin with an example, considering the domain $\mathscr{U} \subset R^2 \setminus (a_1 \cup a_2)$ (a_1 and a_2 are two points in the plane R^2). $\pi_1(\mathscr{U}) = l_1 \cup l_2$ is a free group with two generators l_1 and l_2, which are paths surrounding a_1 and a_2 (Fig. 11). We consider

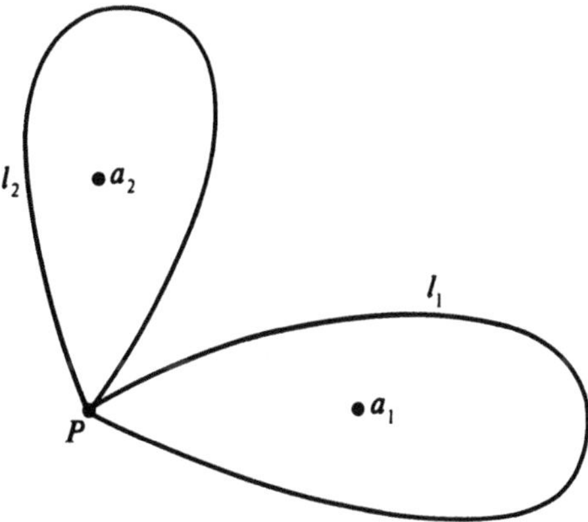

Fig. 11. Fundamental group of domain $\mathscr{U}$. Free group with generators l_1 and l_2

two paths $\gamma_1 = l_1 l_2 l_1^{-1}$ and l_2 which are different but homotopic elements of $\pi_1(\mathscr{U})$. The "paradox" can be explained quite simply: when deforming the path $l_1 l_2 l_1^{-1}$ into l_2, the initial point moves along l_1 bypassing a_1. Such deformations are not allowed by the definition of π_1. Deformations that do not fix the initial point define free-homotopy classes. Free-homotopy classes $\tilde{\pi}_1(M)$ are in a one-to-one correspondence with the conjugation classes of $\pi_1(M, x)$ on a connected manifold. For the commutative group $\pi_1(M, x)$, the conjugation class has only one element, therefore, $\tilde{\pi}_1(M, x) = \pi_1(M)$.

If $\pi_1(M)$ is noncommutative, then in order to find real physical singularities, we should consider conjugation classes of the corresponding disclination.

We list the classes of conjugate elements in the group Q. They are five, namely,

$$e_0, \quad -e_0, \quad \pm e_1, \quad \pm e_2, \quad \pm e_3.$$

The multiplication of two elements of Q is associated with the coalescence of two disclinations. The result depends on the coalescence path.

The non-Abelian nature of the group $\pi_1(M^3)$ is related to a number of the properties of interacting defects we shall study in future subsections.

5.1.3 Defects in Cholesterics

We start with a brief description of structure. Locally, i.e., at distances of the order of a molecular length, a cholesteric resembles a nematic. The positions of the molecules' centers are not correlated and the molecules themselves are oriented along the director **n** axis (this is a local optical axis). However, in contrast to the nematic phase, the cholesteric one has another type of orientation ordering. The state of a cholesteric is not fixed by **n** which varies continuously in space and which describes a spiral curve. If we schematically represent a cholesteric as a collection of planar molecular layers, then from layer to layer the vector **n** continuously rotates and describes a spiral. In this approach, we do not allow for the possibility that rotating molecules leave the "layers". However, for our purposes this simplification is justified. If we consider a macrocrystal in a layer about 100 μm thick, then if light is incident upon a sample, we distinctly discern a spiral structure.

If a cholesteric is taken in a layer between two parallel plates and tangential boundary conditions are given, then the director **n** is determined by the simple formulas

$$n_x = \cos(q_0 z + \varphi), \quad n_y = \sin(q_0 z + \varphi), \quad n_z = 0,$$

where the helical axis points along the z-axis, q_0 is the wave vector, and the angle φ is only fixed by the boundary conditions (see Fig. 12). The spatial period of the spiral L is one-half the helix's pitch, namely,

$$L = \frac{\pi}{|q_0|}$$

(because of the symmetry of the transition $n \to -n$).

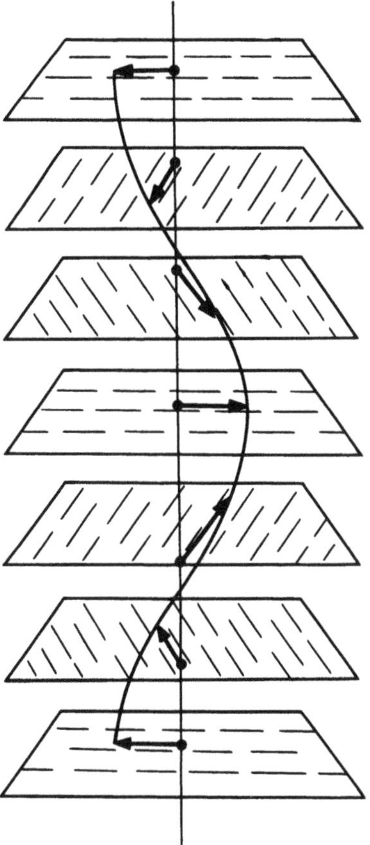

Fig. 12. Cholesteric. Schematic diagram of spiral structure. Several consecutive positions of director **n**

For most cholesterics, L is of order 3,000 Å, a distance much greater than a molecules' length and comparable to the wavelength of light, which accounts for the possible optical observation of the cholesteric's spiral structure.

We now consider the structure of singularities. As for nematics, the problem reduces to the study of the topology of the order parameter's space, or the system's internal state space.

We first define the order parameter. This has already been done for a cholesteric between two parallel plates. In the general case, the parameter is characterized by a frame of three vectors: **n**, the director, **d**, a unit vector pointing in the direction of the spiral's axis, and $\mathbf{l} = [\mathbf{n} \times \mathbf{d}]$, their vector product. To find the complete range of the order parameter, we also have to take into account the system's additional symmetry relative to the transformations

$$\mathbf{n} \to -\mathbf{n}, \quad \mathbf{d} \to -\mathbf{d}, \quad \mathbf{l} \to -\mathbf{l}.$$

Hence, the internal state space M of the cholesteric is isomorphic to $SO(3)/D_2$, which is the same as for a biaxial nematic (the result is obtained in [VM]). Therefore, the homotopy classification of line and point singularities in a cholesteric is similar to a biaxial nematic.

There are no topological stable point singularities in a cholesteric and $\pi_2(M) = 0$, whereas line singularities are described by the classes of conjugate elements of the group $\pi_1(M) = Q$. It has been shown in Subsec. 5.1.2 that the classes consist of the elements $1, -1, \pm e_1, \pm e_2, \pm e_3$.

When classifying the line singularities of a cholesteric, we run up against difficulties due to the discrete structure in the direction of the cholesteric's helical axis. Applying homotopy methods, we should take into account that contracting the contours of one layer onto another requires considerably more energy than deformation in the layer itself. There are hence two types of cholesteric crystals, depending on the relationship between the thickness of a molecular layer and the helix pitch, e.g., in sufficiently thick samples (mixtures of cholesterics with nematics) the helix pitch L is of order 5 μm and the layer's thickness is about 1 μm. If the distance between the layer is unaltered, then we obtain a smectic-like structure, in which case the equilibrium structure consists of equidistant bent surfaces (layers). Such a structure is said to be *confocal.*

A mathematical machinery which can describe such structures is called the *theory of foliations.* The structures can be best seen in smectics with much thinner layers, of order 20 Å. It is obvious that homotopy methods find natural constraints in similar liquid crystals if defects are to be described.

However, homotopy methods for most types are as applicable as for nematics. Additional constraints are only related to the energetics of the processes and to the structure of the order parameter. In particular, the elements of translational symmetry (i.e., shift through the helix pitch) in the cholesteric phase lead to an inequality between the vectors in the order parameter's local frame. We mean the following experimental fact. Singularities in the director field are the most energetically unfavorable. Intuitively, this can be explained by the rotation of the cholesteric's molecules in the ground state (which is not uniform). Therefore, the director either easily "leaves" the singularity line or can be continued by continuity onto it. This property has interesting topological consequences and can be verified experimentally.

Let $I = (-1)$, $\pm e_i$ $(i = 1, 2, 3)$ be the classes of defects in a cholesteric and 1 associated with non-singular lines.

We select a coordinate system e_1, e_2, e_3 in space R^3 (e_1, e_2, e_3 is the canonical orthonormal frame). We orient the order parameter's frame with respect to the basis e_i by pointing the molecular director along the vector e_1, the spiral's vector along the e_3-axis, and the binormal vector along the e_2-axis. We consider the loops γ_i surrounding the $\pm e_i$-axes. In traversing γ_i the order parameter rotates through π. Besides, there exist defects associated with the class I. In traversing the line γ_I, the parameter turns through 2π. It follows from the above argument on the director's behavior that the most frequent defects are of type I and $\pm e_1$, where the director has no singularities.

In the literature a situation, in which one of the frame's vectors has no singularities but the whole frame has is said to be *double topological.*

We consider this phenomenon in more detail, starting with a purely topological analysis of the mapping of internal states, complete space $M_f^3 = S^3/Q$ and the director's degeneracy domain $M_n^2 = RP^2$. We have $M_f^3 \to M_n^2$. Since $M_n^2 \subset M_f^3$ and M_n^2 is the quotient space of M_f^3, then M_f^3 can be regarded as the fiber map over M_n^2

$$SO(3)/D_2 \overset{S_1}{\to} SO(3)/SO(2) \dot{\times} Z_2 \tag{5.14}$$

with fiber S^1.

We consider the segment of the exact sequence of homotopy groups

$$\pi_2(M_f^3) \to \pi_2(M_n^2) \to \pi_1(S^1) \to \pi_1(M_f^3) \to \pi_1(M_n^2) \to \pi_0(S^1) \tag{5.15}$$

or

$$0 \to Z \to Z \to Q \to Z_2 \to 0.$$

We derive from the exactness that

$$\begin{aligned} \{N\} \to \{4N\} &\to 1, \\ \{4N+2\} &\to -1, \\ \{4N+1\} &\to \ \ e_1, \\ \{4N+3\} &\to -e_1 \to 1, \\ & \quad\ \ e_2, \\ & \quad -e_2 \to -1, \\ & \quad\ \ e_3, \\ & \quad -e_3. \end{aligned} \tag{5.16}$$

It can be seen from (5.16) that the singularity lines e_i of the frame (**n**, **d**, **l**) in the kernel of the map $\pi_1(M_f^3) \to \pi_1(M_n^2)$ form the group $Z_4 = \{1, -1, e_1, -e_1\}$. The director is without singularities on the lines.

They are observed experimentally in the director field as thick lines without a singular core and of the order of the helix's pitch. The element $-1 \in \pi_1(M_n^2)$ is associated with singularity lines in the classes $\pm e_2$, $\pm e_3$. They are also singularity lines for the whole order parameter and for the director field.

5.1.3.1 Hopf invariant and disclination loops

Bouligand discovered an interesting structure in a cholesteric in 1974 [Boul]. Two dark linked loops, i.e., singularity lines of the order parameter (Fig. 13), were observed in polarized light. The structure was termed a *double anneau*. It was later shown by a group of French physicists and mathematicians that this structure has the linking number -1 and realizes the Hopf map h: $S^3 \to RP^2$ of degree -1 [BDPPT].

Let us consider this interesting effect in more detail. As has already been noted, lines of I type are absent among the singularity lines of a cholesteric if there are no singularities in the director field **n**.

Fig. 13. Double *anneau* is link of two circles with linking number $k = 1$

These lines may differ in shape, viz., they may be closed. Since I are not singularity lines in $\mathbf{n}$, they are associated with the points $m_0 \in RP^2$.

We consider the covering of RP^2 with the two-dimensional sphere S^2. Each point m_0 is associated with two diametrically opposite points s_0 and $-s_0$ of S^2. Assume that we already have a configuration of double anneau type $I = l_1 \cup l_2$, where l_1 and l_2 are two circles linked together. We specify the field $\mathbf{n}$ on l_1 as directed "upwards" on l_1 and "downwards" on l_2. It is then obvious that the resultant configuration determines a nonsingular configuration of $\mathbf{n}$ which cannot be disentangled (with linking number $k(l_1, l_2) = 1$).

This configuration actually realizes the Hopf map h: $S^3 \to S^2$ with degree 1.

One essential note has to be made. Such a nontrivial configuration can be considered immediately in a nematic with the link of two nonsingularity lines l_1 and l_2 ($k(l_1, l_2) = 1$) forming a nontrivial structure. The Hopf map meanwhile can be obtained as follows. We specify a map of physical space φ: $R^3 \to RP^2$ such that

$$\varphi(\infty) = \text{const.} \tag{5.17}$$

R^3 is thereby compactified into S^3 by means of boundary condition (5.17).

In case of cholesteric, the vector $\mathbf{n}$ in the ground state rotates in space, therefore, a fixed boundary condition for the director cannot be picked. However, this is not necessary for the solution and it suffices to consider a ball D^3 with the curves l_1 and l_2. Assuming that the boundary ∂D^3 is sent into a fixed point in RP^2, we also obtain a correspondence between a linking of l_1 and l_2 and the Hopf map.

In the Bouligand experiments, trivial links with deg $h = 0$ (*anneau eu fuseau* and *anneau en forme de larme*) were observed besides the *double anneau* [Boul].

Since the vanishing of the linking number of two curves is a necessary, but not a sufficient, condition for the disentanglement, the problem of direct disentanglement in greater detail would be of interest. The problem is related to another interesting question of defect crossing and we postpone a solution of both of them until Subsec. 5.1.6. For the present, a brief reprise.

5.1.3.2 The Volterra process and classification of line singularities in cholesterics

Here, I discuss the approach to classification of disclinations in cholesterics on the basis of the Volterra process, which is traditional for the theory of liquid crystals. The classification was obtained by Kleman and J. Friedel in 1969, viz., several years before homotopy theory was applied [K]. A comparison may be useful for the physicist (particularly specializing in liquid crystals) who is used to the classification by the Volterra process. Note that the Volterra process is essentially topological, but based on homology theory ideas.

I begin with a brief summary, following de Gennes [Ge].

Actually, the Volterra process describes operations over an elastic body, which are admissible from the standpoint of internal stresses (or *distortions* if we follow the terminology of Volterra himself). To apply the constructions to liquid crystals, we make the following mental experiment. We take a cholesteric and freeze it to obtain a solid body. We cut the crystal along a surface S bounded by a line l, shift one side of the cut (e.g., S_1) relative to the other, S_2, by a vector $\mathbf{b}$ (analogous to the Burgers vector in the theory of normal crystals [AsM]), and rotate S_1 through an angle ω around an axis t, requiring that, after all the operations, part of S_1 should be consistent with S_2, or abut on S_2 along a certain line.

To fulfil the requirement, the transformations $\mathbf{b}$ and (ω, t) should be in the cholesteric's order parameter symmetry group. Depending on the shift of S_1 relative to S_2, "cavities" or "overlaps" arise in the "body". We add some substance in the former case and remove the extra material in the latter. Since both sides S_1 and S_2 are consistent, the director $\mathbf{n}$ field has no discontinuity on the surface but it can have one on l, which is called a *singularity line*, or *disclination*, in the cholesteric liquid crystal.

We list the allowed operations $(\mathbf{b}, \omega, t)$ in the cholesteric.

(1) Displacement by the vector $\mathbf{b}$ perpendicular to the helical axis. The displacement has no effect on the director and therefore may not be considered at all.

(2) Displacement by $\mathbf{b}$ along the helical axis. The consistency condition yields $b = mL$ (L is the helix's pitch and m an integer or half-integer).

(3) Rotation $\omega = 2\pi m$ about the t-axis parallel to the helical one. Due to the helical symmetry of the spiral, this operation is equivalent to (2).

(4) Rotation $\omega = 2\pi m$ about the t-axis perpendicular to the helical one. The operation is admitted by the symmetry if and only if t is either parallel or perpendicular to the local director.

Operations (1)-(4) are basic and any combination is possible.

In the particular case of a nematic, the unique nontrivial operation is rotation through the angle $\omega = 2\pi m$ about the t-axis perpendicular to that of the unperturbed director $\mathbf{n}$. The position of t in the plane perpendicular to $\mathbf{n}$ is immaterial, since rotations about two parallel axes (t and t') differ only in a translation.

Disclination types. Each type of Volterra operation is associated with a certain disclination class.

1. χ-lines. Disclinations associated with operation (2), or, which is the same, (3), are called *χ-lines.* A χ-line has Frank index $2m$ [Ge]. For $m = n/2$, where

$n = 2k + 1$, a χ-line is topologically stable, while for $m = k$, unstable. A χ-line should not necessarily coincide with its rotation t-axis and can be oriented in an arbitrary way.

2. τ-lines. Let us consider the Volterra process related to operation (4). Two orientations of the rotation of the t-axis are possible. Let the t-axis be perpendicular to the local director $\mathbf{n}$ axis. The corresponding disclination line is called a *τ-line.* The notation τ^- is adopted for the line τ obtained by filling the cavity if S_1 is moved from S_2 and τ^+ by removing "excess" substance if S_1 and S_2 overlap (see Fig. 6.17 in [Ge]).

3. λ-lines. If the rotation axis is parallel to the local director's axis, then there is another type of disclination, a *λ-line*, for which the same remark about separation into λ^+- and λ^--lines holds.

There is an important difference between λ- and τ-lines. The director is continuous on a λ-line and has no core-type singularity. A τ-line does have a core and possesses high energy.

Isolated λ- and τ-lines are considerably less flexible than χ-ones, which can easily be explained on the Volterra process basis. The energy of disclination lines of type λ or τ decreases if the distance between S_1 and S_2 in the neighborhood of l does so. A minimum is attained provided l coincides with the rotation axis. χ-, τ-, and λ-filaments can interact with each other in a rather complicated way. Before considering the familiar experiments, we juxtapose the above classification with the purely topological one considered earlier. This is done in [VM] and is easy to obtain if we notice that the disclinations χ, τ, λ are singularity lines relative to the cholesteric's complete order parameter, but behave differently with respect to the director.

Consider a map φ: $\pi_1(M) \to \pi_1(M_1)$, where M is the cholesteric's internal states space and M_1 the nematic's (see Subsec. 5.1.2). Ker φ is associated with nonsingular configurations relative to the director and the elements of $\pi_1(M)$ sent into -1 with the singular configurations with respect to $\mathbf{n}$. On the basis of this remark and structure of the group $\pi_1(M)$, we have

$$\begin{gathered} 1 \to \chi^{4k}, \quad \tau^{4k}, \quad \lambda^{4k}, \\ -1 \to \chi^{4k+2}, \quad \tau^{4k+2}, \quad \lambda^{4k+2}, \\ \pm e_1 \to \lambda^{2k+1}, \quad \pm e_2 \to \tau^{2k+1}, \quad \pm e_3 \to \chi^{2k+1}. \end{gathered} \tag{5.18}$$

In particular, it follows from the juxtaposition that the transitions of χ^2 into λ^2 and into τ^2 are not topologically forbidden.

However, the transition process $\chi^2 \to \lambda$ is more favorable energetically since the director has no singularity on λ^+- and λ^--lines. The transformation of a topologically unstable line χ^2 into the pair of λ^{+2}- and λ^{-2}-lines is called a *dissociation* and is observed experimentally. It would be interesting to note that (5.18) entails the impossibility of the decomposition of χ^{2k+1} into λ- and τ-lines for topological reasons. Before introducing the topological classification, this was held to be true, but without proof. The topological classification via homotopy methods is more complete. It is impossible to completely separate topologically different singularity types by the Volterra process in the old classification, as seen from (5.18).

The application of topological criteria to actually observed structures is only possible if the energetics of the processes are taken into account, e.g., both possible (though problematic) dissociation processes are mentioned in the de Gennes book, viz., $\chi^1 \to \lambda^+ + \tau^+$ or $\chi^1 \to \lambda^- + \tau^+$. The decompositions are forbidden from the topological viewpoint, since χ^1, λ^1, and τ^1 are in different topological classes. On the other hand, transitions of type $\tau^k \to \tau^{-k}$, $\lambda^k \to \lambda^{-k}$ or dissociation processes $\chi^{4k+2} \to \lambda^{2k+1} + \lambda^{2k+1}$ for large k are allowed topologically and require more energy than transitions between different homotopy classes.

To summarize, we can say that topological methods should go along with estimation of system's energy. Nevertheless, the methods have heuristic importance and enable us to delineate at once the possibilities that arise.

A full topological analysis of the Volterra process is given by Kléman [Kl] (see also the lectures by Poénaru [Po2]).

5.1.4 Two-Dimensional Structures in Nematics

Topological methods enable us to classify nonsingular structures (i.e., textures). Consider the classification of two-dimensional solutions in a uniaxial nematic. Let φ be a map $R^2 \to RP^2$ such that $\varphi(\infty) = m_0 \in RP^2$. The map $\tilde{\varphi}: S^2 \to RP^2$ is thereby defined.

To find a two-dimensional texture, we can, generally speaking, do as follows. The nematic's ground states form the manifold of the minima of the free-energy potential U_{cond}, i.e., the projective plane RP^2. To describe nonuniform states determined by the field $\mathbf{n}$, we have to introduce terms depending on the gradients $(\nabla n)^2 \to F_{\text{grad}}$ into the potential F. The nematic's nonuniform states are thus determined by the Euler-Lagrange equations for the gradient part of free-energy potentials. Geometrically, this means that we specify on RP^2 a metric, defined by F_{grad}, and seek minimal surfaces covering RP^2. In the general case,

$$F_{\text{grad}} = K_1(\text{div } \mathbf{n})^2 + K_2(\mathbf{n}\cdot\text{rot } \mathbf{n})^2 + K_3(\mathbf{n} \times \text{rot } \mathbf{n})^2 \tag{5.19}$$

up to terms of second degree in ∇n, where K_i are the Frank (elastic) constants. This follows from the basic equations of elasticity theory adapted to media with additional symmetry $D_{\infty h}$. Expression (5.19) is basic for nematics' continuum theory and was first obtained by Oseen in 1933, being derived in a more rigorous manner by Ericksen and Frank (see the detail and physical applications in [Ge]).

The constants K_i are associated with the three basic deformation types:

K_1 with $\text{div } \mathbf{n} \neq 0$, *splay*,

K_2 with $\mathbf{n}\cdot\text{rot } \mathbf{n} \neq 0$, *twist*,

K_3 with $\mathbf{n} \times \text{rot } \mathbf{n} \neq 0$, *bend*.

In real nematics, $K_3 \sim K_1 > K_2$.

To find two-dimensional textures with arbitrary Frank constants is a hard problem and completely unsolved. However, there is a so-called *one-constant approximation* $K_1 = K_2 = K_3$ with which a texture's description can be simplified considerably. In this case the qualitative behavior of solutions is in many respects the same as the physically real case of different K_i.

It is not complicated to notice that when $K_1 = K_2 = K_3$ the expression for F_{grad} can be rewritten as

$$F_{\text{grad}} = K(\nabla n)^2 \tag{5.20}$$

(up to surface terms). The two-dimensional textures' description is thereby reduced to minimizing the functional

$$\mathscr{F} = \int K(\nabla n)^2 dV \tag{5.21}$$

of the chiral field, which was considered in Subsec. 4.2.3 in relation to two-dimensional ferromagnets.

It is natural to specify the field $\mathbf{n}(r)$ in a nematic in a two-dimensional domain $\mathscr{U} \subset R^2$ with boundary $\partial\mathscr{U}$ and fixed behavior of $\mathbf{n}(r)$ on $\partial\mathscr{U}$, which is topologically equivalent to the problem of maps $S^2 \to RP^2$ if the boundary condition is selected so that $\mathbf{n}/\partial\mathscr{U} = \mathbf{n}_0$. To find the textures we only have to notice that there is a harmonic mapping (i.e., a minimum of (5.21)) in each homotopy class $\pi_2(RP^2)$.

For this boundary condition we actually reduced finding two-dimensional structures to the description of an instanton map $S^2 \to RP^2$. This problem is already close to the model of a two-dimensional ferromagnet considered in Subsec. 4.2.3, with the natural difference that the "isotopic space" M is RP^2.

I now take a diversion and describe the space of instantons for the chiral model with $M = RP^2$. The problem has a certain methodological interest. First, this is only another example (after $M = S^2$) where the space can be described completely. (If $S^2 \to S^2$, the result is obtained from the explicit construction for meromorphic maps of the Rieman sphere into itself.) For Yang-Mills instantons, the problem seems to be exceptionally difficult. Second, in the solution we use an informative mathematical construction, i.e., Klein surfaces, applying which can be convenient when various structures are studied in physical problems with non-orientable manifolds or manifolds with boundary. The results that follow were obtained by Natanzon and the author in 1978. The physicist who reads the book can skip the subsection and only deal with the final result.

Klein surfaces. We first recall the basic facts (see the detailed treatment in [AG]).

It is well known that algebraic curves in the complex plane are related to compact Rieman surfaces. Similarly, real algebraic curves are connected to Klein surfaces. Like a Rieman surface, a Klein one is a pair of a surface and of an analytic structure on the latter. However, in contrast to a Rieman surface, a Klein surface can be non-orientable or with boundary. The analytic structure of a Klein surface (a *dyanalytic structure*) generalizes the analytic structure of a Riemann surface (a complex analytic structure) to non-orientable surfaces and to those with a boundary.

The simplest example of a Klein surface is the upper half-plane $C^+ = \{z \in C | \operatorname{Im} z \geqslant 0\}$ with the analytic structure induced on it by the complex analytic structure of the complex plane C. Let two sets $A, B \subset C$ be open in C or belong to C^+ and be open in C^+. We call a continuous map $f\colon A \to B$ *dyanalytic* if there exists a set $\tilde{A}$, $C \supset \tilde{A} \supset A$ open in C and a function $\tilde{f}\colon \tilde{A} \to C$ of class C^2 differentiable in the real sense such that $f = \tilde{f}|_A$, with $\partial\tilde{f}/\partial\bar{z} = 0$ or $\partial\tilde{f}/\partial z = 0$ at each point $z_0 \in \tilde{A}$.

Let M^2 be a two-dimensional topological manifold (possibly, non-orientable or with boundary). We call a pair $(\mathscr{U}, \varphi)$, where φ: $\mathscr{U} \to C$ is a homeomorphism onto an open set in C or C^+, a *dyanalytic chart.* The collection of dyanalytic charts $(U, \varphi) = \{(U_j, \varphi_j) | j \in J\}$ such that $M^2 = \bigcup_j \mathscr{U}_j$ and the map $\varphi_j\varphi_k^{-1}$: $\varphi_k(\mathscr{U}_j \cap \mathscr{U}_k) \to \varphi_j(\mathscr{U}_j \cap \mathscr{U}_k)$ is dyanalytic for any k and j is called a *dyanalytic atlas* for M^2.

Two dyanalytic atlases $\mathscr{U}$ and $\mathscr{V}$ for M^2 are said to be *equivalent* if $\mathscr{U} \cup \mathscr{V}$ is also a dyanalytic atlas. The equivalence class of dyanalytic atlases for M^2 is called a *dyanalytic structure* on M^2.

Definition 5.3 The pair of the surface M^2 and the dyanalytic structure on it is called a *Klein surface.*

Definition 5.4 Let M_1 and M_2 be two Klein surfaces. A continuous map f: $M_1 \to M_2$ is called a *morphism* of Klein surfaces if the image of the boundary of M_1 belongs to the boundary of M_2 and, for any point $m \in M_1$,

(1) there is a dyanalytic chart $(\mathscr{U}, \varphi)$, compatible with the dyanalytic structure on M_1, where $\varphi(\mathscr{U}) \subset C^+$,

(2) there is a dyanalytic chart $(\mathscr{V}, \psi)$, compatible with the dyanalytic chart on M_2, where $f(m) \in \mathscr{V}$, $\psi(\mathscr{V}) \subset C^+$, and

(3) there is a dyanalytic map $\hat{f}$: $\varphi(\mathscr{U}) \to C$ such that $f(\mathscr{U}) \subset \mathscr{V}$, $\psi f = \eta\hat{f}\varphi$, where the map η: $C \to C^+$ is given by the equality $\eta(a + ib) = a + i|b|$.

For instance, a Riemann surface is also a Klein surface and a holomorphic or an antiholomorphic mapping of Riemann surfaces is a morphism of Klein surfaces.

The simplest example of a non-oriented Klein surface is the quotient space $S \sim RP^2$, $S = \tilde{C}/\langle\tau\rangle$ of the Riemann sphere $\tilde{C} = C \cup \infty$ relative to the group generated by the antiholomorphic involution τ: $\tilde{C} \to \tilde{C}$ given by the formula $\tau(z) = -1/\bar{z}$. The natural projection p: $\tilde{C} \to S$ is a two-fold Klein map.

Two holomorphic (Klein) maps f_1: $M_1 \to M_2$ and f_2: $M_2 \to M_1$ are said to be *equivalent* if there exist invertible holomorphic (Klein) maps α: $M_1 \to M_1$ and β: $M_2 \to M_2$ such that $f_2\alpha = \beta f_1$.

Compare the concepts of holomorphic and Klein equivalence of mappings in the simplest case of $M_1 = \tilde{C}$.

Let f: $\tilde{C} \to \tilde{C}$ be an n-fold holomorphic map, $m(q)$ the multiplicity index of f at a point $q \in M_1$, and $D_f = \sum_{q \in M_1} (m(q) - 1)_q$ the divisor of f. The degree of D_f is $2(n - 1)$.* For any divisor D of even degree, there exists a holomorphic map f: $\tilde{C} \to \tilde{C}$ such that $D = D_f$. It is not hard to show that two maps g_1 and g_2: $\tilde{C} \to \tilde{C}$ are holomorphically equivalent if and only if the divisors D_{g_1} and D_{g_2} are too, i.e., there exists a bilinear (more exactly, biholomorphic) transformation α: $\tilde{C} \to \tilde{C}$ such

* See the definition of a divisor and its degree, e.g., in [Sp].

that $\alpha D_{g_1} \to D_{g_2}$. The equivalence class of an n-fold holomorphic map $\bar{C} \to \bar{C}$ is thus determined by a collection of $2(n-1)$ (possibly coincident) points on the Rieman sphere if the action of the group Γ of bilinear transformations of the Rieman sphere $\bar{C}$ is not taken into account. Therefore, the manifold of all equivalence classes is isomorphic to the manifold $\bar{C}/\Gamma$ and, in particular, depends on $4(n-1) - 6 = 4n - 10$ real parameters (equal to the dimension of the space of instantons with $M^2 = S^2$). Now, let f: $\bar{C} \to S$ be a Klein map. Since $\bar{C}$ is simply connected, $f = p\bar{f}$, where $\bar{f}$: $\bar{C} \to \bar{C}$ is a holomorphic mapping. Thus the number of folds of f is $2n$, where n is that of $\bar{f}$. If two Klein maps f_1 and f_2: $\bar{C} \to S$ are equivalent, then the associated maps $\bar{f}_1$ and $\bar{f}_2$ are also Klein-equivalent. In other words, there exist conformal (biholomorphic or antibiholomorphic) maps α and β: $\bar{C} \to \bar{C}$ and a Klein equivalence $\tilde{\beta}$: $S \to S$ such that $\bar{f}_2\alpha = \beta\bar{f}_1$ and $p\beta = \tilde{\beta}p$. The latter means that $p\beta p^{-1}$ is a correctly defined map of S into itself and, therefore, β belongs to the centralizer G of an element τ in the group of conformal transformations $H \times \langle\tau\rangle$. Thus, $\bar{f}_2\alpha\bar{f}_1^{-1} \in G$.

It is not hard to see that, conversely, if there exists a Klein equivalence α and β: $\bar{C} \to \bar{C}$ such that $\beta = \bar{f}_2\alpha\bar{f}_1^{-1} \in G$ between two maps $\bar{f}_1$ and $\bar{f}_2$, then the maps f_1 and f_2 are Klein equivalent. Therefore, the equivalence class of a $2n$-fold Klein map $\bar{C} \to S$ is governed by a collection of $2(n-1)$ (possibly coincident) points on the Rieman sphere, given the arbitrary action of the group G of bilinear transformations. Thus, the manifold of all equivalence classes of $2n$-fold Klein maps is isomorphic to $C^{2(n-1)}/G$. Since G is a three-dimensional real group, the real dimension of the manifold (of RP^2-instantons) is $4(n-1) - 3 = 4n - 7$.

5.1.5 Surface Singularities in Nematics and Cholesterics

Here we give the classification of surface singularities via relative homotopy groups [Vo], [Tr]. The usual (absolute) homotopy groups, which carry less information, have been used for the same goal earlier [SPA]. Similar applications were made to superfluid ^{3}He (see Sec. 5.2). Note that in a somewhat different context (but also for ^{3}He) relative homotopy groups appeared earlier in [GM] (see Subsec. 5.2.8.3).

The idea of application of relative homotopy groups is very simple. Let x_0 be a singular point on the surface M^2. Surround x_0 with a closed curve (loop) γ. The point can be isolated singularity or the end of the line singularity inside the spatial configuration Ω bounded by M^2. If x_0 is that endpoint, then x_0 is associated with nonzero elements of the group $\pi_1(V)$; if x_0 is isolated singularity, then with the trivial element $\pi_1(V)$, i.e., the kernel of the map φ_1: $\pi_1(\tilde{V}) \to \pi_1(V)$, where V is the space of internal states of the order parameter given on Ω and $\tilde{V}$ the reduced manifold associated with the space of internal states of the order parameter defined on a domain $\omega \subset M^2$ with x_0.

There can exist singular points on the surface, "plunging" out of Ω. A map φ_2: $\pi_2(\tilde{V}) \to \pi_2(V)$ is thereby defined. The complete classification of singular points on the surface covers all the above possibilities.

We surround a singular point x_0 with a hemisphere S^2_+ so that the boundary $\partial S^2_+ = S^1$ is on M^2. Then the homotopy classes of the maps

$$S^2_+ \to V, \quad S^1 \to \tilde{V}, \quad \text{Map}\,(S^2_+, S^1) \to (V, \tilde{V})$$

classify all the types of singular points on the surface. Map $(S^2_+, S^1) \to (V, \tilde{V})$ coincides with the relative group $\pi_2(V, \tilde{V})$ for whose calculation it is convenient to consider the exact homotopy sequence

$$\pi_2(\tilde{V}) \to \pi_2(V) \to \pi_2(V, \tilde{V}) \to \pi_1(\tilde{V}) \to \pi_1(V) \to \pi_1(V, \tilde{V}) \to \pi_0(\tilde{V}). \tag{5.22}$$

We apply this general argument to the classification of point singularities in a nematic, specifying tangential conditions on its boundary, i.e., selecting the director field to be tangent to M^2. We arrive at the following:

$$V = RP^2, \ \tilde{V} = S^1, \ \pi_2(S^1) \xrightarrow{\varphi_2} \pi_2(RP^2)$$

$$\to \pi_2(RP^2, S^1) \xrightarrow{\psi} \pi_1(S^1) \xrightarrow{\varphi_1} \pi_1(RP^2) \to \pi_1(RP^2, S^1) \to 0, \tag{5.23}$$

i.e.,

$$\begin{array}{ccccccccccc} 0 \to Z \to & X & \to Z \to Z_2 \to & X_1 & \to 0. \\ & \downarrow & & \downarrow & \\ & \pi_1(V, \tilde{V}) & & \pi_1(V, \tilde{V}) & \end{array} \tag{5.24}$$

We now analyze sequence (5.23).

(1) Ker $\varphi_1 = Z$, since each point singularity on a surface with the even Frank index (the degree of the map of the circle S^1) is associated with a removable singularity in Ω. This map $Z \to Z_2$ is known in elementary algebra as a particular case of the map $Z \to Z_p$ into the residue field (for prime p). Here, the subgroup of even integers $\{2n\}$ isomorphic to Z forms the kernel of φ_1.

(2) Since $\pi_2(\tilde{V}) = 0$, $\pi_2(V) = Z$ is embedded in $\pi_2(V, \tilde{V})$ in a one-to-one fashion, $\pi_2(V)$ coincides with Ker ψ and therefore is a normal subgroup of $\pi_2(V, \tilde{V})$.

Hence, $\pi_2(V, \tilde{V})/\text{Ker}\,\psi \sim \text{Im}\,\pi_2(V, \tilde{V}) = Z$, i.e.,

$$\pi_2(V, \tilde{V}) = Z + Z. \tag{5.25}$$

We note that in the general case direct decomposition (5.25) does not hold. Generally speaking, the group $\pi_2(V, \tilde{V})$ is not Abelian and we have to analyze the whole of (5.23).

Formula (5.23) supplies a complete classification of the singular points of a nematic's surface.

We could in the same way consider point surface singularities in a cholesteric. However, we have to take into account the inequality of the vectors of the order parameter's frame. The picture for the director field is the same as in case of nematic. However, for the vector **d** (pointing along the helical axis), certain deformations are forbidden because the helix pitch is finite. On the other hand, the natural boundary conditions specify a field **d** normal to the surface M^2, therefore, there are no point singularities of $\mathbf{d}$ on M^2 at all. The A-phase in superfluid ^{3}He, which we shall consider in Subsec. 5.2.5.1, is a physically more adequate example of a surface singularity in a system with the order parameter close to a cholesteric.

We retrace our steps to exact sequence (5.22). By analogy with point singularities on a surface, we can obtain an existence criterion for line singularities on a surface, only noticing that the line singularity χ in the surface is classified by the set $\pi_0(\tilde{V})$ and accordingly can be either a stable line in the surface or the boundary of a wall in Ω. The walls are classified by the group $\pi_0(V)$. Finally the full group of line surface singularities is $\pi_1(V, \tilde{V})$.

It follows from analyzing sequence (5.24) that $\pi_1(RP^2, S^1) = 0$ for a nematic. A nontrivial example of line singularities is given by the same A-phase of ^{3}He (see Subsec. 5.2.4.1).

5.1.6 Topology of Linked Defects

Here, we discuss the topological aspects of the interaction between several line singularities. Line singularities (linear defects) arise in liquid crystals, superfluid liquids ^{3}He and ^{4}He, neutron stars, etc. On the other hand, analogs of line singularities in magnetohydrodynamics are "frozen-in" magnetic field lines forming intricate configurations. I only confine myself to the mathematical theory of linking effects. The examples from condensed-matter physics are only illustrative. It seems nevertheless that the study of links may be of considerable interest in a number of physical problems which I touch upon at the end of the section.

The problem of line defects interaction can be regarded from different points of view. I discuss two different but related approaches. One was offered by Poénaru and Toulouse in [PT1, 2] and provides that a defect shall move through another. The second approach was considered later [MRe] by Retakh and the author, where the interaction of a linked defect ensemble was studied by knot theory methods. I begin our treatment with the latter work.

5.1.6.1 Linked linear defects

Examples of linked defects in cholesterics have already been discussed in Subsec. 5.1.3.2. From the topological viewpoint, this is the classical theory of links (e.g., see [Ro]) made more complicated by introducing the order parameter, which characterizes the corresponding thermodynamic phase.

Combining the refined apparatus of the theory of links with restrictions imposed by the topology of order parameter space, we succeed in obtaining a number of no-go theorems for links in terms of differential forms.

To construct the theory, we need precise mathematical analogs of the physical concept of a linear defect. In the sequel, we omit the adjective, since only such defects are considered.

Essential defect. A *link* l in S^3 or R^3 is a collection of closed non-self-intersecting curves $l_1, \ldots, l_n$. We fix a base-point $*$ outside l. For $1 \leqslant i \leqslant n$, a loop m_i without self-intersections emanating from $*$ and transversally circumnavigating l_i in the complement to the set of other link curves is called the ith *meridian* m_i of the link

l. This means that l_i crosses the film with boundary m_i at only one point and any other l_j does not cross it. It is obvious that the loop can be selected to be arbitrarily narrow.

Let Φ be a continuous map* of a manifold M into order parameter space V defined outside a set Σ called a *defect.*

Definition 5.5 We call a defect Σ *essential* if Φ cannot be extended continuously to any point in Σ.

We now assume that $M = S^3$ and that $\Sigma = l = (l_1, \ldots, l_n)$ is a link.

Definition 5.6 We call l an *m-essential* defect of Φ if $\Phi(m_i)$ is non-contractible in V for all $i \leqslant n$.

Note that the essentiality of a defect follows from its m-essentiality. In fact, assume that $\Sigma = l$ is not an essential defect. Then there is a point in one of the curves l_j of l, into whose neighborhood Φ is continuously extended. Pick a meridian m_j passing through the neighborhood so that the extension of Φ might be defined on the film spanning m_j. Hence, $\Phi(m_j)$ is contractible and Σ is not m-essential. We have thereby shown that m-essentiality entails essentiality.

On the other hand, in cases important from the physical standpoint, for $M = S^3$ and $\Sigma = l$, an essential defect is m-essential. For an exact formulation, we need the two definitions below.

Definition 5.7 Consider a mapping $\Phi\colon M \setminus \Sigma \to V$. A *deformation* Φ_t is the family of homotopic maps $\Phi_t\colon M \setminus \Sigma \to V$, $t \in [0, 1]$, $\Phi_0 = \Phi$.

Let Φ coincide with Φ_1 outside a small neighborhood of Σ.

Definition 5.8 A defect Σ of a map Φ is said to be *stably essential* if Σ is essential for any map Φ_t.

Theorem 5.4 *Let $M = S^3$ and a link $l = \Sigma$ be a stably essential defect of a map $\Phi\colon S^3 \setminus \Sigma \to V$. Then Σ is m-essential.*

Proof. Assume that Σ is not m-essential, i.e., there exists a curve l_j in the link l, for which $\Phi(m_j)$ is contractible. Hence, the restriction of Φ to m_j is extended to a map φ onto a film σ with boundary m_j.

It is easy to construct a map Φ_1 of the union of the film with $S^3 \setminus \Sigma$, which would extend φ and be homotopic to Φ on $S^3 \setminus \Sigma$. Picking m_j to be sufficiently narrow, we can make the restriction of Φ_1 to $S^3 \setminus \Sigma$ to be the result of the deformation Φ. Since l_j crosses σ, Σ is not essential for Φ_1, contrary to the stability of Σ, a contradiction, which proves the theorem.

The argument used in the proof is also convenient in proving other properties of the defects formed by links l of the curves $l_1, \ldots, l_n$. We recall that the ith parallel of the link is the loop $\bar{l}_i$ obtained as follows: We draw a path γ_i from a base point $*$ reaching a small tubular neighborhood of l_i evading the other curves and then bypass along l_i in the neighborhood, returning to $*$ along γ_i. Further, we assume that the parallel l_i is contractible in $S^3 \setminus l_i$.

Definition 5.9 Let $\Phi\colon S^3 \setminus l \to V$. We call a defect $\Sigma = l$ *strongly essential* for Φ if Σ remains essential for any mapping $\Phi'\colon S^3 \setminus l \to V$ homotopic to Φ.

* We mean only continuous mappings from now on.

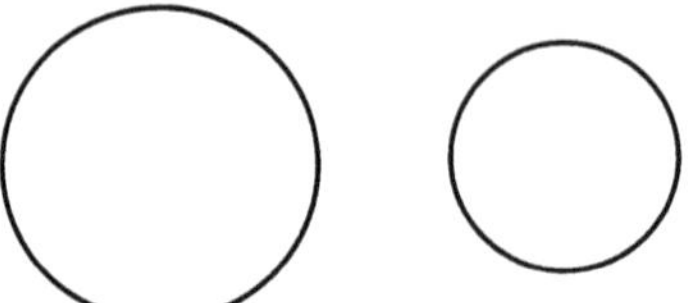

Fig. 14. Trivial link of two circles

Fig. 15. Borromean rings

Fig. 16. Whitehead link

Theorem 5.5 *If $\Sigma = l$ is a strongly essential defect of Φ and a curve l_i of a link l not separable by the homeomorphic image of the two-dimensional sphere from the other curves, then $\Phi(\bar{l}_i)$ is not contractible in V.*

Proof. Assume that $\Phi(\bar{l}_i)$ is contractible in V. The restriction of Φ to $\bar{l}_i$ can be extended to a map ψ of the film σ spanning $\bar{l}_i$. We homotopize Φ into a map Φ' coinciding with ψ on $\sigma \cap (S^3 \setminus \Sigma)$. If l_i is not separable from other curves, then at least one l_j with $j \neq i$ crosses σ. Hence, Σ is not an essential defect of Φ' and by definition not strongly essential of Φ. ■

Examples. Let Φ: $S^3 \setminus l \to V$ be a map into the order parameter space with the defect Σ formed by the link l. We consider four examples.

Example 5.3 (Fig. 13) The simplest is formed by two circles with a Gauss linking number of unity. It is obvious that a parallel of one circle is homotopic to a meridian of the other. The noncontractibility of images of parallels in V follows from the m-essentiality of the defect.

Example 5.4 (Fig. 14) The link l is formed by two unlinked circles, the parallels are contractible in $S^3 \setminus l$, therefore, their images are contractible irrespective of the essentiality of the defect.

Example 5.5 (Fig. 15) A link l is formed by Borromean rings, links of three circles, decomposing if one is cut. A parallel of one circle, say, $\bar{l}_3$, is here homotopic to the commutator $[m_1, m_2] = m_1 m_2 m_1^{-1} m_2^{-1}$ of the meridians of the others. If $\pi_1(V)$ is commutative, then the images of the parallels are obviously contractible in V. However, if $\pi_1(V)$ is not, then $\Phi(\bar{l}_i)$ are homotopic to elements of the commutator, e.g., for a biaxial nematic or cholesteric, $\pi_1(V) = Q$ (Q is the group of eight elements: 1, J, $\pm I_i$, $i = 1, 2, 3$, see p. 257). The commutator Q coincides with the center consisting of 1 and J. Thus, if the defect is formed by Borromean rings and the images of all the parallels and meridians are not contractible in V, then the images of meridians m_1, m_2, and m_3 are homotopic up to relabelling $\pm e_i$, $i = 1$, 2, 3, and $\Phi(\bar{l}_i)$ is homotopic to J for all i.

The Gauss linking numbers of any two Borromean rings vanish. However, the topological properties of a link are distinctly different from those of the ensemble of three unlinked circles.

Example 5.6 (Fig. 16) We consider the familiar Whitehead link of two curves l_1 and l_2 as l. The Gauss number is zero in this case, too. The topological properties of l also differ distinctly from those of the trivial link from Example 5.3. For the Whitehead link, the parallel $\bar{l}_2$ is homotopic to the product of the commutators $[m_1^{-1}, m_2] \circ [m_1, m_2]$ of two meridians m_1 and m_2. Thus, $\Phi(\bar{l}_2)$ is contractible irrespective of the kind of essentiality of the defect for the commutative group $\pi_1(V)$. However, $\Phi(\bar{l}_2)$ is contractible both for a biaxial nematic and a cholesteric.

It follows from the above examples that the topological properties of Φ essentially depend on the structure of its defect which is not determined only by the Gauss linking number. *Higher-order linking numbers* are defined in subsequent sections, enabling us, in particular, to distinguish Borromean rings of a Whitehead link from trivial links.

Various homotopy classes of parallels in terms of those of meridians or of links have been written out in [Ro] and [Por].

5.1.6.2 General theory of links

A. Homotopic and isotopic unlinking. Let $l = (l_1, \ldots, l_n)$ be a link of n curves in S^3. A curve l_i is *homotopically unlinked* from other components if there exists a homotopy h_t of l_i into a point such that the images $h_t(l_i)$ and l_j are disjoint for all t and $j \neq i$.

The link l is homotopically trivial if there exists the same homotopy h_t of each curve in the link to a point such that $h_t(l_i)$ are disjoint for any t and i.

A curve l_i is *isotopically unlinked* from other components of the link if it is separable from them by the homeomorphic image of the two-dimensional sphere S^2. A link l is *isotopically trivial* if each curve is isotopically unlinked from the remaining ones.

From the physical point of view, the disentanglement of a link means its isotopic unlinking. In homotopic unlinking, the image can be a curve with self-intersections and the transition of the curve through a self-intersection point "requires energy". Homotopic unlinking can be regarded as coarser than isotopic unlinking.

Example 5.7 Borromean rings are homotopically unlinked even though each two of its components may be unlinked isotopically.

Example 5.8 The Whitehead link is isotopically unlinked, though homotopically linked. It suffices to "rectify" the first curve without fearing self-intersections (see the details in [Ro], p. 68).

Note that it is possible to have a case where the curve l_1 is homotopically linked with a curve l_2, but l_2 is homotopically unlinked from l_1 ([Ro] p. 60).

The simplest invariant of a link of two oriented curves l_1 and l_2 in S^3 is the Gaussian linking number $k(l_1, l_2)$ determined by the Gaussian integral. For the simplest link (see Fig. 13), $k(l_1, l_2) = 1$, while for the Whitehead link (Fig. 16), $k(l_1, l_2) = 0$. The latter equality is necessary and sufficient for the homotopic unlinking of two curves. Various higher-order linking numbers enable us to solve the problem of the isotopic and homotopic unlinking of l which consists of n curves and are given in point C and Subsec. 5.1.6.4. The definitions are based on generalization of different definitions of the Gaussian linking number and involve the properties of the complement of l in S^3.

B. First-order Milnor numbers. Let $l = (l_1, l_2, l_3)$ be a link in S^3. We once and for all select a base point $*$ in the complement to l. We clarify in which sense the homotopy classes of the meridians $m_1, \ldots, m_n$ are generators of the group $\pi_1(S^3 \setminus l)$. For definiteness, we henceforward shall assume that each curve l is oriented. The orientation of l_i determines that of its parallel and meridian, the parallel inheriting the curve's orientation and the meridian is oriented so that $k(l_i, m_i) = 1$.

Consider the *lower central series* of subgroups $G_1 \supset G_2 \supset \ldots \supset G_n$ of an arbitrary group G, where the subgroups are defined thus: $G_1 = G$ and G_n is the subgroup of G generated by products of the form $xyx^{-1}y^{-1}$, where $x \in G$ and $y \in G_{n-1}$. The minimal subscript s for which G_s consists of only the unit element shows the measure of noncommutativity of G.

If such a value exists, then the central series is said to *terminate*. In particular, if G is commutative, the s is two. If G is the fundamental group of order parameter space for a biaxial nematic, then s is 3. Note that all G_i are normal subgroups of G.

For $G = \pi_1(S^3 \setminus l)$, the homotopy classes of meridians of a link l are generators of the quotient group G/G_s for $s \geqslant 1$. Strictly speaking, Milnor's theorem is valid [Mil1].

Theorem 5.6 *For any q there exists a map α of a free group F of noncommuting variables $x_1, \ldots, x_n$ on $G = \pi_1(S^3 \setminus l)$ so that*

(1) *$\alpha(x_i)$ equals the homotopy class of the meridian m_i,*

(2) *α induces the isomorphism between F/F_q and G/G_q, and*

(3) *if $\alpha(y_i)$ equals the homotopy class of a parallel $\bar{l}_i$, then $[x_i, y_i] = 1$.*

We now define the Milnor numbers. Let $l = (l_1, \ldots, l_n)$ be a link in S^3. In formulating Milnor's theorem, we pick $q = n$. We denote the sum of degrees with which a generator x_i of F is in the decomposition of the element y_i for $i \neq j$ by $\mu(i, j)$, e.g., for Borromean rings, $\mu(1, 3) = \mu(2, 3) = 1 - 1 = 0$. The values $\mu(i, j)$ are called *Milnor numbers of first order of l.* They do not depend on the choice of either the parallels or the meridians.

Milnor has shown that $\mu(i, j) = k(l_i, l_j)$ [Mil1]. The reader should verify this by considering some examples. It was shown in the same paper that for links of two curves l_1 and l_2, the equality $\mu(1, 2) = 0$ is necessary and sufficient for their homotopic unlinking. The Whitehead link shows the insufficiency of the equality for isotopic unlinking.

C. Higher-order Milnor numbers. Let $l = (l_1, \ldots, l_n)$ be a link in S^3. First-order Milnor numbers introduced above are determined by the expression of homotopy classes of parallels in terms of those of meridians. However, the numbers are too coarse a characteristic. In fact, $\mu(1, 2) = 0$ for $n = 2$ and for the trivial link. The same holds for a not isotopically unlinked Whitehead link. For $n = 3$, the linking numbers for any two Borromean rings are zero. However, the rings are not even homotopically unlinked. To distinguish between the above links, Milnor introduced higher-order linking numbers [Mill, 3].

With the notation of the preceding subsection, we construct a homomorphism of a free group F with generators $x_1, \ldots, x_n$ into the multiplicative group of formal power series with integer coefficients in noncommuting variables $X_1, \ldots, X_n$ and the absolute term equal to unity. We set $\theta(x_i) = 1 + X_i$. It is obvious that $\theta(x_i^{-1}) = 1 - X_i + X_i^2 - \ldots$ for $1 \leqslant i \leqslant n$. We now define the Milnor number of order $p - 1$, $p \geqslant 2$. Let us consider a collection of subscripts $1 \leqslant i_1, \ldots, i_p \leqslant n$. Let y_{i_p} be the element sent into the homotopy class of a parallel $\bar{l}_{i_p}$. We denote the coefficient of the monomial $X_{i_1}X_{i_2} \ldots X_{i_p}$ in the formal series of $\theta(y_{i_p})$ by $\mu(i_1, \ldots, i_p)$, which is the $(p - 1)$ Milnor number. It is obvious that the two methods of defining $\mu(i_1, i_2)$ here and above coincide.

We list the Milnor numbers of the second and third order for the above examples. For Borromean rings, $\mu(1, 2, 3) = 1$ and $\mu(2, 1, 3) = -1$. For the Whitehead link, $\mu(1, 2, 2, 1) = \mu(1, 1, 2, 2) = 1$, $\mu(1, 2, 1, 2) = -1$.

Milnor has shown that numbers $\mu(i_1, \ldots, i_p)$ are determined uniquely modulo $\mu(j_1, \ldots, j_q)$, where $j_1, \ldots, j_q$ is an arbitrary tuple of q numbers from $i_1, \ldots, i_p$ for $q < p$ [Mill, 3]. In particular, if $\mu(j_1, \ldots, j_q)$ for all such collections, then $\mu(i_1, \ldots, i_{q+1})$ is determined uniquely. In the general case, the image of $\mu(i_1, \ldots, i_p)$ in the group Z/Z_μ is considered, where μ is the greatest common divisor of $\mu(j_1, \ldots, j_q)$ $(q \leqslant p - 1)$. The image is denoted by $\bar{\mu}(i_1, \ldots, i_p)$ and is called the *Milnor number* of order $p - 1$ for the link l.

Milnor numbers $\bar{\mu}(i_1, \ldots, i_p)$ are linking homotopy invariants of l if there are no equals among the numbers $i_1, \ldots, i_p$. Otherwise, coefficients $\bar{\mu}(i_1, \ldots, i_p)$ are only isotopy invariants. Moreover, a link $l = (l_1, \ldots, l_n)$ is homotopically unlinked if and only if all $\mu(i_1, \ldots, i_p) = 0$ for $p \leqslant n$ and there are no equal numbers in $i_1, \ldots, i_p$. Thus, for homotopic disentanglement of l, it suffices to verify that only finitely many numerical invariants vanish.

But isotopic disentanglement is different. It is obvious that if l is unlinked, then $\mu(i_1, \ldots, i_p) = 0$ for any p-tuple. However, the converse may not hold [Mil3].

Note also that Milnor numbers are related to each other. The simplest relation is of the form

$$\bar{\mu}(i_1, \ldots, i_p) = (-1)^p \bar{\mu}(i_p, \ldots, i_2, i_1).$$

Others are given in [Mill, 3].

Note that for a link of two curves the zero Milnor number of first order entails that of second order.

5.1.6.3 Homotopy properties of maps to order parameter space

In this subsection, we everywhere consider a map Φ: $S^3 \setminus \Sigma$ with defect Σ equal to a link $l = (l_1, \ldots, l_n)$ and study the behavior of the homotopy classes of the images of parallels $\Phi(\bar{l}_i)$ of l in $\pi_1(V)$ depending on the properties of the images of meridians $\Phi(m_i)$ and the Milnor numbers for l. These properties are considered in more detail as the class of spaces V is restricted. The base point in V is taken to be $\Phi(*)$.

(A) The lower central series of the group $\pi_1(V)$ terminates. All known order parameter spaces satisfy the condition and we have the following theorem.

Theorem 5.7 (a) *The homotopy classes of the images of a meridian m_i and parallel $\bar{l}_i$ in $\pi_1(V)$ commute for any i.*

(b) *If the images of all meridians are contractible in V, then those of all parallels are also contractible.*

(c) *If all Milnor numbers for a link l vanish, then the images of all parallels in l are contractible on V.*

Proof. The condition (a) immediately follows from Theorem 5.6. To prove (b), we see that if a parallel $\bar{l}_i$ is contractible in the complement to l, then $\Phi(\bar{l}_i)$ is contractible in V. If $\bar{l}_i$ is not contractible in such a complement, then, by Milnor's theorem mentioned above, we can show that the homotopy class of $\Phi(\bar{l}_i)$ is expressed in terms of those of the images of meridians in $\pi_1(V)$. It is obvious that $\Phi(\bar{l}_i)$ is then contractible.

The statement (c) is proved on the basis of Subsec. 5.1.6.2.

In the next subsection, we show to which restrictions on the linking numbers the calculations of $\pi_1(V)$ for different systems lead. Consider the following examples.

Example 5.9 (1) $\pi_1(V)$ is a commutative group. If $\pi_1(V) = Z$, then the corresponding system is superfluid ^{4}He and the links are formed from Abrikosov vortices, $\pi_1(V) = Z + Z$. A physical example is the A-phase of ^{3}He if the magnetic field and dipole energy F_{dip} are taken into account [VM] (see Sec. 5.2 for the properties of ^{3}He).

(2) $\pi_1(V) = Z_2$, the physical system is a nematic, $V = RP^2$. The superfluid A-phase of ^{3}He has the same fundamental group if spin-orbit interaction of F_{dip} is taken into account. Here, $V \simeq SO(3)$.

(3) $\pi_1(V) = Z_4$. The corresponding system is the A-phase of ^{3}He, $V = S^2 \times SO(3)/Z_2$.

(4) $\pi_1(V) = Q$. The corresponding systems are a biaxial nematic and a cholesteric, $V \simeq SO(3)/D_2$, where D_2 is a two-point group.

(B) The group $\pi_1(V)$ is commutative.

Theorem 5.8 *Let i be fixed, $1 \leqslant i \leqslant n$. If $\mu(i, j) = k(l_i, l_j) = 0$ for $1 \leqslant j \leqslant n$, $j \neq i$, then the image of the parallel $\bar{l}_i$ is contractible in V.*

Proof. Let $G = \pi_1(S^3 \setminus l)$. For any i, we consider the representation of the homotopy class of a parallel $\bar{l}_i$ as the product Π of the homotopy classes of meridians in G/G_n. We can make the sum of degrees with which a homotopy class m_i is in Π zero by a convenient choice of $\bar{l}_i$. The same holds by the theorem for the remaining meridians. Since $\pi_1(V)$ is commutative, the group G_n is sent under the map of homotopy groups induced by Φ into the unit element and $\Phi(\bar{l}_i)$ is homotopic to the product of meridian's images. Since the image of each meridian is involved in the product with the degree sum zero and $\pi_1(V)$ is commutative, $\Phi(\bar{l}_i)$ is contractible in V for any i.

(C) The group $\pi_1(V) = Z_2$. Theorem 5.8 then admits the following refinement.

Theorem 5.9 *Let i be fixed, $1 \leqslant i \leqslant n$. Let a defect Σ be m-essential. A curve $\Phi(\bar{l}_i)$ is contractible in V if and only if*

$$\sum_{j=1}^{i-1} \mu(i, j) + \sum_{j=i+1}^{n} \mu(i, j) \equiv 0 \pmod 2.$$

Proof. By reasoning similar to that in Theorem 5.8, we can show that $\Phi(\bar{l}_i)$ is homotopic to the product

$$\Phi(m_1)^{\mu(i,1)} \ldots \Phi(m_n)^{\mu(i,n)},$$

assuming $\mu(i, i) = 0$.

Let t be a noncontractible loop in V, emanating from the point $\Phi(*)$ and $s = \sum_{j=1}^{n} \mu(i, j)$. It follows from the m-essentiality of the defect l that $\Phi(m_i)$ is homotopic to t for all i. Therefore, $\Phi(\bar{l}_i)$ is homotopic to t^s and the theorem follows.

Example 5.10 For two curves l_1 and l_2, Theorem 5.9 admits the following illustration. If $\Phi(\bar{l}_1)$ is contractible in V, then the linking number $k(l_1, l_2)$ is even. In particular, it follows that if due to the energy conditions $k(l_1, l_2) \leqslant 1$ (and only such links were observed in the Bouligand experiments [Boul]), then $k(l_1, l_2) = 0$ and the curves are homotopically unlinked.

(D) The group $\pi_1(V) = Z_4$. The analog of Theorem 5.9 is then formulated as follows.

Theorem 5.10 *Let t be a loop in V starting at a base point $\Phi(*)$ and serving as a generator of $\pi_1(V)$ and $\Phi(m_j)$ be homotopic to a curve $t^{s(j)}$ for each j, where $0 \leqslant s(j) \leqslant 3$. We put*

$$\mu'(i, j) = s(j)\mu(i, j).$$

Then $\Phi(\bar{l}_i)$ is contractible in V if and only if

$$\Sigma(i, j) \equiv 0 \pmod 4.$$

The *proof* is similar to that of Theorem 5.9.

(E) The case of a biaxial nematic and of a cholesteric, $\pi_1(V) = Q$. Let $\Phi: S^3 \setminus \Sigma \to V$ and $\Sigma = l$. It follows from Theorem 5.7 that the images of a parallel and a meridian of any of the curves l_i homotopically commute in V. Taking into account that $e_i e_j = -e_j e_i$, we obtain that if $\Phi(m_i)$ is homotopic to $\pm e_k$, then $\Phi(\bar{l}_i)$ can be homotopic to only one to the elements $\pm e_k$, J, 1. An arbitrary element of $\Phi(\bar{l}_i)$ is either homotopic to 1, I or to the product $\pm \Pi \Phi(l_j)^{\mu(i,j)}$. Hence, the following is obvious.

Theorem 5.11 *Let $l = (l_1, l_2)$ and a defect $\Sigma = l$ be m-essential. Then $\Phi(\bar{l}_2)$ is contractible if and only if $\mu(1, 2)$ is divisible by four when $\Phi(m_1)$ is homotopic to $\pm e_i$ $(i = 1, 2, 3)$ and if and only if $\mu(1, 2)$ is even when $\Phi(m_1)$ is homotopic to J.*

For a link of three curves, a similar statement is below.

Theorem 5.12 *Let $\Sigma = l = (l_1, l_2, l_3)$ and let $\Phi(m_1)$ and $\Phi(m_2)$ not commute homotopically in V. Then $\Phi(l_3)$ is contractible if and only if $\mu(1, 3) + \mu(2, 3)$ is divisible by four.*

5.1.6.4 Differential forms and higher-order linking numbers

Here, we define cohomology obstructions of different orders to unlinking $l = (l_1, \ldots, l_n)$ in S^3 via differential forms. Poénaru's paper [Po2] contains the remark that in order to distinguish between trivial and Whitehead links we can apply the Sullivan construction* which is also based on the study of differential forms on space $S^3 \setminus l$ [Su]. The Sullivan construction requires a special graded differential algebra. Our method is based on a straightforward calculation of differential forms on $S^3 \setminus l$ and is simple and natural in physical applications.

The cohomology obstructions defined here carry more information than the Alexander polynomials normally used by physicists [Ro], e.g., for a link in Fig. 17 the Alexander polynomial is identically equal to unity, but the tangle is homotopically unlinked. The Stallings conjecture [St] proved in [Tu] and [Por] relates the above cohomology obstructions with higher-order Milnor numbers.

Let $l = (l_1, l_2)$. We span the curve l_2 with a film.

The intersection number for l_1 and the film equals the Gauss linking number $k(l_1, l_2)$, which can be calculated via differential forms.

Recall that a curve l_i in S^3 can be associated with the differential Alexander-dual 1-form u_i (by Theorem 2.14), which is defined in the complement to l_i, closed and characterized by $\int_C u_i = k(c, l_i)$ for any closed curve from the complement. The cohomology class of u_i is determined uniquely.

Now, let B_i ($i = 1, 2$) be the boundary of a tubular neighborhood of l_i not meeting another curve. Then

$$\int_{B_1} u_1 \wedge u_2 = - \int_{B_2} u_1 \wedge u_2 = k(l_1, l_2). \tag{5.26}$$

Fig. 17. The linking of four curves

* Novikov has constructed a theory of multivalued functionals on its basis [No2].

To determine $k(l_1, l_2)$ by density integration, we need a cohomology with compact support. We associate l_i with the class of cohomology with compact supports, Poincaré dual, and defined by a closed 2-form v_i with support outside l_i. The class is determined by the equality

$$\int_Z v_i = \operatorname{ind}(z, l_i).$$

Here ind (Z, l_i) is the intersection number for Z and l_i, where Z is an arbitrary 2-cycle in the complement to l_i.

3-forms $u_1 \wedge v_2$ and $v_1 \wedge u_2$ are defined on the whole sphere and

$$\int_{S^3} u_1 \wedge v_2 = \int_{S^3} v_1 \wedge u_2 = k(l_1, l_2).$$

The tuple of $k(l_i, l_j)$ is one of numerical characteristics of the link $k = (l_1, \ldots, l_n)$. It is also natural to introduce the linking number for the whole of l by the formula

$$\bar{k}(l) = \max_{1 \leqslant i \leqslant j \leqslant n} |k(l_i, l_j)|. \tag{5.27}$$

If l is isotopically unlinked, then $\bar{k}(l) = 0$. However, for the links in Figs. 13, 15, 16, and 17, we should introduce higher-order linking numbers.

Note that if $\bar{k}(l) = 0$, where $l = (l_1, l_2)$, then there exists a 1-form u_{12} on the complement to l and 2-forms v_{12} and v'_{12} with compact support on S^3 such that

$$du_{12} = u_1 \wedge u_2, \quad dv_{12} = v_1 \wedge u_2, \quad dv'_{12} = u_1 \wedge v_2.$$

Assume that $\bar{k}(l) = 0$ for $l = (l_1, l_2, l_3)$. In addition to the above, we define 1-forms u_3 and u_{23} and 2-forms with compact support v_3, v_{23}, v'_{23} and verify by differentiation that the 2-form $\tilde{u}_{123} = u_{12} \wedge u_3 + u_1 \wedge u_{23}$ is closed. The form is defined on the complement to (l_1, l_2, l_3). We also have to check by differentiation that the 3-form

$$\tilde{v}_{123} = v_{12} \wedge u_3 + v_1 \wedge u_{23}, \quad \tilde{u}'_{123} = u_{12} \wedge v_3 + u_1 \wedge v_{23} \tag{5.28}$$

is closed. v_{12} and v_{23} can be picked so that the latter 3-forms are defined on the whole space. The cohomology classes in $H^2(S^3 \setminus l)$ and $H^3(S^3)$, determined by $\tilde{u}_{123}$, $\tilde{v}_{123}$, and v'_{123}, are called the *Massey products* [Masl], [Kr] of cohomology classes $u_1, u_2, u_3, v_1, u_2, u_3$, and u_1, u_2, v_3 denoted by $\langle clu_1, clu_2, clu_3 \rangle, \ldots$, respectively. They do not depend on the choice of u_i, v_j in clu_i, clv_j.

The integrals

$$\int_{B_1} \tilde{u}_{123} = -\int_{B_3} \tilde{u}_{123} = \int_{S^3} \tilde{v}_{123} = \int_{S^3} \tilde{v}'_{123} = k_2(l)$$

have integer values and do not depend on the choice of u_{12}, u_{23}, v_{12}, and v_{23}.

Let us consider sublinks of three curves $l_{ijk} = (l_i, l_j, l_k)$ for $1 \leqslant i < j < k \leqslant n$ of a link $l = (l_1, \ldots, l_n)$ with $n \geqslant 3$. We introduce the linking number

$$\bar{k}_2(l) = \max_{1 \leqslant i < j < k \leqslant n} |k_2(l_{ijk})| \tag{5.29}$$

for l.

For Borromean rings, $k_2(l) = 1$. If l is homotopically unlinked, then the number is zero.

For the link in Fig. 17, $\bar{k}_2(l) = 0$, but the tangle is homotopically unlinked. To characterize such links, we define a linking number of order three.

Let $l = (l_1, l_2, l_3, l_4)$ and $\bar{k}_2(l) = 0$. Then there exist 1-forms u_{123} and u_{234} in the complement to l_1, l_2, l_3 and l_2, l_3, l_4, respectively, 2-forms v_{123} and v_{234} whose compact supports do not intersect with l_1 and l_4, respectively, and such that

$$du_{123} = \tilde{u}_{123}, \quad du_{234} = \tilde{u}_{234}, \quad dv_{123} = \tilde{v}_{123}, \quad dv_{234} = \tilde{v}_{234}.$$

Hence, the 2-form $\tilde{u}_{1234} = u_1 \wedge u_{234} + u_{12} \wedge u_{34} + u_{123} \wedge u_4$ and the 3-forms $\tilde{v}_{1234} = u_1 \wedge v_{234} + u_{12} \wedge v_{34} + u_{123} \wedge v_4$ and $\tilde{v}'_{1234} = v_1 \wedge u_{234} + v_{12} \wedge u_{34} + v_{123} \wedge u_4$ which can be defined everywhere with a convenient choice of v_{123} and v_{234} are closed. Their cohomology classes in $H^2(S^3 \setminus l)$ and $H^3(S^3)$ are called the Massey products of the corresponding 4-tuples (u_1, u_2, u_3, u_4), (u_1, u_2, u_3, v_4), (v_1, u_2, u_3, u_4) and denoted by $(clu_1, clu_2, clu_3, clu_4), \ldots$, respectively. They do not depend on the choice of forms u_i with clu_i and v_j with clv_j.

We put

$$\int_{B_1} \tilde{u}_{1234} = \int_{B_4} \tilde{u}_{1234} = \int_{S^3} \tilde{v}_{1234} = \int_{S^3} \tilde{v}'_{1234} = k_3(l). \tag{5.30}$$

This is an integer not depending on the choice of $u_{12}, u_{23}, u_{123}, \ldots$.

For the four curves in Fig. 17, $\bar{k}_3(l) = 1$. Similarly, the linking number of order three can be defined for a collection of n curves. However, we first make two remarks.

Remark 5.1 The value $k_3(l)$ for $l = (l_1, \ldots, l_4)$ is unaltered under a cyclic permutation of the curves, but can change if the permutation is different, which can be seen if we use the results in [O'].

Remark 5.2 The above constructions enable us to find the third-order numbers for a pair of curves, e.g., in a Whitehead link. Geometrically, it is as if we had several copies of the same curve obtained by small translations. Algebraically, e.g., we assume for a link $l_{1222} = (l_1, l_2, l_2, l_2)$ that $u_2 = u_3 = u_4$ (and, therefore, $u_2 \wedge u_3 = u_3 \wedge u_4 = u_3 \wedge u_4 = 0$). The Whitehead link $l = (l_1, l_2)$ can also be represented as $l_{1212} = (l_1, l_2, l_1, l_2)$, $u_3 = u_1$, $u_4 = u_2$, in which cases

$$k_3(l_{1222}) = 1, \quad k_3(l_{12}l_{12}) = -2.$$

The above enables us to determine the coefficients $k(i_1, \ldots, i_p)$ of order $p - 1$ using a collection of curves $l_{i_1}, \ldots, l_{i_p}$, possibly repeating. With this notation,

$k(l) = k(1, \ldots, n)$. The common linking number of order $p - 1$ can be determined by the formula

$$\bar{k}_{p-1}(l) = \max_{1 \leqslant i_1, \ldots, i_p \leqslant n} |k(i_1, \ldots, i_p)|. \tag{5.31}$$

To find $k(i_1, \ldots, i_p)$, we introduce Massey products of any order [Masl], [Kr].

Let $\omega_1, \ldots, \omega_s$ be s closed differential forms on a manifold M^n, of orders $\alpha_1, \ldots, \alpha_s$, respectively. We will say that the Massey product $\langle cl\omega_1, \ldots, cl\omega_s \rangle$ is defined on their cohomology classes if there exist differential forms ω_{ij} for $0 \leqslant j - i < s - 1$ such that

(1) $\omega_{ii} = \omega_i$,

(2) $d\omega_{ij} = \sum_{k=i}^{k=j-1} \tilde{\omega}_{ik}\omega_{k+1j}, \ \tilde{\omega}_{ik} = (-1)^{\deg \omega_{ik+1}}\omega_{ik}$.

Then the form $\sum_{k=1}^{k=s-1} \omega_{1k}\omega_{k+1s}$ is closed and its cohomology class is called the *value* of $\langle cl\omega_1, \ldots, cl\omega_s \rangle$. For simplicity, we will write $\langle \omega_1, \ldots, \omega_s \rangle$ instead. Generally speaking, a Massey product may have several values. However, all depend only on the cohomology classes of ω_i.

We can consider the closed forms $\omega_1, \ldots, \omega_n$ with compact supports. Then the forms ω_{ij} should be picked with compact supports, too, and the Massey product's values are in the classes of cohomology with compact supports (or in the usual cohomologies for compact manifolds). We call ω_{ij} $(i \neq j)$ *intermediate.*

Here, the Massey product is encountered in two situations. Let $l = (l_1, \ldots, l_n)$ be a link in S^3. In the first situation, the operation $\langle \omega_{i_1}, \ldots, \omega_{i_p} \rangle$ is considered on $S^3 \setminus l$, where ω_{i_p} is the Alexander-dual form u_{i_p} of l_{i_p}, $1 \leqslant i_p \leqslant n$.

In the second situation, $M = S^3$, $\omega_{i_1} = v_{i_1}$, where v_{i_1} is a 2-form with compact support and is Poincaré-dual for the curve l_{i_1}, while the ω_{i_p} are obtained for $2 \leqslant i_p \leqslant n$ from the forms u_{i_p} by making the latter vanish in a small neighborhood of l_i (with a certain liberty, we preserve the same notation u_{i_p} for such forms)*. Similarly, we construct the collection of ω_{i_p} from $u_{i_1}, \ldots, v_{i_p}$.

Theorem 5.13 (a) *Let* $i_1 = 1, \ldots, i_p = n$ *and one of the Massey products* $\langle u_{i_1}, \ldots, u_{i_p} \rangle$, $\langle v_1, \ldots, u_{i_p} \rangle, \ldots, \langle u_{i_1}, \ldots, v_{i_p} \rangle$ *be defined. Then the other two Massey products are defined and*

$$\int_{B_{i_1}} \langle u_{i_1}, \ldots, u_{i_p} \rangle = (-1)^p \int_{B_{i_p}} \langle u_{i_1}, \ldots, u_{i_p} \rangle = \int_{S^3} \langle v_1, \ldots, u_{i_p} \rangle$$

$$= \int_{S^3} \langle u_{i_1}, \ldots, v_{i_p} \rangle = k(i_1, \ldots, i_p),$$

the obtained number being an integer and not depending on the choice of intermediate forms.

* $\omega_{i_p} = \varphi_{i_p} u_{i_p}$, where φ_{i_p} is a smooth function on S^3 and $\varphi_{i_p} = 0$ in a small neighborhood of l_{i_p}.

(b) We call the value obtained in (a) the *linking number* $k(i_1, \ldots, i_p)$ *of order* $p-1$. *Linking numbers of order* q *are defined if and only if all the numbers of smaller orders are zero.*

It follows from the Stallings conjecture, which was proved by Turayev [Tu] and Porter [Por], that $k(i_1, \ldots, i_p) = \mu(i_1, \ldots, i_p)$ if all the linking numbers of smaller orders are zero. Actually, $\langle u_1, \ldots, u_n \rangle$ is a multiple of the cohomology class dual to the path in general position in $S^3 \setminus l$ joining l_1 to l_n.

In reality, a more general statement [Tu], [Por] is true, viz., without having to assume that the linking numbers of smaller orders vanish.

Here, we offer a topological approach to the problem of distinguishing between nontrivial links. The physical applications were basically illustrative and touched upon the restrictions placed on the kinds of links of defects in condensed matter. Nevertheless, we hope that the methods may be useful for other physical problems, e.g., in the study of the statistics of polymer chains or turbulence in superfluid liquids. The effect of linked vortex filaments on turbulence in superfluid ^{4}He was considered in [Schw]. Another example is given by magnetohydrodynamics where the formulas for linking numbers, listed above, play the role of topological conservation laws for frozen-in magnetic field lines of force [MSa].

5.1.6.5 Crossing of defects

Poénaru and Toulouse considered the "crossing" of two defects, or more precisely the possibility that one defect can "move" through another [PT1]. Taking into account that from the physical viewpoint defects are flexible filaments, the problem posed in this way should be refined. The authors suggest the *gedanken* experiment for defect crossing (Fig. 18). It is meanwhile assumed that a defect l_1 moves through

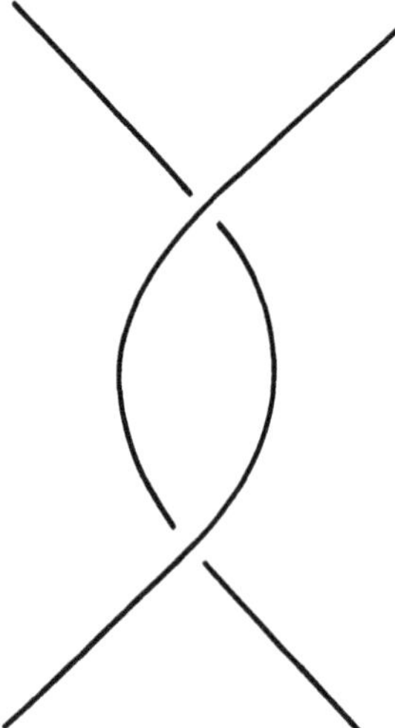

Fig. 18. Crossing of defects

a defect l_2 if in the restructuring process a pair l_2, l_1 unrelated to another defect l_3 is formed. Before passing to exact formulations, we note that the prime objective is the adequate mathematical interpretation of physical phenomena connected to a possible restructuring of defects whose deformations do not require additional energy loss. For instance, given that defects are movable, it is clear that the process of defect linking is energetically more favorable than filament breakdown. It is obvious that if the filaments can be broken, the problem becomes meaningless. Another natural constraint is imposed by filament length: short filaments can simply evade each other.

As will be seen below, defect "crossing" is equivalent to the more comprehensible process of the transition from twisted defects to unlinked ones.

A. Topological model. Let D^3 be a domain in R^3 homeomorphic to the ball D_0^3, $\Sigma = l = (l_1, l_2)$ be a defect in D^3, and V be the order parameter space. There exists a continuous map

$$\varphi\colon D^3 \setminus \Sigma \to V$$

for any point $x \in D^3 \setminus \Sigma$.

It is assumed that V is connected and that $\pi_2(V) = 0$. (This restriction is related to the assumption that there are no stable singular points in the system, otherwise, defects crossing form a stable point singularity.)

We now define which is a topologically admissible surgery of a defect Σ. This is natural to do for an arbitrary defect set, viz., graph Γ. Technically, it is favorable to regard as the defect a tubular neighborhood Σ and not the set itself. Σ is obtained from Γ by inflation and Γ is obtained from Σ by collapse. Note that Γ determines Σ (given an arbitrary isotopy) uniquely, while certainly the converse does not hold.

We will consider the restructuring of tubes Σ similar to those of one-dimensional defects since Σ does not separate space D^3. Because there is a collapse of Σ on Γ, we can define the map

$$\varphi_*\colon \pi_1(D^3 \setminus \Sigma) \to \pi_1(V). \tag{5.32}$$

Given the pair (Σ, φ) of a defect Σ and corresponding map (5.32), a pair (Σ', φ') is said to be *topologically admissible* if it is obtained from (Σ, φ) by the following elementary operations.

(1) Isotopy. Let $\varphi_t\colon D^3 \to D^3$ be an isotopy of D^3. Given a map $\varphi\colon D^3 \setminus \Sigma \to V$, $(\Sigma', \varphi') = (\psi_1\Sigma, \varphi \circ \psi_1)$ is topologically admissible.

(2) Let Σ collapse into a graph Γ and let Γ contain an isolated components Γ' with not more than one endpoint on ∂D^3.

We put $\Gamma' = \Gamma \setminus I$. Let Σ' be a tube of Γ'. We can then extend φ to the map

$$\varphi'\colon D^3 \setminus \Sigma' \to V.$$

(3) Assume that int $D^3 \setminus \Sigma \cap$ int D^3 contains an embedded sphere S^2 separating D^3 into two components so that one contains all the boundary ∂D and the other

is diffeomorphic to the ball D^3. Also, let Σ collapse into a graph Γ so that $D^3 \cap \Gamma$ is a simple closed curve $c \subset \text{int } D$, $\Gamma' = \Gamma \setminus c$, and Σ' an inflation of Γ'. Then φ can be extended to the map

$$\varphi': D^3 \setminus \Sigma' \to V.$$

Here, we use the equality $\pi_2(V) = 0$.

Note that from the physical viewpoint the surgery associated with (2) and (3) means a decrease in core energy proportional to the size of the defect Σ.

(4) Let γ be a simple closed curve in $\partial\Sigma \setminus \partial\Sigma \cap \partial D^3$. Two elements

$$g(\gamma) \in \pi_1(\Sigma), \quad p(\gamma) \in \pi_1(V)$$

can be related to γ (for an inner automorphism) defined as follows:

$g(\gamma)$ is the image of γ under the natural homomorphism $\pi_1(\partial\Sigma) \to \pi_1(\Sigma)$ and $p(\gamma)$ is the homotopy class of $\varphi(\gamma)$.

To formulate (4), we need

Proposition 5.1 1. *If $g(\gamma) = 1 \in \pi_1(\Sigma)$, then γ is the boundary of the embedded disk $D^2 \subset \Sigma$ so that $D^2 \cap \partial\Sigma = \partial D^2 = \gamma$, $D^2 \cap \partial D^3 = \varnothing$ and D^2 transversally intersects $\partial\Sigma$. The disk is determined uniquely (for an arbitrary isotopy). If Σ is cut along the disk D^2, then we obtain a manifold $\Sigma' \subset D^3$ which can be contracted to a graph Γ'. The transition from Σ to Σ' is also an allowed operation* (4).

2. *If we have the additional condition $p(\gamma) = 1 \in \pi_1(V)$, then the map φ can be extended to the map*

$$\varphi': D^3 \setminus \Sigma' \to V,$$

and the surgery $(\varphi, \Sigma) \to (\varphi', \Sigma')$ is an allowed operation.

The *proof* of Proposition 5.1 is based on the Dehn lemma, which was proved by Papakyriakopoulus, and Alexander's theorem [Ce], two classical theorems in the theory of three-dimensional manifolds. We now state the Dehn lemma from which the existence of the disk D^2 follows (the reader will find the proof in [Ro]).

Lemma 5.1 (Dehn) *Let M^3 be a smooth three-dimensional manifold with $\partial M^3 \neq 0$ and let $\gamma \in \partial M^3$ be a simple closed curve such that the embedding of γ in M^3 is null-homotopic. Then there exists a smoothly embedded two-dimensional disk $D^2 \subset M^3$ such that $D^2 \cap \partial M^3 = \partial D^2 = \gamma$.*

Alexander's theorem is necessary for proving the uniqueness of D^2 (for an arbitrary isotopy).

Theorem 5.14 (Alexander) *If S^2 is smoothly embedded in R^3, then there exists a diffeomorphism $\varphi: R^3 \to R^3$ such that $\varphi(S^2) \to S_1^2$, where S_1^2 is a sphere of unit radius.*

We now turn to the last allowed kind of surgery.

(5) Let (Σ, φ) be contracted to a graph Γ and let $\Gamma'' \subset D^3$ be a graph containing Γ and such that the set $\Gamma'' \setminus \Gamma$ belongs to one of the three types:

(a) An isolated component I diffeomorphic to an arc such that $I \cap \Gamma = \varnothing$ and $\partial I \cap \partial D^3$ consists of not more than one point.

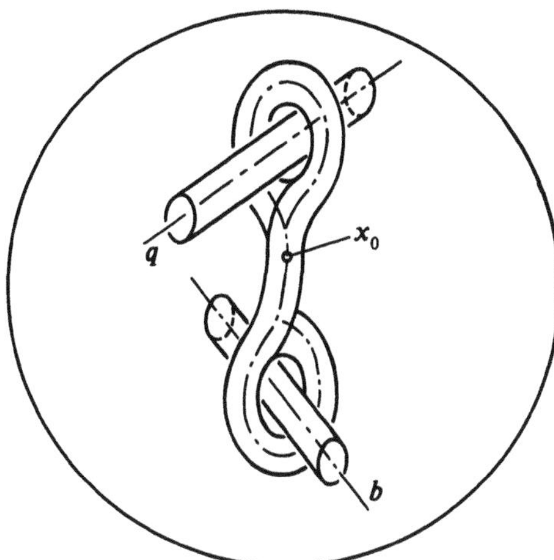

Fig. 19. Structure of element $[\alpha, \beta]$

(b) An isolated component diffeomorphic to a simple closed curve c contained in the domain $(D^3 \setminus \partial D^3) \setminus \Gamma$ and not linked with Γ.

(c) A component $\tilde{I}$ (not necessarily isolated) such that

$$\text{int } \tilde{I} \cap (\partial D^3 \cup \Gamma) = \varnothing, \quad \partial \tilde{I} \subset \partial D^3 \cup \Gamma.$$

We define Σ'' as a tubular inflation Γ'', $\Sigma'' \supset \Sigma$, and φ'' as equal to the restriction of φ to $D^3 \setminus \Sigma''$.

The transition $(\Sigma, \varphi) \to (\Sigma'', \varphi'')$ is an admissible topological surgery and inverse to (1)-(4) operations. From the physical point of view, the transition requires additional energy loss, but has no topological obstructions.

We can now formulate the main result of the subsection.

Let Σ be a tubular neighborhood of a graph Γ. We pick two arcs a and b in Γ. If Γ is inflated into Σ, then a and b arcs are sent into two solid tubes A and B. We consider the arc $q \subset D^3$ joining a and b together (Fig. 19) and not touching ∂D^3 and Γ at other points. We take a fixed point $x_0 \in q$ and pass two closed loops α and β through x_0 bypassing the lines a and b, respectively. Let $[\alpha]$ and $[\beta]$ be the homotopy classes of $\pi_1(D^3 \setminus \Sigma, x_0)$ associated with α and β. We consider the commutator

$$[\alpha, \beta] = [\alpha][\beta][\alpha]^{-1}[\beta]^{-1} \in \pi_1(D^3 \setminus \Sigma, x_0)$$

with image $\varphi_*[\alpha, \beta] = [\varphi_*\alpha, \varphi_*\beta]$ induced by φ and belonging to $\pi_1(V)$.

Theorem 5.15 [PT1] *Defects a and b move through each other if and only if*

$$[\varphi_*\alpha, \varphi_*\beta] = 1. \tag{5.33}$$

The *proof* consists in verifying the following statement. The transition from the configuration in Fig. 19 to the one with a and b interchanged via (1)-(5) is only possible if condition (5.33) holds.

I do not give the proof since after all the preliminaries the reader can obtain it independently or resort to [PT1] (see also the next subsection where Theorem 5.15 is a corollary of Theorem 5.16).

The defect-crossing problem admits generalizations. We can move in different directions, e.g., we can consider a multidimensional analog. Another way is offered in [JaT], where the interaction of two defects twisted with respect to each other n times is considered. The approach is direct generalization of the result obtained by Poénaru and Toulouse.

B. *n*-twisted defects [JaT]. Consider two defects l_1 and l_2 twisted relative to each other n times. How can topological obstructions be found that supply necessary and sufficient conditions for the disentanglement of two defects.

As before, assuming that defects can only be restructured in ambient space in a bounded domain, the problem can be posed as follows:

Let D^3 be a ball in R^3, with sphere S^2 as the boundary ∂D^3. As shown in Fig. 20, we pick four points $(x_1, x_2, x_3, x_4) = X$ on S^2, consider a deficient graph Γ_n consisting of n times twisted lines l_1 and l_2 and a continuous map φ: $S^2 \setminus X \to V$, where V is the space of internal states.

Let $\partial\Gamma_n = X$ and Γ_n be a transversal to S^2. We are interested in two questions. First, can a map φ always be extended to $\tilde{\varphi}$: $D^3 \setminus \Gamma_n \to V$? Second, what is the topology of V, so that we could pass from n-twisted defects to untwisted ones? Below, I use the expressive terminology of Janich and Trebin to call n-twisted defects *n-twists.* It is easy to see then that the defect-crossing problem of Poénaru and Toulouse is equivalent to that of deforming an l-twist into a 0-twist.

We start with preliminary considerations. We fix four points

$$x_1 = (0, \ -\varepsilon, \ \sqrt{1-\varepsilon^2}), \ x_2 = (0, \ \varepsilon, \ \sqrt{1-\varepsilon^2}),$$
$$x_3 = (0, \ -\varepsilon, \ -\sqrt{1-\varepsilon^2}), \ x_4 = (0, \ \varepsilon, \ -\sqrt{1-\varepsilon^2})$$

on the unit sphere $S^2 \subset R^3$ $(\tilde{x}, \tilde{y}, \tilde{z})$ (ε is a fixed number, $0 < \varepsilon < 1$), take $s_0 = (1, 0, 0)$ as the base point, and consider loops emanating from it. a bypasses x_2 in the positive direction, similarly, b bypasses x_1, $a' \to x_4$ and $b' \to x_3$, respectively, also in the positive direction. Let $[g]$ be the homotopy class associated with a loop

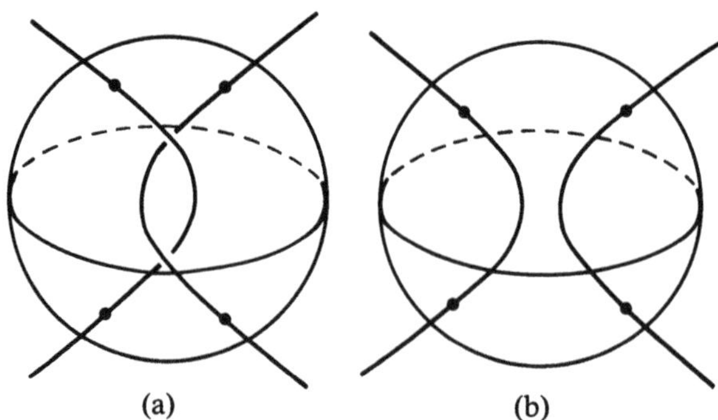

Fig. 20. Defective graph: (a) n-twisted graph Γ_n $(n = 2)$, (b) untwisted graph Γ_n $(n = 0)$

$g \in \pi_1(S^2 \setminus X, s_0)$. The relations between the homotopy classes

$$\begin{aligned} &(1)\ [a][b][a'][b'] = 1,\\ &(2)\ [v] = [b][b'] = [a][a']^{-1},\\ &(3)\ [h] = [a][b] = [a']^{-1}[b']^{-1} \end{aligned} \tag{5.34}$$

are obvious. We specify the map $\varphi: S^2 \setminus X \to V$. Then $[a]$, $[b]$, $[a']$, $[b']$, $[v]$, $[h]$ are associated with the homotopy classes in $\pi_1(V, x_0)$, where $x_0 = \varphi(s_0)$, which we denote by α, β, α', β', ψ, and χ, respectively. The relations

$$\begin{aligned} &(1)\ \alpha\beta\alpha'\beta' = 1,\\ &(2)\ \psi = \beta\beta' = \alpha^{-1}(\alpha')^{-1},\\ &(3)\ \chi = \alpha\beta = (\alpha')^{-1}(\beta')^{-1} \end{aligned} \tag{5.35}$$

follow from (5.34).

We can now formulate the main result.

Theorem 5.16 [JaT] *A map φ can be extended to the map*

$$\tilde{\varphi}: D^3 \setminus \Gamma_n \to V \tag{5.36}$$

if and only if

$$(\alpha\beta)^n = (\alpha'\alpha)(\beta\alpha)^n. \tag{5.37}$$

The proof is based on the following lemma.

Lemma 5.2 *A map φ can be extended to $\tilde{\varphi}: D^3 \setminus \Gamma_0 \to V$ if and only if $\psi = \alpha'\alpha = 1$.*

Proof. (1) If ψ is extended to $D^3 \setminus \Gamma_0$, then it is obviously extended to the disk D^2 bounded by v (hemisphere S^2_+). Since, in contracting the map v on the disk to a point, the map $(\varphi \circ v)$ is also contracted to a point in V, $\psi = [\varphi \circ v] = 1$.

(2) Let $\psi = \varphi_*[v] = 1$. The map φ can be extended to one of the disks bounded by v and separating S^2. There exist only two defects on each hemisphere.

Now we note that φ can be extended from each hemisphere into the ball, evading the line which joins the defect points on the hemisphere together. ■

To prove the theorem, it remains for us to construct the homotopy of the map $\tilde{\varphi}: D^3 \setminus \Gamma_n \to \tilde{\varphi}_1: D^3 \setminus \Gamma_0$.

Let θ be an angle of rotation of the sphere about the $\tilde{z}$-axis. We pick a domain $\mathscr{U} = \{\tilde{x}, \tilde{y}, \tilde{z} | \delta_1 \le z \le \delta_2\}$ in the ball D^3, where δ_1 and δ_2 are two constants, $0 < \delta_1 < \delta_2 < \sqrt{1 - \varepsilon^2}$. We consider a continuous map $\Phi_\theta: D^3 \to D^3$ such that the spherical "cap" over $\mathscr{U}$ $(1 > z > \delta_2)$ rotates through θ, each spherical layer in $\mathscr{U}$ through the angle $\theta(\tilde{z} - \delta_1)/(\delta_2 - \delta_1)$ and the domain placed under $\mathscr{U}$ remains fixed. It is obvious that the transformation $\Phi_{2\pi n}$ sends the 0-twist into the n-twist. Therefore, φ admits the n-twist if and only if $\varphi \circ \Phi_{2\pi n}$ admits the 0-twist.

It follows from Lemma 5.2 that $(\varphi \circ \Phi_{2\pi n}) \circ v = \varphi(\Phi_{2\pi n} \circ v) \sim \gamma_0$, where γ_0 is the trivial loop in $\pi_1(V, x_0)$.

The direct construction of the homotopy $\Phi_{2\pi n} \circ v$ on $S^2 \setminus X$ shows that $\Phi_{2\pi n} \circ v$ is homotopic to $h^n a^{-1} h^{-(n-1)} b^{-1} v$.

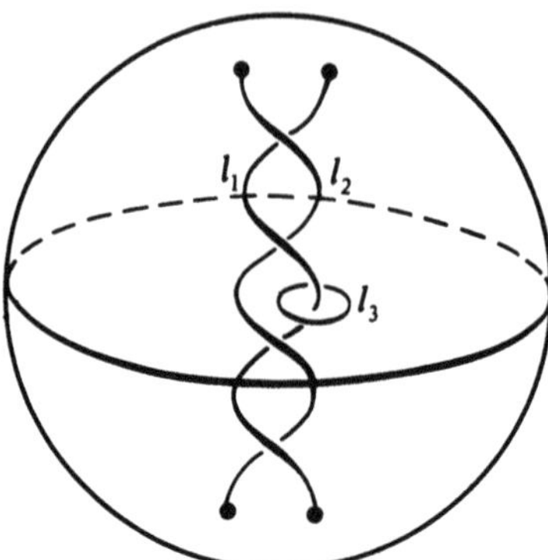

Fig. 21. Link of three defects

Passing to the homotopy classes, we obtain the induced map

$$\varphi_*[h^n a^{-1} h^{-1(n-1)} b^{-1} v] = \chi^n \alpha^{-1} \chi^{-(n-1)} \beta^{-1} \psi. \tag{5.38}$$

Taking into account relations (2) and (3) from (5.34), we obtain the necessary and sufficient condition

$$\chi^n \alpha^{-1} \chi^{-(n-1)} \beta^{-1} \psi = 1 \quad \text{or} \quad (\alpha\beta)^n = \alpha' \alpha (\beta\alpha)^n \tag{5.39}$$

for the n-twist to exist (the result of Poénaru and Toulouse on defect crossing).

Corollary 5.1 *The* 1-*twist can be reduced to the* 0-*twist if*

$$\alpha\beta = \beta\alpha.$$

Other configurations of defects can be considered by the same methods, e.g., a link of two defects l_1 and l_2 and of the third, a closed curve l_3, put on them (Fig. 21; see the particulars in [JaT]).

Remark 5.3 For all the interest in the topological considerations of links in [PT1], [JaT], there are very few real physical systems with noncommutative fundamental groups $\pi_1(V)$ for which the corresponding methods can be applied directly. Only a biaxial nematic can actually be treated, or with considerable reservations a cholesteric. Other systems with non-commutative groups $\pi_1(V)$ are encountered in ordinary crystals and smectics. However, the energies of twisting and shearing deformations are quite different and applying direct homotopy methods leads to far-fetched ideas about defect behavior.

In conclusion, we consider another generalization of the problem of defects, crossing, i.e., a multidimensional analog of the theory of intersections suggested by Poénaru and Toulouse [PT2].

5.1.6.6 Whitehead product and crossing of multidimensional defects

The problem of crossing n-dimensional defects is a direct generalization of the problem for one-dimensional defects in "physical" space R^3. The corresponding

n-dimensional generalizations of defects and topologically possible surgery are given below.

I omit the corresponding "physical" motivation since it is similar to the one-dimensional case.

Definition 5.10 Let M^n be an n-dimensional smooth manifold, a model of physical space, and V the space of internal states. A p-dimensional subset in M^n such that there exists a continuous map $\varphi: M^n \setminus \Sigma^p \to V$ not extendible to Σ^p is called a *p-dimensional defect* Σ^p. Its dimension is determined by the condition $\dim \Sigma^p + q = n - 1$, where q is the number of the ith homotopy group $\pi_i(V) \neq 0$. The tubular inflation $\tilde{\Sigma}$ can be regarded as technically more favorable (naturally, with the same homotopy groups), i.e., the submanifold of codimension zero in M^n and not the set of defects itself. Henceforward, we do not distinguish between Σ and $\tilde{\Sigma}$.

We introduce the allowed operations of defects surgery for a pair (Σ, φ).

1. Isotopy. The transition from (Σ, φ) to (Σ', φ') is allowed, where $\varphi' = \Phi_1 \circ \varphi$, $\Sigma' = \Phi_1\Sigma$, and Φ_t is an isotopy of space M^n.

2. Attaching a handle of index λ. Consider the smoothly embedded ball $D^n = D^\lambda \times D^{n-\lambda} \subset M^n \setminus \operatorname{int} \Sigma$ such that

$$D^n \cap \partial\Sigma = \partial D^n \cap \partial\Sigma = \partial D^\lambda \times D^{n-\lambda}.$$

We define the surgerical defect

$$\Sigma' = \Sigma \cup (D^\lambda \times D^{n-\lambda}), \quad \Phi' = \Phi|M^n \setminus \operatorname{int} \Sigma'.$$

The boundary is

$$\partial(D^\lambda \times D^{n-\lambda}) = \partial D^\lambda \times D^{n-\lambda} \cup D^\lambda \times \partial D^{n-\lambda},$$

where $\partial D^\lambda \times D^{n-\lambda}$ and $D^\lambda \times \partial D^{n-\lambda}$ are identified with respect to the common part, namely,

$$\partial(\partial D^\lambda \times D^{n-\lambda}) \cap \partial(D^\lambda \times \partial D^{n-\lambda}) = \partial D^\lambda \times \partial D^{n-\lambda}.$$

Attaching $D^\lambda \times D^{n-\lambda}$ to Σ' is done along $\partial D^\lambda \times D^{n-\lambda}$. After smoothing, we obtain a manifold Σ'. The operation is similar to the surgery of a manifold by attaching handles familiar from Morse theory [Mil4], where it is proved that any compact manifold Σ^p can be obtained from a point by attaching one by one handles of index λ_i.

3. Negative surgery of index λ. Let D^λ be a ball of dimension λ smoothly embedded in Σ so that $\partial D^\lambda \cap \partial\Sigma$ intersects $\partial\Sigma$ transversally. We extend the embedding of D^λ to the embedding $D^\lambda \times D^{n-\lambda} \subset \Sigma$, identifying D^λ with $D^\lambda \times 0$ (where 0 is the center of $D^{n-\lambda}$), and

$$\partial(D^\lambda \times D^{n-\lambda}) \cap \partial\Sigma = (D^\lambda \times D^{n-\lambda}) \cap \partial\Sigma = \partial D^\lambda \times D^{n-\lambda}.$$

Cutting Σ along D^λ, we define the set $\Sigma' = \Sigma \setminus D^\lambda \times \operatorname{int} D^{n-\lambda}$ (D^λ is determined uniquely) for an arbitrary isotopy Σ'. Assume that $[\varphi|\partial D^\lambda] = 0 \in \pi_{\lambda-1}(V)$. Then there exists an extension of φ to

$$\varphi': M^n \setminus \Sigma' \to V$$

and the pair (Σ', φ') is a topologically possible restructuring of (Σ, φ) by negative surgery of index λ.

The operation is similar to operation 4 for one-dimensional defects.

4. Crossing of handles. Let D^p and D^q be p-dimensional and q-dimensional balls, respectively, embedded in M^n. Their dimensions satisfy the relation

$$p + q = n - 1. \tag{5.40}$$

We now define the crossing of two handles of indices p and q, respectively, for which we have to construct the multidimensional analog of the construction in p. 285.

We mean the set $D^p \times x_0$, where x_0 is the center of the ball D^{n-p}, by D^p and the set $D^q \times y_0$, where y_0 is the center of D^{n-q}, by D^q. We assume that D^p and D^q are embedded in the defect spine so that int D^p and int D^q are open in Γ. We denote the centers of D^p and D^q by p and q, respectively, and consider the curve γ: $[0, 1] \to M^n$, $\gamma(0) = p$, $\gamma(1) = q$ joining the points p and q together, $\gamma(0, 1) \cap \Gamma = \varnothing$. Assume that $\gamma([0, 1])$ intersects D^p and D^q transversally. Assuming that a Riemannian metric exists on M^n, we can assume without loss of generality that γ intersects D^p and D^q orthogonally at p and q, respectively.

We introduce an infinitesimal intersection condition and consider the bundle normal to γ in M^n

$$N(\gamma) \xrightarrow{\theta} R^{n-1}$$

so that $\theta(T_pD^p)$ and $\theta(T_qD^q)$ intersect transversally. Two infinitesimal intersection conditions are homotopic

$$N(\gamma) \underset{\theta_1}{\overset{\theta_0}{\to}} R^{n-1}$$

if there exists a homotopy of the map θ_t such that $\theta_t(T_pD^p)$ and $\theta_t(T_qD^q)$ are transversal for any $t \in [0, 1]$.

For $n \geqslant 4$, there exist four homotopically distinct classes of infinitesimal intersection conditions obtained by combining the reversion of orientation in $\theta(T_pD^q)$ and space R^{n-1} inversion (a nontrivial element of the group $\pi_1(SO(n-1)) = Z_2$ for $n \geqslant 4$). Note that if $p = q = 1$, $n - 1 = 2$, then there are infinitely many such classes with $\pi_1(SO(2)) = Z$.

In addition to infinitesimal conditions, we can introduce intersection conditions by local coordinates.

Definition 5.11 An *intersection condition* is the pair of a local chart for a manifold M^n $(R^n \xrightarrow{\psi} M^n)$ $\{R^n|x^1, \ldots, x^p, y^1, \ldots, y^q, z\}$ and of a map ψ such that

(a) $\psi(R^n)\cap(\Gamma\cup\gamma) = \gamma\cup\tilde{D}^p\cup\tilde{D}^q$, where $\tilde{D}^p$ and $\tilde{D}^q$ are two sufficiently small balls in D^p and D^q, respectively,

(b) $\psi^{-1}(\gamma) = \{x = y = 0, \ -1 \leqslant z \leqslant 1\}$,

(c) $\psi^{-1}(\tilde{D}^p) = \{y = 0, \ z = 1\}$,

(d) $\psi^{-1}(\tilde{D}^q) = \{x = 0, \ z = -1\}$.

Two intersection conditions (ψ_1, R^n) and (ψ_2, R^n) are said to be *isotopic* if the sets $\psi_t(\psi^{-1}(\tilde{D}^p))$ and $\psi_t(\psi^{-1}(\tilde{D}^q))$ under isotopic transformations along γ remain transversal.

It is not hard to see that there exists a one-to-one correspondence between the homotopy classes of infinitesimal and local intersection conditions.

It easily follows from the definitions that it is immaterial which intersection is considered, D_p with D^q or D^q with D^p.

We fix a point x_0 on a segment γ, e.g., $x_0 = \gamma(1/2)$. Consider two sufficiently small balls: $\tilde{D}^{q+1}$ with center at p orthogonal to D^p and $\tilde{D}^{p+1}$ centered at q orthogonal to D^q. We (arbitrarily) orient the boundaries ∂D^{p+1} and $\partial\tilde{D}^{q+1}$ and join them to x_0 along γ. We have thereby defined the elements

$$\alpha_p \in \pi_p(M^n \setminus \Sigma, x_0), \quad \beta_q \in \pi_q(M^n \setminus \Sigma, x_0)$$

for which the Whitehead product $[\alpha_p, \beta_q]$ is defined and we have

$$\varphi_*[\alpha_p, \beta_q] = [\varphi_*\alpha_p, \varphi_*\beta_q] \subset \pi_{p+q-1}(V) = \pi_{n-2}(V).$$

We can now formulate the basic result [PT2].

Theorem 5.17 *For a pair* (Σ, φ) *and a local intersection condition* ψ *specifying an isotopy* $\Sigma \to \Sigma_\psi$, *the following two conditions are equivalent*:

(1) *surgery* (Σ, φ) *in* $(\Sigma\psi, \varphi')$ *by isotopy, negative surgery of index* λ, *attaching handles with indices* 0, 1, *attaching handles with indices* $\lambda \geqslant 2$ (*not touching a neighborhood of the segment* γ);

(2) *the Whitehead product* $[\varphi_*\alpha, \varphi_*\beta] = 1$.

The *proof* of the theorem, suggested by Poénaru and Toulouse, is a natural multidimensional generalization of Theorem 5.15. The reader can either reconstruct it or resort to [PT2]. Another proof can be obtained by using the relation between Massey products and the Whitehead product on the basis of the ideas discussed in Subsecs. 5.1.6.1-5.1.6.4.

Remark 5.4 The technique of surgery applied in this subsection is a particular case of the theory of manifold surgeries used in differential topology. The theory involving Morse surgeries, cobordism theory, Smale "handles", etc., enables us to prove a number of classical theorems in topology.

We now formulate two of them.

Poincaré conjecture (dim $M^n \geqslant 5$) *An n-dimensional smooth simply connected manifold homotopy equivalent to the n-dimensional sphere* S^n *is homeomorphic to* S^n.

***h*-cobordism theorem** *Two manifolds* N_1 *and* N_2 *are said to be h-cobordant if there exists a manifold M such that* ∂M *is* $N_1 \cup N_2$ *and* N_1 *and* N_2 *are deformation retracts of M.*

We recall the definition.

Definition 5.12 A subset $A \subset M$ is called a *retract* if there exists a continuous mapping (or retraction) f: $M \to A$ constant on A (i.e., $f(a) = a$ for any point $a \in A$). If f can be made one of a family of continuous deformations f_t: $M \to M$, $t \in [0, 1]$ so that f_1 is a retraction $M \to A$, then A is called a *deformation retract.*

Theorem 5.18 *If N_1 and (therefore) N_2 are simply connected and h-cobordant,* dim N_1, $N_2 \geqslant 5$, *then M is diffeomorphic to $N \times [0, 1]$. Therefore, N_1 is diffeomorphic to N_2.*

The results with numerous corollaries were first proved by Smale. A remarkably easy treatment is given by Milnor in [Mil6].

The Poincaré conjecture was recently proved by Freedman in dimension four [Fr]. This brilliant result is also based on the surgery technique.

5.2 TOPOLOGY OF SUPERFLUID ^{3}He

Topological methods enable us to investigate quite a rich structure of solutions of equations which describe superfluid ^{3}He.

Recall the main physical concepts related.

Superfluid ^{3}He is a Fermi liquid, i.e., particles with half-integer spin. Therefore, the transition into the superfluid state is only possible by the Cooper particle pairing effect. More precisely, pairing of quasiparticles is meant, i.e., elementary excitations of the original particles. Pitaevsky [Pi] has shown that if the orbital momentum in a system of quasiparticles is nonvanishing, then there is Van der Waals attraction in the system and pairs with nonvanishing orbital momentum are thereby formed with quantum number l.

The value of orbital momentum and transition temperature were not predicted theoretically. However, it was clear that the temperature should be several orders lower than T_c of usual isotope ^{4}He.

A group of researchers from Cornell University discovered the superfluid state of ^{3}He experimentally in 1972. The transition temperature T_c is of order 2.7 μK. Studying superfluid ^{3}He led to the discovery of the remarkable properties of the superfluid liquid.

First, two thermodynamic phases in ^{3}He were discovered. The A- and B-phases were theoretically predicted before [AM], [BW].

The investigations then led to a deeper understanding of phase structure and the phase diagram of the states for practically all temperatures and pressures was studied quite extensively. In particular, another phase A_1 existing in a strong magnetic field was discovered. Nevertheless, there are a number of fundamental problems still unsolved in the study of superfluid ^{3}He.*

1. Phenomenologically, ^{3}He is described by the Ginzburg-Landau equation with the order parameter determined by a composition of experimental and theoretical arguments. The equation has been derived from microscopic theory on the basis of the BCS Hamiltonian [AGR]. However, the mechanism of pairing in ^{3}He is not fully clarified.

* We omit the term "superfluid" below since only this state of ^{3}He is considered in the chapter.

2. There is no complete microscopic theory of the hydrodynamics of ^{3}He although important results have been obtained in this direction [VW].

3. At present, the properties of vortices in rotating ^{3}He [SV], [Thu] are being studied intensively. The identification is mostly by the NMR technique. The phenomenological Ginzburg-Landau equation describes adequately the basic properties of ^{3}He. It turns out to be considerably more complicated than the Ginzburg-Landau equation for superfluid ^{4}He or superconductivity and leads to hopes for applying group-theoretic and topological methods.

5.2.1 Ginzburg-Landau Equation

Here, I discuss the basic physical argument in order to determine the order parameter for superfluid ^{3}He. The reader will find a detailed discussion of the physical properties in [Le] and [VW].

The orbital momentum of a Cooper pair has already been noted to be nonvanishing. The condition $l = 1$ for the Cooper pairs in the superfluid phase of ^{3}He (*P*-wave) is selected for experimental reasons (in particular, taking the magnetic momenta of the pair into account) and is now regarded as a well-established fact. If a Cooper pair ψ has odd orbital momentum, then on the basis of the Pauli principle the spin part of the wave function $\psi = \chi_{\sigma_1\sigma_2}\psi(r_1, r_2)$ should be an even function relative to spin permutation. Since the spin of each particle in the pair is 1/2, the total spin is unity for any odd l (triplet pairing).

We can now determine the macroscopic order parameter for superfluid ^{3}He. As in superconductivity theory, consider the gap $\Delta_{p\sigma\sigma'}$(a scalar function depending on temperature in the case of superconductivity). Here Δ is a symmetric complex 2×2 matrix (a symmetric spinor of rank two) with the three independent elements σ, σ' (particles with spins 1/2) and p (pair momentum),

$$\Delta_{p\sigma\sigma'} = d(\hat{p})\cdot(\tau\cdot i\tau^2)_{\sigma\sigma'} = \begin{pmatrix} -d_1 + id_2, & d_3 \\ d_3, & d_1 + id_2 \end{pmatrix}, \tag{5.41}$$

$$\hat{p} = p/|p|,$$

and $\tau = \{\tau^1, \tau^2, \tau^3\}$, τ^j are Pauli matrices.

The components of the vector d_j determine the amplitude of the order parameter in the states $|j\rangle$ of the complete spin operator S with zero eigenvalues $S_j|j\rangle = 0$, $j = 1, 2, 3$.

Further, if we resolve $d_j(\hat{p})$ in terms of the components of momentum $\hat{p}$ taking into account that $l = 1$ ($\hat{p}^\alpha$, $\alpha = 1, 2, 3$, are the components of the momentum vector of the particles to be paired), then we finally obtain

$$\Delta_{\sigma\sigma'}(p) = \sum_{\alpha, j} A_{\alpha j}\hat{p}_\alpha(\tau^j i\tau^2)_{\sigma\sigma'}, \tag{5.42}$$

$A_{\alpha j}$ being an arbitrary complex 3×3 matrix. If we pass from the momentum representation of p to the coordinate (or x-) representation by means of the Fourier transform, then the dependence $d_j(\hat{p}) = A_{ji}\hat{p}_i$ is sent into $\tilde{d}_j(x) = A_{ik}\hat{x}^k$, where $\hat{x}^k$ are the components of the vector $\hat{x} = r_1 - r_2/|r_1 - r_2|$ and r_1 and r_2 are the radii-vectors of the particles in pairing.

The gap is thereby expressed in terms of the complex 3×3 matrix A_{ij} which can be regarded as the order parameter of superfluid ^{3}He and in whose terms the free-energy density

$$\mathscr{F} = F_{\text{grad}} + U(A) \tag{5.43}$$

can be defined to describe the superfluid state of ^{3}He, where the potential $U(A)$ is selected in the Ginzburg-Landau form (taking the expansion up to the fourth degree), namely,

$$U(A) = \alpha \operatorname{Tr}(AA^+) + \beta_1 |\operatorname{Tr}(AA^t)|^2 + \beta_2[\operatorname{Tr}(AA^+)]^2 + \beta_3 \operatorname{Tr}[(A^+A)(A^+A)^*] \\ + \beta_4 \operatorname{Tr}[(AA^+)^2] + \beta_5 \operatorname{Tr}[(AA^+)(AA^+)^*], \tag{5.44}$$

A^* being the complex conjugate of A.

We have

$$F_{\text{grad}} = \gamma_1 \partial_k A^*_{pi} \partial_k A_{pi} + \gamma_2 \partial_k A^*_{pi} \partial_i A_{pk} + \gamma_3 \partial_k A^*_{pk} \partial_i A_{pi},$$

where α, β_i, and γ_j are constants (the summation is over repeated subscripts).

Writing the free-energy potential in the form of (5.43) is useful since interaction between the system and other fields (e.g., with external magnetic one) or taking additional interactions into account simply leads to the densities of corresponding energies being adjoined to formula (5.43).

Before analyzing (5.44), we make several remarks relative to the validity of the Ginzburg-Landau approach to the strongly interacting system of Fermi-particles which describes ^{3}He. The Landau criterion justifying this holds, since the coherence length $\xi = v_F/(2\pi T_c)$ is much greater than the distance between the particles (where v_F is the Fermi velocity). The form of $U(A)$ in (5.44) is determined by the invariance of the Hamiltonian relative to independent spin and spatial rotations and gradient transformations. Equation (5.44) is the general form of the polynomial $P^4(x)$, invariant under such transformations. $U(A)$ satisfies these requirements if weak dipolar interaction in the system is neglected.

The symmetry group for $U(A)$ is of fundamental importance if we want to describe thermodynamic phases of ^{3}He.

It is easy to see that $U(A)$ is invariant under the transformation group

$$G = SO(3)_R \times SO(3)_L \times U(1), \tag{5.45}$$

where $SO(3)_R$ acts on the spatial degrees of freedom, while $SO(3)_L$ acts on the spin degrees of freedom of the matrix A_{ij} as

$$A_{ij} \to R_{ik} R_{jn} \exp(i\varphi) A_{kn}, \tag{5.46}$$

or, in group-theoretic form:

$$A \to \exp(i\varphi) R_1 A R_2^{-1}, \quad R_1 \in SO(3)_R, \quad R_2 \in SO(3)_L. \tag{5.47}$$

It is obvious that the total potential involving the gradient terms is not invariant under G. However, to determine the thermodynamic phases, it suffices to study $U(A)$. Taking the gradient terms into account is important for textured description (i.e., static distributions of the order parameter; see Subsec. 5.2.7).

That we consider only $U(A)$ means that we confine ourselves to a homogeneous system with infinite characteristic length compared with coherence length. However, the corresponding solutions can have singularities at points, on lines or on surfaces.

We start with the problem of classifying thermodynamic phases.

5.2.2 Phase Structure in ^{3}He and a Neutron Star

5.2.2.1 Phase structure in ^{3}He

Mostly following [GM4], [BoM], we discuss the group-theoretic classification of the phases in ^{3}He. In the next subsection, a physical argument is discussed, enabling us to identify the A- and B-phases of ^{3}He observed experimentally.

The complete solution of the problem of phase classification is reduced to finding the minima of $U(A)$ in 18-dimensional real space, i.e., to the solution of the system

$$\frac{\delta U}{\delta A} = 0, \quad \frac{\delta^2 U}{\delta A} \geqslant 0. \tag{5.48}$$

Generally speaking, the corresponding minima depend on the parameters α and β_i; finding the minima analytically is a hard problem. Using group-theoretic methods, the structure of possible phases can be scanned rather simply. Most importantly, the global description of the space of internal states can be obtained, something necessary for a study of defects and textures.

The basic assumption is as follows. By analogy with field theory, we assume that the whole manifold of the minima of $U(A)$ is separated into the orbits of the gauge group G. The assumption actually implied by this condition is that each nonzero minimum is an analog of the expectation value vacuum $\langle\psi\rangle$ in field theory and is obtained by spontaneous symmetry breakdown. It is obvious that we can only find the necessary conditions for an extremum from the phase group-theoretic classification. The sufficient conditions require that the second relation in (5.48) should be verified and impose restrictions on the coefficients α and β_i (see Remark 5.5).

Let $G = U(1) \times SO(3)_R \times SO(3)_L$ be the gauge group of $U(A)$. Then, following the general scheme in Subsec. 1.3.4, we can immediately list the topological types of manifolds of the minima M_i of $U(A)$. M_i can be in the homogeneous spaces $M_1 = SO(3) \times SO(3)$, $SO(3) \times S^2$, $SO(3) \times U(1)$, $S^2 \times S^2 \times U(1)$, $S^2 \times S^2$,

$S^2 \times U(1)$, $SO(3)$, $U(1)$ of the group G if the discrete subgroups are not accounted for. However, not all of the above manifolds can be a phase of ^{3}He, since $U(A)$ is invariant under certain linear action (5.47).

A description of all the orbits of the group under the action is rather hard. The full proof is given in [BoM] (see also the earlier papers [BM], [J], [GM1, 4]). Since a complete discussion of the work takes us away from the topological emphasis in the book, I only give the main idea. In identifying the orbits with the phases familiar from physics, it is convenient to apply the method of [GM4].

The original problem is to describe the orbits of the group $G = SO(3)_R \times SO(3)_L \times U(1)$ in the space Mat $(3)_C$ of complex 3×3 matrices $A = A_{ij}$ $(i, j = 1, 2, 3)$ relative to (5.47).

Its solution is based on the following reduction. We introduce the space $\{B\} = S^2(C^4)_0$ of complex symmetric matrices with zero trace, $\{B\}$ is isomorphic to $\{A\}$ and decomposed into the direct sum $\{B_1 + iB_2\}$, where $\{B_j\} = S^2(R^4)_0$ $(j = 1, 2)$ is the space of real symmetric 4×4 matrices with vanishing trace. The group

$$G_1 = SO(4): \{B\} \Rightarrow g_1 B g^{-1}, \quad g_1 \in SO(4), \tag{5.49}$$

acts on $\{B\}$ and thereby on B_j.

Lemma 5.3 *The action* (5.47) *of the group* $\tilde{G} = SO(3)_R \times SO(3)_L$ *on* Mat $(3)_C$ *coincides with that of* $SO(4)$ *on* $S^2(C^4)_0$.

Proof. Any irreducible representation of the group $\tilde{G}$ is the tensor product $V_1 \otimes V_2$, where V_j $(j = 1, 2)$ are irreducible representations of $SO(3)_R$ and $SO(3)_L$, respectively. In identifying the covering group $\tilde{G}$ with $SO(4)$, the inverse image of $SO(3)_L$ is identified with the group $SU(2)$, i.e., the subgroup of left multiplications by unit quaternions q, $q \in H \sim R^4$, $N(q) = 1$. $SU(2)$ acts transitively on the unit sphere $S^3 \subset R^4$; therefore, there only exists one quadratic form $SU(2)$-invariant on R^4. It follows that the representation of $SU(2)$ in $S^2(R^4)_0$ has no invariant subspaces, i.e., in the decomposition $S^2(R^4)_0/SO(3)$, dim $V_1^* \geqslant 3$. Since a similar argument also holds for V_2^*, we obtain from the condition dim $S^2(R^4)_0 = 9$ that dim $V_1^* = 3$ and dim $V_2^* = 3$. Therefore, $S^2(R^4)_0$ is the space of the irreducible representation of the group $SO(4)/Z_2$, which is equivalent to Mat $(3)_R$. Since Mat $(3)_C =$ Mat $(3)_R +$ i Mat $(3)_R$, the lemma is thus proved.

We find that the problems of orbit classification for the group G in Mat $(3)_C$ and for $SO(4) \times U(1)$ in $S^2(C^4)_0$ are equivalent.

To identify the phases in ^{3}He, we need explicit formulas describing their equivalence.

We fix the isomorphism between $SO(3)_R$ and $SO(3)_L$, which distinguishes the diagonal subgroup $SO(3)_\Delta$ in $SO(3)_R \times SO(3)_L$. The group $SO(3)_\Delta$ acts on $\{A\}$ by inner conjugation thus:

$$g \in SO(3)_\Delta: gA \Rightarrow gAg^{-1}. \tag{5.50}$$

The space $\{A\}$ is meanwhile separated into the sum of three irreducible representations $\wedge^2 R^3$, $S^2(R^3)_0$, and R^1, or skew-symmetric 3×3 matrices, symmetric matrices with Tr $A = 0$, and scalar matrices λE, respectively. In the representation of $SO(4)$ on R^4, a procedure similar to formula (5.50) is associated with the introduction on R^4 of the structure of the quaternion algebra H. Meanwhile, the action of $SO(3)_\Delta$ on H is identified with inner conjugacy in H relative to the quaternions q with norm $N(q) = 1$

$$q \in S^3, \quad X \subset H, \quad q(X) = qXq^{-1}.$$

The subgroup of quaternion units S^3 factorized relative to $Z_2 = \pm 1$ is the image of $SO(3)_\Delta$.

The space H is decomposed into the direct sum of the space of purely imaginary quaternions isomorphic to R^3 and the space of scalars. The decomposition is invariant under the action of $SO(3)_\Delta$ and specifies the required identification

$$S^2(R^3)_0 \simeq S^2(R^{3\prime})_0, \quad \wedge^2 R^3 = \{A_{ij}\}, \; A_{ij} = 0 \; (i, j = 1, 2, 3), \; A_{i4} = -A_{4i}.$$

The scalar matrices are $A_{ii} = \lambda_i (i = 1, 2, 3, 4), \; -\sum_{i=1}^{3} \lambda_i = \lambda_4$.

The decomposition is of the following form:
Let A be a 3×3 matrix. We have $A = A_s + A_{ss}$, where

$$A_s = \begin{pmatrix} a_{11} & a_{12} & a_{13} \\ a_{21} & a_{22} & a_{23} \\ a_{31} & a_{32} & a_{33} \end{pmatrix}, \quad a_{ij} = a_{ji}, \quad A_{ss} = \begin{pmatrix} 0 & \tilde{a}_{12} & \tilde{a}_{13} \\ -\tilde{a}_{12} & 0 & \tilde{a}_{23} \\ -\tilde{a}_{13} & -\tilde{a}_{23} & 0 \end{pmatrix}, \tag{5.51}$$

are a symmetric matrix and a skew-symmetric one, respectively.

A matrix $B \in S^2(R^4)_0$ is of the form

$$\begin{pmatrix} a_{11} & a_{12} & a_{13} & \tilde{a}_{12} \\ a_{12} & a_{22} & a_{23} & \tilde{a}_{13} \\ a_{13} & a_{23} & a_{33} & \tilde{a}_{23} \\ \tilde{a}_{12} & \tilde{a}_{13} & \tilde{a}_{23} & c \end{pmatrix}, \quad c = -\mathrm{Tr}\, A_s. \tag{5.52}$$

Taking the established isomorphism into account, we study the action of the group $G = SO(4) \times U(1)$ on the space $S^2(C^4)_0$.

Given the matrices $\{B_1, B_2\}$, $B = B_1 + iB_2 \in S^2(C^4)_0$, the commutator $[B_1, B_2] = C$ is a skew-symmetric matrix. The orbits of the group G are essentially related to the structure of commutators and of the corresponding Lie algebra generated by the pairs of B_1, B_2 matrices. Commutator C is invariant under the action of $U(1)$. Therefore, the action of the group $U(1)$ can be as the first step of classification separated from the action of $SO(4)$.

The action of the group $SO(3)_R \times SO(3)_L$ on the space $\wedge^2 R^4$ of skew-symmetric 4×4 matrices is decomposed into the sum $\wedge^+(R^3) + \wedge^-(R^3)$, where $\wedge^+$ is the standard three-dimensional representation of $SO(3)_R$ (a rotation in R^3) and $\wedge^-$ the standard representation of $SO(3)_L$. The commutator $[B_1, B_2]$ thus specifies an equivariant (commuting with the action of G) mapping of the space $\{B\}$ with the full group G action onto the space $\wedge^+ + \wedge^-$ of skew-symmetric matrices, where the action of $U(1)$ is trivial.

Let $SO(4)$ act on the space $\wedge^2 R^4$. The action has a section $R^1_+ \times R^1_- \subset \wedge^+ + \wedge^-$, where the rays R^1_+ and R^1_- of the section are half-lines in $\wedge^+$ and $\wedge^-$ [Br]. The corresponding matrix representatives of such sections in $\wedge^2 R^4$ are the matrices

$$\begin{pmatrix} 0 & \lambda & 0 & 0 \\ -\lambda & 0 & 0 & 0 \\ 0 & 0 & 0 & \lambda \\ 0 & 0 & -\lambda & 0 \end{pmatrix} = R^1_+, \quad \begin{pmatrix} 0 & \mu & 0 & 0 \\ -\mu & 0 & 0 & 0 \\ 0 & 0 & 0 & -\mu \\ 0 & 0 & \mu & 0 \end{pmatrix} = R^1_-, \quad \lambda, \mu > 0. \qquad (5.53)$$

The algebra $\wedge^2 R^4$ is identified with the Lie algebra $so(4)$ of the group $SO(4)$ with $\wedge^+ \simeq so(3)_R$ and $\wedge^- \simeq so(3)_L$.

The description of orbits of the action of $SO(4)$ is based on a study of the structure of an associative algebra $\mathscr{A}(B)$ and the Lie algebra generated by B_1 and B_2.

To classify the orbits of $SO(4)$, it suffices to consider the associative algebra. However, by involving the group $U(1)$, we need a more fine structure, i.e., the Lie algebra Q generated by the matrices B_1 and B_2, $Q = \mathscr{A}(B) \cap \wedge^2 R^4$.

The space $\mathscr{A}(B)$ can be decomposed into the direct sum

$$\mathscr{A}(B) = M + Q, \quad M = \mathscr{A}(B) \cap S^2(R^4)_0. \qquad (5.54)$$

The Lie algebra Q is a subalgebra of $SO(4)$. The decomposition in (5.54) satisfies the relations

$$[M, Q] \subset M, \quad [M, M] \subset Q, \qquad (5.55)$$

which are similar to the familiar containments for the symmetric pair of a Lie algebra [He].

We now classify the symmetric associative algebras generated by $\{B_1, B_2\}$.

As a symmetric algebra, $\mathscr{A}$ is the invariant subalgebra Mat $(4)_R$ of all 4×4 matrices over the field R under the conjugation map (transposition) t

$$X \to X^t, \quad X, X^t \in \text{Mat } (4)_R.$$

It follows in particular that the nilpotent radical (i.e., the largest nilpotent ideal) in $\mathscr{A}$ vanishes, i.e., $\mathscr{A}$ is semi-simple [Sem].

The operation $t: X \to X^t$ induces an anti-automorphism of $\mathscr{A}$, which is generated by the fixed part $\mathscr{A}^t = \mathscr{A}$ of t.

By Wedderburn's classical theorem [We], the complexification $\mathscr{A}_C$ of $\mathscr{A}$ is the sum of simple matrix algebras. Here, $\mathscr{A}_C$ can be one of the algebras or their direct summands

$$
\begin{aligned}
&(1)\ \mathrm{Mat}\ (4)_C,\\
&(2)\ \mathrm{Mat}\ (3)_C + C,\\
&(3)\ \mathrm{Mat}\ (2)_C + \mathrm{Mat}\ (2)_C,\\
&(4)\ \mathrm{Mat}\ (2)_C + C + C,\\
&(5)\ C + C + C + C.
\end{aligned} \tag{5.56}
$$

We have to distinguish the real subalgebras of $\mathscr{A}$ associated with the list (5.56). We omit the proof, which can be found in [BoM], and formulate the final result as

Proposition 5.2 *The associative envelope of a pair* $\{B_1, B_2\}$ *of real symmetric matrices can only be a direct summand of one of the algebras*

$$
\begin{aligned}
&(1)\ \mathrm{Mat}\ (4)_R,\\
&(2)\ \mathrm{Mat}\ (3)_R + R,\\
&(3a)\ \mathrm{Mat}\ (2)_R + \mathrm{Mat}\ (2)_R,\ (3b)\ \mathrm{Mat}\ (2)_C,\\
&(4)\ \mathrm{Mat}\ (2)_R + R + R,\\
&(5)\ R + R + R + R.
\end{aligned} \tag{5.57}
$$

It is now not complicated to find the stability subgroups H_i and thereby the orbits of the group G.

Proposition 5.3 (1) *In* (1) *and* (2) *cases of the Proposition* 5.2, *algebras* $\mathscr{A}$ *have trivial stabilizers and the orbits* Ω *are isomorphic to* $SO(4)/Z_2$.

(2) *In case of* (3a), $H = Z_2$. *H acts on the matrices* $\left(\begin{array}{c|c} a_{ij} & 0 \\ \hline 0 & b_{ij} \end{array}\right)$ *by multiplication by* $\left(\begin{array}{c|c} +1 & 0 \\ \hline 0 & -1 \end{array}\right)$. *Orbits* Ω *are isomorphic to* $SO(4)/Z_2 \times Z_2$.

(3) *In* (3b) *case**, $H \sim S^1 \subset C^* \subset \mathrm{Mat}\ (2)_C$, $\Omega = SO(4)/S^1$.

(4) *In general position, H acts as in* (3a) *and* $H = Z_2$ *and* $\Omega = SO(4)/Z_2$. *However, two subcases are possible, viz.,* (a) *a diagonal embedding* $R_\Delta \subset R + R$, *where* $H = S^1$, $\left(\begin{array}{c|c} 1 & 0 \\ \hline 0 & \lambda \end{array}\right)$, *and* (b) *a space* $\mathscr{A} = \mathrm{Mat}\ (2)_R$. *Therefore, also,* $H \sim S^1$, $H \ni h = \left(\begin{array}{c|c} 1 & 0 \\ \hline 0 & \lambda \end{array}\right)$.

In both cases, $\Omega \sim SO(4)/S^1$. $H \sim S^1$, $H \ni h = \left(\begin{array}{c|c} 1 & 0 \\ \hline 0 & \lambda \end{array}\right)$.

(5) B_1 *and* B_2 *can be diagonalized simultaneously. The stability subgroups H are classified relative to the eigenvalues of matrices* B_i $(i = 1, 2)$. *Let* λ_j $(i = 1, 2, 3, 4)$ *be the eigenvalues of* B_1.

There exist the following H and Ω:

(a) The orbit in general position is associated with matrices with different eigenvalues $\lambda_j \neq \lambda_k$, $H \sim Z_2 \times Z_2 \times Z_2$, and $\Omega \sim SO(4)/Z_2 \times Z_2 \times Z_2$. The singular orbits are associated with repeated eigenvalues.

---

*Here, C^* is the group of complex numbers λ, $\lambda \neq 0$.

(b) $\lambda_1 = \lambda_2 \neq \lambda_3 \neq \lambda_4$. H coincides with the group $SO(2) \times Z_2 \times Z_2$, $\Omega \sim SO(4)/SO(2) \times Z_2 \times Z_2$.

(c) $\lambda_1 = \lambda_2 = \lambda_3 \neq \lambda_4$. $H \sim SO(3) \times Z_2$, $\Omega \sim SO(4)/SO(3) \times Z_2$.

(d) $\lambda_1 = \lambda_2 = \lambda_3 = \lambda_4 = 0$, $H = SO(4)$, $\Omega = *$ (a point).

The description of the Lie algebras of the commutators has a certain mathematical interest. However, we do not dwell on classification of the algebras and pass to the description of the "exterior" group $U(1)$ action (see the proof in [BoM]).

The action of $U(1) = \exp(i\varphi)$. The group acts by rotation

$$B_1 \to B_1 \cos\varphi - B_2 \sin\varphi,$$
$$B_2 \to B_1 \sin\varphi + B_2 \cos\varphi$$

on the pairs of matrices $\{B_1, B_2\}$, $B_1 + iB_2 \in S(C^4)_0$.

$U(1)$ does not act trivially on the space $\{B_1, B_2\}$ if the group is conjugate to the subgroup $K_1 \subset SO(2) \subset SO(4)$.

Let an element $\gamma \in SO(4)/Z_2$ be such that under conjugation it preserves the $\langle B_1, B_2 \rangle$-plane.

We fix the type of rotation γ on the plane. γ can be represented as the composition of two rotations φ and ψ in the planes R^2_φ, R^2_ψ in R^4. On symmetric matrices $S^2(R^2_\varphi)_0$ and $S^2(R^2_\psi)_0$, this specifies rotations through 2φ and 2ψ, respectively, i.e., $1 + R^2_{2\varphi} = S^2(R^2_\varphi)_0$ and $1 + R^2_{2\psi} = S^2(R^2_\psi)_0$. Besides, there exists an action of γ on $R^2_\varphi \otimes R^2_\psi = R^2_{\varphi+\psi} \oplus R^2_{\varphi-\psi}$.

Let L be a subspace of $S^2(R^4)_0$ on which the rotation γ is defined. L can either be two-dimensional and then φ and ψ are arbitrary or $\dim L > 2$ with the following relations between the angles:

(a) $\varphi = \pm\psi$ (since we consider angles to within the sign).

(b) $2\varphi = -\varphi + \psi \Rightarrow \psi = 3\varphi$.

First, consider the special case (b).

L can be reduced to the form

$$\left(\begin{array}{cc|cc} a & b & c & d \\ b & -a & d & -c \\ \hline c & d & & \\ d & -c & \multicolumn{2}{c}{0} \end{array}\right)$$

and possesses the following structure: there exists a subgroup $S^1 \subset S^1 \times S^1$ relative to which the space $\{D, D_1\}$, $D, D_1 \subset L$, is invariant (the $\langle D, D_1 \rangle$-plane rotates through 2φ).

The commutator $[D, D_1]$ is in the Lie algebra of the group $S^1 \times S^1$. The matrices D and D_1 generate the whole algebra Mat $(4)_R$ or Mat $(2)_R$ as an associative algebra.

We retrace our steps to the general case of (a).

Suppose a pair (φ, ψ) specifies a complex rotation on $R^4 \simeq C^2$ and a pair $\{B_1, B_2\}$ lies in the eigenspace of the rotation, i.e., the $\langle B_1, B_2 \rangle$-plane is of the form $\langle B_1, iB_1 \rangle$ relative to a certain complex structure on R^4. Thereby, $B_1, B_2 \in S^2(C^2)_0$ generates Mat $(2)_C$.

We clarify the existence of external symmetries when $\dim L = 2$, in which case γ commutes with the commutator subgroup $[B_1, B_2]$ and acts by the same rotations on the space $\langle[B_1, [B_1, B_2]], [B_2 [B_1, B_2]]\rangle = L$. The matrices B_1, B_2, and $[B_1, B_2]$ either form the Lie algebra $sl(2, R)$ or the nilpotent algebra Y determined by the commutation relations

$$[B_1, B_2] = C \in Q, \; [B_i, C] = 0, \; B_i \subset M, \; \dim M = 2 \; (i = 1, 2). \qquad (5.58)$$

Subalgebras of the form sl (2, R) of Mat $(4)_R$ are associated with various real representations of the group sl (2, R) and are classified as follows:

(1) Irreducible representations in the space of polynomials of degree three.

(2) Representations in the space $R_1^2 + R_2^2$ (acting identically both in R_1^2 and R_2^2).

Case (2a) is also possible where sl (2, R) acts by infinitesimal rotations in R_1^2 and is the identity in the other space.

(3) Representations in $R^3 + R$.

The nilpotent algebra is unique for arbitrary conjugacy in Mat $(4)_R$.

The associative algebra $\langle Y \rangle$ generated by Y is in Mat $(2)_C$ since B_i commutes with complex rotations $[B_1, B_2]$. Thus, in "general" position, the nontrivial external symmetry of space $[B_1, B_2]$ is related only to realizations of the Lie algebras Y and sl (2, R).

In the sequel, we need a realization of certain orbits directly in terms of 3 × 3 matrices. We write out the corresponding orbit representatives. That they are in fact disjoint follows from our classification and can be given explicitly in each case. To denote orbits, we use a physical terminology.

(1) B-phase. The orbit representative is the unit matrix $A_0 = E$ (whose all eigenvalues λ_i are equal).

(2) Planar phase with representative

$$A_0 = \begin{pmatrix} 1 & 0 & 0 \\ 0 & 1 & 0 \\ 0 & 0 & 0 \end{pmatrix}.$$

(3a) Bipolar phase with representative

$$A_0 = \begin{pmatrix} 1 & 0 & 0 \\ 0 & i & 0 \\ 0 & 0 & 0 \end{pmatrix}.$$

(3b) Polar phase with representative

$$A_0 = \begin{pmatrix} 1 & 0 & 0 \\ 0 & 0 & 0 \\ 0 & 0 & 0 \end{pmatrix}.$$

(4a) A-phase with representative

$$A_0 = \begin{pmatrix} 1 & i & 0 \\ 0 & 0 & 0 \\ 0 & 0 & 0 \end{pmatrix}.$$

(4b) β-phase dual to A-phase with representative

$$A_0 = \begin{pmatrix} 1 & 0 & 0 \\ i & 0 & 0 \\ 0 & 0 & 0 \end{pmatrix}.$$

(4c) γ-phase with representative

$$A_0 = \begin{pmatrix} 1 & i & 0 \\ i & 1 & 0 \\ 0 & 0 & 0 \end{pmatrix}.$$

Geometrically, phases (4a) and (4b) are only different in $SO(2) \subset SO(3)_R \times SO(3)_L$, being in $SO(3)_R$ for (a) and in $SO(3)_L$ for (b). However, from the physical point of view, the phases are absolutely different, since there is anisotropy in (a) relative to the orbital variables and relative to spin ones in (b). Experimentally, only the A-phase is observed. We consider it in more detail in Subsec. 5.2.3.

Remark 5.5 From the point of view of the above classification, the orbits are in classes (1) and (2).

We see that there exists another series of orbits, mostly with trivial (discrete) stability subgroups.

The listed phases can only be observed if they are stable and associated with the absolute minima of the Ginzburg-Landau functional.

If the phases correspond to local minima, then they can be interpreted as metastable states. In particular, transitions from one into another can occur with intermediate phases. To find stability criteria for the phases is a technically complicated and unsolved problem. However, calculations in [BM], [J], and [GM1] show that for phases (1), (3a), and (4) there exist domains of coefficients β_i, where the phases can be stable. We have actually investigated the second variation of the Ginzburg-Landau functional in some directions. On the other hand, the bipolar phase is unstable.

In conclusion, we mention another interesting physical object, a neutron star with a superfluid core consisting of several phases. Their classification is based on the same method as for ^{3}He, but is considerably simpler technically.

5.2.2.2 Neutron star

According to modern ideas, the core in a neutron star consists of a neutron superfluid in state $^3P_2$.

A Ginzburg-Landau potential similar to (5.44) is represented as

$$U(A) = (1/3)\alpha(T)\,\mathrm{Tr}\,(A\bar{A}) + \tilde{\beta}_1|\mathrm{Tr}\,A^2|^2 + \tilde{\beta}_2\,\mathrm{Tr}\,(A\bar{A})^2 + \tilde{\beta}_3\,\mathrm{Tr}\,(A^2A^{-2})$$

[SaS], where

$$\tilde{\beta}_1 = \beta_1 + (1/2)\beta_4,\ \tilde{\beta}_2 = \beta_2 + \beta_4,\ \tilde{\beta}_3 = \beta_3 + \beta_5 - 2\beta_4$$

are related to the corresponding coefficients β_j for ^{3}He and A is a complex symmetric

3×3 matrix with Tr $(A) = 0$. The symmetry group of $U(A)$ is $G = SO(3) \times U(1)$, whereas the action

$$G: A \Rightarrow g: A \to \exp(i\varphi) g A g^{-1}, \quad g \in SO(3), \quad A \in S^2(C^3)_0.$$

Proceeding similarly to the classification of phases in ^{3}He, we obtain the following results.

Classification of orbits relative to the action of G. Let $B_1, B_2 \in S^2(R^3)_0$. The classification of associative algebras $\mathscr{A}(B)$ generated by B_1 and B_2 gets strongly simplified. The following cases are possible:

(1) Full algebra Mat $(3)_R$,

(2a) Mat $(2)_R$, (2b) Mat $(2)_R + R$,

(3) $R + R + R$.

Hence, the stability subgroups H for $\mathscr{A}(B)$ can be found easily.

In case (1), H is trivial, in case (2) we have $H = Z_2$, and in (3), B_1 and B_2 are symmetric and commuting, with two possibilities:

(a) The eigenvalues λ_1 and λ_2 of B_1 are different. The stability subgroup $H \sim Z_2 \times Z_2$. (b) $\lambda_1 = \lambda_2$ and the subgroup $H \sim SO(2) \times Z_2$.

The structure of the algebra of commutators also gets simplified considerably. Since the commutators make up a Lie algebra, only cases $Q = so(3)$ or $Q \sim R$ are possible.

In the first case, the representations of Q in M is irreducible, therefore, $\mathscr{A}(B) \sim sl(3, R)$ (see the proof in [BoM]).

In the second case $Q \sim R$ realization of matrices from Q has the standard form

$$\begin{pmatrix} 0 & \lambda & 0 \\ -\lambda & 0 & 0 \\ 0 & 0 & 0 \end{pmatrix}.$$

Accordingly, the symmetric matrices B_1 and B_2 have the form

$$\left(\begin{array}{c|c} A_i & 0 \\ \hline 0 & 0 \end{array}\right), \quad i = 1, 2,$$

where A_i are symmetric 2×2 matrices with vanishing trace. The algebra $\mathscr{A}(B)$ acts by infinitesimal rotations and $\mathscr{A}(B) \sim sl\,(2, R)$.

It remains for us to consider the action of the external group $U(1)$. It is obvious that *a priori* only the case with $Q \sim R$ is possible. In fact, it is easy to verify that the group $U(1)$ specifies the rotation conjugate to the action of the group $SO(2) \subset SO(3)$ on the $\langle B_1, B_2 \rangle$-plane generated by the matrices

$$B_i = \begin{pmatrix} a_i & b_i & 0 \\ b_i & -a_i & 0 \\ 0 & 0 & 0 \end{pmatrix}, \quad i = 1, 2.$$

In all the other cases, $U(1)$ generates additional Z_2-symmetry.

5.2.3 *A*- and *B*-phases of ^{3}He

Here, we study the topology of the *A*- and *B*-phases, which have been observed experimentally.

We start with a description of the space of internal states for the phases on the basis of physical argument. The detailed description of superfluid phases of ^{3}He from the theoretical and experimental points of view is contained in [Le] and [VW].

***A*-phase.** Its principal properties are determined when the magnetic susceptibility χ_A is measured to be equal to χ_n of the ground state of ^{3}He. It will be shown in (5.64) and (5.65) that this admits natural interpretation if we make the assumption that

$$\Delta_{\alpha\beta}(r) = \Delta_{\alpha\beta} \cdot \psi(r) \tag{5.59}$$

about the condensate, where the spin part describes the state with $S = 1$ and the vanishing projection S_z onto a certain fixed $\hat{z}$-axis. The orbital part is associated with the state when $l = 1$ and $l_z = 1$. As in the ordinary theory of superconductivity, we neglect the dependence of ψ on r. Let $\hat{l}$ be the unit vector along which the orbital momentum of the Cooper pairs points. Then the frame $(\hat{l}, u_1, u_2)$, where u_1 and u_2 are the unit vectors orthogonal to $\hat{l}$ $(\hat{l} = u_1 \times u_2)$, determines the orbital part of Δ locally. The orientation of the pair u_1 and u_2 fixes locally the phase of the wave function's orbital part. The anisotropy of the *A*-phase is characterized by $\hat{l}$ and is called the *anisotropy vector.*

As far as the spin part of the wave function Δ goes, Δ is invariant under rotations of the vector $\hat{z}$ whose direction can be taken to be arbitrary.

If we pass from representation (5.41) to (5.42), then the states of the *A*-phase are described by the order parameter

$$\begin{aligned} A_{pi} &= \lambda \hat{z}_i u^p, \quad \lambda = \text{const}, \\ |u_1| &= |u_2|, \quad (u_1 \cdot u_2) = 0, \end{aligned} \tag{5.60}$$

where $u^p = u_1^p + i u_2^p$, $i = \sqrt{-1}$, $\hat{l} = u_1 \times u_2$, and $\hat{z}$ is the unit spin vector.

The *A*-phase's states are also invariant under discrete symmetry, namely, $z \to -z$ and under simultaneous rotation of the pair (u_1, u_2) through π around the $\hat{l}$-axis.

It can easily be obtained from representation (5.60) that the global description of the manifold V_A is $S^2 \times SO(3)/Z_2$.

We have thus obtained the order parameter's description in terms of $\hat{z}$ and l or directly in terms of the matrices A_{pi}.

The equilibrium state of the superfluid phase is determined by the minimum of (5.44) and should be spatially unknown with $F_{\text{grad}} = 0$. However, to describe real distributions in superfluid phases, we should also take into account different additional forces, e.g., the spin-orbit interaction due to the dipole-dipole interaction between the magnetic momenta of the nuclei in ^{3}He, the external magnetic field which orients the action of the walls of the containing vessel, etc. In a macroscopic description of textures in ^{3}He, each interaction F_{grad}, F_{dip}, and F_H has its own characteristic length scale (L_{grad}, L_{dip}, L_H, etc). Their account leads to the reduction of the vacuum manifold V_A.

We now give formulas for the spin-orbit interaction and the energy densities in a magnetic field in the A-phase. The reader can find the details in the remarkable survey [Le].

Dipole energy density F^A_{dip} for the A-phase is determined in terms of $\hat{z}$ and $\hat{l}$ as

$$F_{\text{dip}} = -(2/5)\, g_d(T)(\hat{z}\cdot\hat{l}), \tag{5.61}$$

where g_d is the factor measuring the contributions of the Cooper pairs to dipole energy. We write

$$F_{\text{dip}} \sim |\text{Tr } A|^2 + \text{Tr } (AA^*) - (2/3) \text{ Tr } (AA^*) \tag{5.62}$$

in terms of the matrix A_{pk}.

It follows from formula (5.61) that F_{dip} is minimum, provided $\hat{z} \parallel \pm \hat{l}$. Following [Le], we write the magnetic field's energy density

$$F_H = (-1/2)\chi_{ij}H_iH_j = (-1/2)\chi_n H^2 + F^A_{H}, \tag{5.63}$$

where χ_{ij} is the tensor of magnetic susceptibility.

The magnetic field H only interacts with the spin vector $\hat{z}$. If H points along $\hat{z}$, then the spins of the pair in the plane perpendicular to $\hat{z}$ do not react to the field. Therefore,

$$\chi_{\parallel} = (N_n/N)\chi_n \tag{5.64}$$

along the field is proportional to the number N_n of unpaired quasi-particles, where N is the total number of quasi-particles. If H is perpendicular to $\hat{z}$, then the paired quasi-particles behave relative to the field as ordinary particles and $\chi_{\perp} = \chi_n$.

The dependence on $\hat{z}$ is of the form

$$F^A_H = (1/2)\chi_{\perp\text{-}\parallel}\, (\hat{z}\cdot H)^2, \tag{5.65}$$

therefore, the total of F_H is minimal for $\hat{z} \perp H$.

In terms of the order parameter matrix A_{pi}, formula (5.65) is written as

$$F_H = (1/2)\chi_{\perp\text{-}\parallel}\, A^*_{pi}\, A_{qi}\, H_pH_q. \tag{5.66}$$

***B*-phase.** It is also described by (5.59) provided that the gap energy Δ does not depend on the direction of the spin vector, which leads to triplet states with $S = 1$ and an equiprobable number of pairs with values $S_z = 1, 0, -1$ of the spin projection onto the quantization axis. A similar condition is also fulfilled for the projection $l_z = 1, 0, -1$ of the orbital momentum. The total angular momentum $J = l + S$ for the B-phase vanishes. Thus, the B-phase is characterized by a simultaneous breakdown of spin and orbital symmetry. In terms of the order parameter matrix A_{pi}, the B-phase can be defined as

$$A_{pi} = \lambda \exp (i\varphi)R_{pi},\ R_{pi} \in SO(3), \tag{5.67}$$

where $\lambda = \lambda(T)/\sqrt{3}$ is a numerical factor.

The degeneracy of the B-phase can be partly evaded if dipole interaction F_{dip} and external magnetic field H are taken into account.

Dipole interaction. We parametrize an orthogonal matrix $R_{pi} \in SO(3)$ by the unit vector $\mathbf{n}$ and an angle θ of rotation about $\mathbf{n}$. We have

$$R_{pi} = \delta_{pi} \cos \theta + (1 - \cos \theta) n^p n^i + \varepsilon_{pik} \sin \theta \, n^k. \quad (5.68)$$

Dipole energy is expressed in terms of the traces of the matrix R as

$$F_{\text{dip}} = g_d/5 \, \{[\text{Tr}\,(R)]^2 + \text{Tr}\,(R^2)\}. \quad (5.69)$$

Putting $\text{Tr}\, R = 1 + 2 \cos \theta$ and $\text{Tr}\, R^2 = 1 + 2 \cos 2\theta$, we obtain

$$F_{\text{dip}} = (4/5) g_d [2 \cos^2\theta + \cos \theta]. \quad (5.70)$$

The minimum of F_{dip} is attained if

$$1 + 4 \cos \theta = 0, \quad (5.71)$$

i.e., θ is of order 104°.

Minimizing dipole energy only fixes θ and leaves the rotation axis arbitrary. Thus, the space of internal states for V_B is isomorphic to $S^1 \times S^2$ if F_{dip} is taken into account.

We now give the formula for magnetic field energy in the B-phase. Switching on an external magnetic field H produces the additional contribution

$$F_H = -g_d \left(\frac{\mu H}{\Delta} \right)^2 \left(\mathbf{n} \cdot \mathbf{H}/|\mathbf{H}| \right)^2 \quad (5.72)$$

to free-energy density [Le]. Its minimum is attained when $\mathbf{n} \parallel \pm \mathbf{H}$.

A simultaneous account of dipole interaction and $\mathbf{H}$ leads to the degeneracy of the vacuum manifold

$$V_B = S^1 \times S^0 \quad (5.73)$$

(S^0 is the zero-dimensional sphere).

A description of the structure of linear and point singularities in superfluid phases of ^{3}He can be conveniently carried out by using the homotopy methods already applied to monopoles, instantons, and liquid crystals. However, no less important are the quantities of directly physical meaning, characterizing linear vortex singularities. Such is first of all the superfluid velocity v_s similar in the B-phase to the superfluid velocity in ^{4}He and represented as

$$v_s^B = \frac{\hbar}{2m_3} \nabla \varphi. \quad (5.74)$$

The form of superfluid liquid in the A-phase is more complicated in structure and given below. Note that from the topological viewpoint, the superfluid velocity v_s is homological in character. Therefore, the relation of the homotopy characterization of singularities to the quantization conditions of the flow v_s resembles the relation between homology and homotopy groups.

Before analyzing singularities in the A- and B-phases directly, we study the structure of superfluid velocity there.

I follow the general approach of [GM1, 5]. The method gives a unified idea of superfluid velocities and uses a general construction in chiral models.

It is useful to consider the A- and B-phases as orbits of group G (5.45) with corresponding selection of stability subgroups H_i.

In the London limit, the order parameter matrix A is in an orbit of G, equalling $\{g(r)A_0\}$, where $g(r)$ are elements of G, $r \in R^3$, and A_0 is a fixed point of the orbit.

Following the approach of the theory of chiral fields, we introduce the superfluid velocity as an element of the Lie algebra $\mathcal{G}$ of G

$$v_s = g^{-1}(r)\partial g(r), \tag{5.75}$$

or as the matrix-valued vector

$$\mathbf{v}_s = g^{-1}\partial g = (R_1\partial R,\ R_2^{-1}\partial R_2,\ u^{-1}\partial u), \tag{5.76}$$

where R_1 and R_2 are elements of the subgroups $SO(3)_R$ and $SO(3)_L$, while $u \in U(1)$.

The definition generalizes the usual one of v_s in ^{4}He

$$v_s = i\,\nabla\varphi \exp(i\varphi)\exp(-i\varphi) = i\,\nabla\varphi, \tag{5.77}$$

as the gradient of φ.

Representing v_s as in (5.76) enables us to obtain a good general formula for the quantization of the flux of v_s.

The quantization condition defined in terms of $v_s = g^{-1}\,\partial g$ requires that the integral $I = \oint_\gamma g^{-1}\partial g\, dr$ should not along a closed curve γ depend on the choice of the latter in its homotopy class. Note that I takes values in the Lie algebra of the gauge group G.

Applying Stokes formula, we obtain the relation

$$\begin{aligned} I_j = \oint_\gamma R_j^{-1}\partial R_j\, dr &= \iint_\sigma \left\{\partial_l(R_j^{-1}\partial_k R_j) - \partial_k(R_j^{-1}\partial_l R_j)\right\} dx^k\, dx^l \\ &= \iint_\sigma \left[R_j^{-1}\partial_k R_j,\ R_j^{-1}\partial_l R_j\right] dx^k\, dx^l \quad (j = 1,\ 2), \end{aligned} \tag{5.78}$$

where [ ] denotes the commutator. It is seen that the condition for I_j to be independent of the path is fulfilled if

$$[R_j^{-1}\partial_k R_j,\ R_j^{-1}\partial_l R_j] = 0 \quad (k,\ l = 1,\ 2,\ 3)$$

on the surface σ bounded by γ.

Let us consider a particular case. Let R_j have one direction in bypassing the whole of γ. We then obtain the well-known phase quantization condition, since

$$R_j^{-1}\partial R_j = \partial\varphi_j \begin{pmatrix} 0 & -1 & 0 \\ 1 & 0 & 0 \\ 0 & 0 & 0 \end{pmatrix}. \tag{5.79}$$

By integrating this along γ, we obtain an integral of the gradient of phase, which is an integer multiple of 2π.

We now find expressions for the superfluid velocities in the A- and B-phases. The case of the B-phase is simpler.

***B*-phase.** Since the order parameter A_{pi} is on the orbit $\Omega = G/H = U(1) \times SO(3)$ and $A_{pi} = \exp(i\varphi)RA_0$ (A_0 is the unit matrix), we have

$$v_s = (u^{-1}\partial u,\ R_1^{-1}\partial R_1),\ \exp(i\varphi) = u \in U(1),\ R_1 \subset SO(3). \tag{5.80}$$

Defining the superfluid velocity in the A-phase is slightly more complicated.

***A*-phase.** First, we consider the case of minimal degeneracy $L \ll L_{\text{dip}}$ where dipole interaction can be neglected. It follows from Subsec. 5.2.2 that the order parameter matrix is representable as

$$A = \exp(i\varphi)R_1A_0R_2^{-1},\ R_1 \in SO(3)_R,\ R_2 \in SO(3)_L,$$

$$A_0 = \lambda \begin{pmatrix} 0 & 0 & 0 \\ 0 & 0 & 0 \\ 1 & i & 0 \end{pmatrix},\ \lambda = \text{const}. \tag{5.81}$$

The action of the group $U(1) = \exp(i\varphi)$ is immaterial as it is included in the action of the group $G_1 = SO(3)_R \times SO(3)_L$.

We have

$$\exp(i\varphi)A_0 = R_\varphi^{-1}A_0R_\varphi,\ R_\varphi = \begin{pmatrix} \cos\varphi, & \sin\varphi, & 0 \\ -\sin\varphi, & \cos\varphi, & 0 \\ 0 & 0 & 1 \end{pmatrix}. \tag{5.82}$$

Thus, the action of G on A is G_1:

$$A \to R_1^{-1}A_0R_2.$$

We introduce the chiral velocities

$$v = i\partial R_1^{-1}R_1, \quad w = iR_2^{-1}\partial R_2, \tag{5.83}$$

where v is the spin velocity and w the orbital velocity in the A-phase. Since they are elements of the Lie algebra of the group $SO(3)$, they can be conveniently represented as

$$v = v_a c_a, \quad w = w_a c_a, \tag{5.84}$$

where $(c_a)_{bd} = i\varepsilon_{abd}$, $a = 1, 2, 3$, are generators of the Lie algebra of $SO(3)$. The representation turns out to be useful for finding textures in the A-phase explicitly (see Subsec. 5.2.7).

At distances $L \gg L_{\text{dip}}$, the energy of dipole interaction should be taken into account. The A-phase is then reduced to space $V_A = SO(3)$ and the action of the gauge group G is represented as

$$G: A \Rightarrow R_1^{-1} A_0 R_1. \tag{5.85}$$

The symmetry properties of (5.85) are shown in Subsec. 5.2.7 to enable us to integrate one-dimensional textures in the A-phase with dipole interaction.

The representation for the superfluid velocity v_s^{orb} associated with the orbital part of the order parameter in the A-phase directly in terms of the vectors u_i has been obtained in [MH]. The representation clarifies the physical meaning of superfluid velocity and, in particular, the detail of ^{3}He is explicit, compared with ^{4}He.

We start with the usual definition of velocity for ^{3}He in terms of the order parameter matrix A_{pk} [VW].

Definition 5.13

$$\tilde{v}_s(\mathbf{x}) = \frac{\hbar}{4m_3 i} \operatorname{Tr}\,[A^+ \nabla A - A \nabla A^+], \tag{5.86}$$

where m_3 is the mass of the atom in ^{3}He.

To stress the difference between this definition and that of v_s in (5.83), v_s in (5.86) is labelled with a tilde. Formula (5.86) is the natural generalization of the definition of v_s in ^{4}He, which involves the particle flux density

$$J = \frac{\hbar}{2im_4} (\bar{\psi} \nabla \psi - \psi \nabla \bar{\psi}), \tag{5.87}$$

where ψ is the wave function of superfluid ^{4}He and m_4 the mass of the atom in ^{4}He.

If we regard the superfluid component's density $\psi\bar{\psi} = |\psi|^2 = |\psi_0|^2$ as constant, then (5.86) is similar to the formula $v_s = \dfrac{\hbar}{m_4} |\psi_0|^2 \nabla \varphi$ for ^{4}He, $\psi = \psi_0 \exp(i\varphi)$.

We can show that (5.86) is transformed as velocity in time reversal, space inversions, and Galilean transformations. Definition 5.13 is thereby justified in all phases of ^{3}He.

The Mermin-Ho formula can easily be derived from the definition for $\tilde{v}_s^{\mathrm{orb}}$ in the A-phase. We represent the matrix A_{ij} in terms of the vectors $\hat{z}_i$ and u_j. We have

$$A_{ij}^+ \nabla A_{ji} - A_{ij} \nabla A_{ji}^+ = \hat{z}^j \bar{u}^i \operatorname{grad} (\hat{z}^j u^i) - \hat{z}^i u^j \operatorname{grad} (\hat{z}^j \bar{u}^i).$$

Taking into account that $(\hat{z}^i)^2 = 1$ and $\hat{z}^i \operatorname{grad} \hat{z}^j = 0$, we obtain

$$\bar{u}^i \operatorname{grad} u^i - u^i \operatorname{grad} \bar{i}^1 = A_{ij}^+ \nabla A_{ji} - A_{ij} \nabla A_{ji}^+. \tag{5.88}$$

We express u^i in terms of $u_1^i + iu_2^i$ and have

$$(u_1^i - iu_2^i) \operatorname{grad} (u_1^i + iu_2^i) - (u_1^i + iu_2^i) \operatorname{grad} (u_1^i - iu_2^i) = -2i[u_2^i \operatorname{grad} u_1^i - u_1^i \operatorname{grad} u_2^i]. \tag{5.89}$$

Substituting this into (5.86) we obtain the Mermin-Ho formula

$$\tilde{v}_s^j = -\frac{\hbar}{2m_3} u_2^i \operatorname{grad} u_1^i - u_1^i \operatorname{grad} u_2^i = \frac{\hbar}{2m_3} u_1^i \partial_j u_2^i. \tag{5.90}$$

For the B-phase, (5.86) yields

$$\tilde{v}_s^B = \lambda \operatorname{grad} \varphi$$

for superfluid velocity in notation (5.80). Comparing $\tilde{v}_s$ with the chiral velocity v_s, we see that it has an important relationship. For the B-phase, $\tilde{v}_s^B$ describes changes in the phase of the factor and is one component of chiral velocity v_s^B. In the A-phase case, the irregularity of embedding of the group $U(1)$ in G leads to a more complicated relation between v_s from (5.84) and $\tilde{v}_s$ from (5.90).

5.2.4 Line and Point Singularities in the *A*- and *B*-phases

Line and point singularities in superfluid phases of ^{3}He can be classified by homotopy methods (the corresponding results were obtained in [TK], [VM]; see also the survey [Me2]). Following the general scheme of Subsec. 5.1.1 to find line singularities, it is necessary to calculate $\pi_1(V)$ for the vacuum manifolds of the A- and B-phases if possible degeneracies of V are taken into account when various interactions are present. A number of topological problems then arise.

One is the classification of singularities in liquid and the classification of surface singularities. The local stability of singularities within one homotopy class is quite important. The problem is solved by analytic estimates, requiring a more refined study of the physics of the phenomenon.

5.2.4.1 Line singularities in the *A*-phase

In classifying singularities, it is convenient to use the idea of characteristic lengths of various interactions.

***A*-phase, $L_\xi \ll L \ll L_{\text{dip}}$.** Only taking into account $U(A)$, the manifold has the form

$$V_A = S^2 \times SO(3)/Z_2. \tag{5.91}$$

We show that $\pi_1(V_A) = Z_4$. It follows from the exact sequence of the homotopy groups of the fiber bundle $S^2 \times SO(3) \to S^2 \times SO(3)/Z_2$ that

$$0 \to Z_2 \to \pi_1(V_A) \to Z_2 \to 0. \tag{5.92}$$

It does not suffice to state that the sequence in (5.92) is exact if $\pi_1(V_A)$ should be calculated. It only follows from (5.92) that $\pi_1(V_A)$ is an extension of Z_2 with the group Z_2.

Using the explicit representation of the factorization $S^2 \times SO(3)/Z_2$, we associate each element $\alpha \in \pi_1(V_A)$ with the map $\tilde{\alpha}$: $(z, u_1, u_2) \to (-\hat{z}, -u_1, -u_2)$

specifying the factorization. The mapping can be obtained in two ways: as a rotation through $+\pi$ or $-\pi$ about the vector $l = u_1 \times u_2$ and as a simultaneous rotation of the vector $\hat{z}$ through π about an arbitrary axis perpendicular to $\hat{z}$.

Let α and β be two elements of $\pi_1(V_A)$ associated with rotations through π and $-\pi$, respectively. Obviously, $\beta = \alpha^3$. The element $\alpha^2 = \alpha \cdot \alpha$ is associated with a nontrivial one of the group π_1 $(SO(3))$ since the path γ associated with rotating $\hat{z}$ through 2π is contracted in S^2 to a point. The other relations are obvious. Thus, $\pi_1(V_A) = Z_4$.

Each homotopy class of the group $\pi_1(V_A)$ can be associated with several different types of line singularities. The corresponding solutions are found by the minimum condition for gradient energy $\delta F_{\text{grad}} = 0$, where F_{grad} is defined in (5.44).

Let N be an element of $Z_4(N = 0, 1, 2, 3)$. We write out the distributions of the fields $\hat{z}$ and u for certain solutions representing all homotopy classes of line singularities [VM], [BL], [Bl].

Depending on the detail of the solution, it may be more convenient to use Cartesian $(\hat{x}, \hat{y}, \hat{z})$, cylindrical $(\hat{\varrho}, \hat{\varphi}, \hat{z})$, or spherical coordinates $(\hat{r}, \hat{\theta}, \hat{\varphi})$ (the carat denotes a fundamental vector).

$N = \mathbf{0}$. (a) Homogeneous distribution:

$$u_1 = \hat{x},\ u_2 = \hat{y},\ l = \hat{z},\ \hat{z}_{\text{spin}} = \text{const},\ v_s = 0. \tag{5.93}$$

(b) Vortex with two circulation quanta:

$$u_1 + iu_2 = \exp(2i\varphi)(\hat{x} + i\hat{y}),\ \hat{z}_{\text{spin}} = \text{const},\ v_s = \frac{\hbar}{m_{3\varrho}}\hat{\varphi}. \tag{5.94}$$

(c) Disclination of the field of $\hat{z}_{\text{spin}}$:

$$u_1 = \hat{x},\ u_2 = \hat{y},\ l = \hat{z},\ \hat{z}_{\text{spin}} = \hat{\varrho},\ v_s = 0. \tag{5.95}$$

$N = \mathbf{1}$. (a) Combination of a vortex with one half of the circulation quantum v_s and of disclination $\hat{z}$ with the circulation quantum $n = 1$:

$$u_1 + iu_2 = \exp(\pm i\varphi/2)(\hat{x} + i\hat{y}),\ \hat{z}_{\text{spin}} = \hat{x}\cos\varphi/2 - \hat{y}\sin\varphi/2,\ v_s = \frac{\hbar}{4m_{3\varrho}}\hat{\varphi}.$$

$N = \mathbf{2}$. (a) Disgyrations-disclinations in the field l:

$$u_1 = \hat{\varphi},\ u_2 = \pm\hat{z},\ l = \pm\hat{\varrho},\ \hat{z}_{\text{spin}} = \text{const},\ v_s = 0.$$

(b) Vortex with circulation of v_s quantum ± 1:

$$u_1 + iu_2 = \exp(\pm i\varphi)(\hat{x} + i\hat{y}),\ \hat{z}_{\text{spin}} = \text{const},\ v_s = \pm\frac{\hbar\hat{\varphi}}{2m_{3\varrho}}.$$

(c) Disgyration "hedgehog":

$$u_1 + iu_2 = \theta + i\hat{\varphi}, \quad l = \hat{r}, \quad \hat{z}_{\text{spin}} = \text{const}, \quad v_s = -\frac{\hbar}{2m_3}\frac{\hat{\varphi}}{\varrho}\frac{\cos\theta}{\sin\theta}.$$

(d) "Haired-hedgehog" vortex:

$$u_1 = \hat{\varphi}, \; u_2 = \hat{r}, \; l = \hat{\theta}, \quad \hat{z}_{\text{spin}} = \text{const}, \; v_s = 0.$$

By "folding" in two the vortex, it can be turned into one with two circulation v_s quanta in the A-phase container, or a vortex with an "endpoint".

(e) Vortex with an "endpoint"

$$u_1 - iu_2 = \exp(i\varphi)(\hat{\theta} + i\hat{\varphi}), \quad v_s = \frac{\hbar\hat{\varphi}(1 - \cos\theta)}{2m_3\varrho\sin\theta}. \tag{5.96}$$

$N = 3$. $u_1 + iu_2 = \exp(-i\varphi/2)(\hat{x} + i\hat{y})$, $\hat{z}_{\text{spin}} = \hat{x}\cos\varphi/2 + \hat{y}\sin\varphi/2$,

$$v_s = -\frac{\hbar\hat{\varphi}}{4m_3\varrho}.$$

Singularities within one homotopy class can be deformed into each other. Restructuring a singularity core requires that an energy threshold of order $(L_\xi)^3$ should be overcome.

Analysis of local stability, or estimating the integral $\int F_{\text{grad}}\, d^3r$, carried out in [BL] shows that singularities with $N = 1, 3$ and $N = 2$ (see (a) and (c)) are locally stable, when $N = 2$ (see (b), (d), and (e)) the singularities are unstable and can relax into singularities of the type $N = 2$ (see (a) and (c)) without the threshold.

A-phase, $L_{\text{dip}} \ll L_\xi$. At these distances, the energy F_{dip} of dipole (spin-orbit) interaction should be taken into account. The vector of the spin part of the order parameter $\hat{z}$ is then parallel to the vector l and the manifold V_A degenerates into $V_A^{\text{dip}} \sim SO(3)$. Since $\pi_1(V_A^{\text{dip}}) = Z_2$, there exists only one nontrivial homotopy class $\beta \in Z_2$ which determines a line singularity. The range $\mathcal{U}_1$ of distances $L > L_{\text{dip}}$ is an extension of the domain $\mathcal{U}\,(L_\xi < L < L_{\text{dip}})$. Therefore, it would be interesting to clarify what singularity types survive if dipole interaction is taken into account.

We begin by analyzing the physical situation. The account of the boundary condition $\hat{z} \parallel \hat{l}$ shows that under the degeneracy $V_A \to V_A^{\text{dip}}$ only those solutions which are in the homotopy classes with $N = 0$ and $N = 2$ survive. Distributions (5.93) and (5.94) with $N = 0$ can continuously go from $\mathcal{U}$ to $\mathcal{U}_1$. The vortex solutions with $N = 2$ can for topological reasons continuously go from $\mathcal{U}$ to $\mathcal{U}_1$. However, a solution minimizing F_{grad} in the A-phase if no dipole interaction is taken into account may now not minimize F_{grad} if F_{dip} is considered. It is asserted in [VM] that the minimum of energy in the A-phase when F_{dip} is taken into account is attained in the solution class with $N = 2$ for a vortex with one circulation v_s quantum and with vector l perpendicular to the vortex axis.

Since $\pi_1(V_A^{\text{dip}})$ is embedded in $\pi_1(V_A)$ naturally, it is convenient to label the homotopy classes $\pi_1(V_A^{\text{dip}})$ by the two numbers $N = 0$ and $N = 2$ and add the classes together modulo 4. Singularities with $N = 1, 3$ are absent in $\mathscr{U}_1$, which means from the physical viewpoint that in the transition from $\mathscr{U}$ to $\mathscr{U}_1$ the singularities dissipate, turning into domain walls of thickness of order L_{dip}. The energy threshold to destroy them is of order $F_{dip}L_{dip}^3$. The domain walls can be classified by the group Z_2, but are related to various types of order parameter degeneracy. The classification for ^{3}He can be found in [MV].

Remark 5.6 Transitions of vortex solutions from $\mathscr{U}$ to $\mathscr{U}_1$ if the order parameter degenerates are the simplest illustration of the structure of topological charges as invariants of cobordism classes; this is developed in Subsec. 5.2.8.3.

In fact, since $\mathscr{U}_1$ and $\mathscr{U}$ have the common boundary sphere $L = L_{dip}$, a continuous transition of vortex solutions from $\mathscr{U}_1$ to $\mathscr{U}$ can be regarded as a cobordism J with boundary manifolds determined by the vortex solutions φ in $\mathscr{U}$ and by φ_1 in $\mathscr{U}_1$. The necessary condition for such a cobordism to exist is that the nontrivial homomorphism

$$\pi_1(V_A) \to \pi_1(V_A^{dip}) \Rightarrow Z_4 \to Z_2 \tag{5.97}$$

should be there.

We now pass to the classification of point singularities in the A-phase container.

5.2.4.2 Point singularities in the A-phase

$L_\xi \ll L \ll L_{dip}$. According to the general recipe, we have to calculate

$$\pi_2(V_A) = Z,$$

or $\pi_2(S^2 \times SO(3)/Z_2)$, to classify the point singularities.

Singular points are only determined by the degree of a map of the field $\hat{z}$. As in a nematic, interaction between point singularities is determined by the action of the fundamental group $\pi_1(V_A)$ on $\pi_2(V_A)$. Nontrivial actions relative to the vector $\hat{z}$ are associated with elements α and α^3 of $\pi_1(V_A)$. The actions of α and α^3 are the same on $\hat{z}$. Let $\beta \in \pi_2(V_A)$. The action of α on β,

$$\alpha(\beta) = -\beta,$$

is defined as for a nematic. Therefore, point singularities in the A-phase are determined by the index $|N|$. Depending on the merger path, two identical point singularities may annihilate.

$L \gg L_{dip}$. In this domain, $\pi_2(V_A^{dip}) = \pi_2(SO(3)) = 0$ and there are no point singularities.

In the transition from $\mathscr{U}$ to $\mathscr{U}_1$, so-called *linear solitons* are formed, i.e., singular points in $\mathscr{U}$ dissipate into cylindrical tubes in $\mathscr{U}_1$ [VM].

The linear solitons (cylindrical tubes) are characterized by the same index $|N|$ as the singular points in $\mathscr{U}$ generating the solitons are. They are of thickness $(L_\xi \cdot L_{dip})^{1/2}$. The stability threshold is $F_{dip}\,(L_\xi \cdot L_{dip})^{3/2}$. Soliton addition is naturally equivalent to that of point singularities in $\mathscr{U}$.

5.2.4.3 Line singularities in the *B*-phase

$L_\xi \ll L \ll L_{\text{dip}}$. For these distances, the order parameter is $A_{pk} = \lambda \exp(i\psi)R_{pk}$ (5.67); therefore, $V_B = S \times SO(3)$, $\pi_1(V_B) = Z + Z_2$. Thus, the vortex solutions are characterized by an integer N and by $N_1 \in Z_2$.

The vortex solutions associated with the index N are ordinary vortices in the field grad ψ, $\psi(r) = N\varphi$, where φ is the zenith angle in cylindrical coordinates with the z-axis along the vortex.

The vortex solution with $N = 0$ and $N = 1$, which provides a minimum gradient energy (5.44) for the *B*-phase, namely,

$$\hat{\omega} = \hat{z}, \quad \theta = \pi - \varphi,$$

is given in [VM] where $\hat{z}$ is the direction of the z-axis.

$L \gg L_{\text{dip}}$. Since an account of dipole interaction (5.70) fixes the angle $\theta \simeq 104°$, the manifold V_B degenerates into $V_B^{\text{dip}} = S^1 \times S^2$, $\pi_1(V_B^{\text{dip}}) = Z$. Therefore, only line singularities with N circulation quanta exist in the domain. Something similar to the formation of linear solitons in the *A*-phase occurs to line singularities with $N_1 \neq 0$. In the transition from $\mathcal{U}$ to $\mathcal{U}_1$, "two-dimensional" domain walls (i.e., plane solitons [MV]) are formed, ending on the line singularities with index $N_1 = 1$. The walls are in a one-to-one correspondence with line singularities of $\pi_1(SO(3))$, forming the group Z_2. They are L_{dip} thick, with annihilation energy threshold $F_{\text{dip}}\,(L_{\text{dip}})^3$.

5.2.4.4 Point singularities in the *B*-phase

$L_\xi \ll L \ll L_{\text{dip}}$. $\pi_2(V_B) = 0$. Therefore, there are no point singularities here.

$L \gg L_{\text{dip}}$. The point singularities are those of the field $\hat{\omega}$, which coincide with the degree of a map of $\hat{\omega}$. The simplest solution with $N = 1$ is hedgehog shaped $\hat{\omega} = \hat{r}$.

Since V_B^{dip} is the direct product $S^1 \times S^2$ and $\pi_2(S^1) = 0$, $\pi_1(V_B^{\text{dip}})$ acts trivially on $\pi_2(V_B^{\text{dip}})$. Hence, line singularities do not affect point ones. In the transition from $\mathcal{U}_1$ to $\mathcal{U}$, the point singularities dissipate and $\pi_2(SO(3)) = 0$. Such singularities are said to be *without singular core.*

5.2.5 Surface Singularities

Singularities can exist in ^{3}He both inside the liquid and on the surface. Surface singularities are intimately related to the surface topology and lead to important physical consequences, e.g., to the continuous decay of superfluid current j_s and to other effects observed experimentally. We discuss them at the end of the subsection.

From the mathematical viewpoint, the problem of classifying surface defects is solved by combining differential geometric methods with the simple topological properties of two-dimensional manifolds.

The *A*-phase is of greatest interest. The classification of surface singularities and their relation to spatial structures was given in detail in [Mel] and [AP].

5.2.5.1 Surface singularities in the A-phase [Mel], [AP]

Let $\mathcal{U}$ be a domain in R^3 with boundary $\partial \mathcal{U} = M^2$, an arbitrary compact orientable surface. When classifying singularities we restrict ourselves to the orbital part of the order parameter A_{pi}.

In studies of ^{3}He we commonly use the London approximation. The requirement is not very restrictive and is equivalent to the condition for transitivity for a "gauge" group G on a vacuum manifold V. In terms of the vectors **u** and **l**, the requirement leads to the boundary condition that the vectors $\mathbf{u}_1$ and $\mathbf{u}_2$ are tangent to the surface M^2 and that **l** is thereby normal to M^2. The vector **l** is $\pm\mathbf{n}$ (**n** is the exterior for normal to M^2) and therefore the surface is divided into ranges of positive and negative values $(\mathbf{n}\cdot\mathbf{l})$ separated by line singularities for u. In [Mel] the domains where **l** does not change sign are called *islands* and singularity lines *boundaries.* Our choice of a boundary condition means that the order parameter frame on the islands has only one degree of freedom, rotation through φ about **l**. The manifold of internal states is $V_{\partial\mathcal{U}} = S^1$ and therefore singular points on the surface are point singularities of $\mathbf{u}_1$ or of $\mathbf{u}_2$ fields with quanta of circulation m. The latter singularities can be regarded as terminal points of the quantized vortices within the container.

We shall be interested in two problems. First, what spatial structures generate surface singularities? Second, which is simpler, what restrictions are imposed on the types of surface singularity by the topology of a two-dimensional manifold?

We start with the second question by imposing the condition that the sign of $\mathbf{n}\cdot\mathbf{l}$ does not change on the whole surface. There are no boundaries and so the whole surface is a single island. The immediate answer from Poincaré's theorem (2.5.2) is that

$$\sum_{a_i} \operatorname{ind} \mathbf{v} = \chi(M^2), \tag{5.98}$$

where $\mathbf{v}$ is the tangent field on the surface M^2 and a_i are singular points of $\mathbf{v}$. The field $\mathbf{v}$ can be given differently, e.g., $\mathbf{v}$ can coincide with one of $\mathbf{u}_j$. It will be convenient to define $\mathbf{v}$ as proportional to grad φ, where φ is the angle of rotation of the frame about the l-axis, in which case $\mathbf{v}$ coincides with superfluid velocity $\mathbf{v}_s$ (5.86) on the surface. By relation (5.90) and the Gauss-Bonnet formula (2.63) for a spherical mapping of M^2, with $\mathbf{l} = \mathbf{n}$ selected, (5.98) can be rewritten as

$$\frac{1}{4\pi}\int_{M^2} (1/2)\varepsilon_{ijk}\mathbf{l}\cdot\left(\frac{\partial \mathbf{l}}{\partial x^i} \times \frac{\partial \mathbf{l}}{\partial x^j}\right) dS$$

$$= \frac{m_3}{2\pi\hbar}\int_{M^{2\prime}} \operatorname{rot} \mathbf{v}_s dS' = \chi/2, \tag{5.99}$$

where $M^{2\prime}$ is obtained from the surface M^2 by removing the singular points of the field. Applying the Stokes formula gives

$$\int \operatorname{rot} \mathbf{v}_s dS' = \sum_{a_i} \oint_{\gamma_{a_i}} \mathbf{v}_s d\boldsymbol{\tau} = \frac{\pi\hbar}{m_3}\chi, \tag{5.100}$$

where γ_{a_i} is the infinitesimal contour surrounding a point a_i. Application of Poincaré's theorem thus yields a restriction on the sum of circulation quanta of singular points on the surface, e.g., if there exists one point singularity on the sphere, then the singularity should have a circulation quantum of two. Another important remark is that the unique surface with no singular points is a torus. It is useless to argue from the bulk estimation of the index of a vector field to clarify which is the more real configuration: existence of one singular point of index m or several singular points with bulk index m? To solve the problem, the coupling energy for several point singularities must be estimated and established which spatial configurations they proceed from.

We start with the space $\mathcal{U}$ bounded by the surface homeomorphic to the sphere S^2, first assuming that there are no islands.

It is obvious that singular points on the surface can relax into structures with total circulation quantum two.

We begin by analyzing textures in the class of line singularities such that $N = 0$, i.e., topologically nonsingular solutions.

At first glance, two types of solution sent into point singularities on the surface seem natural:

(1) a point singularity with an adjoint vortex with two quanta of circulation (a 4π-vortex),

(2) two vortices with quanta of circulation $+1$ emanating from a point singularity and moving in opposite directions. Unfortunately, this simple picture is not true for several reasons. One is topological. It follows from the analysis in Subsec. 5.2.4.2 that there are no stable singular points in the orbital part in the A-phase (for $L \ll L_{\mathrm{dip}}$). Therefore, an isolated singular point in the liquid cannot relax into a surface one. A vortex singularity in the liquid must necessarily be either the terminal point of a vortex filament (line singularity) or dissipate into a nonsingular texture. The second reason is the instability of a similar configuration. Nevertheless, spatial textures relaxing into surface singular points can be obtained by a certain modification of vortices of types (1) and (2).

Let us consider a coreless Anderson-Toulouse vortex texture [AT]. We take the domain $\mathcal{U}$ bounded by S^2 and isolate a cylindrical tube $\tilde{\mathcal{U}}$ of radius R in $\mathcal{U}$. Suppose the vortex in (5.94) is given in $\mathcal{U}$ and the vector $\mathbf{l}$ is in the direction of the z-azis, e.g., in the positive direction.

The vortex belongs to the trivial homotopy class $N = 0$ and hence can be deformed into a nonsingular configuration. The corresponding deformation in the cylinder $\tilde{\mathcal{U}}$ can be carried out as follows. We surround the vortex item (the z-axis) by a circle γ of radius R and construct a deformation of $\mathbf{l}$ along any radius $\mathbf{r}$ in the cross-section of the cylinder orthogonal to z such that $\mathbf{l} = f(\varphi(\mathbf{r}))$ and for $\mathbf{r} = 0$ and $\varphi = 0$ we have $\mathbf{l} = \hat{\mathbf{l}}$, while for $\varphi = \pi$ we have $\mathbf{l} = -\hat{\mathbf{l}}$, i.e., $\mathbf{l}$ rotates through π on the circle of radius R. The frame of $\mathbf{u}_1$ and $\mathbf{u}_2$ is then rotated through π on γ, too.

The superfluid velocity $\mathbf{v}_s$ vector described in cylindrical coordinates as $\dfrac{1 + l_z}{2rm_3}\hat{\varphi}$ vanishes on γ.

A continuous deformation of the vortex with two quanta of singularity into a nonsingular configuration is thereby constructed.

We consider another singular configuration: a vortex of type (5.96) with "endpoint", i.e., defined only in the half-space $z < 0$. For $z > 0$, the uniform configuration of the field can be reduced to the form $l = r$. We consider the field $\mathbf{v}_s$ of the vortex. In spherical coordinates (r, θ, φ),

$$\mathbf{v}_s = \frac{(1 - \cos\theta)\hat{\varphi}}{r \sin\theta}, \tag{5.101}$$

where $\mathbf{r}$ is the radius-vector of the point $a = (r, \theta, \varphi)$ measured from the vortex endpoint at the origin. It follows from formula (5.101) that $\mathbf{v}_s$ has a singularity on the line $\theta = \pi$; as $\theta \to 0$, $\mathbf{v}_s \to 0$, i.e., there are no singularities on the upper half-axis $z > 0$. Thus there exists a vortex with endpoint [Bl]. A vortex with endpoint is also called a *vorton*, or a *monopole*, because it is similar to a Dirac monopole. Relation (5.101) is similar to the field of the potential vector of Dirac monopole A_μ (4.3) for an arbitrary replacement of the coefficients $\hbar/2m \to 2mc/e$. However, there is an essential difference. A Dirac monopole is a stable point singularity, whereas a monopole in the A-phase is unstable with a tail and a vortex line singularity. A monopole's energy is proportional to the length of the tail. Therefore, it is favorable to minimize the energy or contract the tail into a point of the boundary of the domain filled with the A-phase. We then obtain a stable surface singularity.

5.2.5.2 Boojum

The physicists's love of neologisms has led to many singularities, of which the commonest is the "Boojum", which successfully encapsulates the enigmatic properties of point singularities.

The term was introduced by Mermin, who borrowed it from Lewis Carroll's poem *The Hunting of the Snark**. The enigmatic Snark hunters are warned:

"...beware of the day,
If your Snark be a Boojum!"

What the "suddenly vanishing Boojum" is remains totally unclear. The mysterious "Boojum" brought about a flood of research papers. There is a very interesting book by Eric Partridge entitled *Lewis Carrol's and Edward Lir's Meaningless Words* but it does not provide a convincing explanation of the term.

We now leave the literary Boojum and consider topological ones.

Two problems are related to the definition.

(1) What spatial structure relaxes into a Boojum with normal boundary conditions for a vector $\mathbf{l}$?

(2) How is a Boojum related to the topology of the surface $\partial\mathcal{U}$ bounding the domain $\mathcal{U}$ filled with the A-phase?

We have actually obtained the answer to the former question in the foregoing subsection for a cylinder or sphere domain. This is either a coreless Anderson-

---

* N. Mermin recounted how the term appeared in "E Pluribus Boojum", *Physics Today*, April, 1981, pp. 46-53, and the funny adventures to legalize it was made.

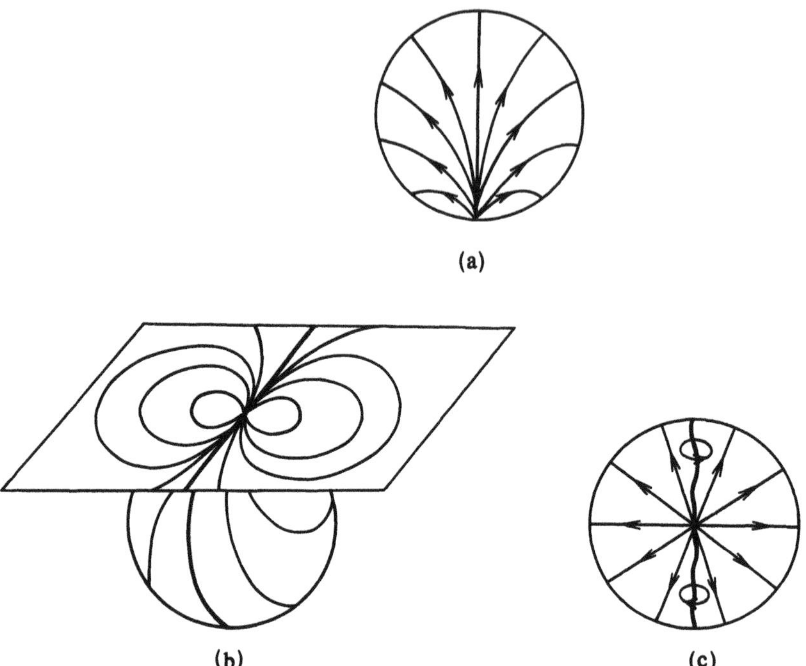

Fig. 22. Surface singularities: (a) Boojum, (b) construction of Boojum field, (c) two surface singularities due to breaking of vortex in a container

Toulouse vortex or a monopole contracted along the tail into a boundary point (Fig. 22*a*, *b*). The consistency of the normal boundary conditions for **l** and with spatial distribution of the field **l** is obvious. The reply to the latter question is contained in Poincaré's theorem since a Boojum on the surface is a zero of the tangent vectors (u_1, u_2). Thus, the necessary condition for a Boojum to exist is that $\chi(\partial\mathscr{U}) \neq 0$ and, therefore, only a torus can have no singular points.

Poincaré's theorem yields a restriction only for the bulk index of a vector field. It can easily be seen that several singular points can exist on the surface. For definiteness, consider the domain bounded by the sphere S^2, e.g., as a matter of principle, two singular points with a circulation quantum 1 may exist (Fig. 22*c*). What is the spatial texture to which they correspond? Obviously, they cannot be terminations of one singular filament since the circulations of the points where they can go onto the surface are of opposite signs.

A natural possibility is a breaking of the two-quanta circulation vortex which is within the container's depth and the formation of two monopole singularities moving in the opposite direction. We call this configuration C_1. It has been shown [AP] that such a configuration is unstable. However, it can be continuously deformed into another with the same behavior on the boundary.

Let C_1 be given in a spherical volume. The vector **l** is in the positive z-axis direction for points above the center O of the sphere with monopole singularities and in the negative direction on the half-axis $z < 0$. The vortex filament in the upper hemisphere defines a positive vortex with circulation 1, while a vortex in the lower hemisphere defines a vortex with circulation -1. The points where the vortices reach the surface have identical quanta of circulation.

We show that a positive vortex can be continuously deformed into a negative when crossing the singularity in the field of **l**. In fact, the transition from the upper hemisphere to the lower one can be made by rotation through π, the vector **l** at 0 being perpendicular to the z-axis. The corresponding singularity in the field of **l** is radial disgyration. The transition from the direction $+z$ at 0 to $-z$ can be carried out continuously by rotating **l** relative to one of the vectors of pair (u_1, u_2). The texture more energetically favorable than a two-vortex one is thereby obtained since infinite gradients of the order parameter of a monopole singularity are transformed into finite disgyration gradients with the configuration type remaining unaltered.

It is evident that a vortex singularity with large values of the circulation quanta can form on the surface. However, the points are less energetically favorable than the above solutions.

5.2.5.3 Islands on a surface

It has been noted (see Subsec. 5.2.5.1) that line singularities or island boundaries can be formed on a surface.

Since the boundary condition for a vector **l** leads to the space of internal states in $O(2)$, it is obvious that line singularities γ, or the lines where **l** changes its sign, are classified by the group $\pi_0(O(2)) = Z_2$. The singularities γ cannot be boundaries of singular two-dimensional domains since $\pi_2(SO(3)) = 0$ (recall that we are in the region $L \gg L_{\text{dip}}$). Therefore, γ are also the lines going into the surface of spatial textures.

A study of surface line singularities is based on the generalization of the Gauss-Bonnet theorem for manifolds with a boundary.

Let γ_i be line singularities of the surface. It is obvious that they can be regarded as disjoint simple arcs. In fact, if two curves do intersect, then the intersection can be eliminated by a small deformation.

For simplicity, consider the domain $\mathscr{U} \subset M^2$ bounded by one curve γ. The argument is similar to arbitrary domains with trivial modifications. We parametrize γ by the arc length, selecting the direction in which the parameter s increases so that the vector product $\mathbf{d} = \mathbf{w} \times \mathbf{n}$ of the unit tangent vector $\mathbf{w}$ and normal $\mathbf{n}$ points to where $\mathbf{n}\cdot\mathbf{l}$ is positive (i.e., $\mathbf{d}$ always points inside the domain).

The formula for the circulation of the field of superfluid velocity $\mathbf{v}_s$ on the boundary $\partial\mathscr{U} = \gamma$, which consists of finitely many simple arcs, can be written as

$$\oint \mathbf{v}_s \, ds = 2\pi m - \iint_{\mathscr{U}} K \, dS, \tag{5.102}$$

where m is the algebraic sum of the quanta of circulation of point vortices in $\mathscr{U}$.

Representing directly the contour integral of $\mathbf{v}_s$ with respect to $\partial\mathscr{U}$ yields

$$\oint \mathbf{v}_s \, ds = \int \left[\frac{d\theta}{ds} \, ds + \mathbf{w} \cdot \left(\mathbf{l} \times \frac{d\mathbf{w}}{ds} \right) ds \right], \tag{5.103}$$

where the term $\mathbf{w} \cdot \left(\mathbf{l} \times \frac{d\mathbf{w}}{ds} \right) = k_g$ is the geodesic curvature of the curve.

If we regard the domain's boundary as consisting of geodesic lines, then $\mathbf{w} \cdot \left(\mathbf{l} \times \frac{d\mathbf{w}}{ds} \right) = 0$. Equating both expressions for $\mathbf{v}_s$ and taking the Gauss-Bonnet formula (2.63) into account, we obtain

$$\frac{1}{2\pi} d\theta + \int \mathbf{w} \cdot \left(\mathbf{l} \times \frac{d\mathbf{w}}{ds} \right) ds = m - \frac{1}{2\pi} \int_{\mathscr{U}} K \, dS = m - \chi(\partial\mathscr{U}), \quad (5.104)$$

where $d\theta$ is the total variation of the angle θ when bypassing the boundary $\partial\mathscr{U}$.

Formula (5.104) is favorable if surface line singularities are analyzed. We start with general properties and adopt the following convention regarding the signs of vortices in $\mathscr{U}$. A vortex ω is regarded as *positive* (*negative*) if the vector $\mathbf{v}_s$ rotates counterclockwise (clockwise) for an observer looking along the vector $\mathbf{n}$ pointing out of the surface. Let $\mathbf{l}$ point in the same direction as $\mathbf{n}$ everywhere except some islands with reverse orientation.

If we introduce a net on a surface associated with a topologically regular decomposition so that the transition from the domain $(\mathbf{n} \cdot \mathbf{l}) > 0$ to $(\mathbf{n} \cdot \mathbf{l}) < 0$ is equivalent to that from one two-face to another, then the total number of the vortices $m_t = m_+ - m_-$ is

$$m_t = E_+ - E_- + (1/(2\pi))(\Delta\theta_+ - \Delta\theta_-), \quad (5.105)$$

where E_+ is the number of positively oriented simplexes and E_- the number of negatively oriented simplexes of decomposition of M^2. It is obvious that the transition from a positively oriented domain into a negatively oriented one (an island) decreases E_+ by 1, since the procedure is equivalent to cutting one face out of E_+. Accordingly, the transition from $\mathscr{U}$ to $\mathscr{U}_+$ increases E_- by 1. Thus, N islands yield the relation

$$E_+ - E_- = \chi(M^2) - 2N. \quad (5.106)$$

It follows from (5.106) that for all surfaces except a sphere with one island, m_t increases in absolute value when new islands are formed. However, if the number of quanta of circulation decreases, then this is an argument for jumps of discontinuity of $\Delta\theta_\pm$ (5.105) to exist.

Following [Mel], we shall show how the account of $\Delta\theta$ occurs. Note that the local argument (which is differential geometric in nature) holds for any type of surfaces. The topology of a surface depends on the Euler characteristic and affects the total number of the vortices.

Let γ be a part of the boundary of an island $\mathscr{U}$. The vector $\mathbf{l}$ outside $\mathscr{U}$ points "upwards" while inside $\mathscr{U}$, $\mathbf{l}$ points "downwards". Consider a small arc τ of the circle, starting at a point $a \notin \mathscr{U}$, ending at a point $b \in \mathscr{U}$ and passing through the A-phase container. The magnitude of $\Delta\theta$ is determined from the change of the frame ($\mathbf{u}_1$, $\mathbf{u}_2$) on the boundary.

Traversing along τ we have

$$\begin{aligned} \mathbf{u}_1 &= (\mathbf{w} \times \mathbf{l}) \cos\theta + \mathbf{w} \sin\theta, \\ \mathbf{u}_2 &= -(\mathbf{w} \times \mathbf{l}) \sin\theta + \mathbf{w} \cos\theta. \end{aligned} \tag{5.107}$$

The vector $\mathbf{l}$ should reverse the direction. If the curve τ has no torsion, then the corresponding transformation $\mathbf{l}$ consists in rotation through $\varphi = \pi$ in the plane containing τ. Such boundaries are said to be *twistfree*. Meanwhile, if $\mathbf{l}$ turns in the same direction as that of traversing τ, then such a boundary is a *circular disgyration*, otherwise, *hyperbolic*.

The divergence in the field of $\mathbf{v}_s$ on the boundary can be eliminated if the phase θ does not depend on $\mathbf{l}$ when moving along τ, i.e., the vectors $\mathbf{u}_1$ and $\mathbf{u}_2$ do not rotate about the local direction of the vector $\mathbf{l}$. For twistfree boundaries, this leads to the condition that a vector $\mathbf{u}_1$ parallel to the boundary before moving along τ remains parallel to it after the traverse. The vector $\mathbf{u}_2$ changes its sign. We can imagine that the final position of the frame ($\mathbf{u}_1''$, $\mathbf{u}_2''$, $\mathbf{l}$) is obtained from the initial one ($\mathbf{u}_1'$, $\mathbf{u}_2'$, $\mathbf{l}$) by rotation about the normal $\mathbf{n}$ through π in the plane with τ, and then there is no jump in θ. Therefore, on all surfaces except a sphere with one island, the number of islands with twistfree boundaries increases that of surface vortices.

The situation is different with a sphere with one island. If the direction of the normal $\mathbf{n}$ is given as outward, then the change of sign of $(\mathbf{n} \cdot \mathbf{l})$ on the island (e.g., on the lower half of the hemisphere) enables us to preserve a constant projection of the vector $\mathbf{l}$ on the z-axis in the exterior domain of the sphere (e.g., where $(\mathbf{n} \cdot \mathbf{l})$ is positive), thereby obtaining a solution for $\mathbf{v}_s$ on the sphere (or the cylinder) in cylindrical coordinates

$$\mathbf{v}_s = \frac{-(1 + l_z)\hat{\varphi}}{2rm_3}. \tag{5.108}$$

The solution is the coreless Anderson-Toulouse vortex already familiar to us. The field of velocity $\mathbf{v}_s$ has no singularity in the domain $\mathscr{D}^3$.

Note that the projection l_z of solution (5.108) vanishes on the equator. If the boundary γ of the island does not pass along geodesic lines, then the vector $\mathbf{v}_s$ has a jump on the boundary. In fact, the change in the tangential component of $\mathbf{v}_s$ is proportional to the curve's geodesic curvature k_g. It follows from this result that, in particular, the system of islands with twistless boundaries is unstable.

It is obvious that it is energetically favorable to contract an island into a point, e.g., a pole. We then obtain the configuration associated with the coreless Anderson-Toulouse vortex texture attaining the surface. Mermin considered twist boundaries in the same way. It follows from (5.106) that considering constant-twist boundaries does not alter the conclusions for twistless boundaries. If boundaries have mixed twist, then we can decrease the number of vortices for surfaces of genus $g > 2$ by

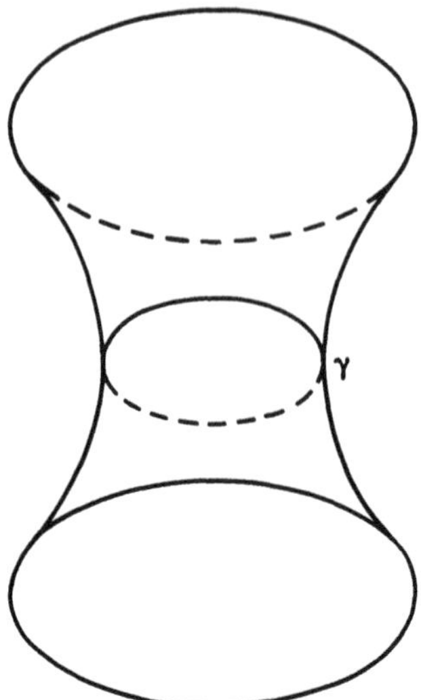

Fig. 23. Linear singularity γ of a surface with an hourglass shape

adding the corresponding number of islands. The contribution to the total energy from the gradient terms then decreases.

Remark 5.7 Note that when comparing topologically admissible solutions with observed textures, the energy of the latter should be taken into account. The simplest restrictions of this sort arise even if the geometry of the manifold in the Gauss-Bonnet formula is taken into account.

As an example, we take a surface $\tilde{M}^2$ in an hourglass shape (Fig. 23). Topologically, it is equivalent to the sphere. However, with contributions to the energy of terms proportional to geodesic curvatures, a surface configuration of an equatorial half-disgyration type may be favorable for $\tilde{M}^2$, a configuration different from a Boojum [AP].

Remark 5.8 We actually encounter the classical mathematical problem of determining the type of a manifold M^n from the asymptotic behavior of the eigenvalues of the Laplacian on M^n. The first term in the asymptotic is known to be proportional to vol (M^n) and is therefore related to the total curvature $K(M^n)$ and the Euler characteristic $\chi(M^n)$: the next term takes the contribution of geodesic curvature into account. In problems when identifying defects in superfluid phases, the corresponding spectral data is obtained from NMR data, which is in turn described by the Leggett equations for the spin and momentum of Cooper pairs [Le], [VW].

We now pass to the physical consequences of surface defects.

5.2.5.4 Boojum and superfluid current decay

Consider the motion of the flux of the superfluid velocity $\mathbf{v}_s$ on the surface bounding an A-phase container. It has been shown that point singularities, i.e., Booja whose bulk index (circulation guantum) is determined by the Euler characteristic, can exist on the surface.

Let the superfluid flow be given in a torus T^2. $\chi(T^2) = 0$ and there can be no Boojum on T^2. However, there can be a random number of Booja with Σ ind $= 0$. We select a point x on the surface T^2 at which a pair of Booja with opposite circulation quanta $\mathbf{v}_s$ are generated. If the Booja b_1 and b_2 are pulled apart a distance L and then one is made to encounter the other after the cycle γ_1 is bypassed n times, then the total circulation quantum of b_1 and b_2 is $1 + (-1 \mp 2n)$.

Thus, the Boojum loses (acquires) two quanta of circulation if a cycle on T^2 is traversed. Encountering the superfluid flow with N quanta of circulation, after traversing the torus along its cycles, a Boojum can completely destroy the superficial superfluid velocity. This "sudden" catastrophe, the collision between a point singularity on the surface and the superfluid flow is what caused Mermin to coin the term Boojum.

The stability of superficial superfluid velocity for domains of arbitrary type is determined solely by the Euler characteristic and the coupling energy necessary for two Booja generation. Normally, there are always some vortices in the container. A similar picture arises if floating islands are considered. However, it has been shown above that it is more energetically favorable for islands to minimize the boundary, i.e., to contract simply connected islands into a point.

5.2.6 Domain Structure in the B-Phase of ^{3}He

In the foregoing, we considered the topological structure of singularities in ^{3}He A- and B-phases. Some interesting nonsingular structures also arise.

The following problem is of interest both from the physical and mathematical viewpoint. Do domain wall type textures exist in ^{3}He? Since order parameter ^{3}He is complicated, such solutions could describe the interactions of various phases or the structure of the core of a concrete superfluid phase. I shall discuss possible applications at the end of the subsection.

From the topological viewpoint, domain wall solutions related to various ranges of the order parameter specify transitions between cobordant manifolds with different topological charges.

Domain wall solutions have been considered in a number of works with the order parameter assuming different values while belonging to a fixed phase of ^{3}He [Mak], [MK1, 2].

Another class of solution is of interest and was first constructed in [GM3]. We consider a spherical domain $\mathscr{U}$ (droplet) filled with B-phase. The boundary $\partial\mathscr{U}$, or the domain wall, is not in the B-phase, while the domain outside is in the phase. The order parameter takes different values inside and outside the droplet.

We now describe the explicit solutions using the Ginzburg-Landau free-energy functional

$$\mathscr{F}_{\mathrm{GL}} = \int d^3x\, F_{\mathrm{GL}} = \int d^3x\, [F_{\mathrm{grad}} + F_{\mathrm{dip}} + U]. \tag{5.109}$$

Here F_{grad} and U are as defined in (5.44) and the dipole interaction F_{dip} as in (5.69).

We assume that a solution is finite in scale L (the droplet radius) and domain wall thickness is of the order of the coherence length L_ξ, $L_\xi \ll L \ll L_{\text{dip}}$. Therefore, we can neglect dipole interaction in the domain. How the interaction is taken into account is shown at the last step of solution construction.

We seek for the solution of the Ginzburg-Landau equation, which is invariant under three-dimensional rotations in R^3. By analogy with the Witten construction of an instanton solution with cylindrical symmetry (see Subsec. 4.2.5), we write the matrix field in the form

$$A_{ij} = f_1 \frac{r^i r^j}{r^2} + f_2 \left(\delta_{ij} - \frac{r^i r^j}{r^2} \right) + f_3 \frac{\varepsilon_{ijk} r^k}{r}, \tag{5.110}$$

where the f_i ($i = 1, 2, 3$) are generally speaking complex functions depending on the radius r. The order parameter in the B-phase can be represented as

$$A_{ij} = (\lambda/\sqrt{3})(\delta_{ij} \cos\theta + n^i n^j (1 - \cos\theta) + \varepsilon_{ijk} n^k \sin\theta), \tag{5.111}$$

where $n^i = r^i/r$ is the unit vector pointing along the rotation axis and θ the rotation angle. Substituting (5.110) into (5.111), we obtain

$$f_1 = \lambda/\sqrt{3},\ f_2 = (\lambda/\sqrt{3}) \cos\theta,\quad f_3 = (\lambda/\sqrt{3}) \sin\theta. \tag{5.112}$$

We select the boundary conditions

$$\begin{aligned} r &= 0,\ f_3 = 0,\ f_j = \lambda/\sqrt{3},\ j = 1, 2, \\ r &= \infty,\ f_3 = 0,\ f_j = \pm\lambda/\sqrt{3},\ j = 1, 2. \end{aligned} \tag{5.113}$$

The Ginzburg-Landau functional (5.109) is invariant under complex conjugation and therefore the Euler-Lagrange equation for it has solutions with real functions. We substitute (5.110) into (5.109) and integrate with respect to angular variables to obtain the functional $\mathcal{F}$ in the form

$$\mathcal{F}_{\text{GL}} = 4\pi \int_0^\infty dr\, F(A),$$

$$F(A) = \gamma_1 (r\dot{f}_1 + 2f_1 - 2f_2)^2 + \gamma_2 (r\dot{f}_2 + f_2 - f_1)^2 + \gamma_2 (r\dot{f}_3 + f_3)^2 + \gamma_2 f_3^2 + r^2 U(A), \tag{5.114}$$

$$\begin{aligned} U(A) = {}& \alpha(f_1^2 + 2f_2^2 + 2f_3^2) + (\beta_1 + \beta_2)(f_1^2 + 2f_2^2 + 2f_3^2)^2 \\ &+ (\beta_3 + \beta_4 + \beta_5)(f_1^4 + 2(f_2^2 + f_3^2)^2) \end{aligned} \tag{5.115}$$

(omitting the term with dipole interaction), where $\dot{f}$ is differentiation with respect to r.

The function U has a maximum (corresponding to the normal phase) at the origin and also two minima (corresponding to the B-phase), viz., the circles given by the equations

$$\begin{aligned} &\alpha + 2(\beta_1 + \beta_2 + \beta_3 + \beta_4 + \beta_5) f_1^2 + 4(\beta_1 + \beta_2)(f_2^2 + f_3^2) = 0, \\ &\alpha + 2(\beta_1 + \beta_2) f_1^2 + 2(2\beta_1 + 2\beta_2 + \beta_3 + \beta_4 + \beta_5)(f_2^2 + f_3^2) = 0. \end{aligned} \tag{5.116}$$

We introduce the notation

$$\sigma_1 = 2\Big(\sum_{i=1}^{5} \beta_i\Big), \quad \sigma_2 = 2(\beta_1 + \beta_2), \; \sigma_3 = 2(2\beta_1 + 2\beta_2 + \beta_3 + \beta_4 + \beta_5). \tag{5.117}$$

The Euler-Lagrange equations $\dfrac{d}{dr}\dfrac{\partial F}{\partial f_i'} - \dfrac{\partial F}{\partial f_i} = 0$ after quite tedious, but straightforward, calculations take the form

$$\gamma_1 f_1'' - 2\,[\alpha + 2\sigma_2(f_2^2 + f_3^2)]\, f_1 - 2\sigma_1 f_1^3 = 0, \tag{5.118}$$

$$\gamma_2 f_2'' - 4[\alpha + \sigma_2 f_1^2 + \sigma_3 f_3^2]\, f_2 - 4\sigma_3 f_2^3 = 0, \tag{5.119}$$

$$\gamma_2 f_3'' - 4[\alpha + \sigma_2 f_1^2 + \sigma_3 f_2^2]\, f_3 - 4\sigma_3 f_3^3 = 0. \tag{5.120}$$

Since solutions on scales $L \gg L_\xi$ are considered, the terms of orders L^{-1} and L^{-2} are omitted in the final form of (5.118)-(5.120). Equations (5.119) and (5.120) in f_2 and f_3 coincide in this approximation.

Equations (5.118)-(5.120), which describe one-dimensional textures, are formally analogous to the system describing the motion of a three-dimensional particle in a field with potential $\tilde{U} = -U$, where U is as in (5.115). The analogy was considered in [FV].

Together with boundary conditions (5.113), equations (5.118)-(5.120) have the explicit solution

$$f_1 = -\sqrt{c_1}\tanh\left[\sqrt{\frac{\sigma_1 c_1}{\gamma_1}}\,(r - L)\right], \tag{5.121}$$

$$c_1 = -\alpha/\sigma_1 - (\sigma_2/\sigma_1)(f_2^2 + f_3^2), \tag{5.122}$$

$$f_{2,3} = -\sqrt{c_{2,3}}\tanh\left[\sqrt{\frac{\sigma_3}{\gamma_2}\,c_{2,3}}\,(r - L)\right], \tag{5.123}$$

$$c_{2,3} = -\alpha/\sigma_3 - (\sigma_2/\sigma_3)\, f_1^2 - f_{3,2}^2. \tag{5.124}$$

Naturally, we assume that

$$c_{1,2,3} \geqslant 0, \; \sigma_1, \; \sigma_3 > 0. \tag{5.125}$$

The necessary condition for (5.125) to hold is

$$\sigma_2 + \sigma_3 > 0.$$

Taking the boundary conditions of (5.113) into account, we find that the solution of (5.121)-(5.125) describes a spherical domain containing the B-phase with the order parameter

$$A_{ij} = (\lambda/\sqrt{3})I = -(\alpha/2)(3\beta_1 + \beta_2 + \beta_3 + \beta_4 + \beta_5). \tag{5.126}$$

Outside the spherical domain,

$$A_{ij} = -(\lambda/\sqrt{3})I, \tag{5.127}$$

where I is the unit matrix.

The domain wall itself does not belong to the B-phase.

The solution in (5.121)-(5.125) admits an interesting topological interpretation. Consider the open ball $\mathscr{D}^3$ bounded by the surface S^2 in space R^3 and the domain Ω complementary to $\mathscr{D}^3$. Equations (5.121)-(5.125) define the map

$$\varphi\colon (R^3, \mathscr{D}^3, \Omega) \to (A, S_1^1, S_2^1). \tag{5.128}$$

Space R^3 is mapped into R^{18} so that $\mathscr{D}^3$ is sent onto S_1^1 and the exterior domain Ω onto the circle S_2^1. Both circles are in the disjoint union of the minima of potential $\mathscr{U}$. The solution is the natural generalization of the usual φ^4-kink (see Subsec. 4.2.1). From the topological viewpoint, (5.128) is a cobordism. We delay until the end of the chapter our discussion of the general construction of topological charges based on the cobordism theory concept.

Let us see how the effect of dipole energy can be considered in the construction. For $L \gg L_{\text{dip}}$, the minimization of dipole interaction energy fixes the angle $\theta = \theta_0 = \cos^{-1}(-1/4)$, but leaves the direction of rotation axis $\mathbf{n}$ arbitrary. Suppose that we have somehow fixed the direction of $\mathbf{n}$, e.g., by switching on a weak magnetic field. We denote by R_θ the matrix of rotation through θ relative to $\mathbf{n}$. The order parameter A_{ij} is determined by dipole energy and is $A_{ij} = (\lambda/\sqrt{3})R_\theta$. We apply the rotation operator R_0 to the spherically symmetric solution

$$A_{ij} = A_{iq}(R_\theta(\mathbf{n}))_{qj} \tag{5.129}$$

(see Eqs. (5.121-5.125) and (5.128)).

Since the Ginzburg-Landau functional $\mathscr{F}_{\text{GL}}$ (5.109) is invariant in spin space under rotation, we obtain a domain wall for A_{ij} accounting for dipole interaction.

A realization of this suggested solution in a real physical situation requires additional research.

A more important question is associated with the domain structure's stability. We can assume that the stability is determined by excess pressure inside the droplet. We clarify the mechanism with the usual assumption that the domain wall is of zero thickness and separates the B-phase into two domains.

In our situation, the domains are associated with two different components S_1^1 and S_2^1 in the order patameter range.

This is likely that the stabilization mechanism is determined by the condition that the pressure difference outside and inside the droplet should only be due to surface energy.

Another problem of interest touches upon the stability of a solution relative to complex perturbations of the functions f_i. Solution (5.121)-(5.125) and (5.128) was obtained assuming that the minima of U are reached in real solutions. However, complex perturbations of the order parameter may take the solutions off the fixed phase.

In the B-phase, f_i have the special form $f_i = \exp(i\varphi)\tilde{f}_i$ (the $\tilde{f}$ are real). The nontrivial phase factor leads to currents $\nabla\varphi$, which can hardly be compensated in the fixed B-phase. Therefore, in the general case, textures starting in the B-phase but transforming into others are the most likely [Sa].

Lately, experiments on rotating ^{3}He have shown that the core of the B-phase consists of different superfluid phases [SV].

In this connection, a study textures which describe transitions between different phases seems to be an important and interesting problem.

5.2.7 One-Dimensional Textures in the A- and B-Phases of ^{3}He

Here, we find exact solutions of the Ginzburg-Landau equation that describe one-dimensional textures in the A- and B-phases of ^{3}He, which depend on one spatial coordinate. The treatment is based on [GMN].

From the physical viewpoint, such solutions describe the static distributions in parallel-plate geometry. Note that similar solutions are called *planar textures* in physical literature.

We start with the most complicated case of the A-phase.

5.2.7.1 One-dimensional textures in the A-phase

We need representations

$$\mathscr{F} = \int d^3x \, F_{\mathrm{GL}} = \int d^3x \, [F_{\mathrm{grad}} + F_{\mathrm{dip}} + F_H + U] \tag{5.130}$$

for Ginzburg-Landau free-energy potential, where F_{grad}, F_{dip}, F_H, and U are given by (5.44), (5.62), and (5.66).

For uniform spatial configurations, $F_{\mathrm{grad}} = 0$ and the vacuum manifolds, V_A and V_B, are determined by the minimization of potential $U(A)$.

The order parameter A_{ij} in the A-phase is written in the ordinary form (5.60).

We now pose the problem precisely. Let A_{ij} depend on the spatial coordinate z, $F_{\mathrm{grad}} \neq 0$. We consider our system in the London limit (in which the description by the Ginzburg-Landau functional necessarily holds). This means that the order parameter assumes its values in the fixed phase of ^{3}He (the A-phase here) and that spatial changes in A_{ij} are determined by the minimization in (5.130) with an account of boundary conditions. The analogy between this problem and chiral field theory can be noticed easily. However, there is one essential difference. In the theory of chiral fields, the field of A_{ij} takes values in homogeneous space G/H and the chiral field's Lagrangian determines the bi-invariant metric on G/H. In our problem, the metric on G/H is given by the gradient term F_{grad}, generally speaking ($\gamma_1 \neq \gamma_2 \neq \gamma_3$)

not invariant under the action of the group G. The constants γ_i are always positive and in the weak-coupling approximation $\gamma_1 \sim \gamma_2 \sim \gamma_3$ [Le], whereas bi-invariance (in the case of principal chiral fields) requires that $\gamma_1 + \gamma_3 = 0$ (see (5.133) below).

Planar textures in the A-phase admit a mechanical analogy with the motion of a top with variable momenta of inertia in which the coordinate part is time. Apparently, this is the first nontrivial example of the kind.

Equations for the A-phase. We consider the cases (a) $L \ll L_{\text{dip}}$ and (b) $L \gg L_{\text{dip}}$ separately, where L and L_{dip} are the characteristic and dipole lengths, respectively.

(a) $L \ll L_{\text{dip}}$. Here, dipole energy can be neglected and the order parameter

$$A = \exp(i\varphi) R_1^{-1} A_0 R_2, \quad A_0 = \begin{pmatrix} 0 & 0 & 0 \\ 0 & 0 & 0 \\ 1 & i & 0 \end{pmatrix}, \quad i = \sqrt{-1}$$

is selected in the most general form of (5.81).

Note that any value of A for the A-phase can be written as

$$A = R_1^{-1} A_0 R_2, \tag{5.131}$$

since the formula

$$e^{i\varphi} A_0 = R_\varphi^{-1} A_0 R_\varphi, \text{ where } R_\varphi = \begin{pmatrix} \cos\varphi & \sin\varphi & 0 \\ -\sin\varphi & \cos\varphi & 0 \\ 0 & 0 & 1 \end{pmatrix} \tag{5.132}$$

holds.

We introduce chiral velocities

$$v = i\partial_z R_1 R_1^{-1}, \quad w = iR_2^{-1}\partial_z R_2$$

(see (5.83)) which assume values in the Lie algebra $so(3)$ of the group $SO(3)$. Therefore,

$$v = v_a c^a, \quad w = w_a c^a,$$

where c^a ($a = 1, 2, 3$) are generators of $so(3)$ with matrix elements $(c^a)_{bd} = i\varepsilon_{abd}$. Using v and w and their coordinates v_a and w_a, we can write (5.44) in the form

$$F_{\text{grad}} = I_{ab}(A) w_a w_b + \chi_{ab} v_a v_b, \tag{5.133}$$

$$I_{ab} = \gamma_1 (c^a A^+ A c^b)_{ii} + (\gamma_2 + \gamma_3)(c^a A^+ A c^b)_{33},$$

$$\chi_{ab} = \gamma_1 (A^+ c^a c^b A)_{ii} + (\gamma_2 + \gamma_3)(A^+ c^a c^b A)_{33}.$$

Note that v and w are determined by (5.83) in the same form as the angular velocities of a three-dimensional solid are. As seen from the expanded form, the matrices I_{ab} and χ_{ab} alter for different values of A, thus

$$\chi_{ab}(A) = |\lambda|^2(2\gamma_1 + (\gamma_2 + \gamma_3)|u^3|)(\delta_{ab} - \hat{z}^a\hat{z}^b),$$
$$I_{ab}(A) = \begin{pmatrix} \gamma_1|u^3|^2 + \hat{\gamma}|\lambda|^2, & -\hat{\gamma}u^1\bar{u}^2, & -\gamma_1 u^1\bar{u}^3 \\ -\hat{\gamma}(\bar{u}^1 u^2), & \gamma_1|u^3|^2 + \hat{\gamma}|u^1|^2, & -\gamma_1 u^2\bar{u}^3 \\ -\gamma_1\bar{u}^1 u^3, & -\gamma_1\bar{u}^2 u^3, & \gamma_1(|u^2|^2 + |u^3|^2) \end{pmatrix}, \tag{5.134}$$

where u^i and $\hat{z}^j$ are the complex and real vectors of the order parameter $A_{ji} = \lambda\hat{z}^j u^i$ and $\sum_{i=1}^{3} \gamma_i = \hat{\gamma}$.

It is convenient to define the scalar product of two complex 3×3 matrices as

$$\langle X|Y\rangle = \gamma_1(X^+Y)_{ii} + (\gamma_2 + \gamma_3)(X^+Y)_{33}. \tag{5.135}$$

Hence,

$$I_{ab}(A) = \langle Ac^a|Ac^b\rangle, \quad \chi_{ab}(A) = \langle c^aA|c^bA\rangle. \tag{5.136}$$

Since, under rotation transformations, order parameter matrices are transformed as

$$R_{1,2} \to R_{1,2}(1 + i\theta^{(1,2)} + \ldots),$$

v_a and w_a and the order parameter are transformed as

$$v_a \to v_a + \partial_z\theta_a^{(1)} + \varepsilon_{abc}v_b\theta_c^{(1)},$$
$$w_a \to w_a + \partial_z\theta_a^{(2)} + \varepsilon_{abc}w_b\theta_c^{(2)},$$
$$A \to A - i\theta_a^{(1)}c^aA + i\theta_a^{(2)}Ac^a.$$

We obtain the "motion", or texture, equations in the form

$$\nabla M^a_{\text{spin}} - (i\langle c^ac^bA|c^dA\rangle + \text{c.c.})v_bv_d = 0, \tag{5.137}$$
$$\nabla M^a_{\text{orb}} - (i\langle Ac^ac^b|Ac^d\rangle + \text{c.c.})w_bw_d = 0, \tag{5.138}$$

where

$$M^a_{\text{spin}} = \frac{\partial F_{\text{grad}}}{\partial v_a}, \quad M^a_{\text{orb}} = \frac{\partial F_{\text{grad}}}{\partial w_a},$$
$$\nabla M^a_{\text{spin}} = \partial_z M^a_{\text{spin}} - \epsilon_{abc}v_bM^c_{\text{spin}},$$
$$\nabla M^a_{\text{orb}} = \partial_z M^a_{\text{orb}} + \varepsilon_{abc}w_bM^c_{\text{orb}},$$

and c.c. means complex conjugate.

They are similar to the Euler equations for a top. Since there are no cross-terms in v and w in (5.137) or (5.138), we can assume that two three-dimensional tops are available, interacting with each other, as follows from (5.134). The second terms in (5.137) and (5.138) appear because the tops have variable moments of inertia. We can transform the equations into Hamiltonian form by calculating the Poisson brackets

$$H = F_{\text{grad}}, \ \partial_z M^a_{\text{spin(orb)}} = [M^a_{\text{spin(orb)}}, H],$$

$$[M^a_{\text{spin}}, M^b_{\text{spin}}] = \varepsilon_{abc} M^c_{\text{spin}}, \ [A_{pi}, A_{qj}] = 0,$$

$$[M^a_{\text{orb}}, M^b_{\text{orb}}] = -\varepsilon_{abc} M^c_{\text{orb}}, \ [M^a_{\text{orb}}, M^b_{\text{spin}}] = 0,$$

$$[M^a_{\text{spin}}, A] = -ic^a A, \ [M^a_{\text{orb}}, A] = -iAc^a.$$

There are six conservation laws or system integrals:

1. Spin current j^a_{spin}, $a = 1, 2, 3$, generated by rotations in the spin space (neglecting dipole energy).
2. The momentum along the z-axis, i.e., the "Hamiltonian" of the system H.
3. Superfluid current

$$j_m = \langle A | Ac^a \rangle w_a + \text{c.c.}$$

generated by the gauge transformation $A \to \exp(i\varphi)A$.

4. M^3_{orb} generated by rotation about the z-axis

$$A \to AR_\psi, \ R_\psi = \begin{pmatrix} \cos\psi & -\sin\psi & 0 \\ \sin\psi & \cos\psi & 0 \\ 0 & 0 & 1 \end{pmatrix}$$

in orbital space.

Note that the components of j^a_{spin} coincide with the spin part of the momentum M^a_{spin} $(a = 1, 2, 3)$.

It is easy to see that the integrals

$$H = F_{\text{grad}}, \ j_m, \ M^3_{\text{orb}}, \ M^3_{\text{spin}}, \ (M^a_{\text{spin}})^2 \tag{5.139}$$

are in involution, i.e., the Poisson bracket of any pair vanishes. Therefore, the system is completely integrable. Nevertheless, real integration requires additional treatment of the order parameter's symmetry (see below).

Since magnetic energy is represented by

$$F_H = g_H |H_p A_{pi}|^2 \tag{5.140}$$

(see (5.66)), two integrals of spin currents vanish when the field H is switched on and the system is not completely integrable any more.

(b) $L \gg L_{\text{dip}}$.

Here, only the superfluid velocity w is used. We have

$$F_{\text{grad}} = I_{ab}(A) w_a w_b,$$
$$I_{ab}(A) = \langle [A, c^a] | [A, c^b] \rangle,$$

where $\langle | \rangle$ is the scalar product defined in (5.133).

We assume that there is no magnetic field. By analogy with (5.138), we obtain the Euler equations for M^a in the form

$$\nabla M^a + (i\langle [A, c^b] | [[A, c^a], c^d] - i\langle [[A, c^a], c^b] | [A, c^d] \rangle) w^b w^d = 0, \quad (5.141)$$

$$M^a = \partial F_{\text{grad}} / \partial w_a,$$
$$\nabla M^a = \partial_z M^a + \varepsilon_{abc} w_b M^c.$$

An external magnetic field (5.140) can also be included into the free-energy density $F_{\text{grad}} + F_H$.

F_H generates terms on the right-hand side of (5.141). We only consider the case of a magnetic field parallel to the z-axis.

Using the same argument as we did when deriving integrals (5.139), we obtain three integrals in involution

$$H = F_{\text{grad}} + F_H, \ j_m, \ M^3$$

and our system is completely integrable.

To find explicit solutions, we resort to Eulerian angles and again consider the cases (a) $L \gg L_{\text{dip}}$ and (b) $L \ll L_{\text{dip}}$.

(a) $L \gg L_{\text{dip}}$. The order parameter is $A = R^{-1} A_0 R$ and A_0 has been defined in (5.81).

The free-energy density U is invariant under the transformation

$$A \to \exp(i\varphi) R_\psi^{-1} A R_\psi, \quad (5.142)$$

where R_ψ is rotation about the z-axis through ψ. The transformation can be written as

$$\exp(i\varphi) R_\psi^{-1} A R_\psi = (R_\varphi^{-1} R R_\psi)^{-1} A_0 (R_\varphi^{-1} R R_\psi). \quad (5.143)$$

It immediately follows that the symmetry in (5.142) is generated by two commuting one-dimensional subgroups of $SO(3)$, which act on $SO(3)$ as

$$R \to R_\varphi^{-1} R R_\psi,$$

where φ and ψ are arbitrary. We conclude from the Eulerian form of the rotation matrix

$$R = R_\varphi^{(z)} R_\theta^{(x)} R_\psi^{(z)} \quad (5.144)$$

(where $R_\varphi^{(z)}$ and $R_\psi^{(z)}$ are rotations about the z-axis through φ and ψ and $R_\theta^{(x)}$ is a rotation through θ about the x-axis, respectively) that φ and ψ can be eliminated from the system's integrals by the transformations in (5.144) with the appropriate choice of φ and ψ. Therefore, the coefficients of free-energy density only depend on the angle θ. Thus, we obtain

$$\begin{gathered} F_{\text{grad}} = I_{11}^2\dot{\theta}^2 + I_{22}\sin^2\theta\dot{\psi}^2 + I_{33}(\dot{\psi}\cos\theta + \dot{\varphi})^2 - 2I_{23}\sin\theta(\dot{\psi}\cos\theta + \dot{\varphi})\dot{\psi}, \\ I_{11} = |\lambda|^2(3\gamma_1 + \gamma_2 + \gamma_3),\ I_{12} = I_{13} = I_{21} = I_{31} = 0, \\ I_{22} = |\lambda|^2(2\gamma_1 + \gamma_2 + \gamma_3 + \gamma_1\cos^2\theta + (\gamma_2 + \gamma_3)\sin^2\theta\cos^2\theta), \qquad (5.145) \\ I_{23} = |\lambda|^2(\gamma_1 + (\gamma_2 + \gamma_3)\sin^2\theta)\sin\theta\cos\theta, \\ I_{33} = |\lambda|^2(2\gamma_1 + \gamma_1\sin^2\theta + (\gamma_2 + \gamma_3)\sin^4\theta), \end{gathered}$$

where the dot means differentiation with respect to z. Using cyclic variables φ and ψ, we can write (5.145) in the form

$$\begin{aligned} F_{\text{grad}} = I_{11}\dot{\theta}^2 + \frac{1}{4\sin^2\theta I(\theta)}\{ & j_m I_{33} - 2j_m M^3(I_{33}\cos\theta + I_{23}\sin\theta) \\ & + (M^3)^2(I_{33}\cos^2\theta + 2I_{23}\sin\theta\cos\theta + I_{22}\sin^2\theta)\}, \qquad (5.146) \\ & I(\theta) = I_{22}I_{33} - I_{23}^2. \end{aligned}$$

Since F_{grad} is a motion integral, we write (5.146) as

$$I_{11}\dot{\theta}^2 + \frac{1}{4\sin^2\theta I(\theta)}\Phi(\theta), \qquad (5.147)$$

where $\Phi(\theta)$ is the function in braces in (5.146).

The magnetic energy given by (5.140) can be involved if the magnetic field points along the z-axis, since

$$F_H = 2g_H H^2|\lambda|^2\cos^2\theta$$

is invariant under symmetry (5.143) and equation (5.147) with magnetic energy is

$$I_{11}\dot{\theta}^2 + \frac{\Phi(\theta)}{4\sin^2\theta I(\theta)} + 2g_H H^2|\lambda|^2\cos^2\theta = E. \qquad (5.148)$$

To solve the above equation, we have to take the inverse of the hyperelliptic integral

$$z = \pm I_{11}^{1/2}\int \frac{2\sin\theta I^{1/2}(\theta)}{\{4\sin^2\theta I(\theta)(E - 2g_H|\lambda|^2H^2\cos^2\theta) - \Phi(\theta)\}^{1/2}}d\theta$$

$$= \mp 2I_{11}^{1/2} \int \frac{I(t)}{\{[4(1-t^2)(E - 2g_H|\lambda|^2H^2t^2)I(t) - \Phi(t)]I(t)\}^{1/2}} dt$$

$$= \mp 2I_{11}^{1/2} \int \frac{I(t)}{[P_{12}(t)]^{1/2}} dt, \quad t = \cos\theta, \tag{5.149}$$

where $I(t)$ and $\Phi(t)$ are polynomials of orders four and six, respectively.

Similarly, the solutions

$$\varphi = \mp I_{11}^{1/2} \int \frac{\left[\frac{K_5(t)}{1-t^2} + K_4(t)\right]}{\sqrt{P_{12}(t)}} dt,$$

$$\psi = \mp I_{11}^{1/2} \int \frac{\tilde{K}_5(t)}{(1-t^2)\sqrt{P_{12}(t)}} dt \tag{5.150}$$

can be written. Here the polynomial $P_{12}(t)$ is the same as in (5.149), K_4, K_5, and $\tilde{K}_5$ are polynomials in t, of degrees four and five.

In the important case of a zero superfluid current $j_m = 0$, essential simplifications arise. As a matter of fact, $\Phi(t)$ is a polynomial in $t^2 = s$. We omit the corresponding simple calculations.

(b) $L \ll L_{\text{dip}}$ can be considered as above. By considering the texture in the absence of a magnetic field, all the Eulerian angles except θ can be eliminated and the coefficients can be represented as

$$\chi_{ab} = |\lambda|^2\{\gamma_1 + (\gamma_2 + \gamma_3)\sin^2\theta\}(\delta_{ab} - \delta_{a3}\delta_{b3}),$$

$$I_{11} = \gamma_1 + (\gamma_2 + \gamma_3)\cos^2\theta, \ I_{12} = I_{21} = I_{31} = I_{13} = 0,$$

$$I_{22} = \gamma_1 + \gamma_2 + \gamma_3 + \gamma_1\sin^2\theta, \ I_{23} = -\gamma_1\sin\theta\cos\theta, \ I_{33} = \gamma_1 + \gamma_1\cos^2\theta.$$

The quantities

$$\varphi_{\text{spin}}, \ \psi_{\text{spin}}, \ (M_{\text{spin}})^2 = \text{const}$$

are preserved.

Therefore, the gradient energy can be written as

$$F_{\text{grad}} = I_{11}\dot{\theta}^2 + I_{22}\sin^2\theta\dot{\psi}^2 - 2I_{23}\sin\theta\dot{\psi}(\dot{\psi}\cos\theta + \dot{\varphi}) + I_{33}(\dot{\psi}\cos\theta + \dot{\varphi})^2$$
$$+ \frac{M_{\text{spin}}^2}{4|\lambda|^2}[\gamma_1 + (\gamma_2 + \gamma_3)\sin^2\theta]^{-1}.$$

Eliminating the cyclic coordinates φ and ψ enables us to obtain an equation in θ. We confine ourselves to the simple case of $M_{\text{orb}}^3 = 0$, $M_{\text{spin}} = 0$. Like (5.149) and

(5.150), we obtain

$$z = \pm \int \frac{(\gamma_1 + (\gamma_2 + \gamma_3)s)I(s)\, ds}{(s(\gamma_1 + (\gamma_2 + \gamma_3)s)[P_3(s)]^{1/2}},$$

$$\varphi = \pm \frac{j_m\gamma_1}{2} \int \frac{(\gamma_1 + (\gamma_2 + \gamma_3)s)\, ds}{(1 - s)(s(\gamma_1 + \gamma_2 + \gamma_3)s)[P_3(s)]^{1/2}}$$

$$\mp j_m \frac{2\gamma_1 + \gamma_2 + \gamma_3}{2} \int \frac{(\gamma_1 + (\gamma_2 + \gamma_3)s)\, ds}{(s(\gamma_1 + (\gamma_2 + \gamma_3)s)[P_3(s)]^{1/2}},$$

$$\psi = \mp \frac{j_m\gamma_1}{2} \int \frac{(\gamma_1 + (\gamma_2 + \gamma_3)s)\, ds}{(\gamma_1 + (\gamma_2 + \gamma_3)s)[P_3(s)]^{1/2}(1 - s)},$$

$$s = t^2,$$

where

$$I(s) = |\lambda|^2(\gamma_1(2\gamma_1 + \gamma_2 + \gamma_3) + \gamma_1(\gamma_2 + \gamma_3)s),$$

$$P_3(s) = I(s)[4E(1 - s)I(s) - \Phi(s)],$$

$$\Phi(s) = (2\gamma_1 + \gamma_2 + \gamma_3)|\lambda|^2 j_m - (\gamma_2 + \gamma_3)|\lambda|^2 j_m s.$$

The planar textures in the B-phase can be described by the same method, but with considerable simplifications.

5.2.7.2 One-dimensional textures in the *B*-phase

Let us assume that the characteristic length L of a texture is much less than dipole length L_{dip}. Then the order parameter is

$$A = (\lambda/\sqrt{3})R \exp(i\varphi),$$

where R is the rotation matrix and λ is a complex number. We again use the velocities

$$w = iR^{-1}\partial_z R, \quad v = -\partial_z\varphi.$$

It is easy to see that the gradient part of the free-energy density can be written as

$$F_{\text{grad}} = (I/2)(w_1^2 + w_2^2) + (J/2)w_3^2 + (m/2)v^2,$$

$$J = \frac{2|\lambda|^2}{3} 2\gamma_1, \; I = (2/3)|\lambda|^2(2\gamma_1 + \gamma_2 + \gamma_3),$$

$$m = (2/3)|\lambda|^2(3\gamma_1 + \gamma_2 + \gamma_3).$$

Note that the Lagrangian in the above equation is the same as of a symmetric top of mass m and moments of inertia I and $\dot{J}$.

The velocities v and $w = (w_1, w_2, w_3)$ are those of the center of masses and of the top (angular).

The technique of integration of a symmetric top should now work in the study of planar textures.

Note the interesting case arising if dipole energy is considered. We add the term

$$F_{\text{dip}} = g_d (\cos\theta + 2\cos^2\theta)$$

to the free energy, where θ is the order parameter's rotation angle. If we assume that the rotation axis is perpendicular to the plates with superfluid liquid in between, then there exists a solution first found in [MK1] and the Euler equations are reduced to

$$J\ddot{\theta} + g_d \sin\theta(1 + 4\cos\theta) = 0.$$

The solution can be found in terms of ellyptic integrals.

Remark 5.9 The study of rotating ^{3}He enables us to observe experimentally which, according to the NMR data, of the theoretically feasible textures are realized, e.g., in addition to the lattice of ordinary singular vortices (5.94), a lattice of Anderson-Toulouse vortices in the A-phase is feasible without singular cores. The different possibilities are related to how the additional magnetic fields, container rotation's velocity, etc., are accounted for. The velocity $\mathbf{v}_s$ is potential in the B-phase and like in ^{4}He we have to expect a lattice of ordinary singular vortices. However, nonsingular domain-wall type textures can exist (see Subsec. 5.2.6).

Intensive studies of rotating ^{3}He are under way now in a number of countries; we expect new results of interest. There is already a vast literature on the subject. The reader is referred to the survey [SV] with a list of original works. Note also that this requires analytic estimates of the energy of the obtained textures more than topological studies. Nevertheless, the existence in the B-phase of vortices with cores from the normal and superfluid components is of paramount importance [SV]. Observations of A- and B-phase cores are as interesting. All the experimental results emphasize the importance of topological studies of A- and B-phase interactions (in particular, those involving cobordism theory to describe transitions between them).

5.2.8 General Construction of Topological Charges. New Examples

5.2.8.1 Topological charges for singularities in general relativity theory

In the general theory of relativity, topological methods are mostly used to describe the global structure of space-time, e.g., the Penrose-Hawking theorems on singularities in cosmological models. A discussion of the profound results requires a good knowledge of general relativity theory [HE]. Here, we consider other singularity types similar to line and point ones in ^{3}He and liquid crystals. The results are due to [Bya].

We assume that the following condition is fulfilled:

The Einstein (or Lorentz) metric g_{ik} with signature 2 $(+ + + -)$ is given in a four-dimensional Minkowski space-time $\check{M}^4$.

There are two types of singularities.

(a) Line singularities: g_{ik} and the topology of the manifold $\check{M}^4$ are singular on a certain line or in its ε-neighborhood (with respect to a Riemannian metric).

(b) Surface singularities: g_{ik} and the topology of $\check{M}^4$ are singular on a certain surface or in its ε-neighborhood.

Line and surface singularities in $\check{M}^4$ naturally generalize point and line singularities in R^3. An orthonormal frame $e_1, \ldots, e_4$ can be defined at each point of $\check{M}^4$, which is regular in the sense of g_{ik}. Any other frame is obtained from $e_1, \ldots, e_4$ by the action of the Lorentz group L. The set of all the frames over regular points in $\check{M}^4$ forms a fiber bundle ξ with a structure group L.

We now define the topological charge for each singularity type.

We begin with surface singularities. Take a surface σ intersecting another surface σ_1 in general position, i.e., at a point p_0. We assume that S^1 is an arbitrary circle in the regular part of σ_1, surrounding p_0.

We specify an arbitrary orthonormal frame $e_1, \ldots, e_4$ at a point $s \in S^1$. Consider a continuous transformation $g_s\{e_i\}$ of the frame along S^1 from s into s, e.g., the transformation can be carried out by a parallel translation of g_{ik}. The new frame e_j' is obtained by a transformation

$$L \ni A: \{e_i\} = e_i \to e_j' = A_j^i e_i$$

from the Lorentz group.

The matrix A_j^i is one of the four connected components of the group. The set of them is itself a group (under multiplication) isomorphic to $Z \oplus Z$.

We define the topological charge Q of a two-dimensional singular surface σ as an element of the group $\pi_0(L) = Z_2 \oplus Z_2$ associated with A_j^i. It is obvious that Q is determined uniquely, i.e., it does not depend on the choice of the matrix A_j^i. $Q = (0, 1)$ or $Q = (1, 0)$ is nontrivial, which means that $\check{M}^4$ is non-orientable in the neighborhood of σ. However, $\check{M}^4$ can also be orientable in the neighborhood, but have non-trivial charge Q. Then $Q \in Z_2 \oplus Z_2$, $Q = (1, 1)$. We call $Q = (0, 0)$ *trivial.*

Concerning line singularities, let f: $S^2 \to \check{M}^4$ be an arbitrary map of S^2 into a regular neighborhood of a singular line l.

We consider a fiber bundle ξ over $f(S^2)$, which can be regarded as that over S^2 due to the regularity of f.

We already know the general classification of fiber bundles over spheres that the group $\pi_1(G)$ classifies bundles over S^2 with structure group G. In our case ξ is determined by the structure group L, $\pi_1(L) = \pi_1(L_0) = Z_2$, where L_0 is the connected component of the unit element of L. Recall that the first Stiefel-Whitney class is $w_1(S^2) = H^2(S^2, \pi_1(L_0)) = H^2(S^2, Z_2) = Z_2$. We now define the topological charge Q as an element of the group $w_1(S^2) = Z_2$. The charge is nontrivial if ξ is.

An example of g_{ik} with singularities and nontrivial topological charges can be obtained by gluing together the local charts which cover space-time. The Einstein equation then holds

$$R_{ij} - (1/2)g_{ij} R = T_{ij} \tag{5.151}$$

(where R_{ij} is the Riemannian curvature tensor, R Ricci curvature, and T_{ij} the momentum energy tensor), since no restrictions are imposed on T_{ij} except $T_0^0 > 0$. For the given solution, (5.151) can be regarded as the definition of T_{ij}. The appearance of singularities is not contrary to the metric being Lorentz at infinity since only the total topological charge should be conserved. Several singularities can make up for each other. We stress that singularities cannot be eliminated by altering the coordinate system and can appear or vanish only pairwise.

Another type of charge can be defined in general relativity theory in a similar way to the integral formula for the degree of a map in chiral theories.

Assume that the metric $g_{ik} \to \check{g}_{ik}^M$ at infinity is Minkowski, viz., space-like regions are isolated (where the metric is negative definite).

Space $\check{M}^4$ is filled with matter with a distribution of 4-speeds u^i. We require that a limit should exist, of directions n of velocities $(n^i = u^i/(u^i \cdot u^i)^{1/2})$ when moving towards infinity along the rays. This means that there should exist a continuous map given by the field of n^i of a two-dimensional sphere S_∞^2 of infinite radius to the unit sphere S_1^2. The topological charge

$$Q = \frac{1}{4\pi} \int \varepsilon_{\alpha\beta\gamma} n^\gamma \frac{\partial n^\beta}{\partial x^i} \frac{\partial n^\alpha}{\partial x^i} \varepsilon^{ij} dx^i dx^j \tag{5.152}$$

is then defined as the degree of the map $S_\infty^2 \to S_1^2$ (see Subsec. 2.5.1).

If a model with the momentum energy tensor $T_{ij} = \varepsilon u^i u^j$ and $u^i u^i = 0$ is considered as the simplest hydrodynamic type, where $i = 1, 2, 3$, ε is energy density, and $\varepsilon \neq 0$, then Q is a Lorentz invariant.

Line singularities in the general theory of relativity are characterized by the first Stiefel-Whitney class w_1. The existence of a global spin structure on manifolds is determined by the second Stiefel-Whitney class w_2.

5.2.8.2 Spinors and w_2

Recall the definition of the spinor group Spin (n) as a two-sheeted covering of the group $SO(n)$. The ordinary definition is related to the existence of a Clifford algebra Cl_n defined as follows. Select an n-dimensional vector space R^n with basis $e_1, \ldots, e_n$ so that the scalar product $\langle x, x\rangle$ of the vector $x = x^i e_i$ by itself may equal the sum of squares $(x^i)^2 + \ldots + (x^n)^2$. The quantities e_i then satisfy the relations

$$e_i e_j + e_j e_i = 2\delta_{ij}. \tag{5.153}$$

The associative algebra Cl with (5.153) is called a *Clifford algebra.* Cl_n is of finite dimension 2^n. In fact, using the commutation relations, it is not hard to show that the basis for the algebra is formed by monomials of the form $e_{i_1 i_2 \ldots i_m} = e_{i_1} e_{i_2} \ldots e_{i_m}$ with subscripts $i_1 < i_2 < \ldots < i_m$, $0 \leqslant m \leqslant n$, $e_0 = 1$.

Hence, dim $Cl_n = 1 + {}^nC_1 + \ldots + {}^nC_n = 2^n$.

Cl is decomposed into the sum $Cl^+ + Cl^-$, where Cl^+ is the subalgebra spanned by linear combinations of $e_{i_1 \ldots i_m}$ of an even number of elements. Its dimension is 2^{n-1}. As an algebra, it is generated by elements of the form $e_i e_j$ and by the identity. Linear combinations of $e_{i_1 \ldots i_m}$ associated with subsets $i_1, \ldots, i_m$ containing an odd number of elements form the subspace of Cl, which is complementary to Cl^+. CL^+ and Cl^- are called *spaces of even and odd degree*, respectively.

Consider the set of invertible elements $D = \{d\}$ from Cl such that $dxd^{-1} \subset R^n$ for all $x \in R^n$. D is a multiplicative group in Cl. We can show that

$$\varrho : d \to dxd^{-1} \tag{5.154}$$

is a homomorphism of D to $O(n)$. The preimage of the subgroup $SO(n)$ under the homomorphism is a subgroup of Cl^+ and is called the *spin group* Spin (n).

It follows from (5.154) that the center of Spin (n) consists of the elements ± 1. Sincee $\pi_1(SO(n)) = Z_2$, Spin (n) covers $SO(n)$.

The definition of a global spin structure on a manifold is reduced to the following topological question. Let M^n be a real compact manifold with a tangent bundle ξ given on it. When can ξ be reduced to a fiber bundle with structure group Spin (n)? After the spin bundle has been constructed, the existence of a global spin structure is equivalent to that of a section of the bundle.

We now state the fundamental theorem on the existence of a spin bundle on a manifold M^n.

Theorem 5.19 (1) *Let M^n be a compact manifold. A spin bundle η exists on it if and only if $w_2(\xi) = 0$.*

(2) *Different spin structures on M are classified by elements $w_1(\xi) \in H^1(M^n, Z_2)$.*

We outline a *proof* based on the Milnor approach [Mil5] (see also [Hi]).

Consider the principal bundle $P \to M^n$ with fiber $SO(n)$. The fiber bundle $P_1 \to M^n$ with fiber $G =$ Spin (n) is a non-trivial two-sheeted covering of P.

Recall that two-fiber bundles over a connected manifold M^n are characterized by the mapping

$$\text{Map}\,(\pi_1(M^n), Z_2) \simeq H^1(M^n, Z_2). \tag{5.155}$$

For the group $SO(n)$, $n \geqslant 3$, $H^1(SO(n), Z_2) = Z_2$; therefore there exist two fiber bundles on the space $SO(n)$: the trivial one associated with the group $SO(n) \times Z_2$ and generated by an element $e \in H^1(SO(n), Z_2)$ and the nontrivial one which is associated with an element $\alpha \in H^1(SO(n), Z_2)$.

We cover P with P_1. If such a covering does exist, then

$$i_*(\sigma) = \alpha, \tag{5.156}$$

where $\sigma \in H^1(E(\xi), Z_2)$ and i_* is the map $H^1(E(\xi), Z_2) \to H^1(SO(n), Z_2)$ induced by the embedding of the fiber $SO(n)$ over a point $x \in M^n$ to space $E(\xi)$.

For proof, we use the existence of Serre exact sequence for the cohomology groups of fiber spaces.

Lemma 5.4 (Serre) *Let $\xi = (E, F, B)$ be a fiber space with a connected base space B and connected fiber F. Then there exists an exact sequence*

$$H^0(F, Z_2) \to H^1(B, Z_2) \to H^1(E, Z_2) \to H^1(F, Z_2) \to H^2(B, Z_2). \quad (5.157)$$

Note that $H^0(F, Z_2) = 0$, since F is connected.
We write (5.157) for the fiber bundle $E \to M^n$ as

$$0 \to H^1(M, Z_2) \xrightarrow{p_*} H^1(E, Z_2) \xrightarrow{i_*} H^1(SO(n), Z_2) \xrightarrow{\tau} H^2(M, Z_2), \quad (5.158)$$

where p_* is the map induced by the projection p: $E \to M^n$. Since (5.158) is exact, (5.156) is equivalent to $\tau(\alpha) = 0$. However, $\tau(\alpha) \in H^2(M, Z_2)$ means $w_2(\xi)$. ■

The second statement of the theorem easily follows from the exactness of (5.158). Let σ_1 and σ_2 be two different spin structures on M^n. Then $i_*(\sigma_1 - \sigma_2) = 0$. It follows from the exactness of (5.158) that there exists an element $\beta \in H^1(M, Z_2)$ such that $p_*(\beta) = \sigma_1 - \sigma_2$.

If an element σ_1 is fixed, then all the remaining spin structures are classified by elements of the group $H^1(M, Z_2)$, i.e., by $w_1(\xi)$. ■

Corollary 5.2 *If a spin structure η exists on a simply connected manifold, then the structure is unique.*

Example 5.11 Manifolds CP^n are spin if and only if n is odd.

Remark 5.10 The Serre lemma is stated with considerably more generality for a p-connected base and a q-connected fiber. The proof is based on the theory of spectral sequences, which is outside the scope of the book. The theory of spectral sequences, a powerful method for computing homology and cohomology groups, is treated in a number of books (e.g., see [MT]).

Remark 5.11 If $w_2(\xi) \neq 0$, then there is no spin structure on M^n. However, there can exist generalized spin structures with complex spin group $\mathrm{Spin}^C(n)$, i.e., the quotient group $\mathrm{Spin}^C(n) \times U(1)$ factorized relative to the subgroup Z_2 generated by the element $(\varepsilon, -1)$, where ε generates the kernel of the projection $\mathrm{Spin}(n) \to SO(n)$ (the latter being the center of Spin (n)) and $-1 \in U(1)$.

M^n is called a $\mathrm{Spin}^C(n)$-manifold if the characteristic class of the tangent bundle associated with the principal one with group $\mathrm{Spin}^C(n)$ is the first Chern class $c_1(M^n) \in H^2(M^n, Z)$ reduced mod 2.

We define the homomorphism

$$\varphi\colon \mathrm{Spin}^C(n) \to U(1)\colon (x, u) \to u^2, \; x \in \mathrm{Spin}\,(n), \; u \in U(1),$$

in turn defining a certain line complex bundle ξ_C^1 associated with $U(1)$. We can show that c_1 is the first Chern class of the bundle [Hi].

The importance of the result is that, by endowing ξ_C^1 with a Hermitian structure, the Dirac operator $\not\partial$ can be constructed on M^n. The operator acts in the spin bundle E_C associated with the spin representation of $\mathrm{Spin}^C(n)$.

Four-dimensional manifolds M^4 with the structure group $\mathrm{Spin}^C(4) = \mathrm{Spin}\,(4) \times_{z_2} U(1)$ are of the particular interest from this point of view. The group is the semidirect product of Spin (4) and $U(1)$, the latter being a normal subgroup of $\mathrm{Spin}^C(4)$. The equivalence relation for $\mathrm{Spin}^C(4)$ is written as

$$(x, u) = (-x, -u), x \in \mathrm{Spin}\ (4) \simeq SU(2) \times SU(2), u \in U(1).$$

It has been proved by Massey [Mas3] that all four-dimensional manifolds are Spin^C-ones and that the class w_2 admits an efficient integral representation in terms of the Chern class $c_1(w_2 \simeq c_1(\mathrm{mod}\ 2))$. The class w_2 is calculated in terms of generalized spin field strengths in [IP].

Because of lack of the space, I am forced to leave the topic here. The reader may find interesting applications of spin manifold theory to quantum gravity in [I].

5.2.8.3 Topological charges and cobordism theory

The topological invariants describing vortices, kinks, instantons, and monopoles specify certain rules for selecting solutions and act as conservation laws.

The reason for topological charge has had simple underlying mathematical results, though our physical interpretation of charges may be different.

The general mathematical construction of topological charges can be based on cobordism theory.

To motivate the construction, we start by analyzing the topology of monopoles and instantons, two types of classical solutions.

To determine solutions of the Euler equations for the Lagrangian

$$L_{\mathrm{YMH}} = -(1/4)(F_{\mu\nu})^2 + |\nabla_\mu \varphi|^2 - U(\varphi),$$

it is necessary to specify the asymptotic conditions at infinity. Two types of condition are normally used. The radial condition for a stationary problem (for monopole theory) is that the field W_μ is invariant under the action of the group $SO(3)$. The field is defined on the two-dimensional sphere of infinite radius, centered at (0, 0, 0) in R^3. If we assume that the solution vanishes at the center, then we obtain a 'tHooft-Polyakov monopole. The radial condition for instantons in R^4 determines the field on S^3_∞, invariant under $SO(4)$.

We see that the asymptotic conditions are determined in both cases by the values of the field on a submanifold in Euclidean space. We call such a submanifold *asymptotic* and assume that the values of fields given on an asymptotic submanifold are localized in the space of degenerate vacua (in isotopic space as in the example of a monopole). Since the field φ is given on the whole domain Ω bounded by the asymptotic submanifold $\partial\Omega$, the map

$$f: (\Omega, \partial\Omega) \to (J, V) \tag{5.159}$$

arises, where J is an isotopic and V a vacuum submanifold.

Another type of asymptotic condition is obtained if we require that φ should be constant at infinity and then have to consider the space with an asymptotic condition, compactified by the point at infinity. The stationary problem in R^3 hence leads to a manifold $\Omega \simeq S^3$ and the corresponding problem for instantons to $R^4 \cup \infty = S^4$. A similar situation arises for one-dimensional-kink type solutions in $\lambda\varphi_2^4$ (Subsec. 4.2.1). Here, we have to adjoint two points

$$\varphi(\infty) = a \text{ and } \varphi(-\infty) = -a$$

to the number line.

We consider now the general boundary-value problem. We assume that Ω lies in "physical" space R^n (the cases of R^3 and R^4 naturally being of most interest). Let Ω^n be a domain in R^n with boundary $\partial\Omega^n$ consisting of several submanifolds $P_1^{n-1}, \ldots, P_k^{n-1}$. The field φ given on Ω determines a mapping to isotopic space J. We require that the boundary components should be mapped by f into degenerate vacuum submanifolds $(V_1, \ldots, V_n) \subset J$ and thereby obtain the cobordism

$$f\colon (\Omega^n, P_1^{n-1}, \ldots, P_k^{n-1}) \to (J, V_1, \ldots, V_n), \tag{5.160}$$

where V_i are vacuum submanifolds of V.

It is the topological invariants of classes of continuous mappings that determine topological charges.

The topological constraints yield necessary conditions for mapping existence.

Example 5.12 Consider a stationary problem with manifold Ω^3 of the form $S^2 \times I$ (I is the unit interval). The isotopic space J is $S_1^2 \times S_2^2$. Let V_1 and V_2 be S_1^2 and S_2^2, respectively. We show that there is no continuous map

$$F\colon (\Omega^3, P_1^2, P_2^2) \to (S_1^2 \times S_2^2, S_1^2, S_2^2), \tag{5.161}$$

i.e., the boundary problem has no solutions.

Proof. Assume that such a mapping exists. The boundary $S_2 \times I$ consists of the two spheres $S^2 \times 0$ and $S^2 \times 1$. The former is continuously mapped to S_1^2, the latter to S_2^2. It follows from the existence of F that

$$f_i\colon S^2 \to S_1^2 \times S_2^2,\quad f_i(S^2) = S_i^2 \quad (i = 1, 2)$$

are homotopic, since $f_t(x) = F(x, t)$, $F(x, i-1) = f_i(x)$, where $F\colon \Omega \to S_1^2 \times S_2^2$ is defined by the field.

But f_1 and f_2 are in different elements of the homotopy group $\pi_2(S_1^2 \times S_2^2) = Z \oplus Z$ and therefore they cannot be homotopic.

Example 5.13 The above topological concepts admit an interesting interpretation when describing phases of superfluid ^{3}He. Assume that the container is filled with superfluid ^{3}He, so that it is in one phase in one part (e.g., in the A-phase) and in the other phase in the other part (in the B-phase). Generally speaking, the shapes of the parts filled with the A- and B-phases can be rather complicated droplets, elongated fibers, etc. The order parameter A_{pi} defines a map of a domain Ω^3 in R^3 to isotopic space J (the space R^{18} of all complex 3×3 matrices) and of the domains Ω_A and Ω_B filled with the A- and B-phases to the vacuum manifolds V_A and V_B associated with change in the order parameter domain in the A- and

B-phases. Assuming that Ω_A and Ω_B are sufficiently simple and that the boundary of Ω is smooth, we arrive at the map

$$f\colon (\Omega^3, \Omega_A, \Omega_B) \to (J, V_A, V_B). \tag{5.162}$$

A study of this problem in general seems quite complicated. Certain conditions for the interfaces between the A- and B-phases, which follow from the conservation laws for currents, enthalpy, etc., have been obtained in [GM5]. Following [GM1], we show how the quantization conditions arise in phase transitions.

For simplicity, we consider the following situation. Let the tube of the A-phase be surrounded by the B-phase (see Fig. 24). We pick the surface Π bounded by a closed curve l in the B-phase domain. Let Π intersect the tube in the A-phase. Assuming that Π is the disk D^2, we obtain the map

$$f\colon (D^2, S^1) \to (J, V_B),$$

which determines an element γ of the relative homotopy group $\pi_2(J, V_B) \sim \pi_1(V_B) = Z$. The element γ is the topological index (charge) of the transition

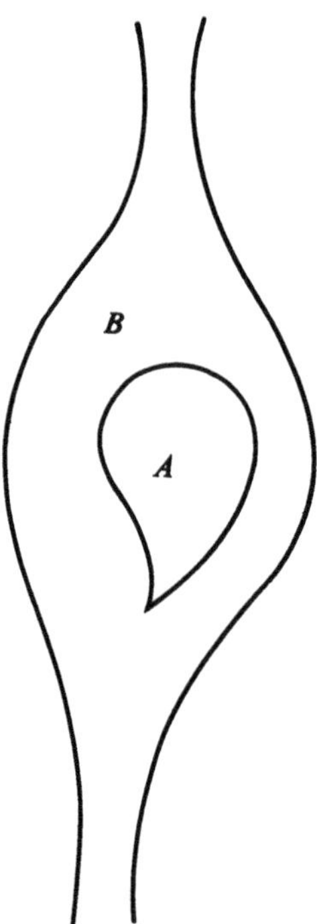

Fig. 24. Droplet of the A-phase surrounded by the B-phase

of the A-phase into the B-phase. The isomorphism $\pi_2(J, V_B) \underset{*}{\sim} \pi_1(V_B)$ here enables us to determine γ immediately in terms of the group $\pi_1(V_B)$.* In the general case, it is necessary to introduce relative homotopy groups. Consider the example due to Mermin from this point of view [Me3].

In the container with the B-phase, there is a tube with superfluid ^{3}He and inside that the core or droplet of the A-phase (see Fig. 24). Taking a closed curve S^1 inside the B-phase embracing the tube and the sphere S^2 inside the A-phase core, we obtain the map

$$f\colon (\Omega, S^1, S^2) \to (J, V_B, V_A),$$

where Ω is the volume of the whole domain in R^3, which is filled with ^{3}He. The map can be deformed into

$$\tilde{f}\colon (\tilde{\Omega}, S^1 \times R, S^2) \to (J, V_B, V_A),$$

where $\tilde{\Omega}$ is the cylinder $D^2 \times R^1$, $S^1 \times R^1$ is the cylinder's boundary and S^2 is the boundary of the ball (droplet) in the A-phase, punched out by the cylinder.

The map $\tilde{f}$ is determined by those of the boundaries $S^1 \times R^1 \to V_B$ and $S^2 \to V_A$. The topological index γ is given by a pair of numbers (m, n), $m \in \pi_1(V_B)$, $n \in \pi_2(V_A)$, which just determine the quantization conditions in the general case of transitions of the A-phase into the B-phase.

Exercise Using the ideas and techniques in this subsection, classify solitons and domain walls in the A- and B-phase in terms of relative homotopy groups (cf. [MV]).

---

* Recall that line singularities are described in helium and liquid crystals in terms of $\pi_1(\Theta_B)$. This situation is a particular case of the problem.

Chapter 6
Conclusion

One cannot encompass the limitless
Koz'ma Prutkov*

The book is coming to a close, but I recognize how many important results remain untouched. This is not surprising, as a detailed description of the latest applications of topology to physics would require many heavy volumes. Nevertheless, it is desirable that the reader should get some idea about the tendencies in modern research while remaining within the framework of elementary topology of the text. We shall discuss certain results, possible prospects and unsolved problems. The treatment is necessarily a summary. Clearly the choice of themes reflects the tastes and work of the author.

6.1 TOPOLOGY OF GAUGE FIELDS

6.1.1 Wu-Yang Copies

The basic measurable quantity in classical electrodynamics was the electromagnetic field strength $F_{\mu\nu}$, while the vector potential A_μ was a derived quantity. In quantum electrodynamics, the remarkable discovery by Aharonov-Bohm [AB] (see the current state of the art in [SK]) has led to a change in standpoint. The macroscopic behavior of quantum effects, including the Meissner effect (see Ch. 3), is due to the explicit dependence of the system's wave function

$$\varphi = \exp i \int A_\mu \, dx^\mu \tag{6.1}$$

on the vector potential (where $\hbar = c = 1$). All the phenomena are related to quantum mechanic systems with the Abelian gauge group $G = U(1)$.

Wu and Yang observed that vector potentials W_μ play a defining role even at the classical stage of non-Abelian gauge theory [WY2]. They constructed important

*Koz'ma Prutkov, a pseudonym used by several writers, including Aleksei Konstantinovich Tolstoy, is the literary predecessor in Russia of Bourbaki. Works attributed to Koz'ma Prutkov (circa 1860) include epigrams, parodies, comedies, etc. The aphorisms have become an organic part of modern Russian.

examples of potentials (or connections) which are not gauge equivalent and possess the same strength (curvature form).

Example 6.1 (Wu-Yang) Let W_μ be the gauge field $SU(2)$ over R^4 and let W_μ take values in $SU(2)$. We select structural constants c_{ij}^k in $SU(2)$ so that $c_{12}^3 = c_{23}^1 = c_{31}^2 = 1$. We set the field

$$(W_0^k)_{\mathscr{U}} = 0,\ (W_i^k)_{\mathscr{U}} = \varepsilon_{ikj}\frac{x^j}{r^2}(\lambda - 1),\ i,\ j,\ k = 1,\ 2,\ 3, \tag{6.2}$$

where λ is a numerical factor and $\mathscr{U}$ is a domain in R^3 not containing the origin or any point of the z-axis.

Calculating the strength of the field $F_{\mu\nu}$ by (3.8), we obtain

$$(F_{ij}^l)_{\mathscr{U}} = \varepsilon_{ijk}\frac{x^k x^l}{r^4}(1 - \lambda^2),\ i,\ j,\ k,\ l = 1,\ 2,\ 3. \tag{6.3}$$

$F_{\mu\nu}$ satisfies equation (3.20) with source I which does not vanish if $\lambda \neq 0, \pm 1$, namely,

$$I_0^k = 0,\ I_i^k = \varepsilon_{ikj}\frac{x^j}{r^4}\lambda(1 - \lambda^2). \tag{6.4}$$

On the other hand, we consider the vector potential W field of Dirac's magnetic monopole given at the origin in R^3 and with one nonzero (e.g., a third) isospin component

$$\begin{gathered}(W_\mu^1)_\mathscr{V} = (W_\mu^2)_\mathscr{V} = (W_0^3)_\mathscr{V} = (W_r^3)_\mathscr{V} = (W_\theta^3)_\mathscr{V} = 0,\\ (W_\varphi^3)_\mathscr{V} = \frac{-(1 - \lambda^2)(1 - \cos\theta)}{r\sin\theta},\end{gathered} \tag{6.5}$$

where r, θ, φ are spherical coordinates in R^3.

The field $\tilde{F}$ associated with W in (6.5) is of the form

$$(\tilde{F}_{ij}^3)_\mathscr{V} = \varepsilon_{ijk}\frac{x^k}{r^3}(1 - \lambda^2),\ (\tilde{F}_{ij}^1)_\mathscr{V} = (\tilde{F}_{ij}^2)_\mathscr{V} = 0. \tag{6.6}$$

Acting by the group $G = SU(2)$ in isospin space, we can transform $\tilde{F}$ into F (see (6.3)), i.e., after the gauge transformation with $g \in SU(2)$ we have $(\tilde{F})_{\mathscr{V}'} = (\tilde{F})_{\mathscr{U}}$. Under this transformation, W in (6.5) is sent into a field W' which is not gauge equivalent to W in (6.2), since $W_\mathscr{V}$ has no source. In explicit form, potential $W_{\mathscr{V}'}$ is given by the formulas

$$(W_0^k)_{\mathscr{V}'} = 0,\ (W_i^k)_{\mathscr{V}'} = (1/r^2)\left[-\varepsilon_{ikj}x^j + \lambda^2\hat{\varphi}^i x^k\frac{\sin\theta}{1 - \cos\theta}\right],$$

where $\hat{\varphi}^i$ is the ith component of the spatial unit vector in the φ direction. It is obvious that the result is local in nature because it is valid in any neighborhood of the origin in R^3. The example admits a natural generalization to the following geometric formulation.

Let (P, p, G, M) be the principal bundle with a base space M and a structure group G, let ω be the connection form and let Ω be the curvature form of the bundle P. Two gauge equivalent forms ω and $\omega' = g\omega g^{-1} + \partial g g^{-1}$, $g \in G$, have gauge equivalent connection forms $\Omega' = g\Omega g^{-1}$. The Wu-Yang problem can be formulated as the conditions on the bundle P under which the form Ω governs the form ω uniquely to within the gauge equivalent form. The most complete results in the solution were obtained in [Mos] with references to the prior results of Deser, Drechsler, Gu, and Yang.

To formulate the Mostow result, we need a definition from holonomy theory [KN1].

Let $\Omega_u(X, Y)$ be the curvature form's value on the horizontal vectors X and Y given at a point u. We denote by $m_0(u)$ the subspace in ($\mathcal{G}$ the Lie algebra in G) generated by all elements of the form $\Omega_u(X, Y)$.

We introduce the ith-order covariant differentiation operator

$$D_i\Omega = D_1 \circ D_{i-1}\Omega, \quad D_0\Omega = \Omega, \quad D_1 = D.$$

We can define covariant differentiation D_I of order I, where I is the multiindex as $I = (i_1, \ldots, i_r)$, $0 \leqslant r \leqslant n$, $n = \dim M^n$,

$$D_I = D_{i_1} D_{i_2} \ldots D_{i_r}\Omega.$$

We apply the operator D_1 to $\Omega_u(X, Y)$ and consider the space m_1 obtained as the union of m_0 and all $\mathcal{G}$-valued functions of the form $D_1\Omega$, $m_1 \supset m_0$.

We define spaces m_k inductively: m_k is the union of m_{k-1} and of the space of all functions of the form

$$D_{i_1 \ldots i_k}\Omega.$$

We can prove that the union of all m_k forms a subalgebra in $\mathcal{G}$ [KN1] and we denote it by $\mathcal{G}'$.

Definition 6.1 The connected subgroup $H'(u)$ in G generated by $\mathcal{G}'$ is called the *infinitesimal holonomy group* at the point $u \in P$.

We now formulate the Mostow result in the particular case of simply connected base space M^n [MSh]. The general case is also considered there.

Theorem 6.1 *Let ω and ω' have coincident curvature forms and their covariant derivatives*

$$D_I\Omega = D_I'\Omega$$

for all points $u \in P$, where D' is covariant differentiation with respect to ω'.

If $\dim \mathcal{G}'(u) = \text{const}$, *then ω and ω' are gauge equivalent.*

The constancy of $\dim \mathcal{G}'(u)$ means that the group $H'(u)$ is isomorphic to the local holonomy group $H^* = \cap H^0(u)$. The intersection is of all connected open neighborhoods of a point $x \in M^n$, $x = p(u)$.

The above results lead to the following natural question: What is the structure of the set of potentials $\{ W_\mu \}$ (connections) which are uniquely (modulo gauge potentials) defined by the strength (curvature form)? The global theorem supplying the answer as applied to fiber bundles over M^n is given in [MSh].

Theorem 6.2 *Let (P, G, M^n) be a principal bundle with a semi-simple group G, $\mathscr{W}^*$ be the set of potentials uniquely defined by the strength, and $\mathscr{W}$ be the set of all potentials.*

Space $\mathscr{W}^*$ forms an open dense set in the space of all potentials $\mathscr{W}$ equipped with the C^∞-topology of Whitney [GG]. However, the operator pr $\mathscr{F}/\mathscr{W}^*$ associating the field $F_{\mu\nu}$ with its potential W_μ is not continuous.

The study of the global structure of the space of gauge potentials is of great interest in Yang-Mills field quantization.

6.1.2 Orbit Space

In a quantization of gauge fields by the path-integral method [FS], the ability to select a representative continuously depending on a point in the base space is always assumed. The condition is called the *choice of gauge.*

The existence of a local gauge provides for the applicability of the path-integral method within the framework of perturbation theory, i.e., Faddeev-Popov quantization. However, there exists no global gauge, as is shown in [Gr], in which the Coulomb gauge ($\partial_k W_k = 0$, $k = 1, 2, 3$) is taken as an example. The result has been generalized by Singer, who proved topologically [Si] that no global gauge could be introduced. The result is based on the study of the topology of the gauge transformation group.

Let (P, G, M) be a principal bundle with the Yang-Mills G-connection W_μ, $\mathscr{W}$ be the space of all connections $\{ W_\mu \}$, $\mathscr{W}$, being an affine space. The group $\mathscr{G}$ acts as

$$W_\mu \to gW_\mu g^{-1} + \partial_\mu g g^{-1}, \; g \in \mathscr{G}, \tag{6.7}$$

on the fields W_μ.

$\mathscr{G}$ is the set of all continuous functions on M with values in G and called the *gauge transformation group*[*]. $\mathscr{G}$ is infinite-dimensional.

Definition 6.2 The set of equivalence classes of space $\mathscr{W}$ relative to action (6.7) is called the *orbit space $\mathscr{O}$.*

$\mathscr{O}$ is an analog of the configuration space of a dynamic system since any physical quantity is gauge invariant and therefore constant on an orbit of $\mathscr{G}$.

Proving that there is no global gauge relative to $\mathscr{G}$ is complicated; however, the crux of the matter can be seen from the following particular example.

Let $\mathscr{G}_0$ be a subgroup of $\mathscr{G}$ consisting of all functions from $\mathscr{G}$ which take the value $e \in G$ at a fixed point of the base space M ($g \in \mathscr{G}_0 \Rightarrow g(x_0) = e$, $g(x) \in G$).

The choice of $\mathscr{G}_0$ as a gauge group is important from the physical standpoint since it is normally relative to $\mathscr{G}_0$ that the gauge is taken. $\mathscr{G}$ differs from $\mathscr{G}_0$ by gauge

---

[*] Strictly speaking, $\mathscr{G}$ is the group of automorphisms of the fiber bundle P. However, we employ the more usual terminology and stick to the elementary style of the treatment.

transformations

$$0 \to \mathcal{G}_0 \to \mathcal{G} \to G \to 0 \tag{6.8}$$

from G.

Below we fix G, assuming $G = SU(n)$ for definiteness. However, the argument remains valid for any semi-simple Lie group.

In contrast to $\mathcal{G}$, the group $\mathcal{G}_0$ acts on $\mathcal{W}$ freely. In fact, it follows from the condition $\mathfrak{g}W = W$, $\mathfrak{g} \in \mathcal{G}_0$, that

$$\nabla \mathfrak{g} = \partial_\mu \mathfrak{g} + [\mathfrak{g}, W] = 0.$$

Since $\mathfrak{g}(x_0) = e$, $\mathfrak{g} = e$.

Since $\mathcal{G}_0$ acts on $\mathcal{W}$ freely, the triple

$$\mathcal{W} \overset{\mathcal{G}_0}{\to} \mathcal{W}/\mathcal{G}_0 \tag{6.9}$$

is a fiber bundle.

Constructing the section s in (6.9) is equivalent to finding a global gauge relative to $\mathcal{G}_0$. It follows from the theory of fibrations (see Subsec. 1.4.2) that the existence of s is equivalent to decomposition into the direct product

$$\mathcal{W} = \mathcal{W}/\mathcal{G}_0 \times \mathcal{G}_0. \tag{6.10}$$

The affine space $\mathcal{W}$ is contractible, i.e., $\pi_i(\mathcal{W}) = 0$ (for all i).

The impossibility of constructing s follows thereby from $\pi_i(\mathcal{G}_0) \neq 0$ for some i. We see that the global gauge is related to the homotopy properties of $\mathcal{G}_0$.

Example 6.2 We calculate $\pi_i(\mathcal{G}_0)$ in several special cases.

(1) $M = R^n$ is a contractible space. Therefore, $\pi_i(\mathcal{G}_0) = 0$ and a gauge does exist. For instance, the gauge conditions for fiber bundles on R^4 are given as

$$W_0 = 0,\ W_1 = 0\ (x^0 = 0),\ W_2 = 0\ (x^0 = 0,\ x^1 = 0),$$
$$W_3 = 0\ (x^0 = 0,\ x^1 = 0,\ x^2 = 0).$$

(2) The physically interesting situation of $M \sim S^3$ or $M \sim S^4$. The spheres S^3 and S^4 are obtained by the compactification of $R^3 \cup \infty$ and $R^4 \cup \infty$, respectively, and $x_0 = \infty$. $M = S^3$, the principal bundle over S^3 with $G = SU(N)$ is also trivial. However, $\pi_i(\mathcal{G}_0) = \pi_i(\Omega^3(SU(N)) \simeq \pi_{i+3}(SU(N))$, where $\Omega^3(SU(N))$ means the functional space of mappings $\varphi: S^3 \to G$ such that $\varphi(x_0) = e \in SU(N)$. Since even $\pi_3(SU(N)) \neq 0$, it follows that there is no gauge.

(3) $M = S^4$. $\pi_i(\mathcal{G}_0) = \pi_i(\Omega^4(SU(N)) \neq \pi_{i+4}(SU(N))$ is also a case where a gauge cannot be constructed. Note that a gauge is not related to the triviality or nontriviality of the fibering. We know that fiber bundles over S^4 are classified by the Pontrjagin class $p_1(\pi_3(G) = Z)$. There exist countably many non-equivalent fiber bundles, the gauge being neither in the trivial $p_1 = 0$ nor in non-trivial bundles $p_1 \neq 0$. Hence, the absence of a gauge is not related to instanton existence.

We retrace our steps to the full group $\mathcal{G}$. We only outline the main stages of the proof of Singer's theorem in the general case (for detail see [Si]).

$\mathscr{G}$ does not act on $\mathscr{W}$ freely. In particular, $\mathscr{G}$ contains the subgroup $\tilde{Z}_N$ of the space of constant functions with values in the center Z_N of the group $SU(N)$. Indeed, functions g taking values in Z_N do not alter the field W_μ, while continuous functions on M with values in a finite group must be constant. However, the group $\mathscr{G}^* = \mathscr{G}/\tilde{Z}_N$ does not act on $\mathscr{W}$ freely either since there exist stability subgroups containing Z_N. We can show that the stability subgroups commute with the holonomy groups $H(W)$ and are classified by the latter. In particular, connections with the maximal holonomy group $H_{\max}(W)$ are associated with the fields on which $\mathscr{G}^*$ acts freely.

Definition 6.3 Fields W with $H(W) = H_{\max}(W)$ are said to be *irreducible.*

The space of irreducible fields $\mathscr{W}^*$ possesses the following properties:

(a) $\mathscr{W}^*$ is open and dense in $\mathscr{W}$, while $\mathscr{W}/\mathscr{G}^*$ is open and dense in $\mathscr{W}/\mathscr{G}$. $\mathscr{W}^*$ is the fiber bundle over $\mathscr{W}/\mathscr{G}^*$ with the structure group $\mathscr{G}^*$.

(b) $\pi_i(\mathscr{W}^*) = 0$.

Hence, the absence of the global gauge for $\mathscr{W}^*$ relative to $\mathscr{G}^*$ follows immediately; since $\mathscr{W}^*$ is a subset of $\mathscr{W}$, there is no gauge for $\mathscr{W}$ either.

We should bear in mind that the openness and denseness of $\mathscr{W}^*$ in $\mathscr{W}$ hold for any functional class of gauge potentials. Even a local gauge exists only in Sobolev spaces L_k^p satisfying the condition

$$p(k+1) \geqslant n,$$

where $n = \dim M^n$ [Uh2], [NR].

Note that it's not easy to study the case of equality analytically. It is natural that Singer's theorem is also valid for this potential class [NR], [So3].

That there is no global gauge points to the complicated structure of the orbit space. The principal question is distinguishing a domain $\mathscr{V}$ in $\mathscr{W}$, where a gauge does exist. It has been shown in [Gr] that the situation in which the orbit intersects the cross-section (i.e., the gauge surface) repeatedly is typical. This phenomenon is called the *Gribov ambiguity.* Another case in which the orbit does not meet the gauge surface seems impossible. There are a number of results in this field, but no complete proof [NR], [So3].

It would be interesting to describe the geometry of $\mathscr{V}$ as it depends on the choice of gauge and on the potential's class, but with certain general properties. The Coulomb gauge, for example, specifies a plane P orthogonal to the vanishing potential's orbit. $\mathscr{V}$ consists of points where P intersects the orbit nearest to the vanishing potential. The boundary $\partial\mathscr{V}$ is called the *Gribov horizon.* It follows from the construction that $\mathscr{V}$ is convex. The convexity inside the horizon is also proved for other gauges, e.g., for the background one, in which case instead of a vanishing potential an irreducible potential is chosen. Irreducible potentials lie on the horizon.

The discovery of Gribov ambiguities stimulated the study of the general structure of the orbit space.

We note two general results related to the differential geometric structure of $\mathscr{O}$.

(1) It has been shown in [BV1] that the sectional curvature of $\mathscr{O}$ is non-negative and that the Ricci tensor is positive definite.

(2) $\mathscr{O}$ is non-compact [Ve].

The introduction of orbit space to field theory enables us to regard the dynamics of the Yang-Mills fields as the motion of a "particle" across the infinite-dimensional Riemannian manifold $\mathcal{O}$. In particular, the approach enables us to obtain infinitely many integrals of motion in involution [BV2]. It should be noted that the complete integrability of the Yang-Mills equations does not follow. Moreover, there exist a solid basis for the opposite conclusion, to which, in particular, finite-dimensional versions of the Yang-Mills equation point.

Recently, models with gauge potentials that only depend on the time component have been studied in detail. The Lagrangian L_{YM} (3.1.1) then reduces to a finite-dimensional system. We write the Lagrangian as

$$L = (1/2)\,\mathrm{Tr}\,(\dot{a} + a_0 a)^t(\dot{a} + a_0 a) - U(a), \tag{6.11}$$

assuming $G = SO(3)$ for definiteness, when a is a real 3×3 matrix, a_0 is skew-symmetric and L is obtained from the $SU(2)$-Lagrangian L_{YM} if we identify W_i^b with a_{bi} and associate W_0^b with the matrix $(a_0)_{bc} = -\varepsilon_{bcd} W_0^d$, assuming W_i only depend on time. The potential is

$$U = \mathrm{Tr}\,(a^t a) - \mathrm{Tr}\,(a^t a)^2. \tag{6.12}$$

L is invariant under gauge transformations

$$a \to \tilde{u}a, \quad a_0 \to \tilde{u}^t a_0 \tilde{u} + \tilde{u}^t \dot{\tilde{u}}, \tag{6.13}$$

where $\tilde{u}$ is the matrix function of time with values in $SO(3)$. Thus, the system in (6.11) is a complete finite-dimensional analog to a Yang-Mills system on the orbit space Mat $(3)_R/SO(3)$.

It has been shown in [So4] that there exist analogs of Gribov ambiguities in (6.11) and the stratification of orbit space has been studied.

The system (6.11) was investigated in the particular case of two-dimensional space-time, from the point of view of integrability of a finite-dimensional Yang-Mills equation. The most interesting is the proof of the nonintegrability [Zi], [NS] and stochasticity of the model [MST]. The results put in doubt the complete integrability of a Yang-Mills equation in M^4.

A study of the topology of orbit space is exceptionally interesting for two other important areas of studies, which we only mention in passing.

One is related to the description of the space of instanton and monopole solutions. In particular, it was shown in [At3] that instantons over R^4 with a gauge group G can be identified with those in two-dimensional chiral theory where the gauge group is a loop space ΩG.

Taking into consideration the Nahm construction of monopoles from instantons [N], we can thus obtain the (incomplete) description of the monopole space.

Another branch of study was started by Taubes [Tau2], who constructed a non-trivial (non-constant) solution in the Yang-Mills-Higgs theory $SU(2)$ (with $M \sim R^3$) in the vacuum sector (zero topological charge Q). The solution is a saddle point in the action functional and requires a refined topological technique (the Ljusternik-

Schnirelmann theory). It was shown later in Taubes' paper that solutions of this type also existed in the monopole sectors ($Q \neq 0$) [Tau3]. In a work by Manton [Man], topologically non-trivial vacuum solutions in the Weinberg-Salam model were investigated. Such solutions can be treated as high-energy resonances observed at large energies, of the order of TeV. The paper contains important heuristic observations without any rigorous proof of the solution's existence. The approach continues to develop [FH], [Bu], but the most interesting results seem yet to follow.

We confine ourselves to the above fragments of the vast area of topological applications to field theory.

6.2 TOPOLOGY OF CONDENSED MATTER

6.2.1 Liquid Crystals

The applications of topology to the theory of liquid crystals can be classified into two groups. One is mostly a description of textures, point, and line defects. The classification of the latter by homotopy groups is only applicable to certain types of liquid crystal and should be supplemented with differential geometric methods (such as the study of caustics) and the analysis of crystallographic symmetry groups. The space of mesomorphic phases promises quite an interesting application of low-dimensional topology.

The other approach is related to the study of phase interaction, structure transitions, and hydrodynamics. When investigating these problems, topology abuts on the theory of dynamic systems. This sphere of research is open to numerous applications. I restrict myself to several examples illustrating the viewpoint.

A. A new type of liquid crystals, *discotics*, was discovered in 1977 [CSS]. They comprise a system of flexible columns whose centers in the cross-section form a regular crystal lattice. In its properties, the discotic is similar to a liquid in one direction and to a two-dimensional crystal in the other two directions. Possible transitions of a discotic into the nematic, isotropic, and other phases can be described by the Ginzburg-Landau theory [KM]. The interesting experimental results of Bouligand [Bou2] show that complicated disclinations can be observed in discotics together with various kinds of walls and other textures. Although the order parameter space is well known, the two-dimensional lattice does not enable us to apply purely homotopy methods. It is interesting to completely classify textures and defects in discotics. The problem is solved for a number of special cases [K2].

B. Another example of a liquid crystal with a poorly understood defect structure is the well-known *blue phase*. Discovered in the first experiments on liquid crystals (Reinitzer, 1882), the blue phase is still the most enigmatic structure. Recently, progress in understanding the blue phase has been made (indeed, three phases have been discovered) [BD].

Several words about its structure. The blue phase is a modification of a cholesteric crystal and is observed between a cholesteric and an isotropic fluid of temperature of order from 0.1 to 1 K. In contrast to cholesterics with a director, which describes a helix, a tensor order parameter, which characterizes the molecular-orientation correlation, is spatially periodic.

We confine ourselves to discussing the formulation of a problem on defect and texture description in one model of the blue phase. Within the framework of continual theory, elastic energy density can be written as [MSAB]

$$F = K_1(\text{div }\mathbf{n})^2 + K_2(\mathbf{n}\cdot\text{rot }\mathbf{n} + q_h)^2 + K_3(\mathbf{n}\times\text{rot }\mathbf{n})^2 + (K_2 + K_4)[\text{Tr }(\nabla\mathbf{n})^2(\text{div }\mathbf{n})^2], \tag{6.14}$$

where $\mathbf{n}$ is the director, $q_h = 1/\lambda_h$, and λ_h is the helix's pitch.

An interesting aspect of the potential F is that it involves a surface term proportional to $\text{Tr }(\nabla\mathbf{n})^2 - (\text{div }\mathbf{n})^2$.

Contrary to the usual situation, the surface term should be taken into account, since surface integration involves an integral over the surface disclination, which contributes to the total energy in proportion to the volume. A complete analysis of defects in the blue phase remains an unsolved problem. However, numerical results show that the structure of singularities is quite complicated [BD]. The topological analysis of disclinations in the blue phase given by (6.4) has been recently carried out [PDDu], [PDuD].

We now pass to the third class of liquid crystals.

C. It was mentioned in Subsec 5.1.1 that *smectics* are ideal for the theory of foliations. We mean only one class A of smectics [Ge], in which the director is orthogonal to the layers. If we assume that the distance between the layers is constant, which is associated to a measured foliation, then it is easy to see that only

$$F = K_1(\text{div }\mathbf{n})^2 \tag{6.15}$$

remains in (5.19), since rot $\mathbf{n} = 0$, i.e., only lateral bending is possible.

By varying the functional

$$\int_S (F)^2 dS,$$

where S is the layer's surface, we can show that a minimum is attained on confocal conics, *Dupin cyclides* [Geu].

It would be of interest to study possible configurations for a type-C smectic whose director is at an angle to the layers. Poénaru initiated the study of point singularities in smectics with foliations [Po1].

The reader can get acquainted with the instability in smectic liquid crystals within the framework of a simple dynamic system in [DGP].

6.2.2 ^{3}He

Experiments on rotating helium have drawn the attention of researchers of superfluid ^{3}He. Vortices in the A- and B-phases of ^{3}He are of greatest interest. The structure of a vortex is complicated. It has a core, which in turn may consist of a normal or any other superfluid phase. Vortices with various symmetries can exist inside the core. Transitions from one phase into another inside the core of the same phase is an unsolved problems.

An example is a core of a B-phase with another phase inside [SV], [Thu]. Modern theoretical achievements are discussed in survey [SV].

Because of a lack of space we do not touch upon other interesting topics which arise in the theory of ^{3}He, but we shall formulate one concrete problem that is closely related to the main text.

All one-dimensional textures in the A- and B-phases have been found in Subsec. 5.2.7. Two-dimensional textures (viz., those depending on two coordinates) would be an interesting subject. The direct generalization of the method encounters great analytical difficulties. The analogy with the chiral model seems quite promising. However, two-dimensional textures have still not been found.

6.2.3 Theory of Links

The theory of links and knots has repeatedly attracted attention due to the discovery of its surprising relationship to other branches of mathematics and to its applications to physics and biology.

We begin with a result having an important application in the theory of ring DNA molecules, i.e., the Călugăreanu-White formula [Wt], [FV], [Ph].

Let γ be a closed smooth curve in R^3 and $\mathbf{v}$ a normal vector field on γ. We assume that the magnitude of a vector $v(t)$ is so small that $v(t)$ intersects γ only at one point. The $\mathbf{v}$ terminals sweep a curve γ_{v} which inherits the orientation of γ, while $\mathbf{v}$ itself inherits a strip embedded in R^3. Let $k(\gamma, \gamma_{\mathrm{v}})$ be the Gauss linking number of γ and γ_{v}. We define the *twist* of $\mathbf{v}$

$$\mathrm{Tw} = (1/2\pi) \int_\gamma \mathbf{v}^\perp \cdot d\mathbf{v} \tag{6.16}$$

in a standard way, where the vector $\mathbf{v}^\perp$ is in the frame $(\hat{\mathbf{t}}, \mathbf{v}, \mathbf{v}^\perp)$, a right-hand system, and $\hat{\mathbf{t}}$ the unit tangent vector to the curve. The twist of the curve is a continuous quantity, the linking number integer is therefore $k \neq \mathrm{Tw}$. It is remarkable that another quantity can be introduced, viz., the writhing number Wr such that

$$k = \mathrm{Tw} + \mathrm{Wr}. \tag{6.17}$$

Wr only depends on γ and its definition is based on the following.

We construct the Gauss map for $\gamma \times \gamma_{\mathrm{v}}$, i.e., we define

$$\varphi: \gamma \times \gamma_{\mathrm{v}} \to S^2, \ \varphi(x, y) = \frac{y - x}{|y - x|} \tag{6.18}$$

for ordered pairs (x, y). Let dS be the element of area on S^2. Then $\varphi_*(dS)$ is induced by φ.

Definition 6.4 The *writhing number* is the integral

$$\mathrm{Wr} = (1/4\pi) \iint_{\gamma \times \gamma_{\mathrm{v}}} \varphi_*(dS). \tag{6.19}$$

The number is a continuous quantity.

Wr is an important quality because it is Wr that is usually measured in experiments with superhelical DNA. The linking number k can also be changed (by topomerase enzymes). The Călugăreanu-White formula thus links two effects (superhelicity and the action of topomerases). The reader interested in the applications of the Călugăreanu-White formula to DNA is referred to the surveys in [FV] and [Ph].

The formula admits a generalization to multidimensions. The role of γ is played by a submanifold embedded in R^n and the intersection number is considered instead of k [Wt].

The Călugăreanu-White formula means that a strip swept around by $\mathbf{v}$ forms a trivial knot, i.e., the linking number is the only topological invariant. It would be interesting to clarify whether analogs of the Călugăreanu-White formula exist when γ and $\gamma_{\mathbf{v}}$ have a trivial Gauss number and non-trivial higher-order linking numbers. The question arises as to whether there are similar formulas for a link ensemble. We note another possible application of the Călugăreanu-White formulas. For instance, in considering linked polymer chains, the integral

$$I_{\gamma_1\gamma_2} = \int \langle k_{\gamma_1\gamma_2}(\varrho, O)\rangle^2 \, d^3\varrho \tag{6.20}$$

is introduced, where $k_{\gamma_1\gamma_2}$ is the linking number of γ_1 and γ_2.

The integral is obtained by averaging $k^2_{\gamma_1\gamma_2}$ over the isometry group for R^3 (where $\boldsymbol{\rho}$ is the displacement and O the orientation of γ_2 with respect to γ_1). We can show that

$$\int \langle k_{\gamma_1\gamma_2}(\varrho, O)\rangle \, d^3\varrho = 0.$$

The interesting formula

$$I = (1/8\pi) \int_0^\infty A_{\gamma_1}(r) A_{\gamma_2}(r) \, dr \tag{6.21}$$

was obtained in [Ph] for plane convex curves, where the functions A_{γ_1} and A_{γ_2} only depend on γ_1 and γ_2.

The results were generalized to curves with twist in [CB] and [Du], with the added term

$$(1/8\pi) \int_0^\infty B_{\gamma_1}(r) B_{\gamma_2}(r) \, dr. \tag{6.22}$$

The quantities $A_\gamma(r)$ and $B_\gamma(r)$ are for the curve of the form

$$\begin{aligned} A(r) &= (1/r) \iint (dr_1 \cdot dr_2)\theta(r - r_{12}), \\ B(r) &= (1/r^2) \iint ([dr_1, \, dr_2] \cdot r_{12})\theta(r - r_{12}), \end{aligned} \tag{6.23}$$

where

$$\theta(x) = \begin{cases} 1, & x > 0, \\ -1, & x < 0, \end{cases} \qquad r_{12} = |\mathbf{r}_1 - \mathbf{r}_2|.$$

Do similar formulas exist for an ensemble of three and more curves? Also, can the result be generalized to higher-order linking numbers?

Another application of the theory of links occurs in multidimensional dynamic systems. I only note three appropriate papers. The structure of periodic solutions in the Lorenz system was described in [BiW]. The system has lately attracted the attention of those interested in dynamic system and turbulence due to the discovery in it of strange-attractor stochastic behavior [LiL], [Wil]. The Lorenz system is described by the Navier-Stokes equations in the *Galerkin approximation*

$$\begin{aligned} \dot{x} &= -10x + 10y, \\ \dot{y} &= \mathrm{Rl}\cdot x - y - xz, \\ \dot{z} &= (-8/3)z + xy, \end{aligned} \tag{6.24}$$

where Rl is the Rayleigh number, equal here to 24.

Birman and Williams have shown that periodic solutions make up a complicated node system. General systems with bifurcations were studied in [U], [Ua] with the discovery of a change in the types of periodic nodal orbits as bifurcation points were crossed.

In conclusion, I cannot help mentioning (but only mentioning) the outstanding recent discovery in a paper by Jones [Jo] of a new polynomial linking invariant $V_L(t)$ which is considerably stronger than the Alexander polynomial. Surprising relationships between the theories of links and von Neumann algebras have been discovered. Similar studies together with discoverd connections to new concepts is physics (topological field theories, quantum groups, etc.) have already led to major achievements in link theory [Wit 4], [V], [BiL], and [At]. Andoubtedly, we are now on the threshold of new discoveries in that developing area.

6.3 HISTORICAL REMARKS

The history of a fundamental branch of science is instructive for a modern researcher. Without attempting in any way to trace the development of topology* I would like to outline certain moments. The reader will recognize more fully the reason for a topological boom in physics.

The birth of topology is usually dated to Leibniz's *Characteristica Geometrica* (1679), where he tried to study the properties of figures in relation to their topological parameters, instead of their metric ones. *Besides the coordinate specification,* he wrote, *we needed other, purely geometric or linear analysis to determine Situs in the way algebra determines quantities.*

It is interesting to note that Leibniz tried to interest Huygens in these ideas; the latter, however, was not curious. This was the end of the first and unsuccessful interaction between topology and physics.

A further development of *Analysis Situs* was made by Euler, who solved the Königsberg bridges problem and derived his formula $V - L + F = 2$ for polyhedrons, where V is the number of vertices, L of edges, and F of faces. Gauss's work (including the Gauss-Bonnet theorem for surfaces) and Listing's *Vorstudien zur Topologie* (1848) must be mentioned as foundations of topology. Listing was the first to introduce the term.

* Recent monograph by Diedonne contains much of the history of topology of the twentieth century.

Works by Riemann in the theory of algebraic functions and Riemann surfaces can also be regarded as topological today. Even though we omit some names, we can surely declare that topology was formed as an independent branch of mathematics by Henri Poincaré. His *Analysis Situs* (1885) and five appendices are the basis for modern topological concepts. Proving the Poincaré conjectures is still considered the highest topological achievement, the latest example being the proof of the Poincaré conjecture in dimension four by Freedman [Fr].

Topological ideas were known by a number of the greatest physicists in the 19th century. In the article on Faraday, in *Nature* in 1873 Maxwell predicted the role of topology in physics:

"It is true that no one can essentially cultivate any exact science without understanding the mathematics of that science. But we are not to suppose that the calculations and equations which mathematicians find so useful constitute the whole of mathematics. Calculus is but a part of mathematics.

The geometry of position is an example of a mathematical science established without the aid of a single calculation. Now Faraday's lines of force occupy the same position in electromagnetic science that lines do in the geometry of position. They furnish a method of building up an exact mental image of the thing we are reasoning about."

It is characteristic of the development of topology that it is associated with universalists, who obtained fundamental results both in physics and mathematics (e.g., Euler, Gauss, Riemann, and Poincaré). With regard to more concrete applications, then we have to mention Volterra's *Sur l'équilibre des corps élastiques multiplement connexes* (1907), in which he offered a method (now bearing his name) [Vol] to describe defects in solids, based on cutting and gluing together. In the process, he constructed a series of homeomorphic three-dimensional manifolds. The construction was published as an appendix to the topology textbook by Lefschetz [Lef].

But, starting in the mid-1920s, the union of topology with physics ceased. In fundamental physics, this break was due to reasons such as the emergence of quantum mechanics and relativity theory, which required other mathematical machinery, while in more specialized branches, say, the theory of liquid crystals, the classification of singularities was done by amateur methods, e.g., the classification of singularities in a nematic by the Frank index, which is the same as the proof of the equality $\pi_1(S^1) = H_1(S^1)$. The basic topological concepts started developing with mid-thirties, being applied to physics only thirty years later. At the time, physical interest concentrated on nuclear structure, the theory of weak interactions, and electrodynamics. Hopf's paper passed totally unnoticed by physicists [Hop]. It was devoted to the homotopy classification of maps $S^3 \to S^2$. It is curious that Dirac's work on monopoles appeared in the same year [Dir].

Only forty years later has the close relation between these papers become clear. It is remarkable that Poincaré himself considered the motion of a charged particle, regarding a monopole as an external source.

Development of topology continued. Differential fiber bundle geometry was created. Ehresmann introduced the notion of connection in a fiber bundle in 1950. Yang-Mills fields appeared in physics in 1954. The first, not very explicit, relationships between topology and physics started in gravitational theory. When studying the structure of space-time, Finkelstein and Misner used homotopy groups (1959) [FM].

The non-linear models of Skyrme appeared in 1959 and included Lagrangians with "kinetic" terms of the fourth degree [Sk]. Finkelstein used homotopy groups in Skyrme models to classify kink solutions (1966) [F]. But elementary-particle physicists were mostly interested in those years in phenomenological theories of strong interactions (Regge's approach). The Skyrme models were not regarded as being realistic. They only attracted attention twenty five years later after Witten's work [Wit2, 3], in which baryons were treated as solitons.

The success in constructing a unified theory of weak and electromagnetic interactions (Weinberg-Salam (1967)) and newly born interest in the attempts of creating a unified theory of interactions involving strong and gravitational ones made the physicists look at the geometric nature of gauge fields in a new way.

A radical change occurred in early seventies.

The discovery of the gauge monopoles in the Georgi-Glashow models by Polyakov and 'tHooft (1974) were based on the earlier work of Wu and Yang [WY1] and led to universal interest in the topology of gauge fields. The topological test of monopole existence was found practically at once [MP2,3], [TFS].

A one-instanton solution of the Yang-Mills equation was found in 1975 [BPST]. An unremitting topological attack on physical theories began.

The algebra-geometry classification of instantons was performed in 1977 [ADHM]. The construction of twistors by Penrose was applied.

A new approach to monopole classification involving earlier instantons was advanced by Nahm [N]. All these results had a strong effect on the development of topology and not only important applications to physics.

With the help of instanton bundles, Donaldson proved the existence of various smooth structures on R^4 in addition to closed simply-connected four-dimensional manifolds. Non-smooth structures were also constructed. After Milnor's proving the existence of various smooth structures on S^7 (1956) the above result was thought to be sensational. No one expected the effect to arise in a low-dimensional manifold. The modern development of quantum-string theory involves the most refined mathematics (such as moduli space). However, this is now not history.

Applications to the theory of condensed matter also turned out to be of certain interest. The methods used earlier in field theory were successfully applied to the theory of condensed state (1976) [TK]. Superfluid ^{3}He, which was only discovered a few years ago, proved to be quite an interesting object for topological applications.

The number of physical systems subjected to topological investigation increases. Complicated topological problems remaining unsolved are connected to the Hall effect [Q].

In a Gibbs lecture in 1971 [Dy], Dyson said "... the marriage between mathematics and physics, which was so enormously fruitful in past centuries, has recently ended in divorce".

Looking at the modern stage of development of mathematics and physics, Dyson was probably mistaken: it was a temporary separation, not a divorce. While not an idyll, normal family life between the marriage partners now continues.

References

[A] Actor, A. Classical solutions of $SU(2)$ Yang-Mills theoreies. *Rev. Modern Phys.*, 51 (1979), pp. 461-525.

[AB] Aharonov, Y. and Bohm, D. Significance of electromagnetic potentials in the quantum theory. *Phys. Rev.*, 115 (1959), pp. 485-491.

[Ad] Adams, J. Vector fields on spheres. *Ann. Math.*, 75 (1962), pp. 603-632.

[ADHM] Atiyah, M., Hitchin, N., Drinfeld, V., and Manin, Yu. Construction of instantons. *Phys. Lett.*, 65A (1978), pp. 185-187.

[AG] Alling, N., and Greenleaf, N. *Foundations of the Theory of Klein Surfaces* (Lect. Notes in Math., 219). Berlin-New York: Springer-Verlag, 1971.

[AGR] Ambegaokar, V., de Gennes, P., and Rainer, D. Landau-Ginzburg equations for an anisotropic superfluid. *Phys. Rev. A*, 9 (1974), pp. 2676-2685.

[AH] Atiyah, M., Hitchin, N. *The Geometry and Dynamics of Magnetic Monopoles.* Princeton: Princeton Univ. Press, 1988.

[AM] Anderson, P., and Morel, P. Generalized Bardeen-Cooper-Schrieffer states and proposed low-temperature phase of liquid He^3. *Phys. Rev.*, 123 (1961), pp. 1911-1934.

[AP] Anderson, P. and Palmer, R. Textures of ^{3}He-A in a sphere: Topological theory of boundary effects and a new defect. In Trickey, S., Adams, E., and Dufty, J. (eds.), *Quantum Fluids and Solids.* New York: Plenum Press (1977), pp. 23-32.

[Ar] Arnold, V. *Geometrical Methods in the Theory of Ordinary Differential Equations.* Berlin: Springer-Verlag, 1978.

[AsM] Ashcroft, N., and Mermin, N. *Solid State Physics.* New York: Holt, Rinehart & Winston, 1976.

[At1] Atiyah, M. *K-Theory.* New York-Amsterdam: W.A. Benjamin, 1967.

[At2] Atiyah, M. *Geometry of Yang-Mills Fields.* Pisa: Acc. Lincei Scuola Normale Superiore, 1979.

[At3] Atiyah, M. Instantons in two and four dimensions. *Comm. Math. Phys.*, 93 (1984), pp. 437-451.

[At4] Atiyah, M. *The Geometry and Physics of Knots.* Cambridge: Cambridge Univ. Press, 1990.

[AT] Anderson, P., and Toulouse, G. Phase slippage without vortex cores: vortex textures in superfluid ^{3}He. *Phys. Rev. Lett.*, 38 (1977), pp. 508-511.

[AW] Atiyah, M., and Ward, R. Instantons and algebraic geometry. *Comm. Math. Phys.*, 55 (1977), pp. 117-124.

[B] Bais, F. The topology of monopoles crossing a phase-boundary, *Phys. Lett. B*, 98 (1981), pp. 437-440.

[BC] Bishop, R., and Crittenden, R. *Geometry of Manifolds.* New York-London: Academic Press, 1964.

[BD] Belyakov, V., and Dmitrienko, V. Blue phase of liquid crystals, *Soviet Phys. Uspekhi*, 28 (1985), pp. 535-562.

[BDFL] Behrends, R., Dreitlen, J., Fronsdal, C., and Lee, W. Simple groups and strong interaction symmetries. *Rev. M. Phys.*, 34 (1962), pp. 1-40.

[BDG] Brand, H., Dörfle, M., and Graham, R. Hydrodynamics parameters and correlation functions of superfluid ^{3}He. *Ann. Phys.*, 119 (1979), pp. 434-479.

[BDPPT] Bouligand, Y., Derrida, B., Poénaru, V., Pomeau, Y., and Toulouse, G. Distortions with double topological character: The case of cholesterics. *J. Physique*, 39 (1978), pp. 863-867.

[Bg] Bogomolny, E. The stability of classical solutions. *Soviet J. Nuclear Phys.*, 24 (1976), pp. 449-454.

[BiL] Birman J., and Lin X. *Knot Polynomials and Vasiliev's Invariants*. Preprint. Columbia University, 1992.

[BiW] Birman, J., and Williams, R. Knotted periodic orbits in dynamical systems-I: Lorenz's equations, *Topology*, 22 (1983), pp. 47-82.

[BL] Bailin, D., and Love, A. Point and line singularities in superfluid ^{3}He. *J. Phys. C*, 11 (1978), pp. 1351-1359.

[Bla] Blaha, S. Quantization rules for point singularities in superfluid ^{3}He and liquid crystals. *Phys. Rev. Lett.*, 36 (1976), pp. 874-876.

[BM] Barton, G., and Moore, M. Some *p*-wave phases of superfluid helium-3 in strong-coupling theory. *J. Phys. C.*, 7 (1974), pp. 4220-4235.

[BoM] Bogomolov, F., and Monastyrsky, M. Geometry of the orbit space and phases of ^{3}He and a neutron star in the *p*-state. *Theoret. and Math. Phys.*, 73 (1987), pp. 1165-1175.

[Bor] Borel, A. Sur la cohomologie des espaces fibrés principaux et des espaces homogènes de groupes de Lie compacts. *Ann. Math.*, 57 (1953), pp. 115-207.

[Bou1] Bouligand, Y. Recherches surles textures des étas mesomorphes. *J. Physique*, 35 (1974), pp. 959-981.

[Bou2] Bouligand, Y. Defects and textures of hexagonal discotics. *J. Physique*, 41 (1980), pp. 1307-1315.

[BP] Belavin, A., and Polyakov, A. Metastable states of two-dimensional isotropic ferromagnets. *JETP Letter*, 22 (1975), pp. 245-247.

[BPST] Belavin, A., Polyakov, A., Schwarz, A., and Tyupkin, Yu. Pseudoparticle solutions of the Yang-Mills equations, *Phys. Lett. B*, 59 (1975), pp. 85-87.

[Bre] Bredon, G. *Introduction to Compact Transformation Groups*. New York-London: Academic Press, 1972.

[Bro] Browder, W. *Surgery on Simply-Connected Manifolds*. New York-Heidelberg: Springer-Verlag, 1972.

[Bt] Bott, R. The Stable homotopy of the classical groups, *Ann. Math.*, 70 (1959), pp. 313-317.

[Bu] Burzlaff, J. Non-contractible hyperloops in gauge models with higgs fields in the fundamental representation. *Lett. Math. P.*, 8 (1984), pp. 459-465.

[BV1] Babelon, O., and Viallet, C. The Riemannian geometry of the configuration space of gauge theories. *Comm. Math. Phys.*, 81 (1981), pp. 515-525.

[BV2] Babelon, O., and Viallet, C. Conserved quantities for the geodesic motion on the configuration space of non-Abelian Yang-Mills theory. *Phys. Lett. B*, 103 (1981), pp. 45-47.

[BW] Balian, R., and Werthamer, N. Superconductivity with pairs in a relative *p*-wave. *Phys. Rev.*, 131 (1963), pp. 1553-1564.

[Bya] Bogoyavlenskiĭ, O. Topological charge of singularity in general theory of relativity. *Lett. Math. Phys.*, 2 (1978), pp. 313-315.

[C] Cartan, E. La Théorie des groupes finis et continus et l'analysis situs, *Mémoir. sci. math. fasc. XLII* (1930).

[CB] Des Cloizeaux, J., and Ball, R. Rigid curves at random positions and linking numbers. *Comm. Math. Phys.*, 80 (1981), pp. 543-553.

[CCR] Catenacci, R., Cornalba, M., and Reina, C. On the energy spectrum and parameter spaces of classical CP^n models. *Comm. Math. Phys.*, 89 (1983), pp. 375-386.

[Ce] Cerf, J. *Sur les Difféommorphismes de la Sphère de Dimension Trois* ($\Gamma_4 = 0$). (Lect. Notes in Math., 53.) Berlin-New York: Springer-Verlag, 1968.

[Ch] Chern, S. *Complex Manifolds Without Potential Theory* (2nd ed.). Berlin, Springer Verlag, 1979.

[CM] Coxeter, H., and Moser, W. *Generators and Relations for Discrete Groups* (3rd ed.). New York-Heidelberg: Springer-Verlag, 1972.

[Co1] Coleman, S. Secret Symmetry: An introduction to spontaneous symmetry breakdown and gauge fields. With discussions. In Zichichi, A. (ed.) *Laws of Hadronic Matter.* New York: Academic Press, 1975.

[Co2] Coleman, S. The uses of instantons. In Zichichi, A. (ed.) *The Whys of Subnuclear Physics.* New York: Plenum Press, 1979.

[CSS] Chandrasekhar, S., Sadashiva, B., and Suresh, K. Liquid crystals of disc-like molecules, *Pramana*, 9 (1977), pp. 471-480.

[DFN] Dubrovin, B., Fomenko, A., and Novikov, S. *Modern Geometry-Methods and Applications.* New York: Springer-Verlag, Part 1, 1984, Part 2, 1985.

[DGP] Dubois-Violette, E., Guazzeli, E., and Prost, J. Smectics: A model for dynamical systems. *Molec. Cryst.*, 114 (1984), pp. 1-17.

[Di] Dieudonné. *A History of Algebraic and Differential Topology 1900-1960.* Boston. Basel: Brikhäuser, 1988.

[Dir] Dirac, P. Quantized singularities in an electromagnetic field. *Proc. Roy. Soc.*, 133 (1931), pp. 601-609.

[Do] Donaldson, S. An application of gauge theory to the topology of 4-manifolds. *J. Differential Geom.*, 18 (1983), pp. 269-316.

[DoK] Donaldson, S., and Kronheimer, P. *The Geometry of Four-Manifolds.* Oxford: Oxf. Univ. Press, 1990.

[Du] Duplantier, B. Linking numbers, contacts and mutual inductances of a random set of closed curves. *Comm. Math. Phys.*, 82 (1981), pp. 41-68.

[Dy] Dyson, F. Missed opportunities. *Bull. Amer. Math. Soc.*, 78 (1972), pp. 635-652.

[EF] Eichenher, R., and Forger, M. More about non-linear σ-models on symmetric spaces. *Nucl. Phys. B*, 164 (1980), pp. 528-535.

[EGH] Eguchi, T., Gilkey, P., and Hanson, A. Gravitation, gauge theories, and differential geometry. *Phys. Report*, 66 (1980), pp. 213-393.

[EL1] Eells, J., and Lemaire, L. A report on harmonic maps. *Bull. London Math. Soc.*, 10 (1978), pp. 1-68.

[EL2] Eells, J., and Lemaire L., Another report on harmonic maps. *Bull. London. Math. Soc.*, 20 (1988), pp. 385-525.

[ES] Eells, J., and Sampson, J. Harmonic mappings of Riemannian manifolds. *Amer. J. Math.*, 86 (1964), pp. 109-160.

[ESt] Eilenberg, S., and Steenrod, N. *Foundations of Algebraic Topology.* Princeton, N.J.: Princeton University Press, 1952.

[EW1] Eells, J., and Wood, J. Maps of minimum energy, *J. London Math. Soc.*, 23 (Apr., 1981), pp. 303-310.

[EW2] Eells, J., and Wood, J. Harmonic maps from surfaces to complex projective spaces. *Adv. Math.*, 49 (1983), pp. 217-263.

[FF] Finkelstein, D. Kinks. *J. Math. Phys.*, 7 (1966), pp. 1218-1225.

[FH] Forgács, P., and Horváth, Z. Topology and saddle points in field theories. *Phys. Lett. B*, 138 (1984), pp. 397-401.

[FM] Finkelstein, D., and Misner, Ch. Some new conservation laws. *Ann. Phys.*, 6 (1959), pp. 230-243.

[Fr] Freedman, M. The topology of four-dimensional manifolds. *J. Diff. Geom.*, 17 (1982), pp. 357-453.

[FQ] Freedman, M., and Quinn, F. *Topology of 4-Manifolds.* Princeton N.J.: Princeton, 1990.

[Fri] Friedel, G. États mésomorphes de la matière, *Ann. Physique*, 18 (1922).

[FS] Faddeev, L.D., and Slavnov, A.A. *Gauge Fields: Introduction to Quantum Theory.* Reading, Mass: Benjamin, 1980.

[FT] Faddeev, L.D., and Takhtajan, L.A. *Hamiltonian Methods in the Theory of Solitons.* Berlin-New York: Springer-Verlag, 1987.

[FU] Freed, D., and Uhlenbeck, K. *Instantons and Four-Manifolds* (2nd ed.). New York: Springer-Verlag, 1990.

[Fu] Fuller, F. Harmonic mappings, *Proc. Nat. Acad. Sci. U.S.A.*, 40 (1954), pp. 987-991.

[FVo] Frank-Kamenetskiĭ, M.D., and Vologodskiĭ, A. V. Topological aspects of polymer physics. Theory and its biophysical applications. *Sov. Phys. Uspekhi*, 27 (1981), pp. 679-696.

[FVu] Fomin, I., and Vuorio, M. Textures and spin currents in the *B*-phase of ^{3}He. *J. L. Temp. P.*, 21 (1975), pp. 271-282.

[G] Glauber, R. Coherent and incoherent states of radiation field. *Phys. Rev.*, 131 (1963), pp. 2766-2788.

[Ge] de Gennes, P. *The Physics of Liquid Crystals.* Oxford: Clarendon Press, 1974.

[Geu] Geurst, J. Continuum theory and focal conic texture for liquid crystals of the smectic mesophase. *Phys. Lett. A*, 34 (1971), pp. 283-284.

[GG] Golubitsky, M., and Guillemin, V. *Stable Mappings and Their Singularities.* New York-Heidelberg: Springer-Verlag, 1973.

[GH] Griffiths, P., and Harris, J. *Principles of Algebraic Geometry.* New York: Wiley-Interscience Publ., 1978.

[Gi] Gibbons, G. Gravitational instantons: Survey. *Lect. Notes in Phys.*, 116. Berlin: Springer-Verlag, 1980, pp. 282-287.

[GM1] Golo, V.L., and Monastyrsky, M.I. Gauge groups and topological invariants of vacuum manifolds, *Ann. Inst. H. Poincaré, Sect. A (N. S.)*, 28 (1978), pp. 75-89.

[GM2] Golo, V.L., and Monastyrsky, M.I. Topological charges and high-dimensional instantons *Ann. Inst. H. Poincaré, Sect. A. (N. S.)*, 29 (1978), pp. 157-167.

[GM3] Golo, V.L., and Monastyrsky, M.I. Domain structure in ^{3}He-B. *Phys. Lett. A*, 66 (1978), pp. 302-304.

[GM4] Golo, V.L., and Monastyrsky, M.I. Gauge group and phases of superfluid ^{3}He. *Lett. Math. P.*, 2 (1978), pp. 373-378.

[GM5] Golo, V.L., and Monastyrsky, M.I. Currents in superfluid ^{3}He. *Lett. Math. P.*, 2 (1978), pp. 379-383.

[GMN] Golo, V.L., Monastyrsky, M.I., and Novikov, S.P. Solutions to the Ginzburg-Landau equations for planar textures in superfluid ^{3}He. *Comm. Cath. Phys.*, 69 (1979), pp. 237-246.

[GO] Coddard, P., and Olive, D. The developments in the theory of magnetic monopoles. *Rep. Prog. Phys.*, 41 (1978), pp. 1357-1438.

[Gr] Gribov, V.N. Quantization of gauge fields. *Nucl. Phys. B*, 138 (1978), pp. 1-19.

[GSW] Green, M., Schwarz, J., and Witten, E. *Superstring Theory.* London-New York: Cambridge University Press, 1987.

[H] Hartman, P. *Ordinary Differential Equations.* New York: John Wiley & Sons, 1964.

[Ha] Harris, B. Some calculations of homotopy groups of symmetric spaces. *Trans. Amer. Math. Soc.*, 106 (1963), pp. 174-184.

[HE] Hawking, S., and Ellis, G. *The Large-Scale Structure of Space-Time.* London-New York: Cambridge University Press, 1973.

[He] Helgason, S. *Differential Geometry and Symmetric Spaces, with an Additional Section by A. Bard.* New York: Academic Press, 1962.

[Hi] Hirzebruch, F. *Topological Methods in Algebraic Geometry* (3rd ed.). New York: Springer-Verlag New York, Inc., 1966.

[HL] Hsiang, W., and Lawson, H. Minimal submanifolds of low cohomogeneity. *J. Diff. Geom.*, 5 (1971), pp. 1-38.

[Ho1] 'tHooft, G. Magnetic monopoles in unified gauge theories. *Nuclear Phys. B*, 79 (1974), pp. 276-284.

[Ho2] 'tHooft, G. Computation of the quantum effects due to a four-dimensional pseudoparticle. *Phys. Rev. D*, 14 (1976), pp. 3432-3450.

[Hop] Hopf, H. Abbildgn. d. 3-dimens. Sphäre auf d. Kugelfläche. *Math. Ann.*, 104 (1930-1931), pp. 637-665.

[Hs] Husemoller, D. *Fibre Bundles.* New York-London-Sydney: MacGraw-Hill Book Co., 1966.

[Hu] Hu, S. *Homotopy Theory.* New York: Academic Press, 1959.

[I] Isham, C. Quantum gravity—An overview. In Isham, C., Penrose, R., and Sciama, D. (eds.) *Quantum Gravity II: A Second Oxford Symposium.* Oxford: Clarendon, 1981.

[IP] Isham, C., and Pope, C. A spinor field representation of the Stiefel-Whitney class. *Phys. Lett. B*, 114 (1982), pp. 137-140.

[IZ] Itzykson, C., and Zuber, J. *Introduction to Quantum Field Theory.* New York: McGraw-Hill, 1980.

[J] Jones, R. Stability of inert *p*-wave phases of superfluid He-3 in strong-coupling theory. *J. Phys. C.*, 10 (1977), pp. 657-669.

[JaT] Janich, K., and Trebin, H. Disentanglement of line defects in ordered media. In Balian, R., *et al.* (eds.), *Physique des Defauts/Physics of Defects.* Amsterdam: North Holland, 1981.

[Jo] Jones, V. A new knot polynomial and von Neumann algebras, *Not. Am. Math. Soc.*, 33 (1986), pp. 219-225.

[JT] Jaffe, A., and Taubes, C. *Vortices and Monopoles.* Boston: Birkhauser, 1980.

[K1] Kléman, M. Points, lines and walls. In *Liquid Crystals, Magnetic Systems and Various Ordered Media.* Chichester: Wiley, 1984.

[K2] Kléman, M. Developable domains in hexagonal liquid crystals. *J. Physique*, 41 (1980), pp. 737-745.

[KM] Kats, E.I., and Monastyrsky, M.I. Ordering in discotic liquid crystals. *J. Physique*, 45 (1984), pp. 709-714.

[KN] Kobayashi, S., and Nomizu, K. *Foundations of Differential Geometry.* New York-London: Interscience Publishers, Vol. 1, 1963, Vol. 2, 1969.

[Kr] Kraines, D. Massey higher products. *T. Am. Math. S.*, 124 (1966), pp. 431-449.

[KS] Klauder, J., and Sudarshan, E. *Fundamentals of Quantum Optics.* New York-Amsterdam: W.A. Benjamin, Inc., 1968.

[L] Langevin R. Feuilletages, énergies et cristaux liquides, *Asterisque*, 107/108 (1983), pp. 201-214.

[La] Lawson, H. *Lectures on Minimal Submanifolds.* Publish or Perish, Inc.: Berkeley, Calif., 1980.

[Le] Leggett, A. A Theoretical description of the new phases of liquid ^{3}He, *Rev. Mod. Phys.*, 47 (1975), pp. 331-413.

[Lef] Lefschetz, S. *Analysis situs et géométrie algébrique.* Paris: Gauthier-Villars, 1924.

[LGP] Lifshits, I., Gredeskul, S., and Pastur, L. *Introduction to the Theory of Disordered Systems.* New York: Wiley, 1988.

[LiL] Lichtenberg, A., and Lieberman, M. *Regular and Stochastic Motion.* New York: Springer-Verlag, 1983.

[LL1] Landau, L., and Lifshitz, E. *The Classical Theory of Fields.* Oxford-New York: Pergamon Press, 1969.

[LL2] Landau, L., and Lifshitz, E. *Statistical Physics.* Oxford-New York: Pergamon Press, 1975.

[LL3] Landau, L.D., and Lifshitz, E.M. *Quantum Mechanics: Non-Relativistic Theory* (3rd ed.). Oxford-New York: Pergamon Press, 1977.

[Ma1] Madore, J. A Classification of SU_3 magnetic monopoles, *Comm. Math. Phys.*, 56 (1977), pp. 115-123.

[Ma2] Madore, J. Geometric methods in classical field theory. *Phys. Report*, 75 (1981), pp. 125-204.

[Mak] Maki, K. Extended objects in superfluid ^{3}He. *Physica B + C*, 90 (1977), pp. 84-93.

[Man] Manton, N. Topology in the Weinberg-Salam theory. *Phys. Rev. D.*, 28 (1983), pp. 2019-2026.

[Mas1] Massey, W. Higher-order linking numbers. In *Conf. on Algebraic Topology, Univ. Illinois at Chicago Circle, Chicago, Ill., 1968.* Chicago: Univ. of Illinois at Chicago Circle, Chicago, Ill., 1969, pp. 174-205.

[Mas2] Massey, W. *Algebraic Topology: An Introduction.* New York: Harcourt, Brace & World, Inc., 1967.

[Mas3] Massey, W. On the cohomology ring of a sphere bundle. *J. Math. Mech.*, 7 (1958), pp. 265-289.

[Me1] Mermin, N. Surface singularities and superflow in ^{3}He. In Trickey, S., *et al.* (eds.) *Quantum Fluids and Solids.* New York: Plenum Press, 1977, pp. 3-22.

[Me2] Mermin, N. Topological theory of defects in ordered media. *Rev. M. Phys.*, 51 (1979), pp. 591-648.

[Me3] Mermin, N. Games to play with ^{3}He-*A*. *Physica B + C*, 90 (1977), pp. 1-10.

[MH] Mermin, N., and Ho, T. Circulation and angular momentum in the *A*-phase of superfluid helium-3. *Phys. Rev. L.*, 36 (1976), pp. 594-597.

[Mi] Michel, L. Minima of Higgs-Landau polynomials. In *Colloq. in the Honour of Antoine Viscontie.* Paris, 1980.

[Mil] Milnor, J. Link groups. *Ann. of Math.*, 59 (1954), pp. 177-195.

[Mil2] Milnor, J. On manifolds homeomorphic to the 7-sphere. *Ann. of Math.*, 64 (1956), pp. 399-405.

[Mil3] Milnor, J. Isotopy of links. In *Algebraic Geometry and Topology. A Symposium in Honor of S. Lefschetz.* Princeton, N.J.: Princeton University Press, 1957, pp. 280-306.

[Mil4] Milnor, J. *Morse Theory.* Princeton, N.J.: Princeton University Press, 1963.

[Mil5] Milnor, J. Spin structures on manifolds, *Enseignement Math.*, 9 (1963), pp. 198-203.

[Mil6] Milnor, J. *Lectures on the h-Cobordism Theorem.* Princeton, N.J.: Princeton University Press, 1965.

[Mil7] Milnor, J. *Topology from the Differentiable Viewpoint.* Charlottesville, Va.: University Press of Virginia, 1965.

[Mim] Mimura, M. Quelques groupes d'homotopie métastables des espaces symmétriques $Sp(n)$ et $U(n)/Sp(n)$. *C.R.Ac.S.*, Sér. A-B, 262 (A-20-A21, 1966).

[Mk1] Maki, K., and Kumar, P. n solitons (planarlike n textures) in superfluid ^{3}He-*B*. *Phys. Rev. B*, 16 (1977), pp. 4805-4814.

[Mk2] Maki, K., and Kumar, P. Composite solitons and magnetic resonance in superfluid ^{3}He-*A*.2. *Phys. Rev. B*, 17 (1978), pp. 1088-1094.

[MLLG] Malthete, J., Liebert, L., Levelut, A., and Galrne, Y. Nematique biaxe thermotrope. *C.R.As.S. II*, 303 (1986), pp. 1073-1076.

[Mo] Monastyrsky, M.I. Coherent states and discrete groups. *Usp. Mat. Nauk*, 31 (1976), pp. 221-222 (in Russian).

[Mos] Mostow, M.A. The field copy problem: To what extent do curvature (gauge field) and its covariant derivatives determine connection (gauge potential)? *Comm. Math. P.*, 78 (1980), pp. 137-150.

[MP1] Monastyrsky, M.I., and Perelomov, A.M. Coherent states and bounded homogeneous domains. *Rep. Math. Phys.*, 6 (1974), pp. 1-14.

[MP2] Monastyrsky, M.I., and Perelomov, A.M. Some remarks on monopoles in gauge field theories. *Preprint ITEP*, 56, Moscow, 1974.

[MP3] Monastyrsky, M.I., and Perelomov, A.M. Concerning existence of monopoles in gauge field theories. *JETP Lett.*, 21 (1975), pp. 43-44.

[MPa] Marciano, W., and Pagels, H. Classical $SU(3)$ gauge theory and magnetic monopoles. *Phys. Rev. D*, 12 (1975), pp. 1093-1095.

[MRa] Michel, L., and Radicatti, L.A. Properties of the breaking of hadronic internal symmetry. *Ann. Phys.*, 66 (1966), pp. 758-783.

[MRe] Monastyrsky, M.I., and Retakh, V.S. Topology of linked defects in condensed matter. *Comm. Math. P.*, 103 (1986), pp. 445-459.

[MSa] Monastyrsky, M.I., and Sasorov, P.V. Topological invariants in magnetohydrodynamics, *Zh. Eksp. Teor. Phys.*, 93 (1987), pp. 1210-1220 (in Russian).

[MSAB] Meiboom S., Sethna, J., Anderson, P., and Brinkman, W. Theory of the blue phase of cholesteric liquid crystals. *Phys. Rev. L.*, 46 (1981), pp. 1216-1219.

[MSh] Mostow, M., and Shnider, S. Does a generic connection depend continuously on its curvature? *Comm. Math. P.*, 90 (1983), pp. 417-432.

[MST] Matinyan, S.G., Savvidi, G.K., and Ter-Arutyunyan-Savvidi, N.G. Classical Yang-Mills mechanics. Non-linear colour fluctuations. *Zh. Eksp. Teo. Phys.*, 80 (1981), pp. 830-838 (in Russian).

[MSt] Milnor, J., and Stasheff, J. *Characteristic Classes* (Ann. of Math. Studies, 76). Princeton, N.J., Princeton Univ. Press, 1974.

[MT] Mosher, R., and Tangora, M. *Cohomology Operations and Applications in Homotopy Theory.* New York-London: Harper & Row, 1968.

[MV] Mineev, V.P., and Volovik, G.E. Planar and linear solitons superfluid ^{3}He. *Phys. Rev. B*, 18 (1978), pp. 3197-3203.

[MZ] Montgomery, D., and Zippin, L. *Topological Transformation Groups.* New York: Interscience Publishers, 1955.

[N] Nahm, W. Construction of all self-dual monopoles by the ADHM method. In Craige, N., *et al.* (eds.), *Monopoles in Quantum Field Theory.* Singapore: World Scientific, 1982, pp. 87-94.

[No1] Novikov, S.P. The Cartan-Serre theorem and inner homologies. *Usp. Mat. Nauk*, 21 (1966), pp. 217-232 (in Russian).

[No2] Novikov, S.P. Hamiltonian formalism and multivalued analog of Morse theory. *Usp. Mat. Nauk*, 37 (1982), pp. 3-49 (in Russian).

[NR] Narasimhan, M., and Ramadas, T. Geometry of $SU(2)$ gauge fields. *Comm. Math. P.*, 67 (1979), pp. 121-136.

[O'] O'Neil, E. Higher-order Massey products and links. *T. Am. Math. S.*, 248 (1979), pp. 37-66.

[O'R] O'Raifeartaigh, L. *Group Structure of Gauge Theories*, Cambridge: Cambridge Univ. Press, 1988.

[Ok] Okun, L. *Leptons and Quarks.* Amsterdam: North-Holland, 1980.

[P] Penrose, R. Structure of space-time. In Dewitt, C., and Wheeler, J. (eds.), *Lectures in Mathematics and Physics Battelle Rencontres.* New York-Amsterdam: W.A. Benjamin, Inc., 1968.

[PDDu] Pansu, B., Dandoloff, R., and Dubois-Violette, E. The double twist connection and the S^3 blue phase. *J. Physique*, 48 (1987), pp. 297-304.

[PDuD] Pansu, B., Dubois-Violette, E., and Dandoloff, R. Disclination in the S^3 blue phase. *J. Physique*, 48 (1987), pp. 305-317.

[Per1] Perelomov, A.M. Coherent states for arbitrary Lie groups, *Comm. Math. P.*, 26 (1972), pp. 222-236.

[Per2] Perelomov, A. *Generalized Coherent States and Their Applications.* New York: Springer, 1985.

[Ph] Pohl, W. The probability of linking of random closed curves. *Lect. Notes in Math.*, 894 (1981).

[Pi] Pitaevski, L.P. On the superfluidity of ^{3}He, *Zh. Eksp. Teo. Phys.*, 37 (1959), pp. 1794-1807 (in Russian).

[Po1] Poénaru, V. Some aspects of the theory of defects of ordered media and gauge fields related to foliations. *Comm. Math. P.*, 80 (1981), pp. 127-136.

[Po2] Poénaru, V. Superalgebras and confinement. In 'tHooft, *et al.* (eds.), *Condensed Matter: Recent Developments in Gauge Theories.* New York: Plenum Press, 1980.

[Poi] Poincaré, H. *Analysis sitûs, École Polyt. J* 1. (1895), pp. 1-121.

[Pol1] Polyakov, A.M. Particle spectrum in quantum-field theory. *JETP Lett.*, 20 (1974), pp. 194-195.

[Pol2] Polyakov, A.M. *Gauge Fields and Strings.* New York: Harwood Acad. Publs., 1987.

[Por] Porter, R. Milnor's $\bar{\mu}$-invariants and Massey products. *T. Am. Math. S.*, 257 (1980), pp. 39-71.

[P-S] Pyatetskii Shapiro, I. *Automorphic Functions and the Geometry of Classical Domains.* New York: Gordon & Breach, 1969.

[PT1] Poénaru, V., and Toulouse, G. The crossing of defects in ordered media and the topology of 3-manifolds. *J. Physique*, 38 (1977), pp. 887-895.

[PT2] Poénaru, V. Topological solitons and graded Lie algebras. *J. Math. Phys.*, 20 (1979), pp. 13-19.

[Q] *The Quantum Hall Effect*, Givin, S., and Prange, R. (eds.), New York-Heidelberg: Springer-Verlag, 1990.

[R] Radcliffe, J. Some properties of coherent spin states, *J. Physics A*, 4 (1971), pp. 313-323.

[Ra] Rajaraman, R. *Solitons and Instantons: an Introduction to Solitons and Instantons in Quantum Field Theory.* Amsterdam: North-Holland, 1982.

[Rh] de Rham, G. *Variétés différentiables, formes courantes, formes harmoniques.* Paris: Hermann et Cie, 1955.

[Ro] Rolfsen, D. *Knots and Links.* Berkeley, Calif.: Publish or Perish, Inc., 1976.

[Sa] Salomaa, M. A phase transition of the moving superfluid ^{3}He *A-B* interface, *J. Physics C*, 21 (1988), pp. 4425-4435.

[Sc] Schwartz, J. *Differentiable Geometry and Topology.* New York: Gordon & Breach, 1968.

[Sch] Schwartz, K. Generation of superfluid turbulence deduced from simple dynamical rules. *Phys. Rev. L.*, 49 (1982), pp. 283-285.

[Sem] *Séminaire "Sophus Lie". Théorie des algèbres de Lie. Topologie des groupes de Lie.* Paris: Secrétariat Mathématique, 1955.

[SaS] Sauls, J., and Serence, J. $^3P_2$ pairing near the transition temperature in neutron-star matter, *Phys. Rev. D.*, 17 (1978), pp. 1524-1528.

[Si] Singer, I. Some remarks on the Gribov ambiguity. *Comm. Math. P.*, 60 (1978), pp. 7-12.

[Sk] Skyrme, T. A non-linear field theory. *Proc. Roy. Soc. London*, Ser. A, 260 (1961), pp. 127-138.

[Ska] Skarzhinsky, V.D. The Aharonov-Bohm Effect: Theoretical calculations and interpretation. *Trudy FIAN*, 167 (1986), pp. 140-161 (in Russian). In Komar A. (ed.) *Group Theory, Cravitation and Elementary Particle Physics*. New York: Nova Sci. Publ. Commack, 1988, p. 176.

[So1] Solovev, M.A. Wu-Yang bundle spaces. *JETP Letter*, 35 (1982), pp. 669-671 (in Russian).

[So2] Solovev, M.A. Dirac monopole as a Lagrangian system in a stratification space. *JETP Lett.*, 39 (1984), pp. 714-716.

[So3] Solovev, M.A. Global gauge in non-Abelian theory. *JETP Letter*, 38 (1984), pp. 504-507.

[So4] Solovev, M.A. On geometry of classical mechanics with non-Abelian gauge symmetries. *Teor. Mat. Fiz.*, 73 (1987), pp. 3-15.

[Sol] *Solitons*. Bullough, K., and Caudrey, P. (eds.). Heidelberg-New York: Springer-Verlag, 1980.

[Sp] Springer, G. *Introduction to Riemann Surfaces*. Reading, Mass.: Addison-Wesley Publishing Company, Inc., 1957.

[SPA] Stein, D., Pisarski, R., and Anderson, P. Boojua in superfluid ^{3}He-A and cholesteric liquid crystals. *Phys. Rev. L.*, 40 (1978), pp. 1269-1272.

[St] Stallings, J. Homology and central series of groups. *J. Algebra*, 2 (1965), pp. 170-181.

[Ste] Sternberg, S. *Lectures on Differential Geometry*. Englewood Cliffs, N.J.: Prentice-Hall, 1964.

[Stee] Steenrod, N. *The Topology of Fibre Bundles*. Princeton, N.J.: Princeton University Press, 1951.

[Sto] Stong, R. *Notes on Cobordism Theory*. Princeton, N.J.-Tokyo: Princeton University Press, University of Tokyo Press, 1968.

[SU] Sacks, J., and Uhlenbeck, K. The existence of minimal immersions of two-spheres. *Ann. of Math.*, 113 (1981), pp. 1-24.

[Su] Sullivan, D. Infinitesimal computations in topology, *Inst. Hautes études. Sci. Publ. Math.*, 47 (1977), pp. 269-331.

[SV] Salomaa, M., and Volovik, G.E. Quantized vortices in superfluid ^{3}He. *Rev. Modern Phys.* 59 (Part 1, 1988), pp. 533-613, *Errata Ibid.* (1988), p. 573.

[SW] Streater, R., and Wightman, A. *PCT, SPIN, and Statistics, and All That*. New York: Benjamin, Inc., 1964.

[Sz] Schweitzer, P. Counterexamples to the Seifert conjecture and opening closed leaves of foliations. *Ann. of Math.*, 100 (1974), pp. 386-400.

[Tam] Tamura, I. *The Topology of Foliations*. Tokyo: Iwanami Shoten, 1976 (in Japanese).

[Tau1] Taubes, C. Arbitrary N-vortex solutions to the 1st-order Ginzburg-Landau equations. *Comm. Math. Phys.*, (1980), pp. 277-292.

[Tau2] Taubes, C. The existence of a non-minimal solution to the $SU(2)$ Yang-Mills equations on R^3. *Comm. Math. Phys.*, 86 (1982), pp. 257-320.

[Tau3] Taubes, C. Min-max theory for the Yang-Mills-Higgs equations. *Comm. Math. Phys.*, 97 (1985), pp. 473-540.

[Tay] Taylor, J. *Gauge Theories of Weak Interactions*. London: Cambridge University Press, 1976.

[TFS] Tuypkin, Yu., Fateev, V., and Schwarz, A. Topologically non-trivial particles in quantum-field theory. *JETP Letter*, 22 (1975), pp. 88-89.

[Th] Thom, R. Quelques propriétés globales des variétés différentiables. *Comment. Math. Helv.*, 28 (1954), pp. 17-86.

[Thu] Thuneberg, E. Identification of vortices in superfluid ^{3}He-B. *Phys. Rev. L.*, 56 (1986), pp. 359-362.

[TK] Toulouse, G., and Kleman, M. Principles of a classification of defects in ordered media. *J. Phys. Lett.*, 37 (1976), pp. 149-151.

[To] Toda, H. *Composition Methods in Homotopy Groups of Spheres* (Ann. of Math. Studies, 49). Princeton, N.J., Princeton Univ. Press, 1962.

[Tr] Trebin, H. The topology of non-uniform media in condensed-matter physics. *Adv. Physics*, 31 (1982), pp. 195-254.

[TT] Tilley, D., and Tilley, J. *Superfluidity and Superconductivity.* New York: Van Nostrand-Reinhold, 1974.

[Tu] Turaev, V.G. The Milnor invariants and Massey products. In *Studies in Topology, II. Zap. nauch. sem. Leningrad. otdel. mat. inst. Steklov.* (*LOMI*), 66 (1976), pp. 189-203 (in Russian).

[TuV] Turaev V. G., and Viro O. Y. State sum invariants of 3-manifolds and quantum 6 J-symbols. Leningrad: Preprint LOMI, 1991.

[U] Uezu, T. Topology in dynamical systems, *Phys. Lett. A*, 93 (1983), pp. 161-166.

[UA] Uezu, T., and Aizawa, Y. Topological character of a periodic solution in 3-dimensional ordinary differential equation system. *Prog. T. Phys.*, 68 (1982), pp. 1907-1916.

[Uh1] Uhlenbeck, K. Removable singularities in Yang-Mills fields, *Comm. Math. Phys.*, 83 (1982), pp. 11-29.

[Uh2] Uhlenbeck, K. Connections with L^p bounds on curvature, *Comm. Math. Phys.*, 83 (1983), pp. 31-42.

[V] Vasiliev, V. Invariants of knots and compliments to discriminants, In Arnold, V. and Monastyrsky, M. (eds.), *Developments of Modern Mathematics.* London: Chapman and Hall, 1992.

[Ve] Vergeles, S. Some properties of orbit space in Yang-Mills theory. *Lett. Math. P.*, 7 (1983), pp. 399-406.

[VM] Volovik, G., and Mineev, V. Investigation of singularities in superfluid He^3 in liquid crystals by homotopic topology methods. *JETP Soviet Physics*, 45 (1977), pp. 1186-1196.

[Vo] Volovik, G. Topological singularities on the surface of an ordered system. *JETP Letter*, 28 (1978), pp. 59-62.

[Vol] Volterra, V. Sur l'équilibre des corps élastiques multipliment connéxes. *Ann. École Norm.*, 24 (1907), pp. 407-517.

[VW] Vollhardt D., Wölfle P. *The Superfluid Phases of* ^{3}*He.* London: Taylor & Francis, 1988.

[VZNS] Vainstein, A., Zakharov, V., Novikov, V., and Shifman, M. ABC of instantons. *Sov. Physics Uspekhi*, 25 (1982), pp. 195-215.

[W] Wallace, A. *Differentiable Topology: First Steps.* New York-Amsterdam: W.A. Benjamin, 1968.

[We] Weyl, H. *The Classical Groups: Their Invariance and Representations.* Princeton, N.J.: Princeton University Press, 1946.

[Wf] Wolf, J. *Spaces of Constant Curvature.* Boston, Mass.: Publish or Perish, Inc., 1974.

[Wh] Whitehead, J. An expression of Hopf's invariant as an integral. *Proc. Nat. Acad. Sci. USA*, 33 (1947), pp. 117-123.

[Whi] Whitney, H. *Geometric Integration Theory.* Princeton, N.J.: Princeton University Press, 1957.

[Wi] Wilczek, F. Geometry and interactions of instantons. In Stump, D., and Weigarten, D. (eds.), *Quark Confinement and Field Theory.* New York: John Wiley & Sons, 1977.

[Wil] Williams, R. The structure of Lorentz attractors. *Lecture Notes in Math.*, 615 (1977), pp. 94-112.

[Wit1] Witten, E. Some exact multipseudo-particle solutions of classical Yang-Mills theory. *Phys. Rev. L.*, 38 (1977), pp. 121-124.

[Wit2] Witten, E. Global aspects of current algebra. *Nucl. Phys. B*, 223 (1983), pp. 422-432.

[Wit3] Witten, E. Current algebra, baryons, and quark confinement. *Nucl. Phys. B*, 223 (1983), pp. 443-444.

[Wit4] Witten, E. Quantum field theory and the Jones polynomial. *Comm. Math. Phys.*, 121 (1989), pp. 351-399.

[Wod] Wood, J. Singularities of harmonic maps and applications of the Gauss-Bonnet formula, *Amer. J. Math.*, 99 (1977), 1329-1344.

[Woo] Woo, G. Pseudoparticle configurations in 2-dimensional ferromagnets. *J. Math. Phys.*, 18 (1977), pp. 1264-1266.

[Wt] White, J. Self-linking and the gauss integral in higher dimensions. *Amer. J. Math.*, 91 (1969), pp. 693-728.

[Wu] Wu, H. *The Equidistribution Theory of Holomorphic Curves* (Ann. of Math. Studies, 64). Princeton, N.J., Princeton Univ. Press, 1970.

[Wuy] Wu, Y. Chern-Simons topological lagrangians in odd dimensions and their Kaluza-Klein reduction. *Ann. Physics*, 156 (1984), pp. 194-211.

[WY1] Wu, T., and Yang, C. Some solutions of the classical isotopic gauge field equations. In Mark, H., and Fernbach, S. (eds.), *Properties of Matter under Unusual Conditions.* New York-London-Sydney-Toronto: Interscience, 1969, pp. 349-354.

[WY2] Wu, T., and Yang, C. Some remarks about unquantized non-Abelian gauge fields. *Phys. Rev. D*, 12 (1975), pp. 3843-3844.

[WY3] Wu, T., and Yang, C. Concept of non-integrable phase factors and global formulation of gauge fields. *Phys. Rev. D*, 12 (1975), pp. 3845-3857.

[YM] Yang, C., and Mills, R. Conservation of isotopic spin and isotopic gauge invariants. *Phys. Rev.*, 96 (1954), pp. 191-195.

[Z] Zakrzewski, W. *Low-Dimensional Sigma Models.* Bristol: Adam Hilger, 1989.

[Ze] Zelobenko, D. *Compact Lie Groups and Their Representations.* Providence, R.I., Amer. Math. Soc., 1983.

[Zi] Ziglin, S.L. Branching of solutions and the non-existence of first integrals. In *Hamiltonian Mechanics.* P.II. *Func. Anal. Appl.*, 17 (1983), pp. 8-23.

[ZMNP] Zakharov V., Manakov, S., Novikov, S., and Pitaevski, L. *The Theory of Solitons.* N.Y.: Plenum, 1984.

Index

GPSR Compliance
The European Union's (EU) General Product Safety Regulation (GPSR) is a set of rules that requires consumer products to be safe and our obligations to ensure this.

If you have any concerns about our products, you can contact us on

ProductSafety@springernature.com

In case Publisher is established outside the EU, the EU authorized representative is:

Springer Nature Customer Service Center GmbH
Europaplatz 3
69115 Heidelberg, Germany

www.ingramcontent.com/pod-product-compliance
Ingram Content Group UK Ltd.
Pitfield, Milton Keynes, MK11 3LW, UK
UKHW051130260726
13967UKWH00010B/2971
9781489924049